Optimization Methods and Applications

Applied Optimization

Volume 52

Series Editors:

Panos M. Pardalos
University of Florida, U.S.A.

Donald Hearn
University of Florida, U.S.A.

The titles published in this series are listed at the end of this volume.

Optimization Methods and Applications

Edited by

Xiaoqi Yang and Kok Lay Teo

Department of Applied Mathematics,
Hong Kong Polytechnic University,
Hong Kong, China

and

Lou Caccetta

School of Mathematics and Statistics,
Curtin University of Technology,
Australia

KLUWER ACADEMIC PUBLISHERS

DORDRECHT / BOSTON / LONDON

A C.I.P. Catalogue record for this book is available from the Library of Congress.

ISBN 0-7923-6866-5

Published by Kluwer Academic Publishers,
P.O. Box 17, 3300 AA Dordrecht, The Netherlands.

Sold and distributed in North, Central and South America
by Kluwer Academic Publishers,
101 Philip Drive, Norwell, MA 02061, U.S.A.

In all other countries, sold and distributed
by Kluwer Academic Publishers,
P.O. Box 322, 3300 AH Dordrecht, The Netherlands.

Printed on acid-free paper

Printed in the Netherlands.

Dr. Nasir U. AHMED

Contents

Preface

This edited book is dedicated to Professor N.U. Ahmed, a leading scholar and a renowned researcher in optimal control and optimization on the occasion of his retirement from the Department of Electrical Engineering at University of Ottawa in 1999.

The contributions of this volume are in the areas of optimal control, nonlinear optimization and optimization applications. They are mainly the improved and expanded versions of the papers selected from those presented in two special sessions of two international conferences. The first special session is Optimization Methods, which was organized by K.L. Teo and X.Q. Yang for the International Conference on Optimization and Variational Inequality, the City University of Hong Kong, Hong Kong, 1998. The other one is Optimal Control, which was organized by K.L. Teo and L. Caccetta for the Dynamic Control Congress, Ottawa, 1999.

This volume is divided into three parts: Optimal Control; Optimization Methods; and Applications. The Optimal Control part is concerned with computational methods, modeling and nonlinear systems. Three computational methods for solving optimal control problems are presented: (i) a regularization method for computing ill-conditioned optimal control problems, (ii) penalty function methods that appropriately handle final state equality constraints, and (iii) a multilevel optimization approach for the numerical solution of optimal control problems. In the fourth paper, the worst-case optimal regulation involving linear time varying systems is formulated as a minimax optimal control problem. The next three papers concern with nonlinear systems and their applications, such as the modified direct gradient descent control of nonlinear systems, the stability of impulsive delay differential equations, and the effect of acceleration on the performance and observability of the true proportional based guidance system.

In the Optimization Methods part, computational methods and algorithms are presented. These include a homogenized cutting plane method for solving the convex feasibility problem, exact algorithms for some hard knapsack problems, asymptotic rates analysis of SQP-type methods of feasible directions for nonlinear programming problems, nonlinear Lagrangian methods in constrained optimization problems, and parallel algorithms for solving large scale nonlinear programming problems.

In the Applications part, a number of practical real world problems are solved using optimization and optimal control techniques. They are: (i) the lobe design of a narrow-band antenna array in signal processing is formulated as a complex Chebyshev optimization problem and is solved by the semi-infinite linear programming technique; (ii) a global optimization method for the optimum pole selection of a digital Laguerre network; (iii) the Bellman principle of dynamic programming is applied to obtain an optimal control policy that maximizes the total distance to be traveled by a solar-powered car; (iv) the personnel task scheduling problem is explored in which a set of tasks with fixed start and end time is assigned to a heterogeneous workforce; and (v) a two-phase heat transfer problem arising from the continuous steel casting process is solved as an optimal control problem.

We would like to express our warmest thanks to a number of referees for their careful and critical evaluation of the manuscripts. They include X. Huang (Chongqing Normal University), J. Lee (The Hong Kong Polytechnic University), D.H. Li (Hunan University), S.J. Li (Chongqing Jianzhu University), A. Lim (The Chinese University of Hong Kong), K.W. Lim (National University of Singapore), Y. Liu (The Hong Kong Polytechnic University), S. Nordholm (University of Karlskrona/Ronneby), B. Vo (Australian Telecommunication Research Institute), S. Wang (The University of Western Australia), W.Y. Yan (Curtin University of Technology), and Y.B. Zhao (Chinese Academy of Sciences).

Xiaoqi Yang, Kok Lay Teo and Lou Caccetta

An Appreciation of Professor N.U. Ahmed

Nasir U. AHMED was born in Bangladesh in 1934. He received his BSc degree in electrical engineering from Dhaka University, Bangladesh in 1956, MSc degree with Magna Cum Laude in electrical engineering from the University of Ottawa, Canada in 1963 and a PhD degree with Summa Cum Laude in electrical engineering also from the University of Ottawa in 1965. He was a Senior Scientific Office in the Pakistan Atomic Energy Commission before he returned to Canada as a Postdoctoral Fellow with the University of Alberta in 1967. He joined the University of Ottawa as an assistant professor in 1968, and was promoted to an associate professor in 1971. Prior to his retirement, he was a professor with the Department of Electrical Engineering of the University of Ottawa since 1976. Professor Ahmed served as the Chairman of Department of Electrical Engineering of the University of Ottawa from 1989 to 1992. He is now a Professor Emeritus of Systems Science in the School of Information Technology and Engineering, and the Department of Mathematics, University of Ottawa.

In view of his outstanding scientific achievements, Professor Ahmed has had many visiting positions at various institutes. He spent a year (1977) at the University of California at Berkeley as a Visiting Professor in the Department of Electrical and Computer Science. He spent the summer months of 1980-81 visiting the Department of Systems Science at the University of California at LA. He spent the winter of 1980-81 as a Visiting Professor at the University of New South Wales, and gave a series of lectures on partial differential equations and control of systems governed by PDE. In the summer of 1988, he was a Visiting Professor at the University of Western Australia. He presented a series of lectures on semigroup theory and its applications to physical and engineering sciences which was subsequently written as a monograph, entitled "Semigroup Theory with Applications to Systems and Control". This monograph was published by Longman Scientific and Technical in 1988. His other invited visitations include Banach Center in Warsaw, Poland; a number of prestigious universities in China ; and many well-known universities in USA and Canada. Professor Ahmed is also active in various professional societies, such as APEO, IEEE, AMS, and SIAM.

Professor Ahmed has been awarded numerous research grants from NSERC and DND Contract. He has graduated 17 PhD students, including K.L. Teo the second editor of this volume, and S. K. Biswas the co-author of the paper, "Worst-case Optimal Regulation of Linear Systems in the Presence of Structured Perturbation". Amongst the authors of the Volume, it is pleasing to note that K.H. Wong, V. Rehbock, H.H. Dam and B. Vo were students of Teo, and G. Peter is currently pursuing his PhD study under the supervision of K.H. Wong. Professor Ahmed has also graduated many masters students, and supervised numerous research associates and post doctoral fellows.

Professor Ahmed has written 5 excellent books, 18 invited papers, over 190 refereed international journal papers, 8 invited articles in edited books and encyclopedia, over 70 conference publications. He was keynote speakers for many important international conferences. He has served as members of editorial board of numerous international journals, including Journal of Dynamics and

Control; Dynamic Systems and Applications; International Journal of Science and Technology; Discussions Mathematicae; Differential inclusions; Mathematical Problems in Engineering: Theory and Applications; Discrete and Continuous Dynamic Systems; Dynamics of Continuous, Discrete and Impulsive Systems; and Communications in Applied Mathematics.

Professor Ahmed has also served as members of Organizing Committee of many International Conferences. His current research areas includes: (i) measure solutions for semilinear evolution equations on Banach spaces and their optimal control; (ii) mathematical modeling (PDE) of the dynamics of suspension bridges and their stability analysis and optimum design; (iii) mathematical modeling of the hemodynamics (hemolysis = red blood cell damage) of artificial heart based on Navier-Stokes equation in 3D and membrane equation in 2D and their optimal control subject to geometric and clinical constraints; and (iv) state space modeling of the dynamics of computer-communication network and their performance analysis.

Professor Ahmed has made many important fundamental contributions in applied functional analysis, stochastic processes, systems and control theory, and practical real world applications of control and optimization. A sample of these contributions are listed as follows:

(1) Developed Fourier analysis on Wiener measure space and introduced the concepts of a vector space for nonlinear functionals of Brownian motion.

(2) Discovered a class of Generalized Functionals of Brownian Motion and constructed a locally convex topological vector space. (This generalizes Hidas generalized functionals of Brownian Motion and opens up the prospect of much wider applications). It has applications in: (i) heat equations with nonlinear (Steffan-Boltzman) boundary conditions; and (ii) modeling turbulence in fluid mechanics. Recently this work was further extended to the space of generalized functionals of Gaussian random fields and evolution equations thereon. These generalized results have applications in the construction of solutions of stochastic Navier-Stokes equations, etc.

(3) Developed existence theorems for optimal controls and necessary conditions of optimality for systems governed by functional-differential equations in finite dimensional spaces.

(4) Developed existence theorems for partially observed diffusions and stochastic functional differential equations in finite dimensional spaces.

(5) Developed optimal control theory for stochastic evolution equations on Banach space with random operator valued coefficients.

(6) Pioneered the development of optimal control theory for general nonlinear evolution equations and differential inclusions on Banach spaces. (This resulted in the book, entitled "Optimal Control of Distributed Parameter Systems", North Holland, 1981, Co-author K.L.Teo)

(7) Introduced a general class of stochastic evolution equations and differential inclusions on Banach space driven by generalized Brownian motion. Application to stochastic partial differential equations with both distributed and boundary noise was demonstrated.

(8) Developed a selection theorem for multivalued maps in reflexive Banach spaces which has been applied by control theorists in infinite dimensional systems.

(9) Developed a general theory for optimization and identification of unbounded operators in evolution equations on Banach Space, which led to the publication of the Monogram "Elements of Finite Dimensional System and Control Theory" (Pitman Research Notes in Mathematics Series, Vol. 184, 1988). (This work provides for the first time a unified approach to inverse problems or identification problems.)

(10) Introduced the concept of differential inclusions for stochastic systems, and subsequently, its applications in control theory. This has been applied to stochastic systems with uncertain parameters or systems with incomplete dynamic models.

(11) Developed nonlinear filter theory for abstract Evolution equations on Hilbert space, which have widespread applications in systems governed by PDE and distributed signal processing. This work was done in collaboration with Professors J.Zabczyk, Polish Academy of Sciences, and Professor M. Fuhrman, Polytechnic University of Milan.

(12) Developed a general theory of partially observed control for nonlinear stochastic differential equations on Hilbert Space. This was an open problems for a decade.

(13) Developed theory of generalized solutions for systems governed by m-times integrated semigroups, which covers the class of distribution semigroups of J.L. Lions. This has much broader applications, including PDE containing non densely defined unbounded operators.

(14) Developed an interesting theory of measure solutions for semilinear evolution equations on Banach spaces both for Deterministic and Stochastic Systems.

(15) Developed a complete stochastic partial differential equation model for suspension bridge, and provided stability analysis.

(16) Introduced the concept of Orlicz stability for general nonlinear systems, covering nonlinearities with exponential growth .

(17) Constructed a complete stochastic dynamic model for large power systems network and computer communication networks, and developed relevant optimization techniques.

(18) Developed a stochastic jump process model for filtering the integrated navigation signals for DND.

(19) Constructed a complete dynamic model for flexible spacecraft based on Hamilton's principle, provided answers to various questions on stability and stabilizability, and developed techniques for constructing stabilizing controls.

(20) Constructed a complete mathematical dynamic model for space stations, and obtained important results on its stabilizing feedback controls. This was presented at JPL-NASA Conference, California , (1989).

(21) Based on Navier-Stokes equation, a complete mathematical model of the hemo-dynamics of artificial heart was constructed. The model was used

to compute optimal controls, which minimize hemolysis (damage to red blood cells).

(22) Developed a rigorous mathematical model for suspension bridge with full analysis and complete illustration by numerical and simulation results.

(23) His latest book is on filtering, entitled " Linear and Nonlinear Filtering for Scientists and Engineers" published by World Scientific Publishers, (1999). This book presents many interesting new results along with general and powerfully numerical techniques for solving nonlinear filtering problems.

The editors of this Volume have had the privilege of knowing Professor Ahmed, personally and professionally. In particular, the second editor has known him for nearly 30 years, first as his thesis supervisor and mentor, and subsequently as his close colleague and friend. Professor Ahmed's work has influenced the scientific activity of many researchers in the area of control and optimization and their applications. More importantly, Professor Ahmed has won the admiration and affection of those who know and benefited from him. Professor Ahmed is not only an outstanding scholar, but more importantly, he is an excellent colleague and friend.

A Publication List of Professor N.U. Ahmed

Professor Ahmed has written 5 books, 8 invited articles in books and encyclopedia, 193 refereed journal publication, 70 refereed conference publications, 18 invited papers, 2 conference presentations without proceedings and 4 contract reports. The total is 300.

BOOKS

1. N.U. Ahmed, and K.L. Teo, "Optimal Control of Distributed Parameter Systems", Elsevier North Holland, New York, 1981 (pages: 443).

2. N.U. Ahmed, "Elements of Finite Dimensional System and Control Theory", Vol. 37, Pitman Monographs and Surveys in Pure and Applied Mathematics, Longman Scientific and Technical, London - New York, 1988 (pages: 421).

3. N.U. Ahmed, "Optimization and Identification of Systems Governed by Evolution Equations on Banach Space", Pitman Research Notes in Mathematics Series, Vol. 184, Longman Scientific and Technical U.K.; John Wiley & Sons, London - New York, 1988 (pages: 187).

4. N.U. Ahmed, "Semigroup-Group Theory with Applications to Systems and Control", Pitman Research Notes in Mathematics Series, Vol. 246, Longman Scientific and Technical U.K.; John Wiley & Sons, London - New York, 1991 (pages: 282).

5. N.U. Ahmed, "Linear and Nonlinear Filtering for Scientists and Engineers", World Scientific Publishers, Singapore, New Jersey, London, Hong Kong (1999), (pages: 256).

INVITED ARTICLES IN BOOKS AND ENCYCLOPEDIA

1. N.U. Ahmed, "Optimal Control of Stochastic Systems", a part of the book "Probabilistic Analysis and Related Topics", (Ed. A.T. Bharucha Reid), Academic Press, Vol. 2, 1979, (pages: 68).

2. N.U. Ahmed, and H.W. Wong, "A Minimum Principle for Systems Governed by Its Differential Equations With Markov Jump Parameters; Differential Games and Control Theory II", Lecture Notes in Pure and Applied Mathematics, Marcel Dekker, Vol. 30, pp. 265-294, 1976.

3. N.U. Ahmed, "Abstract Stochastic Evolution Equations and Related Control and Stability Problems, Control Problems for Systems Described by Partial Differential Equations and Applications", Lecture Notes in Control and Information Sciences, Vol. 97, pp. 107-120, 1986.

4. N.U. Ahmed, "Distributed Parameter Systems" Encyclopedia of Physical Science and Technology, Vol. 4, pp. 362-385, Academic Press, 1987.

5. N.U. Ahmed, "Existence of Optimal Relaxed Controls for Differential Inclusions on Banach Space", Lecture Notes in Pure and Applied Mathematics Nonlinear Analysis and Applications, Marcel Dekker, Vol. 109, pp. 39-49, 1987.

6. N.U. Ahmed, "Identification of Linear Operators in Differential Equations on Banach Space, Operator Methods for Optimal Contract Problems", Lecture Notes in Pure and Applied Mathematics, Marcel Dekker, Vol. 108, pp. 1-30, 1988.

7. N.U. Ahmed, "Stochastic Initial Boundary Value Problems for a Class of Second Order Evolution Equations", Differential Equations and Applications, Proc. of The International Conference on Theory and Applications of Differential Equations, (Ed: A.R. Aftabizadeh), Ohio University, pp. 13-19, 1988.

8. N.U. Ahmed, "Identification of Operators in Systems Governed by Evolution Equations on Banach Space", Springer Verlag Lecture Notes in Control and Information Sciences, Vol. 114, pp. 73-83, 1989.

REFEREED JOURNAL PAPERS

1. N.U. Ahmed and M.S. Khan, "Transistor Characteristic Display Monitor", The Nucleus, Vol.3, No. 6, pp. 107-108, (1966).

2. N.U. Ahmed, "Human Brain: A System of Nonlinear Oscillators", Nuclear Science and Applications,Vol. 2, pp. 26-31, (1966).

3. N.U. Ahmed, "Fourier Analysis on Wiener Measure space", J. Franklin Inst., Vol. 286, No. 2, pp. 143-151, (Aug. 1968).

4. N.U. Ahmed, "An Algebra of Nonlinear Operators on L^2 Spaces", Pak. J. of Scientific and Industrial Research, Vol. 11, No. 4, pp. 348-351, (Oct. 1968).

5. N.U. Ahmed, "Some Novel Properties of Wiener's Canonical Expansion", IEEE Trans. SSC, Vol. 5, No. 2, pp. 140-144, (April 1969).

6. N.U. Ahmed, "A Class of Stochastic Nonlinear Integral Equations on L_p spaces and its Application to Optimal Control", J. Inf. and Control, Vol. 14, No. 6, pp. 512-523, (June 1969).

7. N.U. Ahmed and G.S. Glinski, "A Sequence of Probability Functions on Euclidean Space and the Limit", J. Franklin, Inst. Vol. 288, No. 2, pp. 151-156, (Aug. 1969).

8. N.U. Ahmed, "A Functional Analysis Method for Estimating the Settling Time of Time Varying Systems", J. Engineering Math., Vol. 3, No. 4, pp. 337-343, (October 1969).

9. N.U. Ahmed, "On the $L_p(p \geq 1)$ Stability of a Class of Nonlinear Systems", Proc. IEEE, Vol. 57, No. 10, pp. 1795-1797, (October 1969).

10. N.U. Ahmed, "Optimal Control of a Class of Nonlinear Systems on Hilbert Space Trans. A.C.", IEEE, Vol. 14, No. 6, pp. 711-714, (Dec. 1969).

11. N.U. Ahmed, "Closure and Completeness of Wiener's Orthogonal Set G_n in the Class $L^2(\Omega, B, \mu)$ and its Application to Stochastic Hereditary Differential Systems", J. Inf. Control, Vol. 17, No. 2, pp. 161-1764, (Sept. 1970).

12. N.U. Ahmed, "Comments on an Adaptive Pattern Classification", IEEE Trans. S.M.C., Vol. 12, No. 1, pp. 83-84, (Jan. 1971) (Co-authors V. Gourishankar, J.W. Enerson, Univ. of Alberta).

13. N.U. Ahmed and S.G.S. Shiva, "A Note on the Measure of Self-information", Int. J. of Elect. Eng. Education, Vol. 9, pp. 215-216, (June 1971).

14. N.U. Ahmed, "L_p and Orlicz Stability of a Class of Nonlinear Time Varying Feedback Control Systems", J. Inf. and Control, Vol. 19, No. 2, pp. 114-123, (Sept. 1971).

15. N.U. Ahmed, "On the Stability of a Class on Nonlinear Stochastic Systems", J. Inf. Control, Vol. 20, No. 3, pp. 276-293, (April 1972) (co-author K.L. Teo, Grad. Student).

16. N.D. Georganas and N.U. Ahmed, "Follow-up Controls for Hereditary Systems", Int. J. Systems Sciences, Vol. 4, No. 1, pp. 59-64 (Jan. 1973).

17. N.U. Ahmed, "Optimal Control of Stochastic Dynamical Systems", J. Inf. Control, Vol. 22, No. l, pp. 13-30, (Feb. 1973).

18. S.G.S. Shiva and N.D. Georganas and N.U. Ahmed, "Order Preserving Measures of Information", J. Applied Probability, Vol. 10, No. 3, pp. 666-670, (Sept. 1973).

19. N.U. Ahmed and N.D. Georganas, "Comments on Invariance in Linear Estimators", J. Math. Analysis and Applications, Vol. 44, No. 1, pp. 98-99, (Oct. 1973).

20. N.D. Georganas and N.U. Ahmed, "On a Case of Dynamic Feedback Optimization", Int. J. Systems Sciences, Vol. 14, No. 3, pp. 613-616 (1973).

21. N.D. Georganas and N.U. Ahmed, "Optimal Initial Data and Control for a Class of Linear Hereditary Systems", J. Inf. Control, Vol. 22, No. 4, pp. 399-402, (1973).

22. N.U. Ahmed and N.D. Georganas, "Optimal Control of a Dynamic Aquatic Ecosystem", J. Fisheries Research Board of Canada, Vol. 30, No. 4, pp. 576-579, (1973).

23. N.U. Ahmed and N.D. Georganas, "On Optimal Parameter Selection", Trans. IEEE Vol. AC-18, No. 3, pp. 313-314, (June 1973).

24. N.U. Ahmed, "Strong and Weak Synthesis of Nonlinear Systems on the Systems Space Gl", J. Inf. Control, Vol. 23, No. 1, pp. 71-85, (Aug. 1973).

25. N.U. Ahmed, "On Optimum Phase Locked Loop", Int. J. Electronics, Vol. 36, No. 2, pp. 253-259, (1973).

26. N.U. Ahmed and K.L. Teo, "On Almost Sure and Almost Uniform $L_p(p \geq 1)$ Stability", Int. J. Systems Sc., Vol. 4, No. 5, pp. 803-808, (1973).

27. N.U. Ahmed and K.L. Teo, "Weak Compactness of Conditional Probability Measures and its Application to Stochastic Linear Control Problems", Int. J. Systems, Sc. Vol. 5, No. 1, pp. 67-71 (1973).

28. N.U. Ahmed, "Optimal Control for a Class of Hereditary Dynamical Systems", Int. J. Systems, Sc. Vol. 5, No. 1, pp. 73-82, (1973).

29. N.U. Ahmed and K.L. Teo, "An Existence Theorem on Optimal Control of Partially Observable Diffusions", SIAM Journal on Control, Vol. 12, No. 3, pp. 351-355, (Aug. 1974).

30. N.U. Ahmed, "Optimal Control of a Class of Hereditary Systems with Lagging Control", Int. J. of Systems, Sc. Vol. 5, No. 3, pp. 287-293 (1974).

31. N.U. Ahmed, "Certain Topological Properties of Stochastic Processes Generated by a Family of Stochastic Ito Differential Equations", J. Ricerche di Automatica, Vol. 4, No. 1, pp. 1-13 (July, 1973).

32. N.U. Ahmed and K.L. Teo, "Optimal Feedback Control of Stochastic McShane Differential Systems", J. Applied Probability, Vol. 11, No. 2, pp. 302-309 (June 1974).

33. N.U. Ahmed and K.L. Teo, "Optimal Feedback Control for a Class of Stochastic Systems", Int. J. Systems, Sc. Vol. 5, No. 4, pp. 357-365, (1974).

34. N.U. Ahmed and K.L. Teo, "Stochastic Bang-Bang Control", IEEE Trans. Automatic Control, Vol. 19, No. 1, pp. 73-75 (Feb. 1974).

35. N.U. Ahmed, K.L. Teo and H.W. Wong, "On Optimal Parameter Selection for Parabolic Differential Systems", IEEE Trans. Automatic Control, Vol. 19, No. 3, pp. 286-287, (1974).

36. N.U. Ahmed, "Book Review (Invited by the IEEE Information Dissemination Committee), Optimal Control of Differential and Functional Equations, by J. Warga", Vol. 19, No. 2, pp. 174-175 (1974).

37. N.U. Ahmed, "Existence of Optimal Controls with Frequency Constraints in the Absence of Convexity", Int. J. Systems Sciences, Vol. 5, No. 12, pp. 1165-1178, (1974).

38. N.U. Ahmed, "Existence of Optimal Control for a Class of Hereditary Systems with Lagging Control", J. Information and Control, Vol. 26, No. 2, pp. 178-185, (Oct. 1974).

39. N.U. Ahmed, "Solution of Generalized Stochastic Optimization Problems", Int. J. Systems Sciences, Vol. 6, No. 2, pp. 133-156 (1975).

40. N.U. Ahmed, "Optimal Control of an Economic System Modeled on the Basis of Cyclical Growth Theory", Int. J. of Control, Vol. 25, No. 6, pp. 827-835 (1977) (Co-authors G.J. Jutras, S.H. Yeo).

41. N.U. Ahmed, "Comments on a Selector Theorem in Banach Spaces", J. of Optimization Theory and Applications, Vol. 19, No. 1, pp. 117-118 (1976).

42. N.U. Ahmed and K.L. Teo, "Necessary Conditions for Optimality of Cauchy Problem for Parabolic Partial Differential Systems", SIAM J. of Control, Vol. 13, No. 5, pp. 981-993, (1975).

43. N.U. Ahmed, "Optimal Control of Stochastic Ito Differential Systems with Fixed Terminal Time", Advances in Applied Probability, Vol. 7, No. 1, pp. 154-178, March Issue (1975).

44. N.U. Ahmed and K.L. Teo, "On Optimal Parameter Selection for Stochastic Ito-Differential Systems", Int. J. of Systems Science, Vol. 6, No. 8, pp. 749-754 (1975)

45. N.U. Ahmed and P.H. Chang, "An Optimum Phase-Locked Loop with a Non-Ideal Filter", Int. J. of Electronics, Vol. 38, No. 3, pp. 331-336 (1975).

46. N.U. Ahmed, "An Approximation Theorem for Continuous Functions on L_p Spaces Including Representation", J. Information and Control, Vol. 30, No. 2, pp. 143-149.

47. N.U. Ahmed and M. Aboud, "Optimum Design of a Feedback Controller for Phase-Locked-Loop", IEEE Trans. Comm. Soc., Vol. 23, No. 12, pp. 1521-25 (Dec. 1975)

48. N.U. Ahmed, "A Simple Gradient Algorithm for Least Squares Estimation of System Parameters", Int. J. of Systems Sc. Vol. 7, No. 6, pp. 673-677 (1976).

49. N.U. Ahmed, "Existence of Optimal Band Limited Controls Without Convexity Condition", J. Information and Control, Vol. 31, No. 3, pp. 201-215.

50. N.U. Ahmed and K.L. Teo, "On Optimal Control of Systems Governed by Quasilinear Integropartial Differential Equations of Parabolic Type", Jr. of Math. Analysis and Applications, Vol. 59, No. 1, pp. 33-59, (1977).

51. N.U. Ahmed and A.K. Sen, "Fuel Savings in the Whecon Attitude Control System Using an Asymmetric Controller", IEEE Trans. on Aerospace and Elect. Systems, Vol. AES-12, No. 3, (1976).

52. N.U. Ahmed, "Optimal Control of a Class of Strongly Nonlinear Parabolic Systems", Jr. of Math Analysis and Appl., Vol. 61, No. 1, pp. 188-207, (1977).

53. N.U. Ahmed, "A Note on Satellite Dynamics", IEEE Trans. Aerospace, Vol. AES-12, No. 6, pp. 819-822, (1976) (Co-author A. Guibord).

54. N.U. Ahmed, "Optimal Control of Systems Governed by a Class of Nonlinear Evolution Equations in a Reflexive Banach Space", Jr. Optimization Theory and Applications, Vol. 25, No. 1, pp. 57-81, (1978).

55. H.W. Wong and N.U. Ahmed, "Optimal Control of Precession of Spin Stabilized Satellite Subject to Random Perturbation", IEEE Trans. Aerospace and Elect. Sys., Vol. 13, No. 2, pp. 172-179, (1977).

56. N.U. Ahmed, "Optimal Control of Generating Policies in a Power System Governed by a Second Order Hyperbolic Partial Differential Equation", SIAM J. on Contr. and Opt., Vol. 15, No. 6, pp. 1016-1033, (1977) (also presented at the JACC Conf. 1977).

57. N.U. Ahmed and K.L. Teo, "LP ($p \geq 1$) Stability in the Mean $m (m \geq 1)$ of a Class of Stochastic Volterra Integral Equations", Information Sciences, Vol. 12, pp. 129-150, (1977).

58. N.U. Ahmed and A. Guibord, "Non-Saturating Open Loop and Feedback Controls", IEEE Trans, Aerospace, AES-15, 1, pp. 199-203, (1979).

59. N.U. Ahmed, "Stability Analysis of a Transmission Line with Variable Parameters", IEEE Trans, Circuit and Systems, CAS 25, 10, pp. 878-885, (1978).

60. N.U. Ahmed, "Optimal Availability of Maintainable Systems", IEEE Trans. on Reliability, R-27, 1, pp. 41-45, (1978).

61. N.U. Ahmed and K.L. Teo, "On the Optimal Controls of a Class of Systems Governed by Second Order Parabolic Partial Delay-Differential Equations with First Boundary Conditions", Annali D. Mathematica Pura Ed Applicata, Vol. 7, No. 5, pp. 361-392, (1979).

62. N.U. Ahmed, "Nonlinear Integral Equations on Reflexive Banach Spaces with Applications to Stochastic Integral Equations and Abstract Evolution Equations", Journal of Integral Equations, Vol. 1, No. 1, pp. 1-15, (1979).

63. N.U. Ahmed, "Some Comments on Optimal Control of a Class of Strongly Nonlinear Parabolic Systems", J. of Math. Anal. and Appl. Vol. 68, No. 2, pp. 595-598, (1979).

64. N.U. Ahmed, "Optimal Control of Stochastic Systems", Probabilistic Analysis and Related Topics, (ed. A. T. Bharucha-Reid), Academic Press Series, Vol. 2, pp. 1-68, (1979).

65. N.U. Ahmed, "Stability and Control of a Trap Mode in Synchronous Orbit", IEEE Trans. AES 16, No. 4, pp. 474-480, (1980).

66. N.U. Ahmed and A.K. Sen, "Closed Loop Control of Synchronous Orbit Using Lyapunov Theory", IEEE Trans. AES 17, No. 1, pp. 48-53, (Jan. 1981).

67. N.U. Ahmed, "Stochastic Control on Hilbert Space for Linear Evolution Equations with Random Operator Valued Co-Efficients", SIAM Journal of Control and Optimization, Vol. 19, No. 3, pp. 401-430, (1981).

68. N.U. Ahmed, "Necessary Conditions of Optimality for a Class of Second-Order Hyperbolic Systems with Spatially Dependent Controls in the Coefficients", J. Opt. Th. and Appl., Vol. 38, pp. 423-446, (1982).

69. N.U. Ahmed, "Nonlinear Optimal Feedback Regulation of Satellite Angular Moments", IEEE Trans. AES, 18, pp. 1-10, (1982).

70. N.U. Ahmed, "Stability of a Class of Stochastic Distributed Parameter Systems with Random Boundary Conditions", J. of Math., Analysis & Appl., Vol. 92, No. 1, pp. 274-298, (March 1983).

71. N.U. Ahmed, "Modeling of Flexible Spacecraft and Their Stabilization", International J. of Systems Science, Vol. 16, No. 5, pp. 535-551, (1985) (Co-author S.K. Biswas).

72. N.U. Ahmed, "Sufficient Conditions for Controllability of a Class of Distributed Parameter Systems", J. Systems and Control Letters, Vol. 2, pp. 237-242, (1982).

73. N.U. Ahmed and Y.C. Chen, "An Application of Optimal Control Theory in Fisheries Management", Int J. Systems Sc., Vol. 14, No. 4, 453-462 (1983).

74. N.U. Ahmed, "Optimal Voltage Regulation of Power Systems Under Transient Conditions", Elec. Power Systems Research 6, pp. 71-77 (1983) (Co-author S.K. Biswas).

75. N.U. Ahmed and T.E. Dabbous, "Modeling and Control of Reliability Dynamics of Large Scale Interconnected Power Systems", Int. J. System Sc., Vol. 14, No. 2, pp. 1321-1353, (1983).

76. N.U. Ahmed and M.M. Mittal, "Time Domain Modeling and Digital Simulation of Variable Frequency AC Motor Speed Control Using PPL Technique", IEEE Trans. on Industry Appl., Vol IA-19, No. 2, pp. 174-180 (1983).

77. N.U. Ahmed and S. Roy, "Power System Voltage Regulation Based on Voltage Feedback Using a Dynamic Capacitor", Elect. Power Systems Research, Vol. 6, pp. 81-100 (1983).

78. N.U. Ahmed, "A New Class of Generalized Nonlinear Functionals of White Noise with Application to Random Integral Equations", J. Stock. Analysis and Appl., Vol. 1 (2), pp. 139-158 (1983).

79. N.U. Ahmed, "Properties of Relaxed Trajectories for a Class of Nonlinear Evolution Equations on Banach Space", Siam J. of Control and Optimization, Vol. 21, No. 6, pp. 953-967, (Nov. 1983).

80. N.U. Ahmed and S.K. Biswas, "Existence of Solutions and Stability of a Class of Parabolic Systems Perturbed by Generalized White Noise on the Boundary", J. Stoch. Analysis and Appl., Vol. 2 (4), pp. 347-368 (1984).

81. N.U. Ahmed and T.E. Dabbous, "Non-linear Filtering of Diffusion Processes with Discontinuous Observations", J. Stoch. Analysis and Appl., Vol. 2 (4), pp. 87-106, (1984).

82. N.U. Ahmed and W. Colmeneres, "A Controllability Counter-Example", Int. J. Systems Sci., Vol. 15, No. 8, pp. 837-840, (1984) .

83. N.U. Ahmed, "Generalized Solutions via Integral Equations for Quasi-Linear Non homogeneous Parabolic Boundary Value Problems and Their Stochastic Versions", J. Integral Equations, Vol. 6, pp. 143-158 (1984).

84. N.U. Ahmed, "On the Maximum Principle for Time-Optimal Controls for a Class of Distributed-Boundary Control Problems", J. Opt. Th. and Appl., Vol. 45, No. 1, pp. 147-157 (1985).

85. N.U. Ahmed, "Finite-Time Null Controllability for a Class of Linear Evolution Equations on a Banach Space with Control Constraints", J. Opt. Th. and Appl., Vol. 47, No. 2, pp. 129-158 (1985).

86. N.U. Ahmed, S.S. Lim and K.F. Schenk, "Dynamic Optimal Economic Dispatch Under System Availability Constraints", Elect. Power System Research 9, pp. 11-27 (1985).

87. N.U. Ahmed, "Dynamic Optimal Economic Dispatch Under System Availability Constraints and Transmission Losses", Elect. Power Systems Research, Vol. 9, pp. 63-72 (1985) (Co-author S.S. Lim).

88. N.U. Ahmed and Q. Ahsan, "A Dynamic Model for Generation Expansion Planning", Elect. Power Systems Research, Vol. 9, pp. 79-86 (1985).

89. N.U. Ahmed, and T.E. Dabbous, "Nonlinear Filtering of Systems Governed by Ito Differential Equations with Jump Parameters", Journal of Mathematical Analysis and Applications, Vol. 115, No. 1, pp. 76-92, (1986).

90. N.U. Ahmed, "Abstract Stochastic Evolution Equations on Banach Spaces", J. Stoch. Analysis and Applications, Vol. 3, No. 4, pp. 397-432 (1985).

91. N.U. Ahmed and S.K. Biswas, "Stabilization of Systems Governed by the Wave Equation in the Presence of Distributed White Noise", in IEEE Trans. Auto Contr., Vol. AC-30, No. 9, pp. 1043-1045 (1985).

92. N.U. Ahmed, "Existence of Optimal Controls for a Class of Systems Governed by Differential Inclusions on a Banach Space", J. Opt. Th. and Appl., Vol. 50, No. 2, pp. 213-237, August (1986).

93. N.U. Ahmed, "Stability of a Class of Hybrid Systems Arising in Flexible Space-craft", J. of Optimization Theory and Applications, Vol. 50, No. 1, pp. 83-108, (July 1986) (Co-author S.K. Biswas).

94. N.U. Ahmed, H. Riaz, S.K. Biswas "Stochastic Modeling and Stabilization of Galloping Transmission Lines", J. Electric Power Systems Research, Vol. 10, No. 2, pp.137-143, (1986).

95. N.U. Ahmed, "Stabilizability of Certain Stochastic Systems", International J. of Systems Science, Vol. 17, No. 8, pp. 1175-1185, (1986).

96. N.U. Ahmed and S.K. Biswas, "Optimal Feedback Control of Power Systems Governed by Nonlinear Dynamics", J. of Optimal Control Applications and Methods, Vol. 7, pp. 289-303, (1986).

97. N.U. Ahmed, and Z.Y. Gao, "Feedback Stabilizability of Nonlinear Stochastic Systems with State Dependent Noise", Int. J. of Control, Vol. 45, No. 2, pp. 729-737 (1987).

98. N.U. Ahmed, "Identification of Linear Operators in Differential Equations on Banach Space", in "Operator Methods for Optimal Control Problems" (edited by S.J. Lee), Marcel Dekker Inc., Notes in Pure and Applied Mathematics, Vol. 108, Marcel Dekker Inc., New York (1987).

99. N.U. Ahmed, "Existence of Optimal Relaxed Controls for Differential Inclusions on Banach Space", "Nonlinear Analysis and Applications" (edited by V. Lakshmikantham), Marcel Dekker, Lecture Notes in Pure and Applied Mathematics, Vol. 109, Marcel Dekker Inc., New York, (1987).

100. N.U. Ahmed, "Abstract Stochastic Evolution Equations and Related Control and Stability Problems", Springer-Verlag Lecture Notes on Control and Information Sciences, (Edited by I. Lasiecka, R. Triggiani), Vol. 97, pp. 107-120, (1987).

101. N.U. Ahmed, T.E. Dabbous, H.W. Wong, "Gradient Method for Computing Optimal Controls for Stochastic Differential Equations", Journal of Stochastic Analysis and Applications, Vol. 5, No. 2, pp. 121-150, (1987).

102. N.U. Ahmed, H. Riaz and S.K. Biswas, "Modal Dynamics and Stabilizer Design for Galloping Transmission Lines", J. Electric Power Systems Research, Vol. 12, pp. 175-182, (1987).

103. N.U. Ahmed, T.E. Dabbous and Y.W. Lee, "Dynamic Routing for Computer Queuing Networks", International J. of Systems Science, Vol. 19, No. 6, pp. 967-977, (1988).

104. T.E. Dabbous, N.U. Ahmed, J.C. McMillan and D.F. Liang, "Filtering of Discontinuous Processes Arising in Marine Integrated Navigation Systems", IEEE Trans. on Aerospace and Electronic Systems, Vol. 24, No. 1, pp. 85-102, (1988).

105. N.U. Ahmed and S.S. Lim, "Stabilizing Control for Flexible Spacecraft Under the Influence of Orbital Perturbation", Int. J. of Science and Technology, Vol. 1, No. 2, pp. 5-20, (1988).

106. S.K. Biswas and N.U. Ahmed, "Mathematical Modeling and Control of Large Space Structures with Multiple Appendages", J. Mathematical and Computer Modeling, Vol. 10, No. 12, pp. 891-900, (1988).

107. N.U. Ahmed and P. Li, "On Exponential Stability of Infinite Dimensional Linear Systems with Bounded or Unbounded Perturbations", J. Applicable Analysis, Vol. 30, pp. 175-187, (1988).

108. T. Sellami and N.U. Ahmed, "An Algorithm for Computing Time-Optimal Controls with Examples from Robotics and Ecology", Journal of Mathematical Engineering in Industry, Vol. 2, No.2, pp. 113-125, (1989).

109. N.U. Ahmed and S.S. Lim, "Mathematical Modeling and Stabilization of Large Flexible Spacecraft Subject to Orbital Perturbation", Mathematical and Computer Modeling, Vol. 12, No. 12, pp. 1487-1504, (1989).

110. Peng. Li and N.U. Ahmed, "A Note on Stability of Stochastic Systems with Unbounded Perturbations", Stochastic Analysis and Applications, Vol. 7, No. 4, pp. 425-434, (1990).

111. S.K. Biswas and N.U. Ahmed, "Optimal Control of Large Space Structures Governed by a Coupled System of Ordinary and Partial Differential Equa-

tions", Journal of Mathematics of Control, Signals and Systems, Vol. 2, pp. 1-18, (1989), Temple University, U.S.A.

112. K.L. Teo, N.U. Ahmed and M.E. Fisher, "Optimal Feedback Control for Linear Stochastic Systems Driven by Counting Processes", Journal of Engineering Optimization", Vol. 15, No.1, pp. 1-16, (1990).

113. N.U. Ahmed, "Stochastic Initial Boundary Value Problems for a Class of Second Order Evolution Equations, Differential Equations & Applications", Ohio University Press, Vol. 1, pp. 13-19, (1989).

114. N.U. Ahmed and Peng. Li, "Invariance of Asymptotic Stability of Perturbed Linear Systems on Hilbert Space", Journal of Optimization Theory & Applications, Vol. 68, No. 1, pp. 75-93, (1991).

115. N.U. Ahmed and Peng. Li, "Feedback Stabilization for Perturbed Linear Systems", J. Optimization, D.D.R., Vol. 21, No. 6, pp. 913-924, (1990).

116. N.U. Ahmed, T.E. Dabbous and S.S. Lim, "Linear Filtering for a Class of Jump Processes Arising in Navigation Systems", IMA Journal of Mathematical Control & Information, Vol. 7, pp. 269-292, (1990).

117. P. Li, and N.U. Ahmed, "Feedback Stabilization of Some Nonlinear Stochastic Systems on Hilbert Space", Journal of Nonlinear Analysis: Theory, Methods and Applications, Vol. 17, No. 1, pp. 31-43, (1991) (Co-author P. Li).

118. J.Y. Lin and N.U. Ahmed, "An approach on Controllability Problems for Singular Systems", International Journal of Systems Science, Vol. 22, No. 4, pp. 675-690, (1991).

119. N.U. Ahmed, "Stability of Limit Measures Induced by Stochastic Differential Equations on Hilbert Space", Journal of Computers and Mathematics with Applications (CAMWA), Vol. 22, No. 8, pp. 41-48, (1991).

120. N.U. Ahmed and W. Chouikha, "Congestion Control Using Dynamic Routing and Flow Control", Journal of Stochastic Analysis and Applications, Vol. 10, No. 2, pp. 123-142.

121. N.U. Ahmed and P. Li, "Quadratic Regulator Theory and Linear Filtering under System constraints", Journal of Mathematical Control and Information, Vol. 8, pp. 93-107, (1991).

122. S.S. Lim and N.U. Ahmed, "Modeling and Control of Flexible Space Stations" Journal of Dynamics and Control, Vol. 2, pp. 5-33, (1992).

123. N.U. Ahmed and A. N. Mouadeb, "Optimal Regulators for Linear Systems with no Control Cost", Journal of Dynamics and Control, Vol. 1, pp. 341-355, (1991).

124. N.U. Ahmed, "Nonlinear Evolution Equations on Banach Space", Journal of Applied Mathematics and Stochastic Analysis, Vol. 4, No. 3, pp. 187-202, (1991).

125. N.U. Ahmed and X. Xiang, "Existence of Periodic Solutions for Semilinear Evolution Equations with Time-lags," Journal of Nonlinear Analysis, Theory, Methods and Applications , Vol. 18, No. 11 ,pp. 1063-1070, (1992).

126. P. Li and N.U. Ahmed, "Stabilization of Some Distributed Parameter Systems IEEE Transactions on Automatic Control", Vol. 37, No. 8, pp. 1184-1188, (1992).

127. N.U. Ahmed, "Semilinear and Quasilinear Stochastic Differential Equations in Banach Spaces, Series in Applicable Analysis-Recent Trends in Differential Equations, (Editor, R.P. Agarwal), Vol. 1, pp. 1-14, (1992).

128. N.U. Ahmed and T.E. Dabbous, "Parameter Identification for Partially Observed Diffusions", Journal of Optimization Theory and Applications, Vol. 75, No. 1, pp. 33-50, (1992).

129. N.U. Ahmed, "An Existence Theorem for Stochastic Nonlinear Evolution Equations on Banach Space", Journal of Stochastic Analysis and Applications, Vol. 10, No. 4, pp. 379-385, (1992).

130. N.U. Ahmed, "An Existence Theorem for Differential Inclusions on Banach Space", Journal of Applied Mathematics and Stochastic Analysis, Vol. 5, No. 2, pp. 123-130, (1992).

131. N.U. Ahmed, "Partially Observed Optimal Relaxed Feedback Controls for Infinite Dimensional Systems", IMACS Transactions.

132. N.U. Ahmed and X. Xiang, "Admissible Relaxation in Optimal Control Problems for Infinite Dimensional Uncertain Systems, Journal of Applied Mathematics and Stochastic Analysis, Vol. 5, No.3, pp 227-236, (1992).

133. N.U. Ahmed, "Optimal Control of Hydrodynamic Flow with Possible Applications to Artificial Heart, Dynamic Systems and Applications, Vol. 1, pp. 103-120, (1992).

134. N.U.Ahmed, "Distributed Parameter Systems", Encyclopedia of Physical Science and Technology, Vol. 5, pp. 277-300, (1992).

135. N.U. Ahmed, and J.M. Skowronski, "Stability and Control of Nonlinear Flexible Systems", J. of Dynamic Systems and Applications, Vol.2, No.2, pp. 149-162, (1993).

136. N.U. Ahmed, and X. Xiang, "Properties of Relaxed Trajectories of Evolution Equations and Optimal Controls", SIAM. J. on Control and Optimization, Vol. 31, pp. 1135-1142, (1993).

137. N.U. Ahmed, "Nonlinear Stochastic Differential Inclusions on Banach Space", J. of Stochastic Analysis and Applications, Vol 12, No 1, pp1-10, (1994).

138. N.U. Ahmed and X. Xiang, "Optimal Control of Infinite Dimensional Uncertain Systems", Journal of Optimization Theory and Applications, Vol. 80, No. 2, pp. 261-272, (1994).

139. N.U. Ahmed, "Generalized Functionals of Brownian Motion", Special Issue of the Journal of Applied Mathematics and Stochastic Analysis, Vol. 7, No. III, pp. 247-267, (1994).

140. N.U. Ahmed, "Optimal Relaxed Controls For Nonlinear Infinite Dimensional Stochastic Differential Inclusions, Marcel Dekker, Lecture Notes in Pure and Applied Mathematics, Vol. 180, pp. 1-19, (1994).

141. N.U. Ahmed, and X. Xiang, "Existence of Solutions For a Class of Nonlinear Evolution Equations with Nonmonotone Perturbations", Journal of Nonlinear Analysis, Theory, Methods and Applications, Vol. 22, No. 1, pp. 81-89, (1994).

142. T. Yeap and N.U. Ahmed, "Feedback Control of Chaotic Systems", Journal of Dynamics and Control, Vol. 4, pp. 97-114, (1993).

143. N.U. Ahmed, "Optimal Control of Infinite Dimensional Systems Governed by Integro Differential Equations", Marcel Dekker, Lecture Notes in Pure and Applied Mathematics, Differential Equations, Dynamical Systems and Control Sciences, Vol. 152, pp. 383-402, (1993).

144. N.U. Ahmed, and Kerbal Sebti, "Optimal Control of Nonlinear Second Order Evolution Equations", Journal of Applied Mathematics and Stochastic Analysis, Vol. 6, No. 2, (1993).

145. N.U. Ahmed, and S.M. Radaideh, "Modified Extended Kalman Filtering", IEEE Trans. on Automatic Control, Vol. 39, No. 6, pp. 1322-1326, (1994).

146. N.U. Ahmed and Xinhong Ding, "On Invariant Measures of Nonlinear Markov Processes", Journal of Applied Mathematics and Stochastic Analysis, Vol. 6, No. 4, pp. 385-406, (1993).

147. N.U. Ahmed, "Evolution Equations on Fock Spaces Over Gaussian Random Fields", Dynamic Systems and Applications, Vol. 3, pp. 25-38, (1994).

148. N.U. Ahmed, "A Class of Singularly Perturbed Evolution Systems", J. Applied Mathematics and Stochastic Analysis, Vol. 7, No. 2, pp. 179-190, (1994).

149. N.U. Ahmed, and J. Zabczyk, "Nonlinear Filtering for Semilinear Stochastic Differential Equations on Hilbert Spaces", Publication of The Institute of Mathematics, Polish Academy of Sciences, Vol. 522, pp. 1-12, (1994).

150. Qingxin Zhu and N.U. Ahmed, "Dynamic Programming Equations of Stochastic Optimal Control in Banach Space", Journal of Stochastic Analysis and Applications, Vol. 13, No. 3, pp. 369-387, (1995).

151. N.U. Ahmed, "Compactness in Certain Abstract Function Spaces with Applications to Differential Inclusions", Discussiones Mathematicae: Differential Incusions, Vol. 15, No. 1, pp. 21-28, (1995).

152. N.U. Ahmed, and X. Xiang, "Nonlinear Boundary Control of Semilinear Parabolic Systems", SIAM J. of Control and Optimization, SICON. 34, No. 2, 473-490, (1996).

153. N.U. Ahmed, and X. Xiang, "Necessary Conditions of Optimality for Infinite Dimensional Uncertain Systems", J. Mathematical Problems in Engineering, Vol. 1(2), pp. 179-191, (1995).

154. N.U. Ahmed and Kerbal Sebti, "On Approximate Controllability for Semilinear Systems", J. Nonlinear World, Vol. 2, pp. 53-67, (1995).

155. N.U. Ahmed and X. Ding, "A Semilinear McKean-Vlasov Stochastic Evolution Equation in Hilbert Space", J. Stochastic Processes and Applications, Vol. 60, pp. 65-85, (1995).

156. N.U. Ahmed, "Generalized Solutions for Linear Systems Governed by Operators Beyond Hille-Yosida Type", Publicationes Mathematicae Debrecen, Hungary, Vol. 48, No. 1-2, pp. 45-64, (1996).

157. N.U. Ahmed, S.M. Radaideh, "Identification of Linear Stochastic Systems Based on Partial Information, Journal of Applied Mathematics and Stochastic Analysis, (3), 249-260, (1995).

158. Qingxin Zhu, and N.U. Ahmed, "Some Results on Parabolic Equations on Banach Space", Nonlinear Analysis, Theory and Methods, Vol. 24(9), pp. 1305-1319, (1995).

159. N.U. Ahmed, and T.H. Yeap, "Performance Analysis of Slotted Aloha Protocol for Optical Fiber Local Area Networks with Star Topology", Vol. 13(4), pp. 389-410, (1995).

160. N.U. Ahmed, "Optimal Control of Infinite Dimensional Systems Governed by Functional Differential Inclusions", Discussiones Mathemathematicae Differential Inclusions, Vol. 15, pp. 75-94, (1995).

161. N.U. Ahmed, "Mathematical Problems in Modeling Artificial Heart", Mathematical Problems in Engineering (MPE), Vol. 1(2), pp. 245-254, (1995).

162. N.U. Ahmed, and X. Xiang, "Necessary Conditions of Optimality for a Class of Semilinear Uncertain Systems with Unbounded Monotone Operator", WSSIAA, Vol. 4, Dynamical Systems and Applications, (1995).

163. N.U. Ahmed, "Generalized Functionals of Gaussian Random Fields and Evolutions Thereon", Dynamic Systems and Applications, Vol. 4, pp. 27-46, (1995).

164. N.U. Ahmed, "Optimal Relaxed Controls for Infinite Dimensional Stochastic Systems of Zakai Type", SIAM Journal of Control and Optimization, Vol. 34(5), pp. 1592-1615, (1996).

165. S.K. Biswas, N.U. Ahmed, "Disturbance Rejecting Optimal Regulation of Hyperbolic Systems", Math.Control Signals Systems, Vol. 8, pp. 241-256, (1995).

166. P. Li, S.S. Lim, G. Vukovich and N.U. Ahmed, "Stability and Robustness Analysis of Boundary Control of Flexible Space Structure", Int. J. Systems Science, Vol. 26, No. 10, pp. 1759-1776, (1995).

167. N.U. Ahmed and Tao Wan, "Exact Boundary Controllability of Electromagnetic Fields in General Regions, Dynamic Systems & Applications, Vol. 5, pp. 229-244, (1996).

168. N.U. Ahmed, "Existence and Uniqueness of Measure Valued Solutions for Zakai Equation", Publicationes Mathematicae, Debrecen, Vol. 49, No. 3-4, pp. 251-264, (1996).

169. N.U. Ahmed, S.M. Radaideh, "A Powerful Numerical Technique Solving Zakai Equation for Nonlinear Filtering", Dynamics and Control, Vol. 7, No. 3, (July 1997).

170. N.U. Ahmed, X. Xiang, "Nonlinear Uncertain Systems and Necessary Conditions of Optimality", SIAM Journal Contr. and Optimization, Vol. 35, No. 5, pp. 1755-1772, (1997).

171. N.U. Ahmed and S. Kerbal, "Stochastic Systems Governed by B-evolutions on Hilbert Space", Proceedings of the Royal Society of Edinburgh, U.K 127A, pp. 903-920, (1997).

172. N.U. Ahmed, and J. Zabczyk, "Partially Observed Optimal Controls for Nonlinear Infinite Dimensional Stochastic Systems", DSA, 5, pp. 521-538, (1996).

173. N.U. Ahmed, M. Fuhrman and J. Zabczyk, "Regular Solutions of Filtering Equations in Infinite Dimensions", Journal of Functional Analysis", Vol. 143, No. 1, pp. 180-204, (1997).

174. N.U. Ahmed, "Optimal Control for Linear Systems Described by m-times Integrated Semigroups", Publicationes Mathematicae, Debrecen, Vol. 50, No. 1-2, pp. 1-13, (1997).

175. N.U. Ahmed, "Measure Solutions for semilinear Evolution Equations with Polynomial Growth and their Optimal control", Discussionnes Mathematicae Differential Inclusions, Vol. 17(2), pp. 5-27, (1997).

176. N.U. Ahmed, "Stochastic B-Evolutions on Hilbert spaces", Nonlinear Analysis, Theory, Methods and Applications, Vol. 30(1) pp. 199-109, (1997).

177. N.U. Ahmed, "Linear Systems Governed by Operators Beyond Hille-Yosida Semigroup", Advances in Nonlinear Dynamics, (dedicated to Prof. Leela on her 60th birthday), A volume of the International series of Scientific monographs on Stability and Control (Ed. S.Sivasundaram and A.A.Martynyuk, pp. 197-207, (1997).

178. X. Xiang, and N.U. Ahmed, "Infinite Dimensional Functional Differential Inclusions and Necessary Optimization Conditions", Nonlinear Analysis, Theory, Methods and applications, Vol. 30(1), pp. 429-436, (1997).

179. N.U. Ahmed and H. Harbi, "Stability of Suspension Bridge 1: Aerodynamic and Structural damping", Mathematical Problems in Engineering, Vol. 4, pp. 73-98, (1998).

180. N.U. Ahmed, "Measure Solutions for semilinear Systems with Unbounded Nonlinearities", Nonlinear Analysis : Theory, Methods and Applications, Vol. 35, pp. 487-503, (1998).

181. S. Kerbal Sebti, and N.U. Ahmed, "Optimal Boundary Control of Distributed Systems Involving Dynamic Boundary Conditions", J. Mathematical Problems in engineering, Vol. 3, pp. 387-411, (1998).

182. N.U. Ahmed and H. Harbi, "Mathematical Analysis of Dynamic Models of Suspension Bridge", SIAM Journal of Applied Mathematics, Vol. 58, No. 3, pp. 853-874, (1998).

183. N.U. Ahmed, "Optimal control of Turbulent Flow as Measure Solutions", International Journal of Computational Fluid Dynamics, (IJCFD-1026), pp. 1-13, (1998).

184. N.U. Ahmed, "Relaxed solutions for Stochastic Evolution equations on Hilbert Space with Polynomial Nonlinearities", Publicationes Mathematicae Debrecen, Vol. 54 (1-2) pp. 76-101, (1999).

185. N.U. Ahmed and H. Harbi, "Torsional and Longitudinal Vibration of Suspension Bridge Subject to Aerodynamic Forces", J. Mathematical Problems in Engineering, Vol. 4, pp. 393-421, (1999).

186. N.U. Ahmed and M.A. Rahim, "A Dynamic Model for Crime Control and Policy Evaluation", Mathematical Problems in Engineering, (1999), (accepted)

187. N.U. Ahmed, "A General Result on Measure Solutions for Semilinear Evolution Equations", Nonlinear Analysis: Theory, Methods and Applications, (accepted)

188. N.U. Ahmed, "A General Mathematical Framework for Stochastic Analysis of Suspension Bridges", Nonlinear Analysis, Theory, Methods and Applications Series B, (accepted).

189. N.U. Ahmed, "Measure Solutions for Semilinear and Quasilinear Evolution Equations and Their Optimal Controls", (by invitation) Special Issue, Nonlinear Analysis: Theory, Methods and Applications, (accepted)

190. N.U. Ahmed, "Measure Solutions for Impulsive Systems in Banach Spaces and Their Control", Dynamics of Continuous, Discrete and Impulsive Systems, Vol. 6, pp. 519-535, (1999).

191. N.U. Ahmed, "Optimal Impulse Control for Impulsive Systems in Banach Spaces", International Journal of Differential Equations and Applications, Vol. 1 (1), (2000), pp. 37-52.

192. N.U.Ahmed, "Mathematical Model and Optimal Control of Flow Induced Vibration, of Pipe Lines", Discussiones Mathematicae Differential Inclusions Vol. 19(1), (2000).

193. N.U. Ahmed, "Systems Governed by Impulsive Differential Inclusions on Hilbert Space", Nonlinear Analysis: Theory, Methods and Applications, (accepted)

REFEREED CONFERENCE PAPERS

1. N.U. Ahmed and G.S. Glinski, "A Mathematical Model of Magnetization and Hysteresis", Proc. Nat. Elect. Conf., Chicago 1964, Vol. 20, pp. 807-811

2. N.U. Ahmed and G.S. Glinski, "Properties of a Parametric Family of Symmetric Non-Gaussian Distribution in Euclidean n-Space", 7th Pak. Stat. Conf., Dacca, (Feb. 28-Mar. 2, 1967)

3. N.U. Ahmed, "Estimation of the Settling Time of a Class on Nonlinear Time-Varying Feedback Control Systems", Proc. 6th Allerton Conf. on Circuit and Systems Theory, Illinois, (Oct. 2-4, 1968), pp. 133-141.

4. N.U. Ahmed, "Optimal Control of a Large Class of Stochastic Differential Systems with Finite Memory", Proc. 7th Allerton Conf. on Circuit and Systems Theory, Illinois (Oct. 8-10, 1969), pp. 44-53.

5. N.U. Ahmed, "$L_p (p \geq 1)$ and Orlicz Stability of a Class of Nonlinear Feedback Systems", Proc. 7th Allerton Conf. on Circuit and Systems Theory, Illinois (Oct. 8-10, 1969), pp. 702-709.

6. N.U. Ahmed, "Optimal Control for a Class of Nonlinear Hereditary Dynamical Systems", Proc. 8th Allerton Conf. on Circuit and Systems Theory, Illinois, (Oct. 7-9, 1970), pp. 526-535.

7. N.U. Ahmed, "Constrained Optimization of Nonlinear Filters, and Systems on an Appropriate Systems Space", Proc. 8th Allerton Conf. on Circuit and Systems Theory, Illinois (Oct 7-9, 1970), pp. 111-120.

8. N.U. Ahmed, "A Modified Iterative Procedure for Computing Time-Optimal Controls for Linear Systems", Proc. 5th Annual Princeton Conf. on Information Systems Sciences, Princeton, March 1971, pp. 473-474 (Co-author D.K. Sen, Grad. Student).

9. N.U. Ahmed, G.S. Glinski and B.S. Rao, "Another Application of Optimal Control Theory to Systems with Feel", Proc. 3rd South Eastern Symp. on Systems Theory, Atlanta (April 1971), pp. D4(0-8).

10. N.U. Ahmed and S.G.S Shiva, "Some Results Arising from Information Entropy", Proc. Int. Conf. on Comm. Montreal, (June 1971), pp. 43 (15-18).

11. N.U. Ahmed and K.L. Teo, "On the Stability of a Class of Nonlinear Stochastic Systems with Application to Distributed Parameter Systems", Proc. IFAC Conf. Dist. Parameter Systems, Calgary, (June 1971), pp. 2-6, (1-8).

12. N.U. Ahmed, "Generalized Stochastic Optimization with Applications to Stochastic Control and Optimal Design of Random Systems", Proc. 9th Ann. Allerton Conf. on Circuit and Systems Theory, Illinois, (Oct. 6-8, 1971), pp. 49-57.

13. N.U. Ahmed, "Optimal Bandlimited Controls for Differential Systems without Convexity Condition", Proc. 6th Princeton Conf. Inf. Sc. and Systs., 1972, pp. 686-.

14. N.U. Ahmed, "On the Attainability Function of Stochastic Systems", Proc. 6th Princeton Conf. Inf. Sc. and Systems, 1972, pp. 706-.

15. N.U. Ahmed, "On Optimal Feedback Control of Stochastic Systems and Certain Associated Mathematical Problems", Proc. 10th Ann. Allerton Conf. on Cir. Systems Theory, (Oct. 1972), pp. 532-533.

16. N.U. Ahmed, "An Optimal Phase-Locked Loop", 6th Biennial Symp. on Comm. Theory and Signal Processing, Queen's University, Kingston, Ont., (Aug. 28-30, 1972).

17. N.U. Ahmed and F. Garcia,"Mathematical Model of the Dynamics of a Living Lake and Optimal Control to Prevent Its Eutrophication", Proc. Canadian Conf. on Automatic Control, New Brunswick, (Sept. 24-25, 1973), pp. 7-3 (1-13).

18. N.D. Georganas and N.U. Ahmed,"Optimal Control in Traffic Problems", Proc. 6th Midwest Symposium on Circuit Theory, University of Waterloo, April 1973, pp. XVII 6.1-6.2.

19. N.U .Ahmed and S.H. Yeo, "Optimal Control of a National Economic System Modelled on the Basis of Cyclical Growth Theory", Proc. 8th Ann. Princeton Conf. on Inf. and Systems Sciences (Feb. 1974).

20. N.U. Ahmed and M.A. Aboud, "Synthesis of Optimum Phase Locked Loop Using Optimal Control Theory", 7th Biennial Symp. on Comm., Kingston, Ont. (May 1974)

21. N.U. Ahmed, "An Approximation Theorem for Continuous Functions on L_p Spaces", SIAM 1974 Fall Meeting, Alexandria, VA, Oct. 23-26, 1974, Abstract, SIAM Review, Vol. 17, No. 2, p. 374 (April 1975).

22. N.U. Ahmed, "Optimal Control of Stochastic Systems Governed by Ito Differential Equations", Ann. Meeting of the AM. Math. Soc., 1975, Washington D.C., Jan 23-27, (Invited session).

23. N.U. Ahmed and S.H. Wong, "A Minimum Principle for Systems Governed by Ito Stochastic Differential Systems with Markov Jump Parameters", Second Kingston Conf. on Diff. Games and Control Theory, Rhode Island, June 1976.

24. N.U. Ahmed, "Optimal Control of Generating Policies in a Power System Governed by a Second Order Hyperbolic Partial Differential Equation", presented at the 1977 JACC (Joint Automatic Cont. Conf.), San Francisco, pp. 1339-1344.

25. N.U. Ahmed, "Stochastic Control on Hilbert Space for Linear Evolution Equations with Random Operator Valued Coefficients", 9th Conf. on Stochastic Proc. and their Appl., Evanston, IL, 6-10 Aug. 1979, Abstract in Advances in Appl. Probability, 12, 2, pp. 263-318.

26. N.U. Ahmed, "Stability of a Class of Stochastic Distributed Parameter Systems with Random Boundary Conditions", the 20th IEEE Conf. on Decision and Control, Dec. 1981, San Diego, CA.

27. N.U. Ahmed, "A Class of Generalized Nonlinear Functions of Brownian Motion", 10th Conf. on Stochastic Proc. and Appl., Montreal, Canada, Aug. 23-28, 1981 (Man. page 25).

28. N.U. Ahmed, "Control Theory and Its Prospects for Developing Nations", the 7th Ann. Conf. on Applied Science for Muslim World Development, Purdue University, West Lafayette, Sept. 1981, Vol. II, pp. 361-372.

29. M. Mittal, N.U. Ahmed, "Time Domain Modelling and Digital Simulation of Variable Frequency A.C. Motor Speed Control Using PPL Technique", IEEE's 1981 Industrial Applications Society Annual Meeting, Philadelphia, Oct. 5-9, 1981, pp. 1191-1196.

30. S. Roy and N.U. Ahmed, "Voltage Regulation by Capacitors in a Dynamic System", IEEE-PAS Winter Power Conf. 1982).

31. N.U. Ahmed, "Properties of Relaxed Trajectories for a Class Nonlinear Evolution Equations on a Banach Space", 30 Anniversary Meeting of SIAM, July 19-23, 1982, Stanford University, Stanford, CA, Proc. 21st CDC, Orlando, (1982), pp. 1073-1077, (invited talk).

32. N.U. Ahmed, T.E. Dabbous, S.K. Biswas, "Modelling of Reliability Dynamics of Large Interconnected Power Systems", 4th Internat. Symposium on Large Engineering Systems, June 9-11, 1982, University of Calgary, pp. 173-178.

33. S.K. Biswas and N.U. Ahmed, "Mathematical Modelling and Stabilization of Flexible Space Crafts", SIAM 1983 Fall Meeting, Nov. 7-9, 1983, Norfolk, Virginia

34. N.U. Ahmed and S.K. Biswas, "Stabilization of the Wave Equation in a Bounded Domain by Localized Feedback and Exact Controllability", Proc. 22nd IEEE-CDC, Dec. 14-16, 1983, San Antonio, Texas, 1983.

35. N.U. Ahmed and Y.W. Lee,"A Dynamic Routing for Computer Queueing Networks", Proc. 12th IASTED Int. Conf. on Appl. Simulation and Modelling, Montreal, Quebec, June 3-5, 1985.

36. N.U. Ahmed, "Stability of Distributed Parameter Systems", Int. Conf. on Integral Methods in Science and Engineering, the University of Texas at Arlington, Texas, March 18-21, 1985.

37. N.U. Ahmed, "Stochastic Evolution Equations Perturbed by Generalized White Noise on the Boundary", 15th Int. Conf. on Stochastic Processes and

their Applications, Nagoya, Japan, July 2-6, 1985, Abstract to appear in the Journal of Stochastic Processes and their Applications.

38. N.U. Ahmed, "Nonlinear Filtering of Systems Governed by Ito Differential Equations with Jump Parameters", 15th Int. Conf. on Stochastic Processes and their Applications, Nagoya Japan, July 1-6, 1985, Abstract to appear in the Journal of Stochastic Processes and their Applications.

39. N.U. Ahmed and S.S. Lim, "Modelling and Stabilization of Flexible Spacecraft under the Influence of Orbital Perturbation", The 26th IEEE Conf. on Decision and Control, Los Angeles, California, Dec. 9-11, 1987.

40. N.U. Ahmed, "Identification of Systems Governed by Nonlinear Evolution Equations", SIAM 35th Anniversary Meeting, Oct. 12-15, 1987, Denver Colorado.

41. N.U. Ahmed and S.S. Lim, "Towards Modelling and Control of Large Space Stations", Proc. of 28th IEEE Conference on Decision and Control, Dec. 13-15, 1989, (pp.803-808), Tampa, Florida.

42. N.U. Ahmed and S.S. Lim, "Modelling and Control of Flexible Space Stations (Slew Maneuvers)", Proc. of 3rd Annual Conference on Aerospace Computational Control, Aug. 28-30, 1989, Oxnard, California (to appear).

43. S.S. Lim and N.U. Ahmed, "Vibration Control of Slew Maneuvers for Large Spacecraft", Proc. of Canadian Conference on Electrical and Computer Engineering, Sept. 17-20, 1989, (pp. 650-653), Montreal, Canada.

44. N.U. Ahmed and Peng. Li, "Exponential Stabilization of Perturbed Stochastic Systems", Proceeding of Canadian Conference on Electrical and Computer Engineering, (pp. 203-205) Montreal, 1989.

45. N.U. Ahmed and Peng. Li, "Stabilizing Controllers for Perturbed Distributed Parameter Systems", Proceedings of 28th IEEE-Conference on Decision & Control, (pp. 1446-1447) Tampa, Florida, 1989.

46. N.U. Ahmed, "Computer Aided Design of Optimal Regulators and Filters with Constraints", Proc. 14th IFIP Conference on System Modelling and Optimization, Leipzig, GDR July 3-7, 1989 , Springer Lect. Notes. in Control and Information Sciences, Vol. 143, pp. 894-903, 1990.

47. N.U. Ahmed, "Hydro-Dynamic Problems of the Artificial Heart and its Optimal Control", 9th Canadian Symposiums on Fluid Mechanics, University of Western Ontario, London, Ontario, Canada, June 1990.

48. N.U. Ahmed and Peng. Li, "Optimal Filtering and LQR Theory with Constraints on System and Control Matrices", Proceeding of Canadian Conference on Electrical and Computer Engineering, pp. 10.4.1-10.4.4, Ottawa, Canada, Sept. 1990

49. N.U. Ahmed and S.S. Lim, "Vibration Control of Large Flexible Space Structures in a Noisy Environment", Proceeding of Canadian Conference on Electrical and Computer Engineering, pp. 10.5.1-10.5.4, Ottawa, Canada, Sept. 1990.

50. P. Li, S.S. Lim and N.U. Ahmed, "Robust Boundary Control of Flexible Structures Governed by Hybrid Dynamics", Proceeding of Canadian Conference on Electrical and Computer Engineering, pp. 18.3.1-18.3.4, Ottawa, Canada, Sept. 1990.

51. N.U. Ahmed and Peng. Li, "Exponential Stabilization of Semilinear Evolution Equations", Proc. 29th IEEE-CDC, Vol. 4, pp. 2137-2138, Hawaii, December 5-7, 1990.

52. N.U. Ahmed, "Partially Observed Optimal Relaxed Feedback Controls", 13th Imacs World Congress on Computation and Applied Mathematics, July 1991, Dublin, Ireland.

53. N.U. Ahmed and X. Xiang, "Necessary Conditions for Optimality in a Class of Infinite Dimensional Strongly Nonlinear Control Problems, Proc. 31st .IEEE-CDC, pp. 3100-3105, 1992.

54. N.U. Ahmed and X. Xiang, "Properties of Relaxed Trajectories of Evolution Equations and Optimal Control, Proc.31st .IEEE-CDC, pp. 3126-3128,1992.

55. N.U. Ahmed, and T. Yeap, "New Mathematical Models of an Optical Communication System, Proc. IEEE International Conference on Decision and Control, pp. 1671- 1677, 1993.

56. N.U. Ahmed, "Mathematical Challenges of the Artificial Heart, Proc. of The 9th International Conference on Mathematical and Computer Modeling, University of California, Berkeley, July 1993.

57. N.U. Ahmed, "Existence of Solutions of Nonlinear Stochastic Differential Inclusions on Banach Space, Proc. of the World Congress of Nonlinear Analysis, (Tampa, Florida, August 19-26,) , pp. 1699-1712, 1992.

58. N.U. Ahmed, "Functional Analysis on Wiener Measure Space and Evolution on Fock Spaces", Proc. of Dynamic Systems and Applications, Vol 1, pp. 1-10, 1994.

59. N.U. Ahmed and X. Xiang, "Nonlinear Boundary Control of Semilinear Parabolic Systems", Proc. 34 IEEE-CDC, New Orleans, Dec. 13-15, 1995.

60. N.U. Ahmed, "Stochastic B-evolutions on Hilbert Spaces, The Second World Congress of Nonlinear Analysts, Athens, Greece, July 10-17, 1996. (to appear in the Proc. WCNA-96).

61. N.U. Ahmed and X. Xiang, "Necessary Conditions of optimality for Systems Governed by Differential Inclusions, The Second World Congress of Nonlinear Analysts, Athens, Greece, July 10-17, 1996.

62. N.U. Ahmed, "Partially Observed Stochastic Control on Hilbert Space, Proc. DSA 2 (1996), pp. 19-30.

63. N.U. Ahmed, "Partially Observed Stochastic Control on Hilbert Space, presented at the International Conference on Applied Analysis, July 2-9, Samos, Greece, 1996.

64. S. Kerbal Sebti and N.U. Ahmed, "Necessary Conditions of Optimality for Systems Governed by B-evolutions, Proc. DSA 2 (1996), pp. 293-300.

65. Luis. O. Barbosa and N.U. Ahmed, "Dynamic Modelling of the IEEE 802.6 Medium Access Mechanism, Proc. 18th Biennial Symposium on Communications, Kingston, Ontario, Canada, (June 1996), pp. 21-24.

66. N.U. Ahmed and H. Harbi, "Dynamic Models of Suspension Bridge and their Stability, Proc. IASTED International Conference on Control, Cancun, Mexico, may 28-31,1997.

67. N.U. Ahmed, "Stability and Control of Flexible Structures, Satellites, Space Stations,and Suspension Bridges, Proc. of International Conference in Optimization Techniques and Applications, Perth, Australia, (1998), pp. 3-18.

68. N.U. Ahmed, "Measure Solutions for Infinite Dimensional Stochastic Systems, Proc. of International Conference in Optimization Techniques and Applications, Perth, Australia, (1998), pp. 1239-1246.

69. N.U. Ahmed and H. Harbi, "Stability of Suspension Bridge 11: Aerodynamic VS Structural Damping, Proc. of SPIE, Smart Systems for Bridges, Structures and Highways, Vol. 3325, (1998), pp. 276-289.

70. N.U. Ahmed and H. Harbi, "Stochastic Modeling and Stability of Suspension Bridge", Proceedings of SPIE, Smart Structures and Materials 2000, Smart Systems for Bridges, Structures and Highways, Vol. 3988, pp. 394-399

INVITED PAPERS

1. "Identification of Linear Operators in Differential Equations on Banach Space", presented at the Special Session on Operator Methods in Optimal Control of the American Math. Soc. Meeting, New Orleans, Louisiana, U.S.A., Jan. 1986, A.M.S. Abstracts, Vol. 7, No. 1, p. 150.

2. "Abstract Stochastic Evolution Equations and Related Control and Stability Problems", presented at the IFIP Conf. on Control Systems Governed by Partial Differential Equations, Gainesville, FLA, U.S.A., Feb. 1986 (to appear in the book edited by Prof. I. Lasiecka, published by Optimization Software Inc., published in New York).

3. "Optimal Controls for Differential Inclusions on Banach Spaces", presented at the 7th International Conference on Nonlinear Analysis and Applications, Arlington, Texas, July-Aug. 1986.

4. "On the Stabilization of Flexible Spacecraft with Multiple Appendages", 23rd Annual Meeting of the Society of Engineering Science, State University of New York at Buffalo, N.Y., Aug. 1986.

5. "Identification of Systems Governed by Evolution Equations on Banach Space", IFIP Conf. on Control of Systems Governed by Partial Differential Equations, Santiago de Compostela, Spain, July, 1987.

6. "Modelling & Control of Flexible Space Structures", 3rd Annual Conference on Aerospace: Computational Control, JPL/NASA/NSF/DoD, Los Angeles, California, August 28-30, 1989.

7. "Non-Linear Integral Equations on Banach Spaces", Canadian Math Society: Winter Meeting - December, 1989, Montreal, Canada.

8. "Computer Aided Design of Optimal Regulators and Filters with Constraints", IFIP Conf. on System Modelling and Optimization, Lapzig, DDR, July 1989, (to appear) in Lecture Notes in Springer-Verlag.

9. "Relaxed Controls for Stochastic Boundary Value Problems in Infinite Dimension", IFIP Conf. WG7.2 TC7, Irsee, F.R. Germany, April 1990, in Lecture Notes in Control and Information Sciences, Springer-Verlag,Vol. 149, pp. 1-10, 1991.

10. "Linear Filtering with Constraints and Partially Observed Controls", International Conference on Control and Modelling, July 17-20, 1990, University of Tehran, Iran, Proc. Control and Modelling, pp. 198-203.

11. "Stability and Control of Nonlinear Flexible Systems" Fifth Workshop on Control Mechanics in Honor of the Seventieth Birthday of ANGELO MIELE, January 21-23. 1992, University of Southern California, USA.

12. "Existence of Solutions of Nonlinear Stochastic Differential Inclusions on Banach Spaces", 1st World Congress of Nonlinear Analysts, Tampa, Florida, USA, Aug.19-26, 1992.

13. "Feedback Stabilization of Infinite Dimensional Stochastic Systems Applied to Shuttle Orbiter", SIAM Conference on Control and its Applications, Minneapolis, Minnesota, USA, Sept. 17-19,1992.

14. "Optimal Control of Infinite Dimensional Systems Governed by Integro Differential Equations", At the Joint conference of the AMS (American Mathematics Society) , Special Session on Integro Differential Equations, San Antonio, Texas, USA, Jan 13-16, 1993.

15. "Partially Observed Control of Stochastic Systems on Hilbert Space", At the Joint conference of the AMS (American Mathematics Society) , Special Session on Stochastic Systems and Applications, San Francisco, California, USA, Jan 4-7, 1995.

16. "Functional Differential Inclusions on Banach Spaces and Optimal Control", At the Joint conference of the AMS (American Mathematics Society), Special Session on Nonlinear Dynamics , San Francisco, California, USA, Jan 4-7, 1995.

17. "A Semlinear McKean-Vlasov Stochastic Evolution Equation in Hilbert Space", At the Stochastic Symposium, Banach center, Warsaw, Poland, March 11-21, 1995.

18. "Partially Observed Optimal Controls for Infinite Dimensional Stochastic Systems", At the Stochastic Symposium, Banach center, Warsaw, Poland, March 11-21, 1995.

PAPERS PRESENTED AT CONFERENCES WITHOUT PROCEEDINGS

1. "Mathematical Challenges of the Artificial Heart, Tenth Canadian Symposium on Fluid Dynamics", Saint John, New Brunswick, CANADA, June 4-6, 1992.

2. "An Existence Theorem for Differential Inclusions on Banach Space", Joint AMS Meetings, Baltimore, Jan. 8-11, 1992, (See Abstracts Vol. 13, No. 1, p. 75, 1992).

CONTRACT REPORTS

1. "Research and Design of Optimal Scanning Procedure for Phased Array Radar", DREO, Contract no. OSV 80-00026, (1980-81) (Co-author Y.C. Chen).

2. "Study of Jump Processes in Omega and Loran-C Radio Signals", DREO, Contract no. 05SV97714-4-2566, (1984-85) (Co-author T.E. Dabbous).

3. "Continuation of the Study of Jump Processes in Omega and Loran-C Radio Signals", Interim Report, DREO, Contract no. 24ST.97714-5-2948, (1985-86) (Co-authors S.S. Lim, T.E. Dabbous).

4. "Continuation of the Study of Discontinuous Jump Processes in Omega and Loran-C Radio Signals", Final Report, DREO, Contract no. 24ST.97714-5-2148, (1986-1987) (Co-author S.S. Lim).

I OPTIMAL CONTROL

1 PRACTICAL STABILITY OF IMPULSIVE DELAY DIFFERENTIAL EQUATIONS AND APPLICATIONS TO CONTROL PROBLEMS

George Ballinger and Xinzhi Liu

Department of Applied Mathematics
University of Waterloo
Waterloo, Ontario, Canada N2L 3G1

Abstract: This paper investigates practical stability properties of impulsive delay differential equations. Sufficient conditions for strong practical stability are obtained by using the method of Lyapunov functions and Razumikhin-type techniques. As an application we shall examine the practical stability properties of the LQ regulator problem with a delayed and disturbed control signal and with impulsive perturbations of the state.

Key words: Practical stability, impulsive delay differential equations, control, Lyapunov functions.

X. Yang et al. (eds.), Optimization Methods and Applications, 3-21.

1 INTRODUCTION

From a practical point of view, a physical system is considered to be stable if its state remains within certain bounds of the equilibrium position for all time. Such bounds will depend on the particular physical system and will generally depend on the constraints imposed on the initial values and the system disturbances. The state of a system may be mathematically unstable and yet the system may oscillate sufficiently near an equilibrium that its performance is considered to be acceptable. Many problems fall into this category including the travel of a space vehicle between two points and the problem, in a chemical process, of keeping the temperature within certain bounds. To deal with such situations, the notion of practical stability is very useful (Lakshmikantham & Liu (1993); Liu (1992)). Recently some practical stability results were obtained for LQ regulators with delayed and bounded disturbances in (Hou & Qian (1997)).

Since many evolutionary processes experience abrupt changes of state at certain moments of time, the study of impulsive differential equations as well as impulsive delay differential equations has recently been assuming greater importance (Ballinger & Liu (1999); Lakshmikantham, et al. (1989); Lakshmikantham & Liu (1993); Liu (1992)). In this paper we shall investigate practical stability properties of impulsive delay differential equations. Sufficient conditions for strong practical stability are obtained by using the method of Lyapunov functions and Razumikhin-type techniques. As an application we shall examine the practical stability properties of the LQ regulator problem considered in (Hou & Qian (1997), (1998)) with a delayed and disturbed control signal and with impulsive perturbations of the state.

2 PRELIMINARIES

Let $\Re$ denote the set of real numbers, $\Re_+$ the set of non-negative real numbers and $\Re^n$ the n-dimensional Euclidean space. For $a, b \in \Re$ with $a < b$ and for $S \subset \Re^n$ define

$$\begin{aligned}
PC([a,b],S) = \Big\{\psi : [a,b] \to S \,\Big|\, & \psi(t^+) = \psi(t)\ \forall t \in [a,b),\ \psi(t^-) \text{ exists in } S \\
& \forall t \in (a,b] \text{ and} \\
& \psi(t^-) = \psi(t) \text{ for all but at most a finite number of points} \\
& t \in (a,b]\Big\},
\end{aligned}$$

$$\begin{aligned}
PC([a,b),S) = \Big\{\psi : [a,b) \to S \,\Big|\, & \psi(t^+) = \psi(t)\ \forall t \in [a,b),\ \psi(t^-) \text{ exists in } S \\
& \forall t \in (a,b) \text{ and} \\
& \psi(t^-) = \psi(t) \text{ for all but at most a finite number of points} \\
& t \in (a,b)\Big\}, \text{ and}
\end{aligned}$$

$$PC([a,\infty),S) = \Big\{\psi : [a,\infty) \to S \,\Big|\, \forall c > a,\ \psi|_{[a,c]} \in PC([a,c],S)\Big\}.$$

In these definitions we have used the notation that $\psi(t^+) = \lim_{s\to t^+} \psi(s)$ and $\psi(t^-) = \lim_{s\to t^-} \psi(s)$. These function classes describe functions which are right-continuous on their domain and which are left-continuous except at simple jump discontinuities where they have a left-hand limit. Such functions may have only a finite number of jump discontinuities unless they are defined on a set of the form $[a,\infty)$. In this case they may have simple jump discontinuities at a countably infinite number of points which form an increasing sequence tending to infinity.

Given a constant $r > 0$, which will represent an upper bound on the delays of our system, we equip the linear space $PC_r = PC([-r,0],\Re^n)$ with the norm $\|\cdot\|_r$ defined by $\|\psi\|_r = \sup_{-r\le s\le 0}\|\psi(s)\|$ where $\|\cdot\|$ is the Euclidean norm on $\Re^n$. If $x \in PC([t_0 - r,\infty),\Re^n)$ where $t_0 \in \Re_+$ then for $t \ge t_0$ we define $x_t \in PC_r$ by $x_t(s) = x(t+s)$ for $-r \le s \le 0$. We also define $x_{t^-} \in PC_r$ by $x_{t^-}(s) = x(t+s)$ for $-r \le s < 0$ and $x_{t^-}(s) = x(t^-)$ for $s = 0$.

For any matrix $A \in \Re^{m\times n}$, $\|A\|$ denotes the matrix norm of A induced by the Euclidean vector norms, $\|\cdot\|$, on $\Re^m$ and $\Re^n$. For a square, symmetric, real matrix A, $\lambda_{max}(A)$ denotes the largest eigenvalue of A and similarly $\lambda_{min}(A)$ denotes the smallest eigenvalue of A. For a positive definite (symmetric) matrix $A \in \Re^{n\times n}$, $A^{\frac{1}{2}}$ denotes the square root of A which is defined to be the unique positive definite matrix satisfying $A^{\frac{1}{2}} \cdot A^{\frac{1}{2}} = A$ (Horn & Johnson (1985)).

Consider the following differential equation with bounded delays and disturbances and with impulses at fixed times,

$$x'(t) = f(t, x_t, \eta_t), \quad t \ne \tau_k \tag{2.1a}$$

$$\Delta x(t) = I(t, x_{t^-}, \eta_t), \; t = \tau_k. \tag{2.1b}$$

The state variable $x(t)$ is an n-dimensional vector. Here $x'(t)$ denotes the right-hand derivative of $x(t)$ with respect to t. The function $\eta(t)$ represents an uncertain, bounded disturbance; it is an m-dimensional vector quantity. We assume $\|\eta(t)\| \le \epsilon$ for all t where ϵ is some small positive number. This implies $\|\eta_t\|_r \le \epsilon$ for all t. We define $\Omega = \{\psi : [-r,0] \to \Re^m | \; \|\psi(s)\| \le \epsilon \text{ for all } s \in [-r,0]\}$. The impulse times are assumed to satisfy $0 = \tau_0 < \tau_1 < \tau_2 < \ldots < \tau_k < \ldots$ and $\lim_{k\to\infty}\tau_k = \infty$. We define $\Delta x(t) = x(t) - x(t^-)$. We assume the functionals $f, I : \Re_+ \times PC_r \times \Omega \to \Re^n$ and the function $\eta(t)$ satisfy sufficient conditions (Ballinger & Liu, 1999) to ensure the global existence of solutions of (2.1). Finally we impose an initial condition of the form

$$x_{t_0} = \phi, \tag{2.2}$$

where $\phi \in PC_r$.

By a solution of (2.1) we mean a function $x \in PC([t_0 - r,\infty),\Re^n)$ which satisfies $x(t) = \phi(t - t_0)$ for $t \in [t_0 - r, t_0]$; is piecewise continuous on $[t_0 - r,\infty)$ with discontinuities in (t_0,∞) occurring only at impulse times τ_k; satisfies (2.1a) with its right-hand derivative for all $t > t_0$ except possibly at times $t = \tau_k$; and satisfies (2.1b) at each $t = \tau_k$ for $t > t_0$.

Definition 2.1 *System (2.1) is said to be*

(P1) *practically stable with respect to* (λ, α) *where* $0 < \lambda \leq \alpha$ *if there exists some* $t_0 \in \Re_+$ *such that for all* $\phi \in PC_r$ *with* $\|\phi\|_r \leq \lambda$ *we have* $\|x(t)\| \leq \alpha$ *for all* $t \geq t_0$ *where* $x(t) = x(t, t_0, \phi, \eta)$ *is any solution of (2.1);*

(P2) *uniformly practically stable with respect to* (λ, α) *if (P1) holds for all* $t_0 \in \Re_+$*;*

(P3) *practically quasi stable with respect to* (λ, β, T) *where* $\lambda, \beta, T > 0$ *if there exists some* $t_0 \in \Re_+$ *such that for all* $\phi \in PC_r$ *with* $\|\phi\|_r \leq \lambda$ *we have* $\|x(t)\| \leq \beta$ *for all* $t \geq t_0 + T$*;*

(P4) *uniformly practically quasi stable with respect to* (λ, β, T) *if (P3) holds for all* $t_0 \in \Re_+$*;*

(P5) *strongly practically stable with respect to* $(\lambda, \alpha, \beta, T)$ *where* $0 < \beta < \lambda \leq \alpha$ *if there exists some* $t_0 \in \Re_+$ *such that for all* $\phi \in PC_r$ *with* $\|\phi\|_r \leq \lambda$ *we have* $\|x(t)\| \leq \alpha$ *for all* $t \geq t_0$ *and* $\|x(t)\| \leq \beta$ *for all* $t \geq t_0 + T$*;*

(P6) *uniformly strongly practically stable with respect to* $(\lambda, \alpha, \beta, T)$ *if (P5) holds for all* $t_0 \in \Re_+$*.*

Given a function $V \in C([t_0 - r, \infty) \times \Re^n, \Re_+)$ we define the generalized derivative of V along a particular solution $x(t) = x(t, t_0, \phi, \eta)$ of (2.1) by

$$D^+V(t, x(t)) = \limsup_{h \to 0^+} \frac{1}{h} \left[V(t + h, x(t + h)) - V(t, x(t)) \right], \tag{2.3}$$

for $t > t_0$ and $t \neq \tau_k$. We remark that the generalized derivative of V may equivalently be defined as

$$D^+V(t, \psi(0)) = \limsup_{h \to 0^+} \frac{1}{h} \left[V(t + h, \psi(0) + hf(t, \psi)) - V(t, \psi(0)) \right], \tag{2.4}$$

for $\psi \in PC_r$ providing V satisfies a local Lipschitz condition in its second variable and the composite function $f(t, x_t)$ is right-continuous at each $t \neq \tau_k$ in (t_0, ∞).

3 MAIN RESULTS

In this section we obtain conditions under which system (2.1) is strongly practically stable. Our main tool in developing practical stability results is through the use of Lyapunov functions and Razumikhin-type arguments.

Theorem 3.1 *Let* $V \in C([t_0 - r, \infty) \times \Re^n, \Re_+)$. *Assume that*

(i) *there exist strictly increasing functions* $\nu_1, \nu_2 \in C(\Re_+, \Re_+)$ *with* $\lim_{s \to \infty} \nu_1(s) = \infty$ *and for which*

$$\nu_1(\|x\|) \leq V(t, x) \leq \nu_2(\|x\|), \tag{3.1}$$

for $t \geq t_0 - r$, $x \in \Re^n$*;*

(ii) *there exists a constant* $u_0 \geq 0$ *and a function* $m \in PC([t_0 - r, \infty), \Re_+ \setminus \{0\})$ *which is continuous at each* $t \neq \tau_k$ *in* (t_0, ∞) *and satisfies* $\int_{t_0}^{\infty} m(s) ds =$

∞, such that for every solution $x(t) = x(t, t_0, \phi, \eta)$ of (2.1) we have

$$D^+V(t, x(t)) \le -m(t)\,(V(t, x(t)) - u_0) \text{ for } t \in (t_0, \infty),\ t \ne \tau_k \text{ if } V(t, x(t)) > u_0 \text{ and } V(t, x(t)) \ge V(s, x(s))e^{-\int_{t-t}^{t} m(\tau)d\tau} \text{ for } s \in [t-r, t]; \tag{3.2}$$

(iii) there exist constants $b_k \ge 0$ for $k = 1, 2, \ldots$ with $\sum_{k=1}^{\infty} b_k < \infty$ and for which

$$V(\tau_k, x + I(\tau_k, \psi, \eta_{\tau_k})) \le (1 + b_k)V(\tau_k, x), \tag{3.3}$$

for all $\tau_k > t_0$, $x \in \Re^n$ and $\psi \in PC_r$ with $\psi(0) = \psi(0^-) = x$;
then for all (λ, α, β) satisfying

$$0 < \nu_1^{-1}(Lu_0) < \beta < \lambda \le \nu_1^{-1}(L\nu_2(\lambda)) \le \alpha, \tag{3.4}$$

where $L = \prod_{\{k:\ \tau_k > t_0\}}(1+b_k) < \infty$, there exists a $T > 0$ such that system (2.1) is strongly practically stable with respect to $(\lambda, \alpha, \beta, T)$. More specifically,

$$\|x(t)\| \le \nu_1^{-1}\left(Lu_0 + L(\nu_2(\lambda) - u_0)e^{-\int_{t_0}^{t} m(s)ds}\right), \tag{3.5}$$

for $t \in [t_0 - r, \infty)$.

We shall need the following lemma in the proof of this theorem.

Lemma 3.1 *Let $u, m \in PC([t_0 - r, \infty), \Re_+)$ be such that u and m are continuous at each $t \ne \tau_k$ in (t_0, ∞) and $m(t) > 0$ for all $t \in [t_0 - r, \infty)$. Let $u_0 \ge 0$ be some fixed constant and let $b_k \ge 0$ for $k = 1, 2, \ldots$. Assume*

$$u(\tau_k) \le (1 + b_k)u(\tau_k^-), \tag{3.6}$$

for $\tau_k > t_0$ and

$$D^+u(t) \le -m(t)(u(t) - u_0), \text{ for } t \in (t_0, \infty),\ t \ne \tau_k \text{ if } u(t) > u_0 \text{ and } u(t) \ge \|u_t\|_r e^{-\int_{t-r}^{t} m(s)ds}. \tag{3.7}$$

Then

$$u(t) \le p(t)\left[u_0 + (\hat{u} - u_0)e^{-\int_{t_0}^{t} m(s)ds}\right], \tag{3.8}$$

for $t \in [t_0 - r, \infty)$, where $\hat{u} = \max\{u_0, \|u_{t_0}\|_r\}$ and

$$p(t) = \begin{cases} 1, & t \in [t_0 - r, t_0], \\ \prod_{\{k:\ \tau_k \in (t_0, t]\}} (1 + b_k), & t \in (t_0, \infty), \end{cases} \tag{3.9}$$

where the product is understood to be 1 *if the set* $\{k : \tau_k \in (t_0, t]\}$ *is empty.*

Proof: Define $l(t) = u(t) - p(t)\left[u_0 + (\hat{u} - u_0)e^{-\int_{t_0}^{t} m(s)ds}\right]$. We want to show $l(t) \le 0$ for $t \in [t_0 - r, \infty)$. For $t \in [t_0 - r, t_0]$, $l(t) = u(t) - \left[u_0 + (\hat{u} - u_0)\right.$ $\left.e^{-\int_{t_0}^{t} m(s)ds}\right] \le \|u_{t_0}\|_r - u_0 - (\hat{u} - u_0) \le 0$.

Assume $\tau_{i-1} \le t_0 < \tau_i$ for some i. We first show $l(t) \le 0$ for $t \in (t_0, \tau_i)$. In order to do this we let $\delta > 0$ be arbitrary and show that $l(t) \le \delta$ for $t \in (t_0, \tau_i)$. Suppose for the sake of contradiction that $l(t) > \delta$ for some $t \in (t_0, \tau_i)$. Then let $t^* = \inf\{t \in (t_0, \tau_i) : l(t) > \delta\}$. Then $t^* > t_0$ since $l(t_0) \le 0 < \delta$ and $t^* < \tau_i$. Note that $l(t)$ is continuous on $[t_0, \tau_i)$. For $t \in [t_0, t^*]$, $l(t) \le \delta$ and $l(t^*) = \delta$.

Now $u(t^*) = l(t^*) + \left[u_0 + (\hat{u} - u_0)e^{-\int_{t_0}^{t^*} m(s)ds}\right] \ge \delta + u_0 > u_0$. For $s \in [-r, 0]$,

$$\begin{aligned} u(t^* + s) \le \delta + \left[u_0 + (\hat{u} - u_0)e^{-\int_{t_0}^{t^*+s} m(\tau)d\tau}\right] &\le \delta \\ + \left[u_0 + (\hat{u} - u_0)e^{-\int_{t_0}^{t^*-r} m(\tau)d\tau}\right] & \\ \le \left[\delta + u_0 + (\hat{u} - u_0)e^{-\int_{t_0}^{t^*} m(\tau)d\tau}\right] e^{\int_{t^*-r}^{t^*} m(\tau)d\tau} &= u(t^*)e^{\int_{t^*-r}^{t^*} m(\tau)d\tau}, \end{aligned} \tag{3.10}$$

which implies $\|u_{t^*}\|_r \le u(t^*)e^{\int_{t^*-r}^{t^*} m(s)ds}$. So by (3.7) we have $D^+u(t^*) \le -m(t^*)(u(t^*) - u_0)$. Using this inequality, we find that

$$\begin{aligned} D^+l(t^*) &= D^+u(t^*) + (\hat{u} - u_0)e^{-\int_{t_0}^{t^*} m(s)ds}m(t^*) \\ &\le -m(t^*)\left[u(t^*) - u_0 - (\hat{u} - u_0)e^{-\int_{t_0}^{t^*} m(s)ds}\right] \\ &= -m(t^*)l(t^*) = -m(t^*)\delta < 0. \end{aligned} \tag{3.11}$$

This contradicts the definition of t^*. Thus it must be true that $l(t) \le \delta$ for all $t \in (t_0, \tau_i)$. Letting $\delta \to 0^+$ gives us $l(t) \le 0$ for $t \in (t_0, \tau_i)$. Consequently $l(\tau_i^-) \le 0$.

So far we have proven that $l(t) \le 0$ for $t \in [t_0 - r, \tau_i)$. To prove $l(t) \le 0$ for all $t \in [t_0 - r, \infty)$ we use an induction argument. As our induction hypothesis, suppose $l(t) \le 0$ for $t \in [t_0 - r, \tau_j)$ for some $j \ge i$ and let us prove that $l(t) \le 0$ for $t \in [t_0 - r, \tau_{j+1})$.

To begin with

$$\begin{aligned} l(\tau_j) &= u(\tau_j) - p(\tau_j)\left[u_0 + (\hat{u} - u_0)e^{-\int_{t_0}^{\tau_j} m(s)ds}\right] \\ &\le (1 + b_j)u(\tau_j^-) - (1 + b_j)p(\tau_j^-)\left[u_0 + (\hat{u} - u_0)e^{-\int_{t_0}^{\tau_j} m(s)ds}\right] \\ &= (1 + b_j)l(\tau_j^-) \le 0. \end{aligned} \tag{3.12}$$

by (3.6). Let $\delta > 0$. We want to show $l(t) \leq \delta$ for $t \in (\tau_j, \tau_{j+1})$. Suppose otherwise and define $t^* = \inf\{t \in (\tau_j, \tau_{j+1}) : l(t) > \delta\}$. Then $\tau_j < t^* < \tau_{j+1}$, $l(t^*) = \delta$ and $l(t) \leq \delta$ for $t \in [\tau_j, t^*]$.

Now $u(t^*) = \delta + p(t^*)\left[u_0 + (\hat{u} - u_0)e^{-\int_{t_0}^{t^*} m(s)ds}\right] \geq \delta + u_0 > u_0$. For $s \in [-r, 0]$,

$$\begin{aligned}
u(t^* + s) &\leq \delta + p(t^* + s)\left[u_0 + (\hat{u} - u_0)e^{-\int_{t_0}^{t^*+s} m(\tau)d\tau}\right] \\
&\leq \delta + p(t^*)\left[u_0 + (\hat{u} - u_0)e^{-\int_{t_0}^{t^*-r} m(\tau)d\tau}\right] \\
&\leq \left[\delta + p(t^*)\left(u_0 + (\hat{u} - u_0)e^{-\int_{t_0}^{t^*} m(\tau)d\tau}\right)\right]e^{\int_{t^*-r}^{t^*} m(\tau)d\tau} \\
&= u(t^*)e^{\int_{t^*-r}^{t^*} m(\tau)d\tau}.
\end{aligned} \tag{3.13}$$

Thus $\|u_{t^*}\|_r \leq u(t^*)e^{\int_{t^*-r}^{t^*} m(s)ds}$ and hence by (3.7) we have $D^+u(t^*) \leq -m(t^*)(u(t^*) - u_0)$.

As before we use this inequality to show $D^+l(t^*) < 0$ as follows,

$$\begin{aligned}
D^+l(t^*) &= D^+u(t^*) + p(t^*)(\hat{u} - u_0)e^{-\int_{t_0}^{t^*} m(s)ds}m(t^*) \\
&\leq -m(t^*)\left[u(t^*) - u_0 - p(t^*)(\hat{u} - u_0)e^{-\int_{t_0}^{t^*} m(s)ds}\right] \\
&\leq -m(t^*)\left[u(t^*) - p(t^*)u_0 - p(t^*)(\hat{u} - u_0)e^{-\int_{t_0}^{t^*} m(s)ds}\right] \\
&= -m(t^*)l(t^*) = -m(t^*)\delta < 0.
\end{aligned} \tag{3.14}$$

Again this contradicts the definition of t^* which leads us to conclude that $l(t) \leq \delta$ for all $t \in (\tau_j, \tau_{j+1})$. Letting $\delta \to 0^+$ gives us $l(t) \leq 0$ for $t \in (\tau_j, \tau_{j+1})$. Thus $l(t) \leq 0$ for all $t \in [t_0 - r, \tau_{j+1})$ which proves the induction step.

The above induction argument proves that $l(t) \leq 0$ for all $t \in [t_0 - r, \infty)$ which in turn implies that (3.8) is satisfied for all $t \in [t_0 - r, \infty)$. ■

We now present the proof of Theorem 3.1.

Proof of Theorem 3.1: Let $x(t) = x(t, t_0, \phi, \eta)$ be any solution of (2.1) with $\|\phi\|_r \leq \lambda$ and define $u(t) = V(t, x(t))$ for $t \in [t_0 - r, \infty)$. The assumptions of the theorem ensure that the conditions of Lemma 3.1 are satisfied. Applying Lemma 3.1 therefore gives us

$$V(t, x(t)) \leq p(t)\left[u_0 + (\hat{u} - u_0)e^{-\int_{t_0}^{t} m(s)ds}\right], \tag{3.15}$$

for $t \in [t_0 - r, \infty)$ where $\hat{u} = \max\{u_0, \sup_{s\in[t_0-r,t_0]} V(s, x(s))\}$.

For $t \in [t_0 - r, t_0]$ we have $V(t, x(t)) \leq \nu_2(\|x(t)\|) = \nu_2(\|\phi(t - t_0)\|) \leq \nu_2(\|\phi\|_r) \leq \nu_2(\lambda)$ by (3.1). Also since $\nu_1^{-1}(Lu_0) < \nu_1^{-1}(L\nu_2(\lambda))$ by (3.4) then $u_0 < \nu_2(\lambda)$. Thus $\hat{u} \leq \nu_2(\lambda)$.

Since $p(t) \leq L < \infty$ for all $t \in [t_0 - r, \infty)$ and since $\hat{u} \leq \nu_2(\lambda)$ then (3.15) implies

$$V(t, x(t)) \leq L\left[u_0 + (\nu_2(\lambda) - u_0)e^{-\int_{t_0}^{t} m(s)ds}\right], \tag{3.16}$$

for $t \in [t_0 - r, \infty)$. Applying (3.1) again gives us

$$\|x(t)\| \leq \nu_1^{-1}\left(Lu_0 + L(\nu_2(\lambda) - u_0)e^{-\int_{t_0}^{t} m(s)ds}\right), \tag{3.17}$$

for $t \in [t_0 - r, \infty)$. From (3.17) we get that $\|x(t)\| \leq \nu_1^{-1}(L\nu_2(\lambda)) \leq \alpha$ for $t \in [t_0, \infty)$. Thus $\|x(t)\| \leq \alpha$ for all $t \in [t_0 - r, \infty)$. Since $\beta > \nu_1^{-1}(Lu_0)$ and $\int_{t_0}^{\infty} m(s)ds = \infty$ then (3.17) implies the existence of a $T > 0$ such that $\|x(t)\| \leq \beta$ for all $t \geq t_0 + T$. This completes the proof. ■

We remark that if $u_0 = 0$ in Theorem 3.1 then system (2.1) will be asymptotically stable. Moreover, weaker assumptions on the constants b_k may be imposed that would not require that $\sum_{k=1}^{\infty} b_k$ be finite. Disturbances in the system are what typically cause u_0 to become positive which in turn increases the practical stability parameter β.

4 APPLICATION

To demonstrate the application of the theorems presented in the previous section we examine the LQ regulator problem with a delayed and disturbed control signal and with impulsive perturbations of the state.

First consider the linear system

$$x'(t) = Ax(t) + Bu(t), \ x(t_0) = x_0, \tag{4.1}$$

where $x \in \Re^n$ denotes the state of the system, $u \in \Re^m$ denotes the control signal and $x_0 \in \Re^n$ is the state at the initial time $t_0 \in \Re_+$. The matrices $A \in \Re^{n\times n}$ and $B \in \Re^{n\times m}$ are assumed to be constant with $B \neq 0$ and the pair (A, B) is assumed to be stabilizable (i.e. there exists a matrix $K \in \Re^{m\times n}$ such that $A - BK$ is Hurwitz).

Under these assumptions system (4.1) can be stabilized by finding an appropriate matrix K and then using a linear state feedback controller of the form

$$u(t) = -Kx(t). \tag{4.2}$$

The gain matrix K for which this feedback control will stabilize system (4.1) is not unique. However, a systematic way of finding a unique K is by minimizing the quadratic cost functional defined by

$$v(x_0) = \int_{t_0}^{\infty} \left(x(t)^T Qx(t) + u(t)^T Ru(t)\right) dt, \tag{4.3}$$

where $Q \in \Re^{n\times n}$ and $R \in \Re^{m\times m}$ are constant positive definite matrices. It can be shown (Athans & Falf, 1966) that the input $u(t)$ which minimizes $v(x_0)$ is given by (4.2) where

$$K = R^{-1}B^T P, \tag{4.4}$$

and where P is the unique (constant) positive definite matrix solution of the nonlinear algebraic Riccati equation

$$A^TP + PA - PBR^{-1}B^TP + Q = 0. \tag{4.5}$$

In practice, closed loop systems may experience uncertain delayed perturbations, uncertain bounded disturbances and impulsive perturbations. To take these into account we consider the following modified model

$$x'(t) = Ax(t) + B(u(t - h(t)) + \eta(t)), \quad t \neq \tau_k, \tag{4.6a}$$
$$\Delta x(t) = C_k x(t^-), \qquad t = \tau_k. \tag{4.6b}$$

The function $\eta(t)$ represents an uncertain, bounded disturbance. We assume $\|\eta(t)\| \leq \epsilon$ for all $t \geq t_0$ where $\epsilon > 0$ is constant. We assume there exists a constant $\bar{h} > 0$ such that the delay perturbation function $h(t)$ is bounded by $0 \leq h(t) \leq \bar{h}$ for $t \geq t_0$. The impulse times are assumed to satisfy $0 = \tau_0 < \tau_1 < \tau_2 < \ldots < \tau_k < \ldots$ and $\lim_{k\to\infty} \tau_k = \infty$ and the matrices $C_k \in \Re^{n\times n}$ are constant for $k = 1, 2, \ldots$. The initial condition for system (4.6) is given by

$$x_{t_0} = \phi, \tag{4.7}$$

where $\phi \in PC_{\bar{h}}$.

We wish to investigate the practical stability of the closed loop system (4.6) when subject to the control strategy (4.2)-(4.5). In other words we consider the system

$$x'(t) = Ax(t) - BKx(t - h(t)) + B\eta(t), \quad t \neq \tau_k, \tag{4.8a}$$
$$\Delta x(t) = C_k x(t^-), \qquad t = \tau_k, \tag{4.8b}$$

where K is defined in (4.4).

The functions $h(t)$ and $\eta(t)$ are assumed to be sufficiently smooth to guarantee the global existence of solutions of system (4.8). For example, sufficient conditions are that $h(t)$ and $\eta(t)$ be continuous and that $t - h(t)$ be strictly increasing (Ballinger & Liu (1999)).

We define the following parameters for future use. Let

$$a = \lambda_{min}\left(P^{-\frac{1}{2}}\left(Q + K^T RK\right)P^{-\frac{1}{2}}\right), \tag{4.9}$$

$$d = \lambda_{max}\left(P^{-\frac{1}{2}}(A^T P + PA)P^{-\frac{1}{2}}\right), \tag{4.10}$$

$$c_1 = \epsilon\left\|P^{\frac{1}{2}}B\right\|, \tag{4.11}$$

$$c_2 = \left\|P^{-\frac{1}{2}}K^T RKP^{-\frac{1}{2}}\right\|, \tag{4.12}$$

$$c_3 = \max\{0, d + c_1\} + 2c_2, \tag{4.13}$$

$$b = 2\bar{h}\left(\left\|P^{-\frac{1}{2}}K^T RKAP^{-\frac{1}{2}}\right\| + c_2^2\right), \tag{4.14}$$

$$q = 2\epsilon\left(\left\|P^{\frac{1}{2}}B\right\| + \bar{h}\left\|P^{-\frac{1}{2}}K^T RKB\right\|\right), \tag{4.15}$$

$$b_k = \max\left\{0, \lambda_{max}\left(P^{-\frac{1}{2}}\left(C_k^T PC_k + PC_k + C_k^T P\right)P^{-\frac{1}{2}}\right)\right\}, \quad k = 1, 2, \ldots, \tag{4.16}$$

$$L = \prod_{\{k:\ \tau_k > t_0\}} (1 + b_k), \tag{4.17}$$

$$\hat{L} = \prod_{\{k:\ \tau_k > t_0 + \bar{h}\}} (1 + b_k), \tag{4.18}$$

$$\sigma = \sup_{t > t_0 + \bar{h}} \sum_{\{k:\ \tau_k \in (t - \bar{h}, t)\}} \left\|P^{-\frac{1}{2}}K^T RKC_k P^{-\frac{1}{2}}\right\|, \tag{4.19}$$

and where, if $\sigma < \infty$, we let

$$\hat{b} = b + 2\sigma, \tag{4.20}$$

and we define the function

$$g(\theta) = \theta(a - \hat{b}e^{\bar{h}\theta} - \theta), \tag{4.21}$$

for $\theta \in \Re$. Furthermore, if $a > \hat{b} > 0$ then we define θ_0 to be the unique solution of the equation

$$g'(\theta_0) = a - \hat{b}(1 + \bar{h}\theta_0)e^{\bar{h}\theta_0} - 2\theta_0 = 0, \tag{4.22}$$

and we let

$$u_0 = \frac{q^2}{4g(\theta_0)}. \tag{4.23}$$

The critical point θ_0 of $g(\theta)$ is where $g(\theta)$ attains an absolute maximum. Note that a,c_1,c_2,c_3,b and q are all positive constants.

Theorem 4.1 *Suppose that*

(i) $\sum_{k=1}^{\infty} b_k < \infty$ *(hence* $\hat{L}, L < \infty$*);*

(ii) $\sigma < \infty$*;*

(iii) $a > \hat{b} > 0$*;*

(iv) (λ, α, β) *satisfy*

$$0 < \sqrt{\frac{\hat{L}u_0}{\lambda_{min}(P)}} < \beta \le \lambda \le \hat{\lambda}\sqrt{\frac{\hat{L}\lambda_{max}(P)}{\lambda_{min}(P)}} \le \alpha, \tag{4.24}$$

where

$$\hat{\lambda} = \left\{ \frac{1}{\lambda_{min}(P)} \prod_{\{k:\ \tau_k \in (t_0, t_0+\bar{h}]\}} (1+b_k) \left(e^{c_3 \bar{h}} \lambda_{max}(P)\lambda^2 + \frac{c_1}{c_3}\left(e^{c_3\bar{h}} - 1\right)\right)\right\}^{\frac{1}{2}}. \tag{4.25}$$

Then there exists a $T > 0$ *such that system (4.8) is strongly practically stable with respect to* $(\lambda, \alpha, \beta, T)$*. More specifically,*

$$\|x(t)\| \le \left\{ \frac{1}{\lambda_{min}(P)} \left[\hat{L}u_0 + \hat{L}(\lambda_{max}(P)\hat{\lambda}^2 - u_0)e^{-(t-t_0-\bar{h})\theta_0}\right]\right\}^{\frac{1}{2}}, \tag{4.26}$$

for $t \in [t_0 - \bar{h}, \infty)$*.*

Proof: Define $V(x) = x^T Px$ for $x \in \Re^n$. Then

$$\lambda_{min}(P)\|x\|^2 \le V(x) \le \lambda_{max}(P)\|x\|^2, \tag{4.27}$$

for $x \in \Re^n$. Suppose $x(t) = x(t, t_0, \phi, \eta)$ is any solution of (4.8) with $\|\phi\|_{\bar{h}} \le \lambda$.

For $\tau_k > t_0$ we have

$$\begin{aligned} V(x(\tau_k)) &= x(\tau_k)^T Px(\tau_k) = \left(x(\tau_k^-) + C_k x(\tau_k^-)\right)^T P \left(x(\tau_k^-) + C_k x(\tau_k^-)\right) \\ &= x(\tau_k^-)^T Px(\tau_k^-) + x(\tau_k^-)^T \left(C_k^T PC_k + PC_k + C_k^T P\right) x(\tau_k^-) \\ &= V(x(\tau_k^-)) + \left(P^{\frac{1}{2}}x(\tau_k^-)\right)^T P^{-\frac{1}{2}}\left(C_k^T PC_k + PC_k + C_k^T P\right)P^{-\frac{1}{2}} \\ &\quad \left(P^{\frac{1}{2}}x(\tau_k^-)\right) \\ &\le V(x(\tau_k^-)) + \lambda_{max}\left(P^{-\frac{1}{2}}\left(C_k^T PC_k + PC_k + C_k^T P\right)P^{-\frac{1}{2}}\right) \\ &\quad x(\tau_k^-)^T Px(\tau_k^-) \\ &\le (1+b_k)V(x(\tau_k^-)). \end{aligned} \tag{4.28}$$

Also for $t \ne \tau_k$ in (t_0, ∞) we have

$$\begin{aligned} D^+V(x(t)) &= 2x(t)^T Px'(t) = 2x(t)^T P\left(Ax(t) - BKx(t-h(t)) + B\eta(t)\right) \\ &= 2x(t)^T PAx(t) - 2x(t)^T PBKx(t-h(t)) + 2x(t)^T PB\eta(t) \\ &= x(t)^T(A^T P + PA)x(t) - 2x(t)^T K^T RKx(t-h(t)) + 2x(t)^T \\ &\quad PB\eta(t). \end{aligned} \tag{4.29}$$

We will first show that $\|x(t)\| \leq \hat{\lambda}$ for $t \in [t_0 - \bar{h}, t_0 + \bar{h}]$. In doing this we define $W \in PC([t_0 - \bar{h}, t_0 + \bar{h}], \Re_+)$ by

$$W(t) = \sup_{s \in [t_0 - \bar{h}, t]} V(x(s)), \tag{4.30}$$

for $t \in [t_0 - \bar{h}, t_0 + \bar{h}]$. Note that W is piecewise continuous, with discontinuities occurring only at discontinuous points of $x(t)$, and W is nondecreasing. Therefore $D^+W(t) \geq 0$ for $t \in (t_0, t_0 + \bar{h})$.

If $D^+W(t) > 0$ for some $t \in (t_0, t_0 + \bar{h})$ with $t \neq \tau_k$ then we claim $D^+W(t) = D^+V(x(t))$. To see this first consider that if $V(x(t)) < W(t)$ then by continuity, $V(x(s)) < W(t)$ for all $s \in [t, t+\delta]$ where $\delta > 0$ is sufficiently small. This would imply $W(s) = W(t)$ for $s \in [t, t + \delta]$ which would in turn imply $D^+W(t) = 0$. But since we are assuming that $D^+W(t) > 0$ then we must have $V(x(t)) = W(t)$. Now,

$$\begin{aligned} D^+W(t) &= \limsup_{h \to 0^+} \frac{1}{h}[W(t+h) - W(t)] = \inf_{\delta > 0} \sup_{0 < h < \delta} \frac{1}{h} \Bigg[\sup_{s \in [t_0 - \bar{h}, t+h]} \\ &\quad V(x(s)) - V(x(t)) \Bigg] \\ &= \inf_{\delta > 0} \sup_{0 < h < \delta} \frac{1}{h} \left[\sup_{s \in [t, t+h]} V(x(s)) - V(x(t)) \right]. \end{aligned} \tag{4.31}$$

Let $\delta > 0$ be sufficiently small so that the interval $[t, t + \delta]$ is free of impulses and let $0 < h < \delta$. Let $h_1 \in [0, h]$ satisfy $V(x(t + h_1)) = \sup_{s \in [t, t+h]} V(x(s))$. Since $D^+W(t) > 0$ then $0 < h_1 \leq h$ and so $1/h_1 > 1/h$ which implies

$$\begin{aligned} \frac{1}{h} \left[\sup_{s \in [t, t+h]} V(x(s)) - V(x(t)) \right] &\leq \frac{1}{h_1} [V(x(t + h_1)) - V(x(t))] \\ &\leq \sup_{0 < h_2 < \delta} \frac{1}{h_2} [V(x(t + h_2)) - V(x(t))]. \end{aligned} \tag{4.32}$$

Since this is true for all $0 < h < \delta$ then taking the supremum over all such h gives

$$\begin{aligned} \sup_{0 < h < \delta} \frac{1}{h} \left[\sup_{s \in [t, t+h]} V(x(s)) - V(x(t)) \right] &\leq \sup_{0 < h_2 < \delta} \frac{1}{h_2} [V(x(t + h_2)) - V(x(t))] \\ &\leq \sup_{0 < h_2 < \delta} \frac{1}{h_2} \left[\sup_{s \in [t, t+h_2]} V(x(s)) - V(x(t)) \right], \end{aligned} \tag{4.33}$$

and taking the infemum over $\delta > 0$ reduces these inequalities to

$$D^+W(t) \leq D^+V(x(t)) \leq D^+W(t), \tag{4.34}$$

from which we conclude $D^+W(t) = D^+V(x(t))$.

Therefore if $D^+W(t) > 0$ for some $t \in (t_0, t_0 + \bar{h})$ with $t \neq \tau_k$ then by (4.29)

$$\begin{aligned}
D^+W(t) &= D^+V(x(t)) \\
&= x(t)^T(A^TP + PA)x(t) - 2x(t)^TK^TRKx(t - h(t)) + 2x(t)^TPB\eta(t) \\
&= \left(P^{\frac{1}{2}}x(t)\right)^T \left(P^{-\frac{1}{2}}\left(A^TP + PA\right)P^{-\frac{1}{2}}\right)\left(P^{\frac{1}{2}}x(t)\right) \\
&\quad - 2\left(P^{\frac{1}{2}}x(t)\right)^T \left(P^{-\frac{1}{2}}K^TRKP^{-\frac{1}{2}}\right)\left(P^{\frac{1}{2}}x(t - h(t))\right) \\
&\quad + 2\left(P^{\frac{1}{2}}x(t)\right)^T P^{\frac{1}{2}}B\eta(t) \\
&\leq \lambda_{max}\left(P^{-\frac{1}{2}}\left(A^TP + PA\right)P^{-\frac{1}{2}}\right)x(t)^TPx(t) \\
&\quad +2\left\|P^{-\frac{1}{2}}K^TRKP^{-\frac{1}{2}}\right\| \cdot \left\|P^{\frac{1}{2}}x(t)\right\| \cdot \left\|P^{\frac{1}{2}}x(t - h(t))\right\| \\
&\quad +2\epsilon\left\|P^{\frac{1}{2}}B\right\| \cdot \left\|P^{\frac{1}{2}}x(t)\right\| \\
&= dV(x(t)) + 2c_2\sqrt{V(x(t))}\sqrt{V(x(t - h(t)))} + 2c_1\sqrt{V(x(t))} \\
&\leq (d + c_1)V(x(t)) + 2c_2\sqrt{V(x(t))}\sqrt{V(x(t - h(t)))} + c_1 \\
&\leq \max\{0, d + c_1\}W(t) + 2c_2\sqrt{W(t)}\sqrt{W(t)} + c_1 = c_3W(t) + c_1.
\end{aligned} \tag{4.35}$$

Thus

$$0 \leq D^+W(t) \leq c_3W(t) + c_1, \tag{4.36}$$

for all $t \in (t_0, t_0 + \bar{h})$ with $t \neq \tau_k$. Moreover, by (4.28) we have

$$\begin{aligned}
W(\tau_k) &= \max\left\{\sup_{s\in[t_0-\bar{h},\tau_k)} V(x(s)), V(x(\tau_k))\right\} = \max\{W(\tau_k^-), V(x(\tau_k))\} \\
&\leq \max\{W(\tau_k^-), (1 + b_k)V(x(\tau_k^-))\} \leq \max\{W(\tau_k^-), (1 + b_k)W(\tau_k^-)\} \\
&= (1 + b_k)W(\tau_k^-),
\end{aligned} \tag{4.37}$$

for all $\tau_k \in (t_0, t_0 + \bar{h}]$. Also from (4.27) we have, for $t \in [t_0 - \bar{h}, t_0]$,

$$V(x(t)) \leq \lambda_{max}(P)\|x(t)\|^2 = \lambda_{max}(P)\|\phi(t - t_0)\|^2 \leq \lambda_{max}(P)\lambda^2, \tag{4.38}$$

and so we get

$$W(t_0) \leq \lambda_{max}(P)\lambda^2, \tag{4.39}$$

by the definition of W.

By combining inequalities (4.36), (4.37) and (4.39) we obtain the following upper bound for $W(t_0 + \bar{h})$ (see (Lakshmikantham et al. 1989) for details on solving impulsive differential inequalities)

$$W(t_0 + \bar{h}) \leq \prod_{\{k:\ \tau_k\in(t_0,t_0+\bar{h}]\}} (1 + b_k)\left(e^{c_3\bar{h}}\lambda_{max}(P)\lambda^2 + \frac{c_1}{c_3}\left(e^{c_3\bar{h}} - 1\right)\right). \tag{4.40}$$

Since

$$\|x(t)\| \leq \left(\frac{1}{\lambda_{min}(P)}V(x(t))\right)^{\frac{1}{2}} \leq \left(\frac{1}{\lambda_{min}(P)}W(t_0+\bar{h})\right)^{\frac{1}{2}}, \tag{4.41}$$

for $t \in [t_0 - \bar{h}, t_0 + \bar{h}]$ by (4.27) then we get

$$\|x(t)\| \leq \hat{\lambda}, \tag{4.42}$$

for $t \in [t_0 - \bar{h}, t_0 + \bar{h}]$ where $\hat{\lambda}$ is defined by (4.25).

Next we show that the conditions of Theorem 3.1 are satisfied where the initial time is now taken to be $t_0 + \bar{h}$ and the delay constant r is given by $r = 2\bar{h}$. Conditions (i) and (iii) of Theorem 3.1 are clearly satisfied by (4.27) and (4.28), respectively, along with the assumption that $\sum_{k=1}^{\infty} b_k < \infty$. We define

$$m(t) = \theta_0, \tag{4.43}$$

for all $t \in [t_0 - \bar{h}, \infty)$ where $\theta_0 > 0$ is defined to be the unique solution of (4.22). Thus m satisfies the properties given in condition (ii) of Theorem 3.1.

Starting with equation (4.29) for $D^+V(x(t))$ we now derive an inequality of the form (3.2) given in condition (ii) of Theorem 3.1. Note that $e^{-\int_{t-r}^{t} m(s)ds} = e^{-2\bar{h}\theta_0}$. For $t \neq \tau_k$ in $(t_0 + \bar{h}, \infty)$ we have

$$\begin{aligned}
D^+V(x(t)) &= x(t)^T(A^TP + PA)x(t) - 2x(t)^TK^TRKx(t-h(t)) \\
&\quad +2x(t)^TPB\eta(t) \\
&= x(t)^T(K^TRK - Q)x(t) - 2x(t)^TK^TRKx(t-h(t)) \\
&\quad +2x(t)^TPB\eta(t) \\
&= -x(t)^T(K^TRK + Q)x(t) + 2x(t)^TK^TRK(x(t) - x(t-h(t))) \\
&\quad +2x(t)^TPB\eta(t) \\
&= -\left(P^{\frac{1}{2}}x(t)\right)^T\left(P^{-\frac{1}{2}}\left(K^TRK+Q\right)P^{-\frac{1}{2}}\right)\left(P^{\frac{1}{2}}x(t)\right) \\
&\quad +2x(t)^TK^TRK(x(t) - x(t-h(t))) + 2\left(P^{\frac{1}{2}}x(t)\right)^T P^{\frac{1}{2}}B\eta(t) \\
&\leq -\lambda_{min}\left(P^{-\frac{1}{2}}\left(K^TRK+Q\right)P^{-\frac{1}{2}}\right)x(t)^TPx(t) \\
&\quad +2x(t)^TK^TRK(x(t) - x(t-h(t))) + 2\epsilon\left\|P^{\frac{1}{2}}B\right\|\cdot\left\|P^{\frac{1}{2}}x(t)\right\| \\
&= -aV(x(t)) + 2x(t)^TK^TRK(x(t) - x(t-h(t))) + 2c_1\sqrt{V(x(t))}.
\end{aligned} \tag{4.44}$$

Moreover, if $t \in (t_0 + \bar{h}, \infty)$, $t \neq \tau_k$ and $V(x(t)) \geq e^{-2\bar{h}\theta_0}V(x(s))$ for all $s \in [t - 2\bar{h}, t]$ then

$$\leq \sqrt{V(x(t))}\left(\int_{t-\bar{h}}^{t}\left(e^{\bar{h}\theta_0}\left\|P^{-\frac{1}{2}}K^TRKAP^{-\frac{1}{2}}\right\|\sqrt{V(x(t))}\right.\right.$$

$$\left.+e^{\bar{h}\theta_0}\left\|P^{-\frac{1}{2}}K^TRKP^{-\frac{1}{2}}\right\|^2\sqrt{V(x(t))}+\epsilon\left\|P^{-\frac{1}{2}}K^TRKB\right\|\right)ds$$

$$\left.+e^{\bar{h}\theta_0}\sum_{\{k:\ \tau_k\in(t-\bar{h},t)\}}\left\|P^{-\frac{1}{2}}K^TRKC_kP^{-\frac{1}{2}}\right\|\sqrt{V(x(t))}\right)$$

$$=\bar{h}e^{\bar{h}\theta_0}\left\|P^{-\frac{1}{2}}K^TRKAP^{-\frac{1}{2}}\right\|V(x(t))+\bar{h}e^{\bar{h}\theta_0}\left\|P^{-\frac{1}{2}}K^TRKP^{-\frac{1}{2}}\right\|^2V(x(t))$$

$$+\epsilon\bar{h}\left\|P^{-\frac{1}{2}}K^TRKB\right\|\sqrt{V(x(t))}+e^{\bar{h}\theta_0}\sum_{\{k:\ \tau_k\in(t-\bar{h},t)\}}\left\|P^{-\frac{1}{2}}K^TRKC_k\right.$$

$$\left.P^{-\frac{1}{2}}\right\|V(x(t))$$

$$\leq \frac{b}{2}e^{\bar{h}\theta_0}V(x(t))+\epsilon\bar{h}\left\|P^{-\frac{1}{2}}K^TRKB\right\|\sqrt{V(x(t))}+\sigma e^{\bar{h}\theta_0}V(x(t)). \tag{4.45}$$

By combining this inequality with (4.44) we get

$$\begin{aligned} D^+V(x(t)) &\leq -\left(a-e^{\bar{h}\theta_0}(b+2\sigma)\right)V(x(t))+q\sqrt{V(x(t)} \\ &= -\left(a-\hat{b}e^{\bar{h}\theta_0}\right)V(x(t))+q\sqrt{V(x(t)}, \end{aligned} \tag{4.46}$$

whenever $t \in (t_0+\bar{h},\infty)$, $t \neq \tau_k$ and $V(x(t)) \geq e^{-2\bar{h}\theta_0}V(x(s))$ for all $s \in [t-2\bar{h},t]$.

Next we would like to get an inequality for $D^+V(x(t))$ in the form (3.2) where, ideally, the term u_0 is as small as possible. We made the simplifying assumption that $m(t)$ be constant. As we will see, θ_0 and u_0 defined by (4.22) and (4.23) will allow us to derive the inequality

$$D^+V(x(t)) \leq -\theta_0(V(x(t))-u_0), \tag{4.47}$$

from (3.2) with u_0 minimized.

If we let $z=\sqrt{V(x(t))}$ then (4.46) will imply (4.47) if

$$-\left(a-\hat{b}e^{\bar{h}\theta_0}\right)z^2+qz \leq -\theta_0(z^2-u_0), \tag{4.48}$$

or equivalently,

$$-\left(a-\hat{b}e^{\bar{h}\theta_0}-\theta_0\right)z^2+qz-\theta_0u_0 \leq 0, \tag{4.49}$$

for all $z \geq 0$ where $\theta_0>0$ and $u_0 \geq 0$. For this to be true we need

$$a-\hat{b}e^{\bar{h}\theta_0}-\theta_0>0, \tag{4.50}$$

and

$$u_0 \geq \frac{q^2}{4\theta_0\left(a - \hat{b}e^{\bar{h}\theta_0} - \theta_0\right)}. \tag{4.51}$$

Now, condition (iii) ensures that (4.50) is satisfied providing $\theta_0 > 0$ is sufficiently small. By defining $g(\theta)$ as we did in (4.21) we can find the point which maximizes it. This point is θ_0 which is the solution of $g'(\theta) = 0$. The smallest u_0 can be and yet still satisfy (4.50) and (4.51) is to equal the right hand side of inequality (4.51) with the denominator of that fraction maximized by θ_0.

To verify that (4.46) does indeed imply (4.47) one can start with the fact that

$$\left(2g(\theta_0)\sqrt{V(x(t))} - \theta_0 q\right)^2 \geq 0. \tag{4.52}$$

Expanding and dividing through by $4\theta_0 g(\theta_0)$ gives

$$\frac{g(\theta_0)}{\theta_0}V(x(t)) - q\sqrt{V(x(t))} + \frac{\theta_0 q^2}{4g(\theta_0)} \geq 0, \tag{4.53}$$

which implies

$$\left(a - \hat{b}e^{\bar{h}\theta_0} - \theta_0\right) V(x(t)) + \theta_0 u_0 \geq q\sqrt{V(x(t))}, \tag{4.54}$$

which in turn reduces to

$$-\left(a - \hat{b}e^{\bar{h}\theta_0}\right) V(x(t)) + q\sqrt{V(x(t))} \leq -\theta_0(V(x(t)) - u_0). \tag{4.55}$$

We have thus verified that condition (ii) of Theorem 3.1 is satisfied. Inequality (4.24) gives us

$$0 < \nu_1^{-1}(\hat{L}u_0) < \beta \leq \lambda \leq \hat{\lambda} \leq \nu_1^{-1}(\hat{L}\nu_2(\hat{\lambda})) \leq \alpha, \tag{4.56}$$

where $\nu_1(s) = \lambda_{min}(P)s^2$ and $\nu_2(s) = \lambda_{max}(P)s^2$. Since $\|x(t)\| \leq \hat{\lambda}$ for $t \in [t_0 - \bar{h}, t_0 + \bar{h}]$ then Theorem 3.1 tells us that

$$\begin{aligned} \|x(t)\| &\leq \nu_1^{-1}\left(\hat{L}u_0 + \hat{L}(\nu_2(\hat{\lambda}) - u_0)e^{-\int_{t_0+\bar{h}}^{t}\theta_0 ds}\right) \\ &= \left\{\frac{1}{\lambda_{min}(P)}\left[\hat{L}u_0 + \hat{L}(\lambda_{max}(P)\hat{\lambda}^2 - u_0)e^{-(t-t_0-\bar{h})\theta_0}\right]\right\}^{\frac{1}{2}}, \end{aligned} \tag{4.57}$$

for $t \in [t_0 - \bar{h}, \infty)$. Moreover, system (4.8) is strongly practically stable with respect to $(\lambda, \alpha, \beta, T)$ for some $T > 0$. The proof is therefore complete. ■

In the previous theorem conditions for practical stability were obtained which were dependent on the delay constant $\bar{h}$ as well as the bound ϵ on the disturbance function and the magnitudes of the impulse operators C_k. In general as $\bar{h}, \epsilon$ and the matrices C_k tend to zero then conditions (i), (ii) and (iii) of Theorem 4.1 become satisfied and so the system becomes strongly practically stable with respect to suitable parameters. Theorem 4.1 gives robustness bounds for the feedback controller given in (4.2)-(4.5).

We conclude by examining a particular numerical example.

Example 4.1: Consider the control system

$$x'(t) = Ax(t) + Bu(t), \tag{4.58}$$

where

$$A = \begin{bmatrix} -4 & 3 & 6 & -5 \\ -1 & -3 & 2 & 5 \\ -2 & 2 & 5 & 3 \\ 1 & 1 & -2 & -7 \end{bmatrix} \quad \text{and} \quad B = \begin{bmatrix} -4 & 5 \\ 0 & 0 \\ 1 & 2 \\ 0 & 2 \end{bmatrix}. \tag{4.59}$$

System (4.58) with zero input (i.e. $u(t) = 0$) is not stable since the matrix A has a positive real eigenvalue of 2.9680. Nevertheless, system (4.58) is stabilizable. If we let

$$Q = \begin{bmatrix} 1 & 0 & 0 & 0 \\ 0 & 1 & 0 & 0 \\ 0 & 0 & 2 & 0 \\ 0 & 0 & 0 & 2 \end{bmatrix} \quad \text{and} \quad R = \begin{bmatrix} 3 & 0 \\ 0 & 1 \end{bmatrix}, \tag{4.60}$$

and then solve for P in the Riccati equation (4.5) and calculate K using the formula (4.4) we get

$$P = \begin{bmatrix} 0.2217 & -0.0954 & -0.4402 & -0.2556 \\ -0.0954 & 0.2702 & 0.3824 & 0.2673 \\ -0.4402 & 0.3824 & 2.0941 & 0.6978 \\ -0.2556 & 0.2673 & 0.6978 & 0.6151 \end{bmatrix} \quad \text{and} \tag{4.61}$$

$$K = \begin{bmatrix} -0.4423 & 0.2547 & 1.2849 & 0.5734 \\ -0.2831 & 0.8223 & 3.3830 & 1.3480 \end{bmatrix}. \tag{4.62}$$

Thus using the feedback $u(t) = -Kx(t)$ with this particular choice of K will stabilize system (4.58) and minimize the quadratic performance index (4.3). The eigenvalues of the matrix $A - BK$ are -3.1167, -9.3516 and $-3.8160 \pm 2.6731i$, all of which have negative real parts implying $A - BK$ is Hurwitz.

Suppose that in applying the control law $u(t) = -Kx(t)$ we find that the input signal $u(t)$ is delayed by an amount $h(t) \geq 0$ which can never exceed some small fixed amount $\bar{h} > 0$ and suppose the feedback is disturbed by an amount $\eta(t)$ which is bounded by $\epsilon = 0.01$. Finally, suppose the state $x(t)$ is subjected to small and diminishing impulsive perturbations of the form

$$\Delta x(t) = 10^{-k} x(t^-), \tag{4.63}$$

at integer times $t = k$ for $k = 1, 2, \ldots$. Then we consider the modified system

$$x'(t) = Ax(t) - BKx(t - h(t)) + B\eta(t), \quad t \neq k, \tag{4.64a}$$

$$\Delta x(t) = C_k x(t^-), \qquad t = k, \tag{4.64b}$$

where $0 \leq h(t) \leq \bar{h}$ and $\|\eta(t)\| \leq 0.01$ for all $t \geq t_0 = 0$ and $C_k = 10^{-k}I$ for $k = 1, 2, \ldots$.

Calculating the parameters for our system in order to check the conditions of Theorem 4.1, we get $a = 4.0707$, $d = 7.5233$, $c_1 = 0.0363$, $c_2 = 9.1595$, $c_3 = 25.8786$ and $b = 244.2764\bar{h}$. For $k = 1, 2, \ldots$ we have

$$b_k = \frac{1}{10^{2k}} + \frac{2}{10^k}, \tag{4.65}$$

whose sum $\sum_{k=1}^{\infty} b_k = 0.2323$ is finite. Assuming $\bar{h} < 1$ we have $\hat{L} = L = 1.2371$ and $\sigma = 0.9160$ all of which are finite. In order to satisfy condition (iii) of Theorem 4.1 we need $a > b + 2\sigma$ or $\bar{h} < (4.0707 - 2 \times 0.9160)/244.2764 = 0.0092$. In other words Theorem 4.1 gives us strongly practical stability results in this example if the delay in feedback never exceeds 0.0092. Let us suppose that $\bar{h} = 0.005$. Then $b = 1.2214$, $\hat{b} = 3.0533$ and $q = 0.0757$.

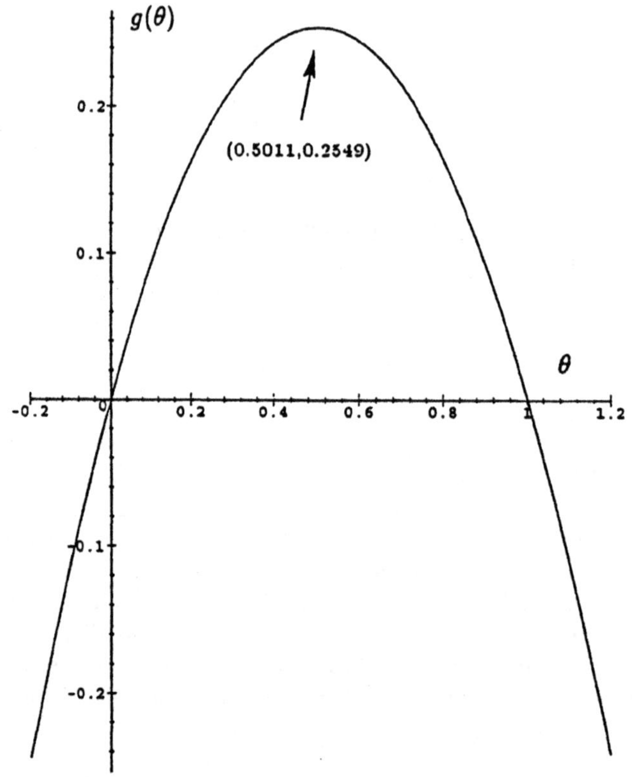

Figure 4.1 Plot of $g(\theta)$ versus θ.

A plot of $g(\theta)$ is given in Figure 4.1 from which it can be seen that the maximum of $g(\theta)$ occurs at $\theta = 0.5011$ where $g'(\theta) = 0$. Thus $\theta_0 = 0.5011$ and at this point $g(\theta_0) = 0.2549$ from which we obtain $u_0 = 0.0056$. Since $\bar{h} = 0.005$ is small, the graph of $g(\theta)$ is approximately given by the equation of the parabola $y = \theta(a - \hat{b} - \theta)$ as long as $|\theta|$ is not too large.

We therefore need $0.3103 < \beta \leq \lambda \leq 24.7231(2.9334\lambda^2 + 0.0001936)^{\frac{1}{2}} \leq \alpha$ in order for condition (iv) of Theorem 4.1 to be satisfied. If these parameters (λ, α, β) are so chosen then according to Theorem 4.1, system (4.58) will be strongly practically stable with respect to $(\lambda, \alpha, \beta, T)$ for some $T > 0$. □

ACKNOWLEDGEMENTS

This research was supported in part by NSERC-Canada which is gratefully acknowledged.

References

Athans, M. and Falb, P. (1966), *Optimal Control*, McGraw-Hill, New York.

Ballinger, G. and Liu, X. (1999), Existence and uniqueness results for impulsive delay differential equations, *Dynamics of Continuous, Discrete, and Impulsive Systems*, Vol. 5, pp. 579-594 .

Horn, R.A. and Johnson, C.R. (1985), *Matrix Analysis*, Cambridge University Press, Cambridge.

Hou, C. and Qian, J., (1997), Technical note: stability criterion for LQ regulators including delayed perturbations, *International Journal of Systems Sciences*, Vol. 28, No. 3, pp. 321-323.

Hou, C. and Qian, J. (1998), Practical stability and exponential estimates of trajectory bounds for retarded systems with bounded disturbances, *Journal of Mathematical Analysis and Applications*, Vol. 223, No. 1, pp. 50-61.

Lakshmikantham, V. (1989), Bainov, D.D. and Simeonov, P.S. (1989), *Theory of Impulsive Differential Equations*, World Scientific, Singapore.

Lakshmikantham, V. and Liu, X. (1993), *Stability Analysis in Terms of Two Measures*, World Scientific, Singapore.

Liu, X. (1992), Practical stabilization of control systems with impulse effects, *Journal of Mathematical Analysis and Applications*, Vol. 166, No. 2, pp. 563-576.

2 A REVIEW OF ILL-CONDITIONING AND REGULARIZATION IN OPTIMAL CONTROL COMPUTATION

Francis Benyah and Les S. Jennings

Center for Applied Dynamics and Optimization
Department of Mathematics and Statistics
University of Western Australia, Australia

Abstract: This article presents a summary of work undertaken to quantify aspects of the ill-conditioning encountered in the computations of optimal control problems. A mathematical machinery for the quantification of aspects of ill-conditioning is developed. This is then used to study how ill-conditioning varies with the size of discretization of control functions, and how it varies with different basis functions. The ill-conditioning associated with two different computational algorithms are also compared. A regularization method is used to obtain a stable and smooth solution. Implementation in the optimal control software MISER3 and computation of test examples involving various constraint types are considered.

Key words: Optimal control, control parametrization, ill-conditioning, regularization.

X. Yang et al. (eds.), Optimization Methods and Applications, 23-44.

1 INTRODUCTION

The computation of solutions to optimal control problems is known to be ill-conditioned. That is, relatively large variations of parameter values are allowed for small variations of constraints and/or objective values. This is in fact a good engineering design feature, in that if the parameters represent in some way the operating parameters of some plant, then as the plant wears out and the optimal parameters need to change, the constraints and/or objective may only change a little.

A summary of work done to qualify and quantify ill-conditioning is presented in this article. It brings together work presented at recent conferences, (eg., Benyah and Jennings (1998a) and Benyah and Jennings (1998b)) and gives an overview of some important considerations when computing solutions to OCPs. The folklore relied upon here is that smaller condition numbers imply better performance for the resulting constrained optimization codes. We demonstrate with a simple test problem, how the condition number increases as the control discretization becomes finer. This phenomenon is well known in numerical analysis (and statistics). However, the relationship between ill-conditioning and the number and type of basis elements is of particular interest, as it shows the natural numerical limits of this procedure.

Optimal control computation algorithms fall under two broad categories, namely, *control parametrization methods*, see section (3), and *state discretization methods*, see section (4). The main vehicle for demonstrating results in this paper is the optimal control software `MISER3`, which uses the control parametrization method, see for example, Teo et al (1991), and Jennings et al (1997). For this method, the control functions are discretized by low order splines; the ordinary differential equations are computed very accurately, and gradients with respect to parameters are computed via costate variables and Hamiltonians. A small nonlinear optimization problem is generated, but gradients are expensive to compute. For the state discretization method the control variables are discretized as well as the state variables, resulting in a large nonlinearly constrained optimization problem, with the discretized state differential equations as equality constraints.

Just how the condition number for discretized optimal control computation is defined is problematic in itself. A method for doing this, in terms of measuring the condition number of a regular non-linearly constrained optimization problem over a finite dimensional space is outlined. This does have some computational drawbacks, as the ill-conditioning gets in the way of being able to make clear decisions about which constraints are active.

Of course whenever ill-conditioning is present, the automatic remedy is to regularize the solution in some way. The most common and well-known form of regularization of ill-posed least-squares problems is *Tikhonov regularization*, see for example, Tikhonov and Arsemin (1997), Groetsch (1984), Engl (1993), and Fierro et al (1997). An important tool which is used in connection with the regularization of discrete ill-posed problems is the so-called *L-curve*. This has been used extensively for determining optimal regularization parameters, in the solution of linear least-squares problems, see for instance Hansen (1992); Hansen (1993), Hansen and O'Leary (1993), and Engl and Grever (1994).

An article on the use of the *L-curve* in determining optimal regularization parameters for OCPs can be found in Benyah and Jennings (1998c).

In sections 2 and 3 the mathematical background to the optimal control code MISER3 Jennings et al (1997) which computes solutions using the control parametrization method is presented. In section 4, we discuss *state discretization methods* for OCPs. In section 5 the perturbation analysis for constrained optimization is presented. In section 7 we discuss regularization of optimal control problems, including test examples.

2 OPTIMAL CONTROL PROBLEM TEMPLATE FOR MISER3

A form of the combined optimal control and optimal parameter selection problem is:

$$\min_{\boldsymbol{u},\boldsymbol{z}} \left\{ G_0(\boldsymbol{u},\boldsymbol{z}) = \phi_0(\boldsymbol{x}(t_f),\boldsymbol{z}) + \int_{t_s}^{t_f} g_0\left(t,\boldsymbol{x}(t),\boldsymbol{u}(t),\boldsymbol{z}\right) dt \right\} \tag{2.1}$$

subject to the state differential equation

$$\dot{\boldsymbol{x}}(t) = \boldsymbol{f}\left(t,\boldsymbol{x}(t),\boldsymbol{u}(t),\boldsymbol{z}\right), \quad \boldsymbol{x}(0) = \boldsymbol{x}^0(\boldsymbol{z}), \tag{2.2}$$

where $\boldsymbol{x}(t) = [x_1(t),\ldots,x_{n_s}(t)]^T \in R^{n_s}$, $\boldsymbol{u}(t) = [u_1(t),\ldots,u_{n_c}(t)]^T \in R^{n_c}$, and $\boldsymbol{z} = [z_1,\ldots,z_{n_z}]^T \in R^{n_z}$ are respectively, the state function vector, the control function vector and system parameter vector; $\boldsymbol{f} = [f_1,\ldots,f_{n_s}]^T \in R^{n_s}$, $\boldsymbol{x}^0 = [x_1^0,\ldots,x_{n_s}^0]^T \in R^{n_s}$; and ϕ_0, $g_0 \in R$. MISER3 is capable of dealing with a wide range of constraints. In the classical optimal control literature, constraints of different types are often treated differently. In the present unified framework, three standard forms of constraints are used. Any constraints involving the state variables or nonlinear in control functions are transformed into the standard canonical form:

$$G_k(\boldsymbol{u},\boldsymbol{z}) = \phi_k(\boldsymbol{x}(\tau_k),\boldsymbol{z}) + \int_{t_s}^{\tau_k} g_k\left(t,\boldsymbol{x}(t),\boldsymbol{u}(t),\boldsymbol{z}\right) dt \quad \geqq \quad 0, \; k = 1,2,\ldots,n_{g_c} \tag{2.3}$$

where n_{g_c} is the total number of canonical constraints, and $\tau_k \in (0,t_f]$ is a known constant and is referred to as the 'characteristic time' associated with the constraint G_k. Note that $\tau_0 = t_f$ for the cost functional.

The other two forms of constraints are firstly those which only involve the system parameters, and secondly those all-time linear constraints involving control functions only. Both of these forms are special in that they do not need costate variables to compute their gradients. They are

$$G(\boldsymbol{z}) \quad \geqq \quad 0, \tag{2.4}$$

and

$$g(t) = \sum_{i=1}^{n_c} \alpha_i u_i(t) + \beta \quad \geqq \quad 0. \tag{2.5}$$

There are likely to be finite upper and lower bounds on all control functions and system parameters. More complicated upper and lower bounds on control which depend on time t usually have to be transformed into a constraint of type (2.3).

$$u_i^L \leq u_i(t) \leq u_i^U, \quad \forall t \in [t_s, t_f], \quad i = 1, \ldots, n_c \tag{2.6}$$

$$z_j^L \leq z_j \leq z_j^U, \quad j = 1, \ldots, n_z. \tag{2.7}$$

3 CONTROL PARAMETRIZATION

The central idea behind MISER3 is the concept of control parametrization, see section 2.6 of Jennings et al (1991). Essentially, each control component $u_i(t)$ is approximated by a zero-th order or first order spline function (that is, a piecewise constant function or a piecewise linear continuous function) defined on a set of knots $\{t_s = t_0^i, t_1^i, t_2^i, \ldots, t_{k_i}^i = t_f\}$, which could be unequally spaced and different for each control. Let $i_0 = 1$ for the case of piecewise constant basis functions and $i_0 = 0$ for piecewise linear continuous basis functions. Formally we write the i-th control as a sum of basis functions with coefficients or parameters $\{\sigma_{ij}, j = i_0, \ldots, k_i\}$

$$u_i(t) = \sum_{j=i_0}^{k_i} \sigma_{ij} B_{ij}(t) \tag{3.1}$$

where $B_{ij}(t)$ is, for piecewise constant control, the indicator function for the j-th interval of the i-th set of knots, or for piecewise linear continuous control, the witches' hat functions:

$$B_{i0}(t) = \begin{cases} (t - t_1^i)/(t_0^i - t_1^i), & t \in [t_0^i, t_1^i], \\ 0, & \text{otherwise}, \end{cases}$$

$$B_{ij}(t) = \begin{cases} (t - t_{j-1}^i)/(t_j^i - t_{j-1}^i), & t \in [t_{j-1}^i, t_j^i], \\ (t - t_{j+1}^i)/(t_j^i - t_{j+1}^i), & t \in (t_j^i, t_{j+1}^i], \quad j = 1, \ldots, k_i - 1, \\ 0, & \text{otherwise}, \end{cases}$$

$$B_{ik_i}(t) = \begin{cases} (t - t_{k_i-1}^i)/(t_{k_i}^i - t_{k_i-1}^i), & t \in [t_{k_i-1}^i, t_{k_i}^i], \\ 0, & \text{otherwise}. \end{cases}$$

These basis functions have finite support and have the added property that the $\sigma_{ij} = u_i(t_j)$. This latter property is important for constraints of the form

(2.6) and hence the direct refinement to (3.4), as well as for the all-time linear constraints of control only. Higher degree splines present further problems as we cannot have both small finite support and $\sigma_{ij} = u_i(t_j)$. We suspect also that the ill-conditioning inherent in optimal control problems could get worse if higher degree splines are used to approximate the controls. Note that the system parameters are constant for all t.

After the control parametrization, the cost functional (2.1) as well as all the constraint functionals (2.3) – (2.7) can be regarded as (somewhat implicit) functions of parameter vector $\boldsymbol{\theta}$. More precisely, the corresponding approximate problem may be written as:

$$\min_{\boldsymbol{\theta}} \widetilde{G}_0(\boldsymbol{\theta}) \tag{3.2}$$

subject to the constraints:

$$\widetilde{G}_i(\boldsymbol{\theta}) = 0, \quad i = 1, \ldots, n_e \qquad \widetilde{G}_i(\boldsymbol{\theta}) \geq 0, \quad i = n_e + 1, \ldots, n_g \tag{3.3}$$

where

$$\boldsymbol{\theta} = [\sigma_{11_0}, \ldots, \sigma_{1k_1}, \sigma_{22_0}, \ldots, \sigma_{2k_2}, \ldots, \sigma_{n_c n c_0}, \ldots, \sigma_{n_c k_{n_c}}, z_1, z_2, \ldots, z_{n_z}]^T.$$

The control parameters and system parameters are bounded by a more general set of upper and lower bounds than (2.6) and (2.7):

$$u_{ij}^L < \sigma_{ij} < u_{ij}^U, \quad j = i_0, \ldots, k_i, \; i = 1, \ldots, n_c, \tag{3.4}$$

$$z_k^L \leq z_k \leq z_k^U, \quad k = 1, \ldots, n_z. \tag{3.5}$$

Clearly, this approximate problem can be regarded as a standard nonlinearly constrained mathematical programming problem. Note that bounds on σ_{ij} automatically put bounds on the control functions for all time as appropriate. This does not happen for higher order spline basis functions, another reason for not considering such basis functions. The problem posed by (3.2)-(3.5) is now a standard constrained nonlinear mathematical programming problem which can be solved by means of a sequential quadratic programming technique. See, for example, Schittkowski (1987), The Mathworks Inc., (1997), and Zhou et al (1998). Like many nonlinear programming techniques, these prefer the analytical gradients of the cost function $\widetilde{G}_0$ as well as the constraint functions $\widetilde{G}_k$, $k = 1, \ldots, n_{gc}$. However, due to the non-explicit dependency of $\widetilde{G}_k$, $k = 0, \ldots, n_{gc}$ on $\boldsymbol{\theta}$ when state functions are present, the computation of gradients of canonical functionals has to be done in a somewhat roundabout way. See Jennings et al (1991), and Teo et al (1991).

4 STATE DISCRETIZATION METHODS

An alternative approach to the method of control parametrization is to discretize both the control and state variables. The differential equations are then approximated by collocation, or by a variety of discretization methods. This transforms the original problem into a constrained nonlinear programming problem with higher dimension.

4.1 *The Method of Consistent Approximations:* RIOTS

RIOTS is a group of programs and utilities, written mostly in C and designed as a Toolbox for MATLAB. It provides an interactive environment for solving a broad class of optimal control problems.

The numerical methods used by RIOTS are supported by the theory in Schwartz and Polak (1996a), Schwartz and Polak (1996b), Schwartz and Polak (1996c), Schwartz (1996a), which uses the method of *consistent approximations* as defined by Polak (1997). In this approach, a solution is obtained as an accumulation point of the solutions of a sequence of discrete-time optimal control problems that are, in a specific sense, consistent approximations to the original continuous time optimal control problem. These approximating problems are obtained by numerically integrating the continuous time system dynamics with one of four Runge-Kutta integration methods, and representing the controls as finite-dimensional B-splines. In this way, the approximating problem can be solved using standard mathematical programming techniques to optimize over the spline coefficients, in addition to any free initial conditions.

The integration proceeds on a (possible non-uniform) mesh that specifies the splines' breakpoints. The solution obtained for one such discretized problem can be used to select a new integration mesh upon which the OCP can be re-discretized to produce a new discrete-time problem that more accuretely approximates the original problem. In practice, only a few such re-discretizations need to be performed to achieve an acceptable solution.

4.2 *Point discretization*

Here the controls are discretized as in section 3, however, special care has to be taken in the case of piecewise constant controls, as the control functions are discontinuous at the chosen points. Equally-spaced (or otherwise) quadrature points are chosen between the control knots, so that the state discretization is finer than the control discretization, to give a vector of state variables $\boldsymbol{x}(t_i)$, at the points t_i, $i = 0, 1, \ldots, N_x$, where N_x is the total number of time points chosen. The state differential equations are then discretized, for instance, using Euler's method

$$\boldsymbol{x}(t_{i+1}) - \boldsymbol{x}(t_i) = (t_{i+1} - t_i) f(t_i, \boldsymbol{x}(t_i), \boldsymbol{\sigma}, z),$$

or using Modified Euler's method,

$$\boldsymbol{x}(t_{i+1}) - \boldsymbol{x}(t_i) = 1/2(t_{i+1} - t_i)\big(f(t_i, \boldsymbol{x}(t_i), \boldsymbol{\sigma}, z) + f(t_{i+1}\, \boldsymbol{x}(t_{i+1}, \boldsymbol{\sigma}, z)\big).$$

All the integrals of the objective and the constraints are replaced by appropriate quadratures over each switching sub-interval. The objective and the constraints are written as functions of the parameters $(\boldsymbol{\sigma}, \boldsymbol{x}(t_0), \boldsymbol{x}(t_1), \ldots, \boldsymbol{x}(t_{N_x}))$. The ode then becomes N_x structured equality constraints. The total number of variables to be optimized over can be large.

4.3 *Direct Collocation: DIRCOL*

DIRCOL is a direct collocation method for the numerical solution of optimal control problems. By a discretization of state and control variables the infinite dimensional optimal control problem is transcribed into a sequence of finite dimensional nonlinearly constrained optimization problems, see von Stryk (1993), von Stryk (1999). First, a discretization of the time interval is chosen:

$$0 = t_1 < t_2 < t_3 < \ldots < t_n = t_f. \tag{4.1}$$

The controls are approximated by piecewise linear interpolating functions between t_j and t_{j+1}; that is,

$$\hat{u}(t) = u(t_j) + \frac{t - t_j}{t_{j+1} - t_j}\left(u(t_{j+1}) - u(t_j)\right), \quad \text{for } t_j < t < t_{j+1}. \tag{4.2}$$

Using Hermite interpolation, each state is approximated by piecewise cubic polynomials $\hat{x}(t_j)$ on each segment, where

$$\hat{x}(t_j) = \sum_{k=0}^{3} c_k^j \left(\frac{t - t_j}{h_j}\right)^k, \quad t_j < t < t_{j+1}, \quad j = 1, \ldots, n-1, \tag{4.3}$$

$$c_0^j = x(t_j), \tag{4.4}$$

$$c_1^j = h_j f_j, \tag{4.5}$$

$$c_2^j = -3x(t_j) - 2h_j f_j + 3x(t_{j+1}) - h_j f_{j+1}, \tag{4.6}$$

$$c_3^j = 2x(t_j) + h_j f_j - 2x(t_{j+1}) + h_j f_{j+1}, \tag{4.7}$$

where $f_j = f(x(t_j), u(t_j), t_j)$, $\quad h_j = t_{j+1} - t_j$. At the grid points $\dot{\hat{x}}_j$ is determined by evaluating the right hand side of

$$\dot{\hat{x}}_j = f(x(t_j), u(t_j), t_j), \; j = 1, \ldots, n.$$

The approximating functions of the states have to satisfy the differential equations at the grid points t_j, $j = 1, \ldots, n$, and at the centres $t_{c,j} = t_{j+1/2} = (t_j + t_{j+1})/2$, $j = 1, \ldots, n-1$, of the discretization interval. The chosen approximations (4.3)—(4.7) of $x(t)$ already satisfy these constraints at t_j. The only remaining constraints in the nonlinear programming problem are

(1) the collocation points at $t_{c,j}$,

$$f(\hat{x}(t_{c,j}), \hat{u}(t_{c,j}), t_{c,j}) - \dot{\hat{x}}(t_{c,j}) = 0, \quad j = 1, \ldots, n-1, \tag{4.8}$$

(2) the inequality constraints at the grid points t_j

$$g(\hat{x}(t_j), \hat{u}(t_j), t_j) \geq 0, \quad j = 1, \ldots, n-1, \tag{4.9}$$

(3) and the initial and endpoint constraint at t_0 and t_n

$$r(\hat{x}(t_0), \hat{x}(t_n), t_n) = 0. \tag{4.10}$$

This scheme is known as *cubic collocation at Lobatto points.*

5 CONDITION NUMBERS FOR CONSTRAINED OPTIMIZATION

Sensitivity to perturbations for optimization problems can be reduced to considering a linear system linearized from the gradient of the Lagrange function. Assume the gradients of the active constraints are linearly independent and that there is a unique minimizer $\boldsymbol{x}^*$. Write the objective $f(\boldsymbol{x}^*)$ and constraints $\boldsymbol{h}(\boldsymbol{x}^*)$ in Taylor series about a computed solution point $\boldsymbol{x}$, where $\boldsymbol{x}^* = \boldsymbol{x} + \boldsymbol{z}$,

$$f(\boldsymbol{x}^*) \approx f(\boldsymbol{x}) + \nabla f(\boldsymbol{x})\boldsymbol{z} + \frac{1}{2}\boldsymbol{z}^T \boldsymbol{F}(\boldsymbol{x})\boldsymbol{z}, \quad \text{and} \quad \boldsymbol{0} = \boldsymbol{h}(\boldsymbol{x}^*) \approx \boldsymbol{h}(\boldsymbol{x}) + \nabla \boldsymbol{h}(\boldsymbol{x})\boldsymbol{z},$$

where $\boldsymbol{F}(\boldsymbol{x})$ is the Hessian of f at $\boldsymbol{x}$. The perturbations to be considered are in those variables used to determine convergence of any optimization algorithm, namely, the constraint values and the Lagrange condition for optimality, in particular, the projected gradient values. For a constrained problem the sensitivity of Lagrange multipliers (and hence the Kuhn-Tucker conditions) to perturbations is also of interest. Let $\boldsymbol{H} = \nabla \boldsymbol{h}(\boldsymbol{x})$, $\boldsymbol{g}(\boldsymbol{x}) = \nabla \boldsymbol{f}(\boldsymbol{x})^T$ and the non-compliance of the Kuhn-Tucker conditions be $\boldsymbol{\zeta}$, where at the computed solution point $\boldsymbol{x}$,

$$\boldsymbol{g}(\boldsymbol{x}) + \boldsymbol{H}^T(\boldsymbol{x})\boldsymbol{\lambda} = \boldsymbol{\zeta}. \tag{5.1}$$

The true Lagrange multipliers are denoted by $\boldsymbol{\lambda}^*$, and are assumed close to the computed value of the Lagrange multipliers $\boldsymbol{\lambda}$. Let $\boldsymbol{\xi} = \boldsymbol{\lambda}^* - \boldsymbol{\lambda}$, and consider the Taylor expansion of $\boldsymbol{g}(\boldsymbol{x}^*)$ about the computed solution point $\boldsymbol{x}$. Then from the Kuhn-Tucker condition, $\boldsymbol{g}(\boldsymbol{x}^*) + \boldsymbol{H}^T(\boldsymbol{x}^*)\boldsymbol{\lambda}^* = 0$, we get

$$\begin{aligned}
\boldsymbol{g}(\boldsymbol{x}) + \boldsymbol{F}(\boldsymbol{x})\boldsymbol{z} + \boldsymbol{H}^T(\boldsymbol{x}^*)\boldsymbol{\lambda}^* &= -\boldsymbol{H}^T(\boldsymbol{x})\boldsymbol{\lambda} + \boldsymbol{\zeta} + \boldsymbol{F}(\boldsymbol{x})\boldsymbol{z} + \boldsymbol{H}^T(\boldsymbol{x}^*)\boldsymbol{\lambda}^* \\
&= \boldsymbol{F}(\boldsymbol{x})\boldsymbol{z} + \left(\boldsymbol{H}^T(\boldsymbol{x}^*) - \boldsymbol{H}^T(\boldsymbol{x})\right)\boldsymbol{\lambda} \\
&\quad + \boldsymbol{H}^T(\boldsymbol{x}^*)(\boldsymbol{\lambda}^* - \boldsymbol{\lambda}) + \boldsymbol{\zeta} \\
&= \left(\boldsymbol{F}(\boldsymbol{x}) + \sum_i \boldsymbol{\lambda}_i \frac{\partial H_i}{\partial \boldsymbol{x}}(\boldsymbol{x})\right)\boldsymbol{z} + \boldsymbol{H}^T(\boldsymbol{x}^*)\boldsymbol{\xi} + \boldsymbol{\zeta} \\
&= \mathcal{O}(\|\boldsymbol{z}\|^2).
\end{aligned}$$

Ignoring second order small quantities,

$$\begin{bmatrix} \boldsymbol{F}(\boldsymbol{x}) + \sum_i \boldsymbol{\lambda}_i \frac{\partial H_i}{\partial \boldsymbol{x}}(\boldsymbol{x}) & \boldsymbol{H}(\boldsymbol{x}^*)^T \\ \boldsymbol{H}(\boldsymbol{x}) & 0 \end{bmatrix} \begin{bmatrix} \boldsymbol{z} \\ \boldsymbol{\xi} \end{bmatrix} = \begin{bmatrix} -\boldsymbol{\zeta} \\ -\boldsymbol{h} \end{bmatrix}. \tag{5.2}$$

This equation now gives the correction to the computed solution vector and Lagrange multipliers in terms of the measurable errors in the Kuhn-Tucker conditions and the constraint compliance. The condition number of the matrix in (5.2) gives a condition number for the optimization problem. To this order of small quantities the (unknown) term $\boldsymbol{H}(\boldsymbol{x}^*)$ can be replaced by $\boldsymbol{H}(\boldsymbol{x}) = \boldsymbol{H}$. Let $\boldsymbol{F} = \boldsymbol{F}(\boldsymbol{x}) + \sum_i \boldsymbol{\lambda}_i \frac{\partial H_i}{\partial \boldsymbol{x}}(\boldsymbol{x})$.

We now consider the various components of the solution to the optimization problem given by (5.2). Factorize the matrix $\boldsymbol{H}$ to $[\boldsymbol{L} \quad \mathbf{0}]\boldsymbol{Q}$, using the QU factorization, where $\boldsymbol{Q}$ is orthogonal and $\boldsymbol{L}$ is lower triangular of full rank as the active constraints are assumed regular. Let $\boldsymbol{w} = \boldsymbol{Q}\boldsymbol{z}$, $\boldsymbol{y} = \boldsymbol{Q}\boldsymbol{\zeta}$ and partition $\boldsymbol{w}$, $\boldsymbol{y}$ and $\boldsymbol{Q}\boldsymbol{F}\boldsymbol{Q}^T$ to the partitioning of $[\boldsymbol{L} \quad \mathbf{0}]$. This gives a transformation of (5.2), where the different components of the solution can be distinguished,

$$\begin{bmatrix} \boldsymbol{F}_{11} & \boldsymbol{F}_{12} & \boldsymbol{L}^T \\ \boldsymbol{F}_{12}^T & \boldsymbol{F}_{22} & \mathbf{0} \\ \boldsymbol{L} & \mathbf{0} & \mathbf{0} \end{bmatrix} \begin{bmatrix} \boldsymbol{w}_1 \\ \boldsymbol{w}_2 \\ \boldsymbol{\lambda} \end{bmatrix} = \begin{bmatrix} -\boldsymbol{y}_1 \\ -\boldsymbol{y}_2 \\ -\boldsymbol{h} \end{bmatrix}. \tag{5.3}$$

As $\boldsymbol{Q}$ is orthogonal, arbitrary perturbations in $\boldsymbol{F}$, $\boldsymbol{H}$ and $\boldsymbol{\zeta}$ are transformed to the same sized arbitrary perturbations in $\boldsymbol{F}_{11}$, $\boldsymbol{F}_{12}$, $\boldsymbol{F}_{22}$, $\boldsymbol{L}$, $\boldsymbol{y}_1$ and $\boldsymbol{y}_2$.

The component $\boldsymbol{w}_1$, regarded as the component of $\boldsymbol{z}$ fixed by the constraints is given by

$$\boldsymbol{L}\boldsymbol{w}_1 = -\boldsymbol{h} \tag{5.4}$$

and so has a condition number of $\chi(\boldsymbol{L})$. The component $\boldsymbol{w}_2$, of $\boldsymbol{z}$, gives the step to the minimizer of f in the subspace free of constraints (in the tangent space), and is given by

$$\boldsymbol{F}_{22}\boldsymbol{w}_2 = -\boldsymbol{y}_2 - \boldsymbol{F}_{12}^T\boldsymbol{w}_1. \tag{5.5}$$

Perturbations in this equation, through $\boldsymbol{w}_1$, are proportional to $\chi(\boldsymbol{L})$, so an approximate condition number for $\boldsymbol{w}_2$ is $\chi(\boldsymbol{L})\chi(\boldsymbol{F}_{22})$, based on arbitrary perturbations in $\boldsymbol{h}$, $\boldsymbol{g}$, $\boldsymbol{F}$ and $\boldsymbol{H}$. But notice that the sub-matrix $\boldsymbol{F}_{12}$ does effect the true condition number.

The Lagrange multipliers are given by

$$\boldsymbol{L}^T\boldsymbol{\lambda} = -\boldsymbol{y}_1 - \boldsymbol{F}_{11}\boldsymbol{w}_1 - \mathbf{F}_{12}\boldsymbol{w}_2, \tag{5.6}$$

but as $\boldsymbol{w}_2$ can have perturbations proportional to $\chi(\boldsymbol{L})\chi(\boldsymbol{F}_{22})$ this gives an approximate condition number for $\boldsymbol{\lambda}$ of $\chi(\boldsymbol{L})^2\chi(\boldsymbol{F}_{22})$. Of course for any particular problem the effect of $\boldsymbol{F}_{11}$ and $\boldsymbol{F}_{12}$ could lessen or worsen this condition number. This shows that the Lagrange multipliers are more likely to be affected by perturbations than the solution point. The extent of coupling between the three components of the solution can be seen in the inverse matrix description of (5.3).

$$\begin{bmatrix} \boldsymbol{w}_1 \\ \boldsymbol{w}_2 \\ \boldsymbol{\lambda} \end{bmatrix} = \begin{bmatrix} \mathbf{0} & \mathbf{0} & \boldsymbol{L}^{-1} \\ \mathbf{0} & \boldsymbol{F}_{22}^{-1} & -\boldsymbol{F}_{22}^{-1}\boldsymbol{F}_{12}^T\boldsymbol{L}^{-1} \\ \boldsymbol{L}^{-T} & -\boldsymbol{L}^{-T}\boldsymbol{F}_{12}\boldsymbol{F}_{22}^{-1} & \boldsymbol{L}^{-T}(\boldsymbol{F}_{12}\boldsymbol{F}_{22}^{-1}\boldsymbol{F}_{12}^T - \boldsymbol{F}_{11})\boldsymbol{L}^{-1} \end{bmatrix} \begin{bmatrix} -\boldsymbol{y}_1 \\ -\boldsymbol{y}_2 \\ -\boldsymbol{h} \end{bmatrix}.$$

If $\boldsymbol{F}_{12}\boldsymbol{F}_{22}^{-1}\boldsymbol{F}_{12}^T - \boldsymbol{F}_{11}$ is small the condition number for $\boldsymbol{\lambda}$ will be smaller than $\chi(\boldsymbol{L})^2\chi(\boldsymbol{F}_{22})$. This also shows which perturbations are most important to reduce. Generally the active constraint equations should be the most accurate,

$\boldsymbol{h} = \mathbf{0}$ being ideal. Any perturbation from $\boldsymbol{h} = \mathbf{0}$ will adversely affect the accuracy of $\boldsymbol{\lambda}$ and to a lessor extent $\boldsymbol{w}_2$. In practice, practitioners may care little about exactly satisfying the constraints. The above analysis determines the inaccuracies that could be expected resulting from arbitrary perturbations in $\boldsymbol{h}$, $\boldsymbol{g}$, $\boldsymbol{F}$ and $\boldsymbol{H}$.

The three scenarios, in the strongly coupled case of (i) $\boldsymbol{L}$ well-conditioned and $\boldsymbol{F}_{22}$ ill-conditioned, (ii) $\boldsymbol{L}$ ill-conditioned and $\boldsymbol{F}_{22}$ well-conditioned, and, (iii) both $\boldsymbol{L}$ and $\boldsymbol{F}_{22}$ ill-conditioned, are of interest as they give an overall ill-conditioned constrained optimization problem, in worsening order. In (i), the solution point and Lagrange multipliers are equally ill-determined through the unconstrained component of the solution. In cases (ii) and (iii) the Lagrange multipliers can be almost impossible to determine accurately as the solution component fixed by the constraints is ill-determined. This will most likely show as non-convergence of Kuhn-Tucker conditions, although this may depend on the algorithm. In cases (i) and (iii) large changes in iterates $\boldsymbol{x}$ are expected for little change in function values or gradient values. In cases (ii) and (iii) large changes in iterates $\boldsymbol{x}$ are expected for small changes in constraint compliance. In all cases large changes in Lagrange multiplier estimates and hence Kuhn-Tucker conditions are expected.

In the light of the foregoing discussions, we compute the condition numbers for (i) the active constraints $\chi(\boldsymbol{L})$, (ii) the projected Hessian $\chi(\boldsymbol{F}_{22})$, and (iii) the Lagrange condition, $\chi(\mathcal{M})$ for the following example.

6 TEST PROBLEM 1

The following example is an optimal control problem with two state variables, one control and an all-time state variable constraint. The original problem is:

Example 6.1

$$\min_{u} \left\{ J(u) = \int_0^1 (x_1^2(t) + x_2^2(t) + 0.005u^2(t))\, dt \right\} \tag{6.1}$$

subject to the dynamics

$$\dot{x}_1(t) = x_2(t), \; x_1(0) = 0,$$
$$\dot{x}_2(t) = -x_2(t) + u(t), \; x_2(0) = -1,$$

and subject to all-time state inequality constraint

$$x_2(t) - 8(t - 0.5)^2 + 0.5 \le 0, \forall t \in [0,\, 1]. \tag{6.2}$$

We replace the inequality constraint (6.2) with

$$x_2(t_i) - 8(t_i - 0.5)^2 + 0.5 \le 0, \tag{6.3}$$

where t_i is chosen at all the knot points of the control function. This is first solved using piecewise constant controls. and then piecewise linear controls, with function values as parameters. In either case the number of control parameters is varied from $n = 4, 8, 16, 32$ to 64.

6.1 *Re-formulation I: Using First Derivative Values as Parameters*

The control is modelled as a piecewise linear function with first derivative values as parameters by the transformation, $u(t) = x_3(t)$, and $\dot{x}_3(t) = v$, where v is a new control function and $u(t)$ is now a state variable. This transforms the original OCP into

$$\min_{v} \left\{ J(v) = \int_0^1 (x_1^2(t) + x_2^2(t) + 0.005x_3^2(t))\, dt \right\} \tag{6.4}$$

subject to the dynamics

$$\begin{aligned} \dot{x}_1(t) &= x_2(t),\ x_1(0) = 0, \\ \dot{x}_2(t) &= -x_2(t) + x_3(t),\ x_2(0) = -1, \\ \dot{x}_3(t) &= v(t),\ x_3(0) = z_1, \end{aligned}$$

subject to

$$x_2(t_i) - 8(t_i - 0.5)^2 + 0.5 \le 0, \tag{6.5}$$

where t_i is chosen at all the knot points of the control function, and z_1 is a system parameter.

If v is approximated by piecewise constant functions as parameters, then $u = x_2$ becomes piecewise linear with first derivative values as parameters. On the other hand if v is approximated by piecewise linear functions then u becomes a piecewise quadratic control with first derivative values as parameters.

6.2 *Re-formulation II: Using Second Derivative Values as Parameters*

The control is modelled as a piecewise quadratic function with second derivative values as parameters by the transformation, $u(t) = x_3(t)$, $\dot{x}_3(t) = x_4$, $\dot{x}_4(t) = v$, where v is a new control function; $u(t) = x_3(t)$ and $x_4(t)$ are now state variables:

$$\min_{v} \left\{ J(v) = \int_0^1 (x_1^2(t) + x_2^2(t) + 0.005x_3^2(t))\, dt \right\} \tag{6.6}$$

subject to the dynamics

$$\begin{aligned} \dot{x}_1(t) &= x_2(t),\ x_1(0) = 0, \\ \dot{x}_2(t) &= -x_2(t) + x_3(t),\ x_2(0) = -1, \\ \dot{x}_3(t) &= x_4(t),\ x_3(0) = z_1, \\ \dot{x}_4(t) &= v(t),\ x_4(0) = z_2, \end{aligned}$$

where z_1 and z_2 are system parameters. The constraint remains unchanged. Approximating the control v with piecewise constant functions makes the original control u piecewise quadratic with second derivative values as parameters.

The graphs in Figure 6.1 show the plot of *condition number* versus *number of parameters* on a `log-log` scale: The number of control parameters is varied from 4, through 8, 16, 32 to 64. An 'o' represents the control as a piecewise constant; a '$\cdot$' represents control as piecewise linear with function values as parameters; a '$+$' represents control as piecewise linear with derivative values as parameters; a '$\times$' represents control as a piecewise quadratic spline with derivative values as parameters; and a '$*$' represents control as a piecewise quadratic spline with second derivative values as parameters.

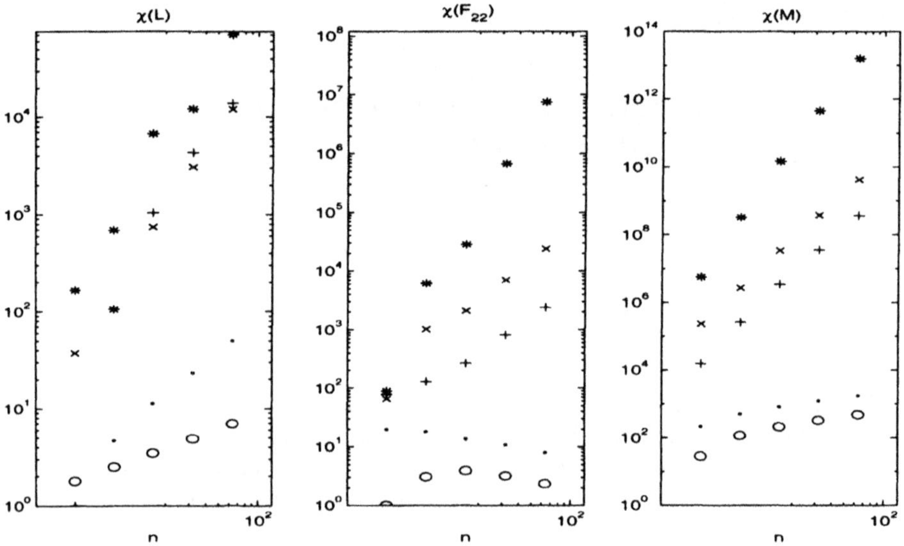

Figure 6.1 A plot of condition numbers for Test Problem 1

From the graph we see that condition numbers associated with controls using derivative values, namely, *pdc* ($+$), *pdl* ($\times$), and *pddc* ($*$) as parameters increase steadily with n. On the other hand, condition numbers associated with controls using function values as parameters, namely, piecewise constant controls and piecewise linear controls, shown on the graph by ($\cdot$), and (o) respectively, remain *small.*

To determine a possible relationship between the condition number χ and the number of parameters n, we find a least-squares fit of the form $\log \chi = \log \alpha + \beta \log n$, for $\chi(L)$, $\chi(F_{22})$ and $\chi(M)$. This suggests a relationship of the form $\chi = \alpha n^{\beta}$ for condition numbers associated with controls using derivative values as parameters. The values of α and β for each of the condition numbers $\chi(L)$, $\chi(F_{22})$ and $\chi(M)$, for Test Problem 1 is shown in Table 6.1.

In Table 6.2 we display the computed condition numbers of active constraints $\chi(\boldsymbol{L})$, the projected Hessian $\chi(\boldsymbol{F}_{22})$, and the Lagrange condition $\chi(\mathcal{M})$, associated with control functions using various derivative values as parameters. The

Table 6.1 **A model for the condition numbers**

Type of smoothness	$\chi(\mathcal{L})$	$\chi(\mathcal{F}_{22})$	$\chi(\mathcal{M})$
pdc	$10^{1.54}n^{1.26}$	$10^{0.69}n^{1.47}$	$10^{2.13}n^{3.58}$
pdl	$10^{2.24}n^{0.92}$	$10^{0.88}n^{1.97}$	$10^{3.25}n^{3.50}$
pddc	$10^{3.15}n^{0.95}$	$10^{-0.44}n^{4.17}$	$10^{3.48}n^{5.43}$

number of control parameters is represented by n; the condition numbers are represented by χ; *pdc* represents control as piecewise linear with first derivative values as parameters; *pdl* piecewise quadratic splines with first derivative values as parameters, and *pddc* piecewise quadratic splines with second derivative as parameters.

Table 6.2 **Computed condition numbers**

n	χ	*pdc*	*pdl*	*pddc*
	$\chi(L)$	$1.455E+02$	$5.003E+02$	$2.260E+03$
4	$\chi(F_{22})$	$3.909E+01$	$8.754E+01$	$9.945E+01$
	$\chi(\mathcal{M})$	$1.384E+04$	$1.803E+05$	$3.949E+06$
	$\chi(L)$	$6.271E+02$	$9.768E+02$	$2.603E+04$
8	$\chi(F_{22})$	$1.082E+02$	$5.168E+02$	$1.963E+03$
	$\chi(\mathcal{M})$	$2.908E+05$	$3.012E+06$	$3.071E+08$
	$\chi(L)$	$9.921E+02$	$4.003E+03$	$2.874E+04$
16	$\chi(F_{22})$	$2.374E+02$	$2.565E+03$	$5.689E+04$
	$\chi(\mathcal{M})$	$3.713E+06$	$3.518E+07$	$1.412E+10$
	$\chi(L)$	$5.035E+03$	$5.411E+03$	$3.730E+04$
32	$\chi(F_{22})$	$7.902E+02$	$7.047E+03$	$7.921E+05$
	$\chi(\mathcal{M})$	$3.972E+07$	$3.792E+08$	$4.703E+11$
	$\chi(L)$	$4.041E+03$	$5.172E+03$	$5.149E+04$
64	$\chi(F_{22})$	$2.349E+03$	$2.185E+04$	$9.296E+06$
	$\chi(\mathcal{M})$	$2.927E+08$	$2.992E+09$	$1.521E+13$

In the next section(s) we examine the degree to which *regularization* can help in making the computation better conditioned.

7 REGULARIZATION OF OPTIMAL CONTROL PROBLEMS

The aim of regularization is to make the computation better conditioned and, at the same time change the value of the objective only slightly.

To regularize an optimal control problem a small amount of a penalty function $P(\boldsymbol{\sigma})$, of the control parameters $\boldsymbol{\sigma}$, is added to the objective $G_0(\boldsymbol{\sigma})$ to create a new objective $G_0^r(\boldsymbol{\sigma})$ given by,

$$G_0^r(\boldsymbol{\sigma}) = G_0(\boldsymbol{\sigma}) + rP(\boldsymbol{\sigma}). \tag{7.1}$$

The penalty function $P(\boldsymbol{\sigma})$ is a measure of some unwanted property of the control function, increasing as the unwanted property becomes worse. The penalty weight r controls the weight given to the minimization of the objective function $G_0(\boldsymbol{\sigma})$ relative to the minimization of $P(\boldsymbol{\sigma})$.

For notational convenience we will consider just one control function, specified by $u(t) = \sigma_i$, $t \in [t_{i-1}, t_i]$ for $i = 1, \ldots, k$, for a piecewise constant approximation; and $u(t_i) = \sigma_i$, for $i = 0, \ldots, k$ for a piecewise linear approximation. Let $h_i = t_i - t_{i-1} > 0$, $i = 1, \ldots, k$.

We consider below, three *orders of regularization* of the control parameters.

7.1 Penalty on Deviation from Zero

Penalties related to the L_2 norm of the control function will take the form:

$$P = \sum_{i=i_0}^{k} w_i \sigma_i^2 \quad \text{or } P = \int_0^1 u^2\, dt = \sum_{i=1}^{k} h_i \sigma_i^2.$$

for some suitable relative weights w_i, $i = i_0, \ldots, k$. Here the penalty is put on distance from the zero function. These are called regularization functionals of *order zero.* Integral forms (on a positive interval length, $h_i > 0$) correspond to particular weighted forms as shown above.

7.2 Penalty on Deviation from A Constant

Penalties related to the L_2 norm of the derivative or first difference of the control function, and the integral form are respectively,

$$P = \sum_{i=i_0+1}^{k} w_i(\sigma_i - \sigma_{i-1})^2, \quad \text{or } P = \int_0^1 (\dot{u})^2\, dt = \sum_{i=1}^{k} \frac{(\sigma_i - \sigma_{i-1})^2}{h_i}.$$

These penalize large changes in control levels, that is, as the penalty weight is increased the control will tend to some constant value if it can. Hence these regularization functionals are said to be of *order one.* The relative weights w_i control local variation in that if w_i is larger than all others then variation close to t_i is penalized more than in other intervals.

7.3 Penalty on Deviation from A Straight Line

Penalties related to the L_2 norm of the second derivative of the control function are only defined as the bounded variation of the derivative of the piecewise linear continuous approximation where $k \geq 2$. However if we compute divided differences based on the control function values at each t_i, we have, if u is smooth enough

$$\ddot{u}_i \approx 2\left(\frac{\sigma_{i+1}-\sigma_i}{h_{i+1}} - \frac{\sigma_i - \sigma_{i-1}}{h_i}\right)/(h_i + h_{i+1}), \quad i_0+1 \le i \le k-1.$$

So possible penalties are: ($k \ge 2$ for a piecewise linear control and $k \ge 3$ for a piecewise constant control.

$$P = \sum_{i=i_0+1}^{k-1} \frac{4w_i}{(h_i+h_{i+1})^2}\left(\frac{\sigma_{i+1}-\sigma_i}{h_{i+1}} - \frac{\sigma_i-\sigma_{i-1}}{h_i}\right)^2$$

If these should be compared to integrals of the second derivative then w_i should be chosen as h_i or better $(h_{i+1}+h_i)/2$. These second derivative penalties penalize behaviour away from a straight line and in particular will prevent chattering controls, to a degree defined by the penalty parameter, much better than first derivative penalties. These regularization functionals are of *order two.*

Remark 7.1 *The choice of the order of the penalty depends on which aspect of the control function we want to penalize. If a highly oscillatory noise component, for example* $a\sin(2\pi kt)$, *exists in the control* u, *then an* order zero *regularization will tend to decrease* a, *the amplitude. An* order one *regularization will tend to decrease* ak; *an* order two *regularization will tend to decrease* ak^2. *We note that for a large* k, *the corresponding amplitude* a, *must be small (since* $a \propto \frac{1}{k}$, *or* $a \propto \frac{1}{k^2}$). *Hence the higher frequencies are attenuated more than the lower frequencies, especially for higher order penalties.*

Table 7.1 **Regularization order 0.**

r	$G_0(\boldsymbol{\sigma}_r)$	$P(\boldsymbol{\sigma}_r)$,	$\left\lvert\frac{G_0(\boldsymbol{\sigma}_r)-G_0(\boldsymbol{\sigma}_0)}{G_0(\boldsymbol{\sigma}_0)}\right\rvert$	χ
10^{-1}	0.487012	8.2420E-04	1.83E+00	2.8995E+00
10^{-2}	0.485534	8.2057E-02	1.82E+00	2.8994E+01
10^{-3}	0.471670	7.5778E+00	1.73E+00	2.8992E+02
10^{-4}	0.377603	4.8126E+02	1.19E+00	2.8971E+03
10^{-5}	0.219458	5.9790E+03	2.75E-01	4.7545E+04
10^{-6}	0.176931	1.7445E+04	2.76E-02	5.1215E+05
$\mathbf{10^{-7}}$	**0.173897**	**2.3880E+04**	**1.00E-02**	**3.9288E+06**
10^{-8}	0.172191	3.0451E+04	1.10E-04	1.6685E+07
10^{-9}	0.172173	3.0780E+04	8.57E-06	2.7435E+07
10^{-10}	0.172172	3.0798E+04	8.02E-07	2.8997E+07
0	0.172172	3.0781E+04	0	2.9182E+07

In Table 7.1 we display the results of a *regularization of order* 0 applied to Re-formulation I (6.1) of Test Problem 1, with 32 control parameters. Corresponding to each penalty weight r, is the objective function for the regularized solution $G_0(\boldsymbol{\sigma}_r)$, the size of the solution (semi)norm $P(\boldsymbol{\sigma}_r)$, the relative error in the objective function $\epsilon_0 = |\frac{G_0(\boldsymbol{\sigma}_r) - G_0(\boldsymbol{\sigma}_0)}{G_0(\boldsymbol{\sigma}_0)}|$, and its condition number χ, (see Benyah and Jennings (1998a), and Benyah and Jennings (1998b)). The last row in the Table gives the value of the objective function for the unregularized solution, $G_0(\boldsymbol{\sigma}_0) = 0.172172$, with corresponding solution (semi)norm $P(\boldsymbol{\sigma}_0) = 3.0781E+04$, condition number $\chi_0 = 2.9182E+07$ and relative error $\epsilon_0 = 0$. From the Table 7.2, we observe that the condition number, as well as the solution norm decrease with increasing penalty weight.

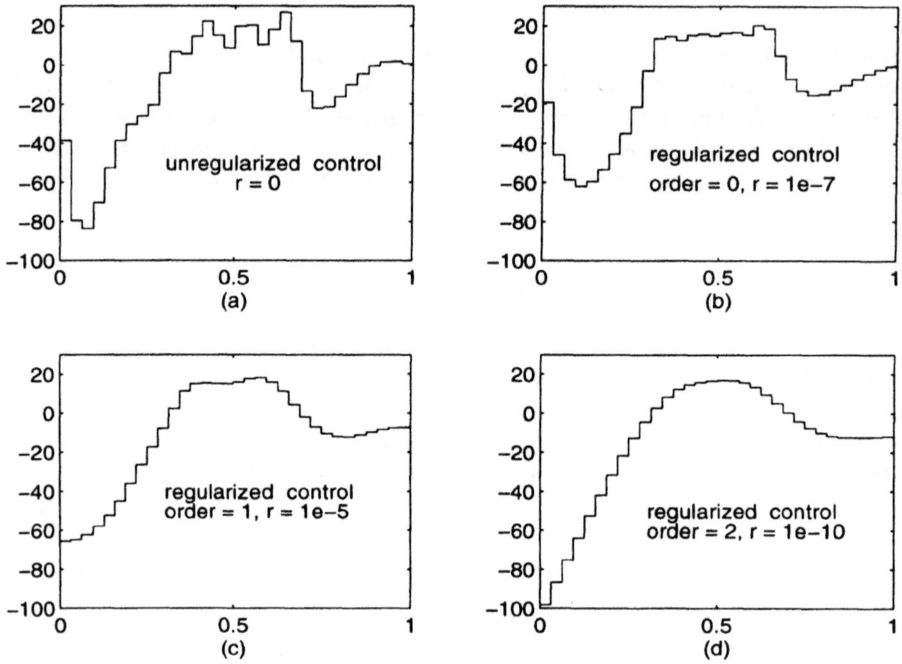

Figure 7.1 control functions for example 6.1

Table 7.2 Summary

regularized:	order = 0	order = 1	order = 2	unregularized
optimal r:	r = 1e-07	r = 1e-05	r = 1e-10	r = 0
$G_0(\boldsymbol{\sigma}_r)$:	0.173897	0.173245	0.173782	0.172172

The graphs in Figure 7.1 (a) show the control function for the unregularized problem; Figures 7.1 (b), 7.1 (c), and 7.1 (d), respectively, show the control functions for the regularization orders 0 (with $r = 1e-7$), order 1 (with $r = 1e-5$), and order 2 (with $r = 1e-10$). An optimal regularization order in each case was obtained using the *discrepancy principle*, given in Benyah and Jennings (1998c), with relative error, in objective function, $\epsilon_0 = 0.01$. This is summarized in Table 7.2.

8 TEST PROBLEM 2: THE CONTAINER CRANE PROBLEM

We now compute and compare the condition numbers of the solution of the *container crane problem*, using two discretization methods, namely, *control parametrization* and *state discretization.* The dynamics of the Container Crane model, see Sakawa and Shindo (1982), are given by:

Example 8.1

$$\min_{\boldsymbol{x},\,\boldsymbol{u}} \left\{ G_0 = \int_0^9 x_3^2(t) + x_6^2(t)\,dt, \right\} \tag{8.1}$$

subject to

$$\begin{aligned}
\dot{x}_1(t) &= x_4(t),\\
\dot{x}_2(t) &= x_5(t),\\
\dot{x}_3(t) &= x_6(t),\\
\dot{x}_4(t) &= u_1(t) + 17.2656x_3(t),\\
\dot{x}_5(t) &= u_2(t),\\
\dot{x}_6(t) &= -(u_1(t) + 27.0756x_3(t) + 2x_5(t)x_6(t))/x_2(t),
\end{aligned}$$

with initial and final state values

$$\begin{aligned}
\boldsymbol{x}(0) &= [0,\ 22,\ 0,\ 0,\ -1,\ 0]^T,\\
\boldsymbol{x}(9) &= [10,\ 14,\ 0,\ 2.5,\ 0,\ 0]^T,
\end{aligned}$$

and subject to the following bounds on the controls (torques of motors), u_1 *and* u_2 :

$$\begin{aligned}
-2.83374 \;\le\; u_1(t) \;\le\; 2.83374,\ \forall t \in [0,\ 9],\\
-0.80865 \;\le\; u_2(t) \;\le\; 0.71265,\ \forall t \in [0,\ 9],
\end{aligned}$$

and subject to bounds on the states x_4 *and* x_5 :

$$\begin{aligned}
|x_4(t)| &\le 2.5,\ \forall t \in [0,\ 9],\\
|x_5(t)| &\le 1.0,\ \forall t \in [0,\ 9].
\end{aligned}$$

This problem was first solved using `MISER3`, which is based on the method of control parametrization, see Benyah and Jennings (1998c). It was then reformulated using a *point discretization method*, see (4.2), and solved by the Optimization Toolbox of `MATLAB`.

8.1 Condition Numbers of the Container Crane Problem

The condition numbers of the solution of the crane problem obtained from the two methods, namely, *control parametrization*, and *state discretization* are displayed in Tables 8.1 and 8.2, respectively. In each table, the number of control parameters n, the objective function value G_0, the condition number of active constraints $\chi(\boldsymbol{L})$, the condition number of the projected Hessian $\chi(\boldsymbol{F}_{22})$, and the overall condition number of the Lagrange condition $\chi(\mathcal{M})$, are given. Note that the ill-conditioning gets worse as the number of control parameters increase.

Table 8.1 Control Parametrization Method

n	G_0	$\chi(\boldsymbol{L})$	$\chi(\boldsymbol{F}_{22})$	$\chi(\mathcal{M})$
4	0.077919	4.21E+02	6.78E+01	2.96E+06
8	0.011288	3.21E+02	2.71E+04	2.33E+07
16	0.010654	3.16E+02	4.71E+04	8.10E+07
32	0.010467	3.14E+02	8.81E+04	2.81E+08

Table 8.2 State Discretization Method

n	G_0	$\chi(\boldsymbol{L})$	$\chi(\boldsymbol{F}_{22})$	$\chi(\mathcal{M})$
4	0.013825	2.12E+03	8.52E+00	1.06E+05
8	0.011551	3.20E+02	4.95E+01	7.90E+03
16	0.010712	1.45E+02	3.30E+04	4.02E+06
32	0.010500	1.38E+02	1.54E+05	2.30E+11

Tables 8.3 and 8.4 display the condition numbers of the *regularized* solution of the Crane problem, using 32 control parameters, from the *control parametrization method* and *state discretization method* respectively. In each Table, the penalty weight r, the corresponding objective function value G_r, the condition number of active constraints $\chi(\boldsymbol{L})$, the condition number of the projected hessian $\chi(\boldsymbol{F}_{22})$, and the overall condition number of the Lagrange condition $\chi(\mathcal{M})$ are given.

9 CONCLUSIONS

We have examined the extent to which the ill-conditioning in optimal control computation can be influenced by the *number* of control parameters, as well as the *smoothness* of the control. We have shown that when derivative values are used to parametrize the controls the condition number χ satisfies approximately, the relationship $\chi \approx \alpha n^\beta$, $\alpha > 0$, $\beta > 0$. In particular, if piecewise

Table 8.3 Control Parametrization Method

r	G_r	$\chi(\boldsymbol{L})$	$\chi(\boldsymbol{F}_{22})$	$\chi(\mathcal{M})$
10^{-1}	0.09265	313.6	3.82E+03	2.56E+04
10^{-2}	0.020481	313.7	1.65E+03	1.96E+03
10^{-3}	0.012837	313.8	2.99E+02	1.81E+03
10^{-4}	0.011662	313.9	1.67E+02	2.36E+03
10^{-5}	0.011113	314.0	1.10E+03	4.61E+04
10^{-6}	0.010803	314.1	3.86E+03	1.75E+05
10^{-7}	0.010598	314.1	1.10E+04	6.17E+05
10^{-8}	0.010519	314.2	1.20E+05	8.90E+06
10^{-9}	0.010576	314.3	4.64E+05	2.56E+07
10^{-10}	0.010521	314.2	2.49E+06	1.10E+08
0	0.010467	314.3	8.03E+04	2.81E+08

Table 8.4 State Discretization Method

r	G_r	$\chi(\boldsymbol{L})$	$\chi(\boldsymbol{F}_{22})$	$\chi(\mathcal{M})$
10^{-1}	0.012382	138.65	6.30E+05	6.63E+05
10^{-2}	0.011500	138.65	4.73E+05	4.82E+05
10^{-3}	0.011020	138.65	3.26E+05	3.37E+05
10^{-4}	0.011631	138.65	6.21E+04	6.41E+05
10^{-5}	0.010620	138.65	7.41E+03	6.55E+05
10^{-6}	0.010547	138.65	4.94E+03	2.56E+05
10^{-7}	0.010508	138.65	4.90E+03	1.15E+06
10^{-8}	0.010560	138.65	4.89E+05	1.12E+07
10^{-9}	0.010512	138.65	2.01E+05	2.63E+07
10^{-10}	0.010499	138.65	2.01E+05	3.45E+07
0	0.010500	138.65	7.55E+03	2.43E+11

linear controls are required, it is better to use the *hat function basis* than to integrate a piecewise constant control function. Also, if second derivative values are used to parametrize the controls then the condition number can increase by as much as n^5. This shows that ill-conditioning gets worse as the number of parameters increase and as the parameters represent higher derivatives.

The results in Tables 8.1 and 8.2 also confirm that the two methods of discretization do not appear to have any significant effect on the degree of ill-conditioning, but this may be compounded by the inaccuracies of the whole process of computing these condition numbers. The state discretization used in this paper is very rudimentary, but it will contain the major part of the condition number for approximating a differential equation by finite differences or collocation.

Regularization, on the other hand, reduces the extent of ill-conditioning to a managable level, (see Tables 8.3 and 8.4). It also results in smoother control functions, (see Figures 7.1, 9.1 and 9.2), without compromising the constraint compliance and objective minimum to any great extent.

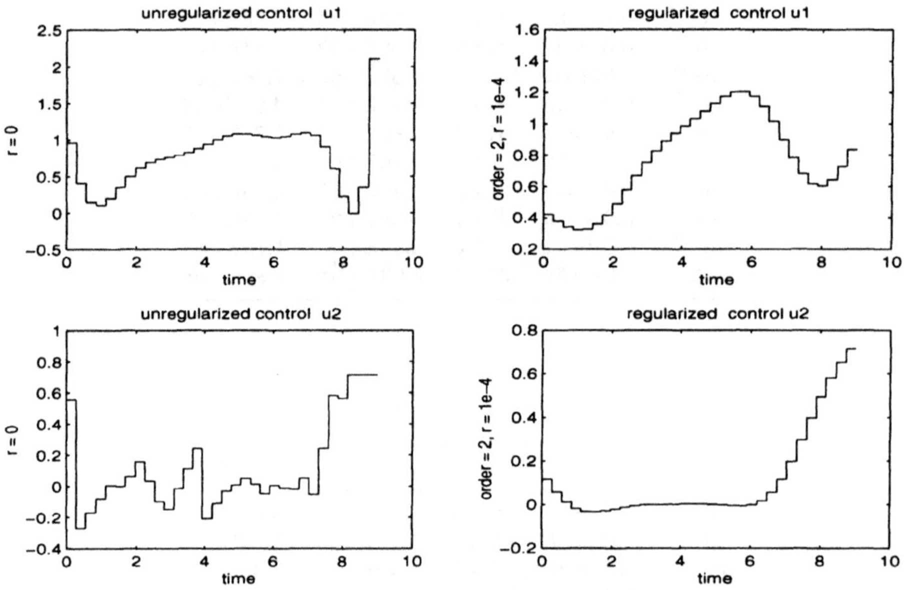

Figure 9.1 `control functions`:(*control parametrization method*)

References

Benyah, F. and Jennings, L. S. (1998), Ill-conditioning in optimal control computation, *Proceedings of the Eighth International Colloquium on Differential Equations* (ed. D. Bainov), (VSP BV, Utrecht, The Netherlands), pp. 81-88.

Benyah, F. and Jennings, L. S. (1998), Regularization in optimal control computation, *International Symposium on Intelligent Automation and Control (ISIAC '98)*, World Automation Congress, Anchorage, Alaska, May 10-14, 1998. TSI press, CDROM pp. ISIAC 091.1-091.6, 1998.

Benyah, F. and Jennings, L. S. (1998), The L-curve in regularization of optimal control computation, *J. Austral. Math. Soc. Ser. B*, Vol. 40, pp. 138-172.

Engl, H. W. (1993), Regularization methods for the stable solution of inverse problems, *Surv. Math. Ind.* Vol. 3, no. 2, pp. 71-143.

Engl, H. W. and Grever, W. (1994), Using the L-curve in determining optimal regularization Parameters, *Numerische Mathematik*, Vol. 69, pp. 25-31.

Groetsch, C. W (1984), *The Theory of Tikhonov Regularization for Fredholm Equations of the First Kind*, Pitman, Boston.

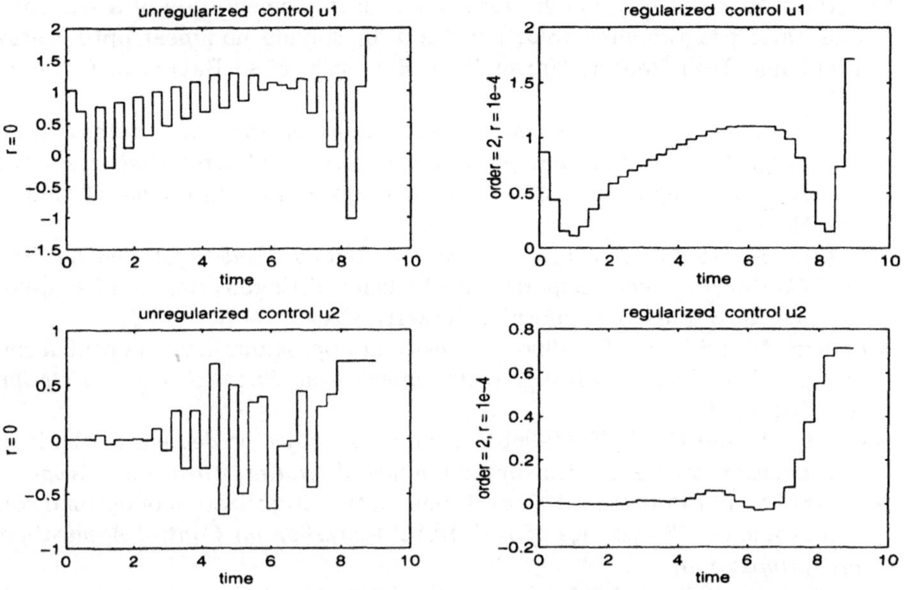

Figure 9.2 `control functions:` (*finite difference method*)

Hansen, P. C. (1992), Analysis of discrete ill-posed problems by means of the L-curve, *SIAM Review*, Vol. 34, pp. 561-580.

Hansen, P. C. (1993), Regularization tools, a MATLAB package for analysis and solution of discrete ill-posed problems, Tech. Report UNIC-92-0, Danish Computing Center for Research and Education.

Hansen, P. C. and O'Leary, D. P. (1993), The use of the L-curve in the regularization of discrete ill-posed problems, *SIAM J. Sci. Comput.*, Vol. 14, pp. 1487-1503.

Fierro, R. D. and G. H. Golub, and P. C. Hansen and D. P O'Leary, Regularization by truncated total least squares, *SIAM J. Sci. Comput.*, Vol. 18, no. 4, pp. 1223–1241.

MATLAB's Optimization TOOLBOX, Mass. U.S.A, 1997.

Jennings, L. S., Fisher, M. E., Teo, K. L. and Goh, C. J. (1991), MISER3: Solving optimal control problems: —an update, *Advances in Engineering Software*, Vol. 13, no. 4, pp. 190-196.

Jennings, L. S., Fisher, M. E., Teo, K. L. and Goh, C. J. (1997), ***MISER3**: Software for solving optimal control problems*, `http://cado.maths.uwa.edu.au`.

Polak, E. (1997), *Optimization: Algorithms and Consistent Approximations* , Springer-Verlag, New York.

Sakawa, Y. and Shindo, Y. (1982), Optimal control of container cranes, *Automatica*, Vol. 18, pp. 257-266.

Schittkowski, K. (1987), NLPQL version 2.3: Implementation of a sequential quadratic programming (SQP) method for solving nonlinear optimization problems, Tech. report, Universitaet Bayreuth, 8580 Bayreuth, Germany, F.R.

Schwartz, A. (1996a), *Theory and Implementations of Numerical Methods Based on Runge-Kutta Integration for Solving Optimal Control Problems: Ph.D Dissertation*, Department of Electrical Engineering, University of California, Berkeley.

Schwartz, A. (1996b) *RIOTS Version 1: A Matlab Toolbox for Solving Optimal Control Problems*, Department of Chemical Engineering and Computer Science, University of California, Berkerly, CA.

Schwartz, A. and Polak, E. (1996a), Consistent approximations for optimal control problems based on Runge-Kutta integration, *SIAM J. Control Optim.*, Vol. 34, no. 4.

Schwartz, A. and Polak, E. (1996b), A family of projected descent methods for optimization problems with simple bounds, *J. Optim. Theory and Appl.*.

Schwartz, A. and Polak, E. (1996c) Runge-Kutta discretization of optimal control problems, *Proceedings of 10th IFAC Workshop on Control Applications of Optimization.*

Stryk, O. von (1993), DIRCOL: *Numerical solution of optimal control problems by direct collocation, Optimal Control - Calculus of Variations, Optimal Control Theory and Numerical Methods, International Series of Numerical Mathematics 111* (ed. R. Bulirsch et. al.), (Birkhauser, Basel), pp. 129-143.

Stryk, O. von (1996), PAREST: A multiple shooting code for optimization problems in differential-algebraic equations, Tech. Report No. TUM-M9616, Fakultat fur Mathematik, Technische Universitat Munchen, Germany.

Stryk, O. von (1999), DIRCOL: User's Guide, Version 2.0, Tech. Report, Mathematisches Institut, Techinisches Universitat Muenchen, Muenchen 2, Germany.

Teo, K. L., Goh, C. J. and Wong, K. H. (1991), *A Unified Computational Approach to Optimal Control Problems*, Longman Scientific and Technical.

Tikhonov, A. N. and Arsemin, V. Y. (1997), *Solutions of ill-posed problems*, Winston & Sons, Washington D.C.

Zhou, J. L., Tits, A. and Lawrence, C. T. (1998), *FFSQP version 3.7b: A fortran code for solving optimization problems*, Tech. Report SRC-TR-92-107r5, University of Maryland, College Park, MD 20742, USA.

3 WORST-CASE OPTIMAL REGULATION OF LINEAR SYSTEMS IN THE PRESENCE OF STRUCTURED PERTURBATIONS

Saroj K. Biswas

Department of Electrical & Computer Engineering
Temple University
Philadelphia, PA19122

and M. Bala Subrahmanyam

Lockheed Martin Electronics & Missiles
Orlando, FL 32819

Abstract: Worst-case optimal regulation of linear time varying systems in the presence of both parameter perturbations as well as exogenous inputs is considered. The control problem is formulated as a state space minimax problem. Necessary conditions for the optimal solution are derived for the worst disturbance, the worst parameter perturbation, and the optimal controller. The necessary conditions are given in terms of a nonlinear two-point-boundary-value problem. A method is also presented for a numerical solution of this two-point-boundary-value problem. The results are illustrated by examples.

Key words: Worst-case control, H_∞ control, structured uncertainty, optimization, minimum principle.

X. Yang et al. (eds.), Optimization Methods and Applications, 45-64.

1 INTRODUCTION

An inevitable feature of practical control systems is the dynamic shift in behavior due to parameter or environmental perturbations. In recent years, considerable research efforts have been devoted to the H_∞ optimal control problems of such uncertain dynamical systems. Although the original formulation of the H_∞ method was in terms of frequency domain terms, time domain extensions in state space terms have been developed leading to Riccati type equations for feedback design [Doyle et.al., (1989), Francis (1987), Glover (1988), Khargonekar (1988), Peterson (1987), Ravi et.al (1989), Schmitendorf (1988), Stoorvogel (1992), Subrahmanyam (1995), Zhou et.al. (1988)]. It is known that under certain assumptions H_∞ suboptimal controllers can be synthesized via the solution of two algebraic Riccati equations. For time varying systems, similar Riccati differential equations have been developed by various authors, such as, Chen (1993), Ravi et.al (1989), Subrahmanyam (1993), Subrahmanyam (1995), Tadmor (1990). For a comprehensive introduction to the subject, the reader is referred to Doyle et.al., (1989). Linear quadratic differential game theory Basar and Bernhard (1995), Limebeer et.al. (1992), Yaesh et.al. (1993) has also been used in order to establish the existence of a H_∞ controller along with its structure.

Perturbations affecting the performance of a system can be broadly classified into two categories: a) environmental perturbations appearing as an extraneous force, and b) perturbations of the system parameters. Although systems with parameter perturbations can be mathematically represented as systems with unknown external inputs, such a transformation requires state and control dependent input matrices leading to additional complexities in the control design and analysis. This type of transformation has been used in Biswas (1990), and Biswas (1988) for robust stabilization of linear systems in the presence of Gaussian type structured perturbations. References Subrahmanyam (1993), Subrahmanyam (1994) investigated the parameter perturbation problems using variational formulation with respect to the H_∞ optimal and suboptimal solutions.

In this paper we investigate a direct approach to solving the H_∞ type optimal control problem in the presence of parameter perturbations in the system matrices as well as exogenous input disturbances. The H_∞ control problem in the presence of norm bounded parametric perturbations has also been considered in Keqin (1994), however, an additional matching condition on the uncertain matrices has been assumed. A similar result [Xie et.al. (1992)] allowed uncertainties to appear only in the system matrix. In this paper (as well as in Biswas (1993)), we do not assume any matching condition on parameter uncertainties, and both system and control matrices are assumed to contain uncertain parameters. For the parameter perturbations, we consider two cases: a) L_∞-bounded perturbations, and b) L_2-bounded perturbations.

We formulate the control problem as a minimax problem of a performance criterion. The ratio of the disturbance energy to the controlled system energy is taken as the performance criterion. This problem is equivalent to the state space approach of the H_∞ control problem. The method of analysis used in this paper parallels that of references Subrahmanyam (1993), Subrahmanyam (1995), Tadmor (1990), Chen (1993). We develop the necessary conditions that the optimal controller and the worst exogenous disturbance must satisfy along with the conditions for the worst variations of the system parameters. These necessary conditions are given in terms of a nonlinear two-point-boundary-value problem (TPBVP). Because of nonlinearity, derivation of a Riccati type equation is not expected. Instead a numerical method is suggested for the solution of the TPBVP, and used to compute the numerical results presented in the paper.

The optimal control computed through the solution of the two-point-boundary-value problem is an open loop control. Furthermore, this control is optimal with respect to the computed worst-case disturbances under the assumed minimax criterion. In reality, since disturbances cannot be predicted, the designer designs the controller based on the worst-case situation. In particular, the designer computes the worst-case disturbance as well as the corresponding optimal controller for an acceptable level of performance. The implication is that, in case the actual disturbance is milder than the worst-case, the performance of the controlled system will be better.

The paper is organized as follows: Section 2 introduces the minimax control problem. Necessary conditions of the optimal solution are developed in Section 3. Section 4 presents a method for numerical solution of the necessary conditions for the minimax solution which is used in Section 5 in numerical examples. The paper is concluded in Section 6 with some remarks.

Notations: Throughout the paper we shall use $'$ to denote the transpose of a matrix or a vector, and $\langle \cdot, \cdot \rangle$ to denote the usual scalar product in the Euclidean space R^n. Given the matrices M and N of an identical size, we use $M \circ N$ to denote the element-wise product of the matrices M and N, for example, the ij-th entry of $M \circ N$ is given by

$$(M \circ N)_{ij} = M_{ij} N_{ij}$$

Clearly, $M \circ N$ has the same size as M or N.

2 PROBLEM STATEMENT

Consider the time varying linear system in R^n, defined over a finite time interval $[0, T]$, given by

$$\begin{aligned} \dot{x} &= \widetilde{A}(t)x + \widetilde{B}(t)u + E(t)v \\ x(0) &= x_0 \end{aligned} \tag{2.1}$$

where $x(t) \in R^n$ and $u(t) \in R^m$ are the state vector and the control vector, respectively. Suppose that the system is perturbed by an exogenous disturbance $v(t) \in R^s$ through the matrix E which is assumed to be known and fixed. We also consider the initial state $x_0 \in R^n$ as an initial disturbance.

We assume that the elements of the system matrix $\widetilde{A}(t)$ and the control matrix $\widetilde{B}(t)$ are randomly perturbed with respect to their nominal values in the sense that

$$\begin{aligned} \widetilde{A}(t) &= A(t) + \Delta A(t), \\ \widetilde{B}(t) &= B(t) + \Delta B(t) \end{aligned} \tag{2.2}$$

where $A(t)$ and $B(t)$ are the nominal matrices, and $\Delta A(t)$ and $\Delta B(t)$ are perturbation matrices. For the perturbation matrices, we shall consider two cases: Case I for perturbations with magnitude bounded between certain upper and lower limits, and Case II for perturbations without any strict magnitude bounds but bounded in L_2 sense.

2.1 *Case I: L_∞ Perturbation of Parameters*

In many engineering applications perturbation of system parameters can be considered to be bounded between certain upper and lower limits. Without any loss of generality we assume that

$$\begin{aligned} |\Delta A_{ij}(t)| &\leq \Delta A_{ij}^{\max}, \\ |\Delta B_{ij}(t)| &\leq \Delta B_{ij}^{\max} \end{aligned} \tag{2.3}$$

where M_{ij} represents the ij-th element of the matrix M, and $M_{ij}^{\max}$ is its largest perturbation. In case M_{ij} does not have a perturbation, we set $\Delta M_{ij}^{\max} = 0$.

The problem we investigate in this paper is a minimax control problem of achieving the best possible state regulation in the presence of maximum disturbances with minimum control efforts. For a mathematical formulation of the control problem, we introduce the performance index

$$J_\infty(u, v, x_0, \Delta A, \Delta B) = \frac{\frac{1}{2}\langle x_0, S_0 x_0\rangle + \frac{1}{2}\int_0^T \langle v, R_v v\rangle\, dt}{\frac{1}{2}\langle x(T), S_1 x(T)\rangle + \frac{1}{2}\int_0^T \{\langle x, Qx\rangle + 2\langle x, Nu\rangle + \langle u, R_u u\rangle\}\, dt} \tag{2.4}$$

where R_u and R_v are positive definite (time varying) matrices. In this setup, the initial state x_0 is also considered as a perturbation, and we use the positive definite matrix S_0 as the corresponding weighting matrix. Regulation of the final state is very important in many applications which is introduced in this performance criterion through the terminal cost with S_1 taken as a positive semidefinite matrix, appearing in the denominator of (2.4). In addition, the matrix Q is taken as a positive semidefinite matrix, and the matrix N is such that the matrix $\begin{bmatrix} Q & N \\ N' & R_u \end{bmatrix}$ is positive semidefinite. The various matrices in the performance index (2.4) have similar significance as in a conventional optimal control problem. It is also easily seen that this setup is equivalent to an output regulation problem with the output equation given by

$$y(t) = C(t)x + D(t)u$$

with known matrices for $C(t)$ and $D(t)$.

We now introduce the minimax problem of finding the optimal control $u \in L_2(0,T)$, the worst exogenous disturbance $v \in L_2(0,T)$, the worst initial disturbance $x_0 \in R^n$, and the worst parameter perturbations $\Delta A(t)$ and $\Delta B(t)$, and a scalar λ such that

$$\lambda = \min_{\substack{v \in L_2(0,T);\, v \neq 0 \\ x_0 \in R^n;\, x_0 \neq 0 \\ |\Delta A(t)| \leq \Delta A^{\max} \\ |\Delta B(t)| \leq \Delta B^{\max}}} \max_{u \in L_2(0,T)} J_\infty(u, v, x_0, \Delta A, \Delta B) \tag{2.5}$$

2.2 *Case II: L_2 Perturbation of Parameters*

Determination of *a priori* bound on the perturbation of the system parameters is not possible in many applications. In such cases, it is necessary to design controllers that can accommodate the largest amount of perturbations. For these problems, we assume that the uncertain matrices $\Delta A(t) \in L_2(0,T)$ and $\Delta B(t) \in L_2(0,T)$, and introduce the performance criterion:

$$J_2(u, v, x_0, \Delta A, \Delta B) = \frac{\frac{1}{2}\langle x_0, S_0 x_0\rangle + \frac{1}{2}\int_0^T \{\langle v, R_v v\rangle + \sum_{i,j\in\mathcal{P}_\mathcal{A}} R^A_{ij}\Delta A^2_{ij} + \sum_{i,j\in\mathcal{P}_\mathcal{B}} R^B_{ij}\Delta B^2_{ij}\}\, dt}{\frac{1}{2}\langle x(T), S_1 x(T)\rangle + \frac{1}{2}\int_0^T \{\langle x, Qx\rangle + 2\langle x, Nu\rangle + \langle u, R_u u\rangle\}\, dt} \tag{2.6}$$

where

$$\mathcal{P}_\mathcal{A} = \{\text{indices } i, j \text{ for which } \Delta A_{ij} \text{ is a perturbation}\}$$

$$\mathcal{P}_\mathcal{B} = \{\text{indices } i, j \text{ for which } \Delta B_{ij} \text{ is a perturbation}\}$$

and R^A_{ij} and R^B_{ij} are positive weighting factors. It is noted that not every element of ΔA or ΔB is possibly a perturbation so that we carry out the minimax optimization only over the set of entries of ΔA and ΔB that are perturbed. The other terms of the cost function (2.6) have the same significance as in (2.4).

The corresponding minimax problem is then defined as the problem of finding the optimal control $u \in L_2(0,T)$, the worst exogenous disturbance $v \in L_2(0,T)$, the worst initial disturbance $x_0 \in R^n$, and the worst parameter perturbations $\Delta A(t)$ and $\Delta B(t)$, and a scalar λ such that

$$\lambda = \min_{\substack{v \in L_2(0,T);\, v \neq 0 \\ x_0 \in R^n;\, x_0 \neq 0 \\ \Delta A \in L_2(0,T);\, \Delta A \neq 0 \\ \Delta B \in L_2(0,T);\, \Delta B \neq 0}} \max_{u \in L_2(0,T)} J_2(u, v, x_0, \Delta A, \Delta B) \tag{2.7}$$

The minimax optimization problems (2.5) and (2.7) are equivalent to the state space formulation of the H_∞ control problem. We note that the performance index (2.4) (also (2.6)) is the ratio of the disturbance energy to the controlled energy, and the H_∞ norm of the control system (2.1) - (2.2) is given by $\frac{1}{\sqrt{\lambda}}$. The quantity λ is interpreted as the disturbance rejection capacity of the system. A large λ implies a better controller in the sense that the system will be able to tolerate larger amount of perturbations for the same controlled performance. A small λ means that the system is too sensitive to disturbances; despite the effects of the control, the state trajectory is not close to the desired trajectory even in the presence of small amount of perturbations.

As part of the design procedure, one computes the worst-case disturbance and the corresponding controller for the optimal state regulation. We note that this controller is optimal for the worst-case disturbance in the sense of the minimax criteria (2.5) or (2.7). Such worst-case design is necessary since disturbances cannot be predicted. In reality, if the disturbances are stronger than the computed worst-case magnitude, the controlled performance will be degraded compared to the optimal. Similarly, a better performance is expected in the presence of milder disturbances.

3 OPTIMAL SOLUTION

In this section we derive the necessary conditions that must be satisfied by the optimal controller, the worst case disturbances, and the worst parameter variations. For this purpose we follow a method similar to that given in Subrahmanyam (1995).

3.1 Case I: L_∞ Perturbation of Parameters

Let $\lambda > 0$ be the solution of the minimax problem (2.5). Then it is easy to see that the cost function

$$\begin{aligned}\tilde{J}_\infty(u, v, x_0, \Delta A, \Delta B) &= \frac{1}{2}\langle x_0, S_0 x_0\rangle + \frac{1}{2}\int_0^T \langle v, R_v v\rangle \, dt \\ -\lambda\Big\{\frac{1}{2}\langle x(T), S_1 x(T)\rangle &+ \frac{1}{2}\int_0^T \{\langle x, Qx\rangle + 2\langle x, Nu\rangle + \langle u, R_u u\rangle\}\, dt\Big\} \quad (3.1)\end{aligned}$$

is nonnegative, and has the minimum $\tilde{J}_\infty = 0$. Hence the minimax problem of the performance criterion (2.4) is equivalent to that of the performance criterion (3.1). For this purpose, we use the Pontryagin's minimum principle. We first find a saddle point $(u, v, x_0, \Delta A, \Delta B)$ with respect to the above criterion (3.1) subject to the system dynamics (2.1).

For a given $u(t)$, let the perturbation $(v, x_0, \Delta A, \Delta B)$ minimizes (3.1). We do not assume any constraint on the disturbances v and x_0, however, the parameter perturbations ΔA and ΔB are assumed to satisfy the bound (2.3). The following lemma characterizes the perturbations $(v, x_0, \Delta A, \Delta B)$.

Lemma 3.1 *Given the system* (2.1), *for a given* $u(t)$, *if* $(v, x_0, \Delta A, \Delta B)$ *minimizes* (3.1), *then there exists an absolutely continuous vector* $\eta(t) \in R^n$ *such that*

$$\begin{aligned}\dot{\eta} &= \lambda Q x + \lambda N u - (A + \Delta A)'\eta \\ \eta(T) &= -\lambda S_1 x(T)\end{aligned} \quad (3.2)$$

and

$$\begin{aligned} v &= -R_v^{-1}E'\eta \\ x_0 &= -S_0^{-1}\eta(0) \end{aligned} \tag{3.3}$$

$$\begin{aligned} \Delta A &= -\Delta A^{\max} \circ \text{sign}\{\eta x'\} \\ \Delta B &= -\Delta B^{\max} \circ \text{sign}\{\eta u'\} \end{aligned} \tag{3.4}$$

Proof: By Pontryagin's minimum principle, there exists an adjoint state $\eta(t)$ such that the Hamiltonian

$$\begin{aligned} H(v, x_0, \Delta A, \Delta B) &= \frac{1}{2}\langle v, R_v v\rangle - \frac{\lambda}{2}\{\langle x, Qx\rangle + 2\langle x, Nu\rangle + \langle u, R_u u\rangle\} \\ &\quad +\langle \eta, (A+\Delta A)x + (B+\Delta B)u + Ev\rangle \end{aligned} \tag{3.5}$$

is minimized almost everywhere on $[0, T]$ satisfying

$$\begin{aligned} \dot{\eta} &= -\frac{\partial H}{\partial x} = \lambda Qx + \lambda Nu - (A+\Delta A)'\eta \\ \eta(T) &= -\lambda S_1 x(T). \end{aligned} \tag{3.6}$$

The necessary condition (3.3) for v is obtained from satisfaction of $\frac{\partial H}{\partial v} = 0$, and that for x_0 follows from the transversality condition. Similarly, minimizing the Hamiltonian (3.5) with respect ΔA and ΔB within the given bounds, one obtains the parameter variations ΔA and ΔB. □

We emphasize here that the equation (3.4) must be understood as an element-wise equality as mentioned earlier, for example, for the ij-th element of ΔA, we have

$$\Delta A_{ij} = -\Delta A_{ij}^{\max}\text{sign}\{(\eta x')_{ij}\} = -\Delta A_{ij}^{\max}\text{sign}\{\eta_i x_j\}.$$

For a given $(v, x_0, \Delta A, \Delta B)$, the maximizing controller u for the cost (3.1) is obtained in a similar manner. We present the result in the following lemma:

Lemma 3.2 *Consider the system* (2.1) *with the performance criterion* (3.1). *Then for a given* $(v, x_0, \Delta A, \Delta B)$, *if* u *maximizes* (3.1), *there exists an adjoint variable* $\psi(t) \in R^n$ *such that*

$$\begin{aligned} \dot{\psi} &= -Qx - Nu - (A+\Delta A)'\psi \\ \psi(T) &= S_1 x(T) \end{aligned} \tag{3.7}$$

and the optimizing control is given by

$$u(t) = -R_u^{-1}[N'x + (B+\Delta B)'\psi] \tag{3.8}$$

Proof: The proof follows from minimization of the cost

$$J(u) = \frac{1}{2}\langle x(T), S_1 x(T)\rangle + \frac{1}{2}\int_0^T \{\langle x, Qx\rangle + 2\langle x, Nu\rangle + \langle u, R_u u\rangle\}\, dt$$

with respect to the control u, subject to the dynamics (2.1). □

Simultaneous solution of the equations (3.2) - (3.4) and (3.7) - (3.8) then yields a saddle point solution $(u, v, x_0, \Delta A, \Delta B)$ for the problem (2.5) as stated in the following theorem:

Theorem 3.1 *Consider the system* (2.1) *and the performance measure* (3.1). *Then a saddle point solution* $(u, v, x_0, \Delta A, \Delta B)$ *of the problem is characterized by the solution of the two-point-boundary-value problem*

$$\begin{bmatrix} \dot{x} \\ \dot{\psi} \end{bmatrix} = \begin{bmatrix} \tilde{A} - \tilde{B}R_u^{-1}N' & -\tilde{B}R_u^{-1}\tilde{B}' + \lambda E R_v^{-1}E' \\ -(Q - NR_u^{-1}N') & -(\tilde{A} - \tilde{B}R_u^{-1}N')' \end{bmatrix} \begin{bmatrix} x \\ \psi \end{bmatrix} \tag{3.9}$$

with the boundary conditions

$$S_0 x_0 - \lambda\psi(0) = 0 \tag{3.10}$$
$$S_1 x(T) - \psi(T) = 0. \tag{3.11}$$

The worst parameter variations are determined by

$$\begin{aligned} \Delta A &= \Delta A^{\max} \circ \operatorname{sign}\{\psi x'\} \\ \Delta B &= \Delta B^{\max} \circ \operatorname{sign}\{\psi u'\}. \end{aligned} \tag{3.12}$$

The optimal controller and the worst exogenous disturbance are given by

$$u = -R_u^{-1}[N'x + (B + \Delta B)'\psi] \tag{3.13}$$
$$v = \lambda R_v^{-1}E'\psi \tag{3.14}$$

Proof: From (3.2) and (3.7), it follows that

$$\begin{aligned} \tfrac{d}{dt}(\eta + \lambda\psi) &= -(A + \Delta A)'(\eta + \lambda\psi) \\ (\eta + \lambda\psi)(T) &= 0 \end{aligned} \tag{3.15}$$

so that

$$\eta(t) + \lambda\psi(t) = 0 \tag{3.16}$$

for almost all $t \in [0, T]$. The proof then follows by combining the results of Lemma 3.1 and Lemma 3.2 and substituting $\eta(t) = -\lambda\psi(t)$. □

We note that λ remains as an unknown parameter in the boundary value problem (3.9) - (3.14). In fact, this boundary value problem can have a non-trivial solution for countably many values of λ. The smallest positive value of λ satisfying (3.9) - (3.14) is the optimal λ corresponding to the worst perturbations $(v, x_0, \Delta A, \Delta B)$ and the optimal control u.

Theorem 3.2 *Let* λ *be the smallest positive value for which the boundary value problem* (3.9) - (3.14) *has a non trivial solution* $\{x, \psi\}$ *with* $\frac{1}{2}\langle x(T), S_1 x(T)\rangle +$

$\frac{1}{2}\int_0^T \{\langle x, Qx\rangle + 2\langle x, Nu\rangle + \langle u, R_u u\rangle\}\,dt > 0$. *Then* λ *is the minimax value of the cost* (2.4), $\{x, \psi\}$ *is the optimal pair,* u *is the optimal control, and* $(v, x_0, \Delta A, \Delta B)$ *are the worst perturbations.*

Proof: It is clear from Lemma 3.1 and Lemma 3.2 that if $(u, v, x_0, \Delta A, \Delta B)$ is a saddle point, then (3.9) - (3.14) are satisfied with λ being the minimax value of the (2.4). Now suppose that $\{x, \psi\}$ satisfies (3.9) - (3.14) for some λ such that

$$\frac{1}{2}\langle x(T), S_1 x(T)\rangle + \frac{1}{2}\int_0^T \{\langle x, Qx\rangle + 2\langle x, Nu\rangle + \langle u, R_u u\rangle\}\,dt > 0.$$

From the second equation of (3.9) and (3.13), we have

$$-\int_0^T \langle x, \dot{\psi}\rangle\,dt = \int_0^T \{\langle x, Qx\rangle + \langle x, Nu\rangle + \langle x, (A + \Delta A)'\psi\rangle\}\,dt \tag{3.17}$$

Integrating the left hand side of the above equation by parts and using the boundary conditions (3.10), (3.11) and the system dynamics (2.1), we have

$$\begin{aligned} -\int_0^T \langle x, \dot{\psi}\rangle\,dt &= -\langle x, \psi\rangle\Big|_0^T + \int_0^T \langle \dot{x}, \psi\rangle\,dt \\ &= -\langle x(T), S_1 x(T)\rangle + \tfrac{1}{\lambda}\langle x_0, S_0 x_0\rangle \\ &\quad + \int_0^T \{\langle (A + \Delta A)x, \psi\rangle + \langle (B + \Delta B)u, \psi\rangle + \langle Ev, \psi\rangle\}\,dt \end{aligned} \tag{3.18}$$

Note that from (3.13) and (3.14), we have

$$\begin{aligned} (B + \Delta B)'\psi &= -R_u u - N'x \\ E'\psi &= \tfrac{1}{\lambda} R_v v \end{aligned} \tag{3.19}$$

Using the above equations in (3.17), we obtain

$$\begin{aligned} \int_0^T \{\langle x, Qx\rangle + \langle x, Nu\rangle\}\,dt = &\; -\langle x(T) S_1 x(T)\rangle + \tfrac{1}{\lambda}\langle x_0, S_0 x_0\rangle \\ &+ \int_0^T \{\tfrac{1}{\lambda}\langle v, R_v v\rangle - \langle u, R_u u\rangle - \langle u, N'x\rangle\}\,dt \end{aligned} \tag{3.20}$$

leading to

$$\frac{\frac{1}{2}\langle x_0, S_0 x_0\rangle + \frac{1}{2}\int_0^T \langle v, R_v v\rangle\,dt}{\frac{1}{2}\langle x(T), S_1 x(T)\rangle + \frac{1}{2}\int_0^T \{\langle x, Qx\rangle + 2\langle x, Nu\rangle + \langle u, R_u u\rangle\}\,dt} = \lambda \tag{3.21}$$

Thus the value of the performance index (2.4) associated with the saddle point solution $(u, v, x_0, \Delta A, \Delta B)$ is λ. Hence if $\{x, \psi\}$ is a solution of the boundary value problem (3.9) - (3.14) for the smallest positive parameter λ such that the denominator of (2.4) is nonzero, then u given by (3.13) is the optimal control and v, x_0 and $\Delta A, \Delta B$ given by (3.14), (3.10), (3.12), respectively, are the worst-case perturbations. □

This completes the minimax solution of the problem (2.5). As shown in Theorem 3.2, the optimal control and the worst-case disturbance and parameter variations are characterized by a nonlinear two-point-boundary-value problem. Because of nonlinearity, further simplification of these necessary conditions is not expected. However, the two-point-boundary-value problem (3.9) - (3.14) can be solved using appropriate numerical techniques. This will be discussed in Section 4.

3.2 Case II: L_2 Perturbation of Parameters

We investigate the minimax problem (2.7) for the cost function (2.6) subject to the dynamics (2.1). In this case we do not assume any bound on the magnitude of the parameter variations, but assume that $\Delta A \in L_2(0,T)$ and $\Delta B \in L_2(0,T)$. This problem is of extreme practical significance if it is not possible to establish any bound on the parameter variations; in such cases one designs a control system that is capable of regulating the system in the presence of the largest perturbations.

In order to find the minimax solution of the problem (2.7) we essentially follow the same procedure as in the previous case. Note that the performance criterion (2.6) contains two additional terms in its numerator corresponding to the parameter perturbation ΔA and ΔB as compared to the performance criterion (2.4). In this case one wishes to find the largest possible perturbations with minimum control efforts.

We assume that λ is the minimax solution of the problem (2.7). Consider the performance criterion

$$\begin{aligned}\tilde{J}_2(u, v, x_0, \Delta A, \Delta B) &= \frac{1}{2}\langle x_0, S_0 x_0\rangle \\ &+\frac{1}{2}\int_0^T \{\langle v, R_v v\rangle + \sum_{i,j\in\mathcal{P}_A} R^A_{ij}\Delta A^2_{ij} + \sum_{i,j\in\mathcal{P}_B} R^B_{ij}\Delta B^2_{ij}\}\, dt \\ &-\lambda\{\frac{1}{2}\langle x(T), S_1 x(T)\rangle + \frac{1}{2}\int_0^T \{\langle x, Qx\rangle + 2\langle x, Nu\rangle + \langle u, R_u u\rangle\}\, dt\}\end{aligned} \tag{3.22}$$

Since λ is the optimal solution, it is clear that the function $\tilde{J}_2$ defined above is nonnegative, and has the minimum $\tilde{J}_2 = 0$. Utilizing the Pontryagin's minimum principle, we first find a saddle point $(u, v, x_0, \Delta A, \Delta B)$ with respect to the criterion (3.22) subject to the system dynamics (2.1). The following lemma characterizes $(v, x_0, \Delta A, \Delta B)$ for a given u.

Lemma 3.3 *Consider the system* (2.1) *and the performance criterion* (3.22). *Then for a given* $u(t)$, *if* $(v, x_0, \Delta A, \Delta B)$ *minimizes* (3.22), *then there exists a*

$\eta(t) \in R^n$ such that

$$\begin{array}{l}\dot{\eta} = \lambda Qx + \lambda Nu - (A + \Delta A)'\eta \\ \eta(T) + \lambda S_1 x(T) = 0 \\ \eta(0) + S_0 x(0) = 0\end{array} \tag{3.23}$$

and

$$\begin{array}{rcl} v & = & -R_v^{-1}E'\,\eta \\ \Delta A_{ij} & = & -\frac{1}{R_{ij}^A}\,\eta_i\,x_j, \quad i,j \in \mathcal{P}_A \\ \Delta B_{ij} & = & -\frac{1}{R_{ij}^B}\,\eta_i\,u_j, \quad i,j \in \mathcal{P}_B \end{array} \tag{3.24}$$

Proof: The proof follows from minimization of the Hamiltonian

$$\begin{array}{rl} H(v, x_0, \Delta A, \Delta B) = & \frac{1}{2}\langle v, R_v v\rangle + \frac{1}{2}\sum_{i,j\in\mathcal{P}_A} R_{ij}^A \Delta A_{ij}^2 + \frac{1}{2}\sum_{i,j\in\mathcal{P}_B} R_{ij}^B \Delta B_{ij}^2 \\ & -\frac{\lambda}{2}\{\langle x, Qx\rangle + 2\langle x, Nu\rangle + \langle u, R_u u\rangle\} + \langle \eta, (A + \Delta A)x + (B + \Delta B)u + Ev\rangle \end{array} \tag{3.25}$$

along the optimal trajectory almost everywhere on $[0, T]$. □

Next we find the maximizing controller u for a given $(v, x_0, \Delta A, \Delta B)$. We present the result in the following lemma:

Lemma 3.4 *Consider the system* (2.1) *with the performance criterion* (3.22). *Then for a given $(v, x_0, \Delta A, \Delta B)$, if u maximizes* (3.22), *there exists an adjoint variable $\psi(t) \in R^n$ such that*

$$\begin{array}{rcl} \dot{\psi} & = & -Qx - Nu - (A + \Delta A)'\psi \\ \psi(T) & = & S_1 x(T) \end{array} \tag{3.26}$$

and the optimizing control is given by

$$u(t) = -R_u^{-1}[N'x + (B + \Delta B)'\psi] \tag{3.27}$$

□

Simultaneous solution of the equations (3.23) - (3.24) and (3.26) - (3.27) then yields a saddle point solution $(u, v, x_0, \Delta A, \Delta B)$ for the problem (2.7). This is presented in the next theorem.

Theorem 3.3 *Consider the system* (2.1) *and the performance measure* (3.22). *Then a saddle point solution $(u, v, x_0, \Delta A, \Delta B)$ of the problem is characterized by the solution of the two-point-boundary-value problem*

$$\begin{bmatrix} \dot{x} \\ \dot{\psi} \end{bmatrix} = \begin{bmatrix} \tilde{A} - \tilde{B}R_u^{-1}N' & -\tilde{B}R_u^{-1}\tilde{B}' + \lambda E R_v^{-1}E' \\ -(Q - NR_u^{-1}N') & -(\tilde{A} - \tilde{B}R_u^{-1}N')' \end{bmatrix} \begin{bmatrix} x \\ \psi \end{bmatrix} \tag{3.28}$$

with the boundary conditions

$$\begin{aligned} S_0x(0) - \lambda\psi(0) &= 0 \\ S_1x(T) - \psi(T) &= 0. \end{aligned} \tag{3.29}$$

The worst parameter variation of ΔA *and* ΔB *are given by*

$$\begin{aligned} \Delta A_{ij} &= \tfrac{\lambda}{R^A_{ij}}\,\psi_i\,x_j, \quad i,j \in \mathcal{P}_A \\ \Delta B_{ij} &= \tfrac{\lambda}{R^B_{ij}}\,\psi_i\,u_j, \quad i,j \in \mathcal{P}_B. \end{aligned} \tag{3.30}$$

The optimal controller and the worst exogenous disturbance are given by

$$\begin{aligned} u &= -R_u^{-1}[N'x + (B + \Delta B)'\psi] \\ v &= \lambda R_v^{-1}E'\psi \end{aligned} \tag{3.31}$$

Proof: In light of (3.23) and (3.26), we have $\eta(t) = -\lambda\psi(t)$. The proof then follows easily from Lemma 3.3 and Lemma 3.4. □

As in the case of L_∞ bounded perturbations, the parameter λ is unknown in the two-point-boundary-value problem (3.28)-(3.31). The smallest value of this parameter is indeed the minimax value of the cost (2.6) as stated in the next theorem.

Theorem 3.4 *Let* λ *be the smallest positive value for which the boundary value problem* (3.28) - (3.31) *has a non trivial solution* $\{x, \psi\}$ *with* $\frac{1}{2}\langle x(T), S_1x(T)\rangle + \frac{1}{2}\int_0^T \{\langle x, Qx\rangle + 2\langle x, Nu\rangle + \langle u, R_uu\rangle\}\,dt > 0$. *Then* λ *is minimax value of the cost* (2.6), *and* $\{x, \psi\}$ *is the optimal pair.*

Proof: Similar to that of Theorem 3.2. □

The main difference between Theorem 3.2 and Theorem 3.4 is that in the case of L_∞-bounded perturbations, the worst variation of $\Delta A(t)$ and $\Delta B(t)$ are due to jump type variations of the elements taking values from the vertices of the $\Delta A^{\max}$ and $\Delta B^{\max}$ polytope, respectively. On the other hand, in the case of L_2 type perturbations, the worst variations occur due to the elements of $\Delta A(t)$ and $\Delta B(t)$ determined by the appropriate equations given in Theorem 3.3. In both cases the minimax solution is characterized by a two-point-boundary-value problem that contains λ as an unknown parameter. The next section investigates a method for numerical solution of this two point boundary value problem.

4 COMPUTATION OF DISTURBANCE REJECTION CAPACITY

In both Theorem 3.1 and Theorem 3.3 we observe that the minimax solution of the worst case optimal regulation problem is determined by a nonlinear two-point-boundary-value problem of size $2n$ where n is the size of the state vector. In this two-point-boundary-value problem, the parameter λ is an unknown and appears in both the dynamic equations as well as the boundary conditions. Because of nonlinearity further simplification of the TPBVP or derivation of a Riccati type equation is not expected. Nevertheless, we can solve the TPBVP using the Newton's method as discussed below. We discuss the method by referring to the appropriate equations in Theorem 3.1. Numerical method for solving the TPBVP of Theorem 3.3 is essentially the same.

The method starts by assuming a value of the parameter λ and the unknown initial condition $x(0)$. Denote the two unknowns as

$$z = \begin{bmatrix} \lambda \\ x(0) \end{bmatrix} \tag{4.1}$$

Because of (3.10), we have $\psi(0)$ so that we can solve the set of differential equations (3.9) (with (3.12) and (3.13) substituted as necessary) forward in time. Clearly, the solutions $x(t)$ and $\psi(t)$ will be functions of z. Furthermore, since the initial guess for z is nonoptimal, the boundary condition (3.11) at the final time is not expected to be satisfied. Define an error function

$$E(z) = \alpha_1 \|S_1 x(T;z) - \psi(T;z)\|^2 + \alpha_2 \lambda^2 \tag{4.2}$$

where α_1 and α_2 are some positive constants. Note that the error function contains λ^2 since we need to find the minimum value of λ for which the TPBVP has a nontrivial solution, and this minimum value is the optimal disturbance rejection capacity. In this study we have used the Matlab macro `fmins` to minimize $E(z)$ subject to the TPBVP given in Theorem 3.1 or Theorem 3.3.

In practice, convergence of iterations to a solution is somewhat dependent upon the initial guess for z. It is also possible that iterations may converge to one of the local minima corresponding to any one of the λ's. It is reminded that the two-point- boundary-value problem (3.9)-(3.14) has a nontrivial solution for countably many values of λ. A large weight α_2 in (4.2) is helpful in this regard to lead the iterations to the smallest value of λ.

While the parameter λ is used as a quantitative measure of performance, the state trajectory of the controlled system provides further details of the controlled performance. In addition, the trajectory of the worst-case disturbance can be considered as the worst-case bound of disturbance. Any disturbance stronger than this bound will degrade the performance of the controlled system beyond an acceptable level.

An alternate procedure [Subrahmanyam (1995)] based on the nominal system can also be used to initialize the iteration. Consider the nominal system, i.e., with $\Delta A = 0$ and $\Delta B = 0$, and let $\Phi(t, \tau)$ be the (nominal system) state transition matrix corresponding to the system (3.9) (with $\Delta A = 0$ and

$\Delta B = 0$):

$$\Phi(t,\tau) = \begin{bmatrix} \Phi_{11}(t,\tau) & \Phi_{12}(t,\tau) \\ \Phi_{21}(t,\tau) & \Phi_{22}(t,\tau) \end{bmatrix} \tag{4.3}$$

Then we have

$$\begin{aligned} x(t) &= \Phi_{11}(t,0)x(0) + \Phi_{12}(t,0)\psi(0) \\ \psi(t) &= \Phi_{21}(t,0)x(0) + \Phi_{22}(t,0)\psi(0) \end{aligned} \tag{4.4}$$

Then utilizing (3.10), one obtains

$$\begin{aligned} x(T) &= [\Phi_{11}(T,0) + \tfrac{1}{\lambda}\Phi_{12}(T,0)S_0]x(0) \\ \psi(T) &= [\Phi_{21}(T,0) + \tfrac{1}{\lambda}\Phi_{22}(T,0)S_0]x(0) \end{aligned} \tag{4.5}$$

These equations, in association with (3.11), then gives

$$(\lambda[S_1\Phi_{11}(T,0) - \Phi_{21}(T,0)] + [S_1\Phi_{12}(T,0) - \Phi_{22}(T,0)]S_0)x(0) = 0$$

or equivalently,

$$\begin{bmatrix} S_1 & -I \end{bmatrix} \begin{bmatrix} \Phi_{11}(T,0) & \Phi_{12}(T,0) \\ \Phi_{21}(T,0) & \Phi_{22}(T,0) \end{bmatrix} \begin{bmatrix} \lambda I \\ S_0 \end{bmatrix} x(0) = 0 \tag{4.6}$$

Then in order for a nontrivial solution for the TPBVP to exist, it is necessary that

$$\det\left| \begin{bmatrix} S_1 & -I \end{bmatrix} \Phi(T,0) \begin{bmatrix} \lambda I \\ S_0 \end{bmatrix} \right| = 0 \tag{4.7}$$

Note that the state transient matrix Φ depends on λ. The equation (4.7) is solved for the smallest λ_{nominal} which is the disturbance rejection capacity of the nominal system, i.e., for $\Delta A = 0, \Delta B = 0$. The smallest λ can be easily identified by ploting the determinant as given in (4.7) as a function of λ. The perturbed system, i.e., when $\Delta A \neq 0$ and $\Delta B \neq 0$, is expected to have a disturbance rejection capacity close to (in fact, less than) λ_{nominal}. Hence λ_{nominal} makes the most natural choice for initialization of the Newton's iteration for the perturbed system.

5 EXAMPLES

Example 5.1 Consider the system described by

$$\dot{x} = \tilde{a}x + \tilde{b}u + v, \qquad x(0) = x_0 \tag{5.1}$$

For the L_∞-type perturbations, we consider the performance index

$$J_\infty(u, v, x(0), \Delta a, \Delta b) = \frac{\frac{1}{2}x_0^2 + \frac{1}{2}\int_0^1 v^2\, dt}{\frac{1}{2}\int_0^1 (x^2 + u^2)\, dt} \tag{5.2}$$

We assume that the parameters $\tilde{a} \in [-0.9, -1.1]$ and $\tilde{b} \in [0.9, 1.1]$ so that the nominal parameters are $a = -1$, $b = 1$ with the corresponding perturbations $\Delta a^{\max} = 0.1$, $\Delta b^{\max} = 0.1$.

The two-point-boundary-value problem corresponding to (3.9) is given by

$$\begin{bmatrix} \dot{x} \\ \dot{\psi} \end{bmatrix} = \begin{bmatrix} \tilde{a} & -\tilde{b}^2 + \lambda \\ -1 & -\tilde{a} \end{bmatrix} \begin{bmatrix} x \\ \psi \end{bmatrix}, \quad \begin{bmatrix} x(0) \\ \psi(1) \end{bmatrix} = \begin{bmatrix} x_0 \\ 0 \end{bmatrix} \tag{5.3}$$

where $\tilde{a} = -1+0.1 \ \mathrm{sign}(x\psi)$. We first assume that $\Delta b = 0$. Then the optimal λ was found by numerically solving (5.3) for a nontrivial solution, and was found to be $\lambda = 1.8763$. The corresponding worst parameter was $\tilde{a} = -0.9$. We note here that in the absence of any parameter variations, the optimal solution is $\lambda = 2$. This result shows that due to variations in the parameter $\tilde{a}$, the performance of the system has degraded. Next we consider both $\tilde{a}$ and $\tilde{b}$ as perturbations. In this case, using (3.12) we have the worst perturbations are $\Delta a = 0.1$ and $\Delta b = -0.1$ so that $\tilde{a} = -0.9$ and $\tilde{b} = 0.9$ are the worst parameters. The optimal value of λ is $\lambda = 1.8282$ indicating a further deterioration of performance. We also notice that since this is a one dimensional system, the parameters $\tilde{a}$ or $\tilde{b}$ did not change with time since the sign of $x\psi$ remained the same over the entire time interval $[0, 1]$.

For the L_2-type perturbations, the performance index is taken as

$$J_2(u, v, \Delta a, \delta b) = \frac{\frac{1}{2}x_0^2 + \frac{1}{2}\int_0^1 \{v^2 + (\Delta a)^2 + (\Delta b)^2\}\, dt}{\frac{1}{2}\int_0^1 (x^2 + u^2)\, dt} \tag{5.4}$$

In this case the perturbation parameters Δa and Δb are bounded in the L_2 sense. The two-point-boundary-value problem obtained from Theorem 3.3 was solved to obtain the worst-case λ. Simulation results indicated that the worst-case variations of Δa and Δb are time varying functions over the time horizon of control. Table 5.1 summarizes the results.

Table 5.1 Worst-case Performance for System 5.1

Case	Perturbation	Perturbation Type	Optimal λ
5.1.1	$\Delta a = 0, \Delta b = 0$	Nominal	2.0
5.1.2	$\Delta a \neq 0, \Delta b = 0$	L_∞	1.8763
5.1.3	$\Delta a \neq 0, \Delta b \neq 0$	L_∞	1.8282
5.1.4	$\Delta a \neq 0, \Delta b = 0$	L_2	1.9910
5.1.5	$\Delta a \neq 0, \Delta b \neq 0$	L_2	1.9893

Example 5.2 Next we consider the two dimensional system

$$\begin{bmatrix} \dot{x}_1 \\ \dot{x}_2 \end{bmatrix} = \begin{bmatrix} 0 & 1 \\ \tilde{a} & 0 \end{bmatrix} \begin{bmatrix} x_1 \\ x_2 \end{bmatrix} + \begin{bmatrix} 0 \\ \tilde{b} \end{bmatrix} u + \begin{bmatrix} 1 \\ 0 \end{bmatrix} v \tag{5.5}$$

where $\tilde{a} = \bar{a} + \Delta a$ and $\tilde{b} = \bar{b} + \Delta b$ with the nominal parameters $\bar{a} = -1$, and $\bar{b} = 1$.

The performance index for the L_∞-type perturbation is taken as

$$J_\infty(u, v, x(0), \Delta a, \Delta b) = \frac{\frac{1}{2}\langle S_0 x(0), x(0)\rangle + \frac{1}{2}\int_0^5 \langle v, v\rangle dt}{\frac{1}{2}\langle x(5), S_1 x(5)\rangle + \frac{1}{2}\int_0^5 \{\langle x, x\rangle + \langle u, u\rangle\} dt} \tag{5.6}$$

The corresponding performance index for the L_2-case has the form

$$J_2(u, v, x(0), \Delta a, \Delta b) = \frac{\frac{1}{2}\langle S_0 x(0), x(0)\rangle + \frac{1}{2}\int_0^5 \{\langle v, v\rangle + R_a(\Delta a)^2 + R_b(\Delta b)^2\} dt}{\frac{1}{2}\langle x(5), S_1 x(5)\rangle + \frac{1}{2}\int_0^5 \{\langle x, x\rangle + \langle u, u\rangle\} dt} \tag{5.7}$$

For the simulations, we took $S_0 = 10 * I$, and $S_1 = 10^4 * I$, where I is an identity matrix. The weights in the cost (4.2) are $\alpha_1 = 10,000$ and $\alpha_2 = 100$.

Using the method discussed in Section 4, we compute the disturbance rejection capacity λ. For this example, we consider both L_∞-type as well as L_2 type perturbation of Δa and Δb. For the L_∞-perturbations, we consider $\Delta a^{\max} = 0.1$ and $\Delta b^{\max} = 0.1$, respectively, whereas in the L_2-case they are bounded in the L_2 sense. Table 5.2 shows the disturbance rejection capacity of the various cases:

Table 5.2 Worst-case Performance for System 5.5

Case	Perturbation	Perturbation Type	Optimal λ
5.2.1	$\Delta a = 0, \Delta b = 0$	Nominal	0.41139
5.2.2	$\Delta a \neq 0, \Delta b = 0$	L_∞	0.39343
5.2.3	$\Delta a \neq 0, \Delta b \neq 0$	L_∞	0.34248
5.2.4	$\Delta a \neq 0, \Delta b = 0$	L_2	0.41001
5.2.5	$\Delta a \neq 0, \Delta b \neq 0$	L_2	0.39828

The results indicate deterioration of controlled performance in the presence of parameter perturbations. For the L_∞-type perturbations, Case 5.2.2 and Case 5.2.3, the worst-case perturbation for Δb was found to be -0.1 for the entire time horizon; however, the worst case perturbation for Δa switched from one extreme $+0.1$ to the other -0.1. Figures 5.1 - 5.3 show the results for the Case 5.2.3, i.e., when both Δa and Δb are considered L_∞-type perturbations. Table II also indicates that the L_2-type perturbations reduce the disturbance rejection capacity compared to the nominal system. Simulation results show that in this case the worst-case parameters for both Δa and Δb are actually time varying functions over the entire time horizon. It can also be seen from Table 1.2 that the optimal λ for the L_2-type perturbation is larger than that for the L_∞-type perturbation. This is expected since a L_2-type variation is *milder* than a bang-bang L_∞-type variation. This means that the system will tolerate

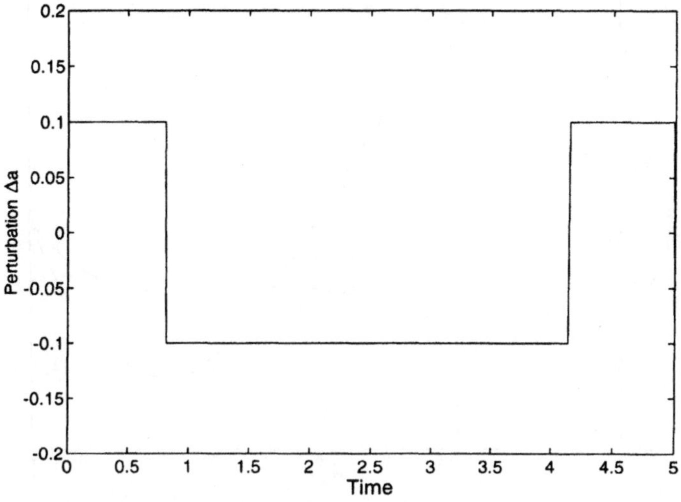

Figure 5.1 Worst-case Perturbation Δa

stronger L_2-type perturbations compared to L_∞-type perturbation without any degradation of performance.

6 CONCLUSIONS

We present the necessary conditions for the worst parameter perturbations and the worst exogenous disturbances for linear time varying systems. The necessary conditions are given by a nonlinear two-point-boundary-value problem. The worst variations of the system matrix occur due to the parameters taking values from the vertices of the polytope of perturbations in the case of L_∞-type perturbations. For L_2-type perturbations, worst perturbations are determined by certain equations. The necessary conditions represented by the two-point-boundary-value problem can be solved using numerical techniques.

References

Basar, T. and Bernhard, P. (1995), H_∞ *Optimal Control and Related Minimax Design Problems*, Birkhauser, Berlin.

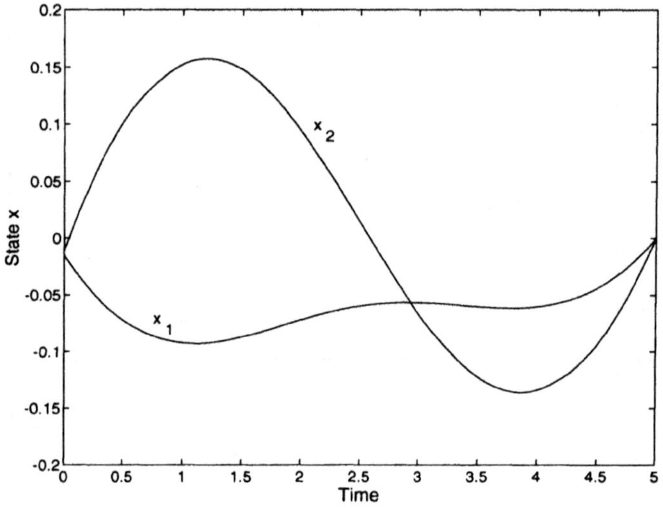

Figure 5.2 Worst-Case State Trajectory

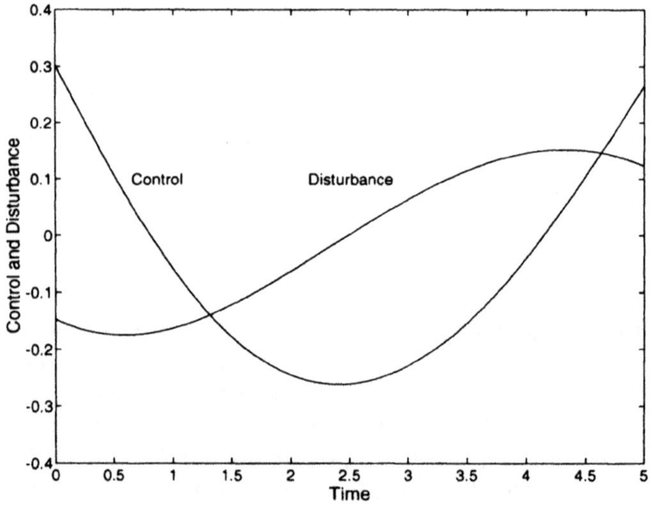

Figure 5.3 Worst-Case Control and External Disturbance

Biswas, S.K. (1990), Robust stabilization of linear systems in the presence of unbounded uncertain parameters with application to yaw stabilization of a satellite, *Proceeding of the American Control Conference*, San Diego, pp. 1326-1331.

Biswas, S.K. (1988), Output feedback stabilization of linear systems in the presence of parametric uncertainties, *Proceedings of the 20th Southeastern Symposium on System Theory*, University of North Carolina at Charlotte, pp. 293-297.

Biswas, S.K., and Subrahmanyam, M.B. (1993), Worst-case optimal control of linear systems in the presence of parameter perturbations, *Proceedings of the American Control Conference*, San Francisco, pp. 2444-2449.

Chen, S.B. (1993), The robust optimal control of uncertain systems: state space method, *IEEE Trans. on Auto. Control*, Vol. AC 38, pp. 951-956.

Doyle, J, Glover, K., Khargonekar, P. and Francis, B. (1989), State-space solutions to standard H_2 and H_∞ control problems, *IEEE Trans. on Automatic Control*, Vol. AC 34, pp. 831 - 847.

Francis, B. and Doyle, J. (1987), Linear control theory with an H_∞ optimality criterion, *SIAM Jour. on Control and Optimization*, Vol. 25, pp. 815-844.

Glover, K. and Doyle, J. (1988), State space formulae for all stabilizing controllers that satisfy an H_∞ norm bound and relations to risk sensitivity, *Systems and Control Letters*, Vol. 11, pp. 167-172.

Keqin, G. (1994), H_∞ control of systems under norm bounded uncertainties in all system matrices, *IEEE Trans. on Auto Control*, Vol. AC 39, pp. 1320-1322.

Khargonekar, P., Petersen, I., and Rotea, M. (1988), H_∞ optimal control with state feedback, *IEEE Trans. on Automatic Control*, Vol. AC 33, pp. 786-788.

Limebeer, D., Anderson, B.D.O., Khargonekar, P., and Green, M. (1992), A game theoretic approach to H_∞ control for time varying systems, *SIAM J. on Control and Optm.*, Vol. 30, pp. 262-284.

Petersen, I. (1987), Disturbance attenuation and H_∞ optimization: a design method based on the algebraic Riccati equation, *IEEE Trans. on Automatic Control*, Vol. AC 32, pp. 427-429.

Ravi, R., Nagpal, K.M., and Khargonekar, P. (1989), H_∞ control of linear time varying systems: a state space approach, *SIAM Journal on Control and Optimization*.

Schmitendorf, W.E. (1988), Designing stabilizing controllers for uncertain systems using the Riccati equation approach, *IEEE Trans. on Auto Control*, Vol. AC-33, pp. 376-379.

Stoorvogel, A. (1992), *The H_∞ Control Problem: A State Space Approach*, Prentice Hall, Englewood Cliffs, NJ.

Subrahmanyam, M.B. (1993), Synthesis of suboptimal H_∞ controllers over a finite horizon, *International Journal of Control*, Vol. 57, pp. 351-364.

Subrahmanyam, M.B. (1995), *Finite Horizon H_∞ and Related Control Problems*, Birkhauser, Boston.

Subrahmanyam, M.B. (1994), Finite horizon H_∞ control with parameter variations, *International Journal of Robust and Nonlinear Control*, Vol. 4, pp. 631-643.

Tadmor, G. (1990), Worst-case design in the time domain: the minimax principle and the standard H_∞ problem, *Mathematics of Control, Signals, and Systems*, Vol. 3, pp. 301-324.

Xie, L. and de Souza, C. (1992). Robust H_∞ control of linear systems with norm bounded time varying uncertainty, *IEEE Trans. Auto. Control*, Vol. 37, pp. 1188-1191.

Yaesh, I. and Shaked, U. (1993). Game theory approach to finite time horizon optimal estimation, *IEEE Trans. of Auto. Control*, Vol. AC 38, pp. 957-963.

Zhou, K. and Khargonekar, P., (1988), An algebraic Riccati equation approach to H_∞ optimization, *System and Control Letters*, Vol. 11, pp. 85-91.

4 TRUE PROPORTIONAL NAVIGATION SYSTEM WITH ACCELERATION SATURATION CONSTRAINT

Cheng-Chew Lim and Mingyan Li

Department of Electrical and Electronic Engineering
Adelaide University
Australia

Abstract: The effects of acceleration saturation on the performance and observability of the true proportional navigation based guidance system are investigated. We derive conditions for achieving effective interception when the closed loop system is subject to acceleration saturation or partial saturation during the pursuit, and present findings on how the constraint affects system observability.

Key words: Proportional navigation, saturation constraint, observability, limited measurements.

X. Yang et al. (eds.), Optimization Methods and Applications, 65-80.

1 INTRODUCTION

Short range guidance systems use proportional navigation guidance law extensively. The control is effective to use but difficult to analyse due to its highly nonlinear closed loop behaviour. Approaches have been proposed to solve the nonlinear target-pursuit motion equation, and to derive conditions for the system capturability, most analyses are however carried out under the assumption that the pursuer is able to provide adequate acceleration without saturation constraints. In reality, an achievable pursuer acceleration is limited by such constraints as actuator limits, angle-of-attack constraints at high altitudes, and the pursuer's structure at low altitudes.

While the acceleration saturation of guidance systems is an important issue, it has not been fully addressed, except some studies mainly via simulations as in Murtaugh and Criel (1966), Zarchan (1995), and Ohlmeyer (1996). Rigorous analytical study on the influence of acceleration saturation on system performance and observability has not, however, been reported in the literature.

We study analytically the impacts of acceleration saturation on system performance in terms of target interception and system observability. The guidance law is based on true proportional navigation (TPN). In developing a framework for analysing the observability and performance, a mathematical model is firstly established for a pursuer with acceleration constraints to intercept a manoeuvring target. Conditions for effective interception are then derived for non-saturation, partial saturation, and saturation of acceleration delivered by the actuator. We demonstrate that these conditions can be used in computing the most favorable launch conditions, and in predicting the occurrence of target interception. Analysis reveals that interception can be accomplished under the saturation mode if the derived interception conditions are satisfied.

We then investigate the impacts of saturation constraints on the observability under closed loop. This investigation is motivated by the wide spread use of bearings-only measurements in short range guidance systems to minimise cost as well as jamming interference from an intelligent target, (see for example Hepner and Geering (1990) and Lim and Li (1999)) . In theses systems, state variables unavailable but needed in the guidance law are often estimated by an extended Kalman filter. The conditions on how saturation constraints influence the system observability are derived.

Simulation studies are conducted and numerical results are given to confirm the generality of interception conditions in catering for a range of manoeuvring target models.

2 SYSTEM MODEL

The system equations governing the target-pursuit motion take the form

$$\ddot{R} - R\dot{\sigma}^2 = A_{TR} - A_{PR} \tag{2.1}$$

$$R\ddot{\sigma} + 2\dot{R}\dot{\sigma} = A_{T\sigma} - A_{P\sigma} \tag{2.2}$$

where R is the relative range with initial value R_0; $\dot{\sigma}$ is the line-of-sight (LOS) angular rate with initial value $\dot{\sigma_0}$; A_{TR} and $A_{T\sigma}$ are the target acceleration components along the LOS and normal to the LOS, respectively; A_{PR} and $A_{P\sigma}$ are the pursuer acceleration components along the LOS and normal to the LOS, respectively. Figure 2.1 shows the engagement geometry.

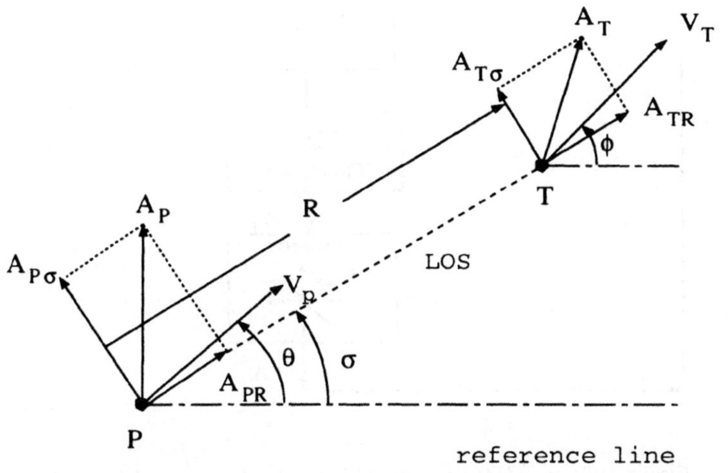

Figure 2.1 Engagement with pursuer heading angle θ and target heading angle ϕ

To account for the finite maximum acceleration capability of the pursuer, the maximum acceleration that a pursuer can provide is incorporated in the model. Figure 2.2 shows the block diagram of a proportional navigation based guidance system. The acceleration amplitude constraint is modelled by a non-linear saturation function at the output of the actuator which is taken as a unity gain component.

We use true proportional navigation (TPN) guidance. The control is to nullify the LOS angular rate aiming to drive the LOS angle between the pursuer and the target to a constant value, so that the pursuer and the target are on a "collision course". Under such condition, an interception can be accomplished without requiring further guidance. The expression for the pursuer acceleration A_P is given as

$$A_{PR} = 0; \qquad A_{P\sigma} = -N_1 \dot{R}\dot{\sigma} \tag{2.3}$$

where N_1 is a positive constant.

When A_P is constrained by certain maximum amplitude due to saturation effects, it takes the form

$$A_{PR} = 0; \qquad A_{P\sigma} = \begin{cases} -N_1 \dot{R}\dot{\sigma} & \text{when } \mid N_1 \dot{R}\dot{\sigma} \mid \leq A_{max} \\ -\text{sign}(\dot{R}\dot{\sigma}) \cdot A_{max} & \text{otherwise} \end{cases} \tag{2.4}$$

where A_{max} is the pursuer acceleration saturation constraint. In modelling the manoeuvring target, we assume that the target manoeuvre acceleration is

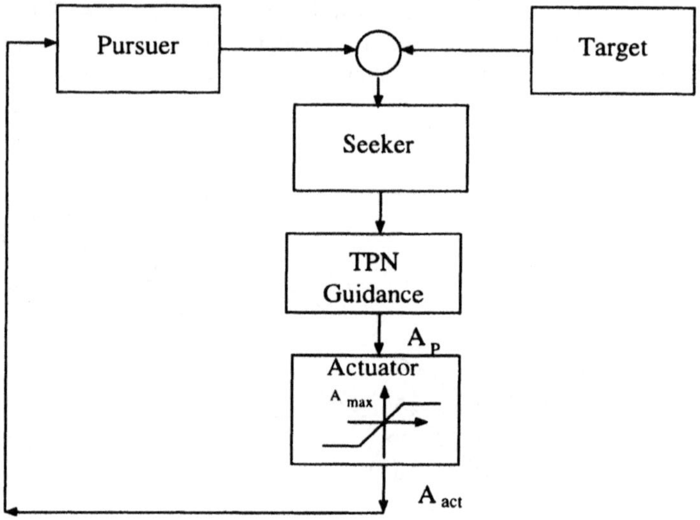

Figure 2.2 Guidance system with acceleration saturation constraint

proportional to the range rate $\dot{R}$ and normal to the line-of-sight. The target manoeuvre acceleration is expressed, as in Yuan and Chern (1992a), as

$$A_{TR} = 0; \qquad A_{T\sigma} = -c\dot{\sigma}_0\dot{R} \tag{2.5}$$

where c is a non-negative constant of the target manoeuvre acceleration. This constant is directly proportional to the manoeuvrability of the target. When $c = 0$, (2.5) reduces to a non-manoeuvring target model.

Substituting (2.4) and (2.5) into (2.1) and (2.2), we obtain the equations represent a pursuer tracking a manoeuvring target under TPN, with the pursuer acceleration subject to the saturation constraint A_{max}, as

$$\ddot{R} - R\dot{\sigma}^2 = 0 \tag{2.6}$$

$$R\ddot{\sigma} + 2\dot{R}\dot{\sigma} = \begin{cases} N_1\dot{R}\dot{\sigma} - c\dot{\sigma}_0\dot{R} & \text{when } \mid N_1\dot{R}\dot{\sigma} \mid \leq A_{max} \\ \text{sign}(\dot{R}\dot{\sigma}) \cdot A_{max} - c\dot{\sigma}_0\dot{R} & \text{otherwise} \end{cases} \tag{2.7}$$

3 SATURATION CONSTRAINT ANALYSIS

The effects of acceleration saturation constraints are studied by first deriving conditions for the occurrence of target interception based on (2.6) and (2.7).

In considering saturation nonlinearity as given in the block diagram shown Figure 2.2, the pursuer acceleration after constrained by saturation effects may operate in a normal, partial saturation, or saturation mode. The operations can be categorised into four cases as shown in Figure 3.1 – Figure 3.4.

The conditions required for a target interception under these four operating regions are now investigated.

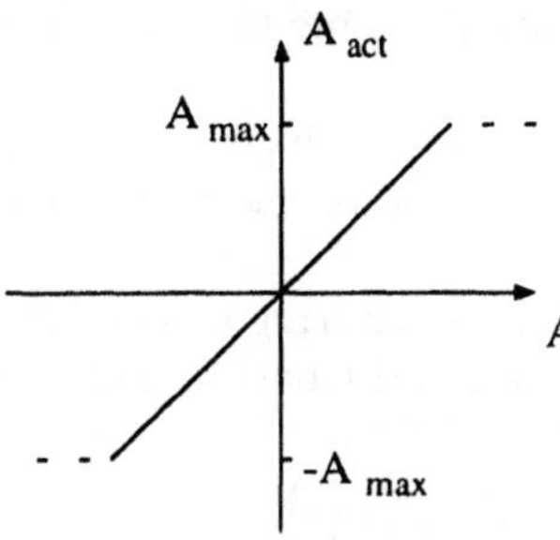

Figure 3.1 Case I: Non-saturation mode

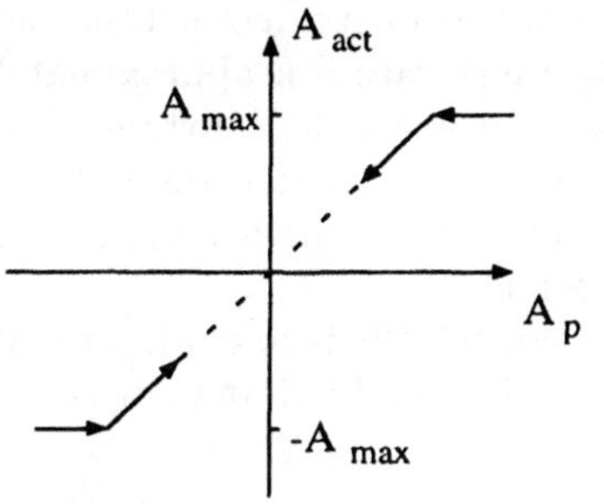

Figure 3.2 Case II: Saturation followed by non-saturation mode

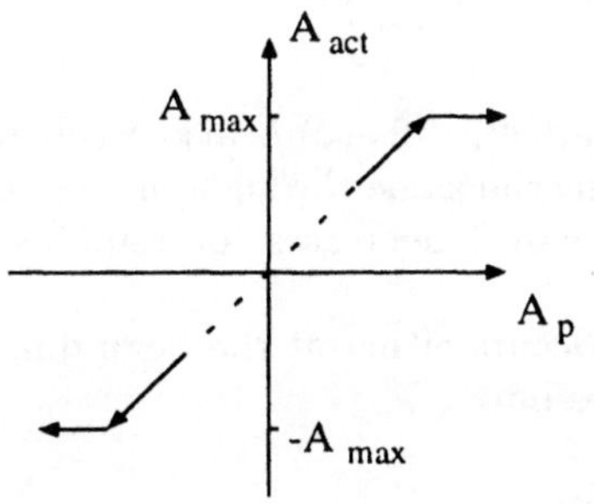

Figure 3.3 Case III: Non-saturation followed by saturation mode

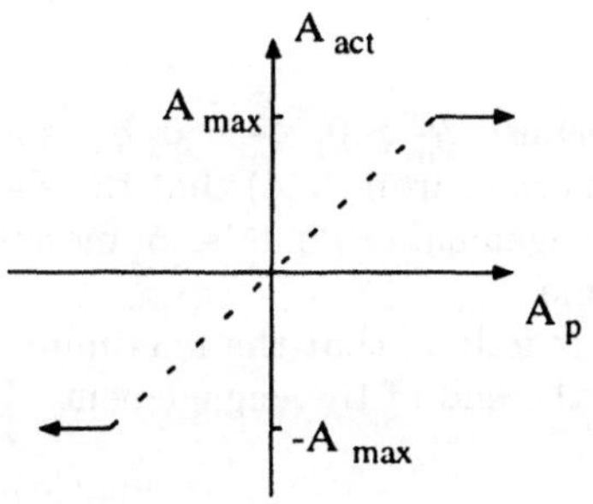

Figure 3.4 Case IV: Saturation mode throughout

3.1 Case I: Non-saturation mode

Consider Figure 3.1 when the commanded acceleration of the pursuer under TPN guidance never exceeds the saturation constraint A_{max} throughout the entire engagement.

Theorem 3.1 *A necessary and sufficient condition for a non-saturated pursuer acceleration under TPN throughout the entire engagement, when the pursuer acceleration is subject to the saturation constraint A_{max}, is*

$$\max(N_1, \frac{N_1 c}{N_1 - 2}) \leq b \tag{3.1}$$

where

$$b = \frac{A_{max}}{|\dot{R}_0 \dot{\sigma}_0|}. \tag{3.2}$$

Proof: Under TPN, there is no acceleration component along the LOS. The relative velocity vector thus remains closely aligned with the LOS, and hence the range rate $\dot{R}$ is approximately a negative constant during the entire pursuit, (see, for example, Murtaugh and Criel (1966)).

Since the magnitude of TPN guidance law is given as $|N_1 \dot{R} \dot{\sigma}|$, the variable $\dot{\sigma}$ determines whether the commanded acceleration exceeds the saturation constraints.

When TPN is to guide a pursuer to intercept a manoeuvring target described by (2.5), the LOS angular rate is obtained by Yuan and Chern (1992a) as

$$\dot{\sigma} = \dot{\sigma}_0 \left(\frac{N_1 - 2 - c}{N_1 - 2} \left(\frac{R}{R_0} \right)^{N_1 - 2} + \frac{c}{N_1 - 2} \right). \tag{3.3}$$

Differentiating (3.3) yields

$$\ddot{\sigma} = \dot{\sigma}_0 (N_1 - 2 - c) \left(\frac{R}{R_0} \right)^{N_1 - 3} \left(\frac{\dot{R}}{R_0} \right). \tag{3.4}$$

Because $\frac{R}{R_0} > 0$, $\frac{\dot{R}}{R_0} < 0$, $\dot{\sigma}_0$ is a constant, and $(N_1 - 2 - c)$ is also a constant, it is clear from (3.4) that the sign of $\ddot{\sigma}$ remains the same throughout the entire engagement. That is, $\dot{\sigma}$ monotonically increases, decreases, or remains the same.

It follows that the maximum value of $|\dot{\sigma}|$ occurs either at the beginning, or at the end of the engagement. This translates into

$$\begin{aligned} \max |\dot{\sigma}| &= |\dot{\sigma}_0| \quad \text{or,} \\ \max |\dot{\sigma}| &= |\dot{\sigma}_f| = \tfrac{c}{N_1 - 2} |\dot{\sigma}_0| \end{aligned}$$

where $\dot{\sigma}_f$ is the LOS angular rate at the final interception, and is obtained by substituting $\frac{R}{R_0} = 0$ into (3.3). The maximum pursuer acceleration commanded by TPN is either at the beginning of the pursuit, i.e., $N_1 |\dot{R}_0 \dot{\sigma}_0|$, or at the end, i.e., $\frac{N_1 c}{N_1 - 2} |\dot{R}_f \dot{\sigma}_0|$, where $\dot{R}_f$ is the final range rate.

With $\dot{R}$ approximately a negative constant, we have $\dot{R}_0 \approx \dot{R}_f$. Then, if

$$\max \left(N_1 |\dot{R}_0 \dot{\sigma}_0|, \quad \frac{N_1 c}{N_1 - 2} |\dot{R}_0 \dot{\sigma}_0| \right) \leq A_{max} = b |\dot{R}_0 \dot{\sigma}_0|,$$

the acceleration issued by TPN is non-greater than the saturation constraints. It follows that if inequality (3.1) is satisfied, then saturation will not occur. Conversely, if saturation does not occur, then condition (3.1) should be satisfied. □

Theorem 3.2 *The conditions for effective interception of a manoeuvring target, when a pursuer operates with non-saturated acceleration throughout the entire engagement under TPN, are*

$$N_1 > 3 \tag{3.5}$$

$$C^2 > \frac{1}{N_1 - 1} + \frac{c^2 + 2c}{N_1(N_1 - 1)} \tag{3.6}$$

with

$$C = \frac{\dot{R}_0}{R_0 \dot{\sigma}_0}. \tag{3.7}$$

Proof: From (3.3), $N_1 > 2$ is necessary to prevent $\dot{\sigma}$ from becoming infinity when R approaches zero at the final course of pursuit. Otherwise, an infinite acceleration is required in the final engagement.

From (3.4), it follows that $N_1 > 3$ can prevent the LOS angular acceleration $\ddot{\sigma}$ from approaching infinity as $R \to 0$. Therefore, $N_1 > 3$ should hold to avoid an infinite acceleration.

The final closing speed is obtained, as in Yuan and Chern (1992a), as

$$\dot{R}_f = \dot{R}_0 \cdot \sqrt{1 - \frac{1}{C^2}\left(\frac{1}{N_1 - 1} + \frac{c^2 + 2c}{N_1(N_1 - 1)}\right)} \tag{3.8}$$

To obtain a real $\dot{R}_f$, condition (3.6) must be satisfied. □

3.1 Remark Inequality (3.6) determines the ranges for $\dot{\sigma}_0$, R_0, and $\dot{R}_0$ to capture a target. The *capture area* for TPN based systems is thus defined by (3.6) and (3.7).

3.2 Remark Conditions (3.5) and (3.6) are necessary conditions for effective interception.

3.2 Case II: Saturation followed by TPN

This mode of operation is illustrated in Figure 3.2 where a maximum thrust is applied during the initial phase of an engagement due to insufficient available acceleration. The target-pursuit motion equations (2.6) and (2.7) under this scenario are

$$\ddot{R} - R\dot{\sigma}^2 = 0 \tag{3.9}$$

$$R\ddot{\sigma} + 2\dot{R}\dot{\sigma} = \text{sign}(\dot{R}\dot{\sigma}) \cdot A_{max} - c\dot{\sigma}_0\dot{R} \tag{3.10}$$

The interception conditions for this case are as follows.

Theorem 3.3 *A TPN guidance system with initial pursuer acceleration saturation must satisfy the following conditions in order to intercept a manoeuvring target*

$$N_1 > b > c + 2 \tag{3.11}$$

$$N_1 > 3 \tag{3.12}$$

$$C^2 > \frac{b^2k^2 + 2cbk^2 + c^2k^2N_1}{N_1^2(N_1 - 1)} \tag{3.13}$$

where b and C are given in (3.2) and (3.7), respectively, and

$$k = \sqrt{\frac{N_1(b-c-2)}{N_1(b-c)-2b}}. \tag{3.14}$$

Proof: As there is no acceleration component along the LOS, the range rate $\dot{R}$ is taken as approximately a negative constant during the course.

At the beginning, a maximum acceleration A_{max} is applied because of the saturation, we have

$$N_1|\dot{R}_0\dot{\sigma}_0| > A_{max} = b|\dot{R}_0\dot{\sigma}_0|,$$

that is,

$$N_1 > b. \tag{3.15}$$

In addition,

$$\begin{aligned} \text{sign}(\dot{R}\dot{\sigma})A_{max} &= \text{sign}(\dot{R}_0\dot{\sigma}_0)\cdot b\cdot|\dot{R}_0\dot{\sigma}_0| \\ &= -\text{sign}(\dot{\sigma}_0)|\dot{\sigma}_0|b(-\dot{R}_0) \\ &\approx b\dot{R}\dot{\sigma}_0 \end{aligned} \tag{3.16}$$

Solving (3.9) and (3.10) with (3.16) yields the LOS angular rate

$$\dot{\sigma} = \dot{\sigma}_0\left(\frac{(c+2-b)}{2}\left(\frac{R}{R_0}\right)^{-2} + \frac{(b-c)}{2}\right). \tag{3.17}$$

Differentiating (3.17) gives

$$\frac{\ddot{\sigma}}{\dot{\sigma}_0} = (b-c-2)\left(\frac{R}{R_0}\right)^{-3}\left(\frac{\dot{R}}{R_0}\right).$$

Since $\frac{R}{R_0} > 0$, and $\frac{\dot{R}}{R_0} < 0$, we have $\frac{\ddot{\sigma}}{\dot{\sigma}_0} < 0$, if $b > c+2$. This means that if $\dot{\sigma}_0 > 0$, $\dot{\sigma}$ will decrease because $\ddot{\sigma} < 0$.

On the other hand, if $\dot{\sigma}_0 < 0$, $\dot{\sigma}$ will increase due to $\ddot{\sigma} > 0$. Both cases give the same consequence that $|\dot{\sigma}|$ will decrease from the very beginning.

When $|\dot{\sigma}|$ reduces to a point at which $A_{max} = |N_1\dot{R}\dot{\sigma}|$, the mode changes from saturation to non-saturation. At the transition,

$$\dot{\sigma} = \frac{b}{N_1}\dot{\sigma}_0 \tag{3.18}$$

$$R = kR_0 \tag{3.19}$$

where k is given in (3.14), and $k \in (0,1) \subset \mathbf{R}$ under the interception condition (3.11), i.e. when $N_1 > b > c+2$.

Equation (3.18) is derived because of $|N_1\dot{R}\dot{\sigma}| = A_{max} \approx |\dot{\sigma}_0 b\dot{R}|$, and (3.19) is obtained by substituting (3.18) into the left-hand side of (3.17).

Note that after the transition from a saturated mode to an unsaturated mode, we must have new system initial values for the unsaturated TPN based system. The new initial relative range, given in (3.19), is denoted as R_0^*. The new initial LOS angular rate, given in (3.18), is denoted as $\dot{\sigma}_0^*$.

The pursuit motion equations for non-saturation acceleration mode are

$$\ddot{R} - R\dot{\sigma}^2 = 0 \tag{3.20}$$

$$R\ddot{\sigma} + 2\dot{R}\dot{\sigma} = N_1\dot{\sigma}\dot{R} - c\dot{\sigma}_0\dot{R} \tag{3.21}$$

Solving (3.20) and (3.21) yields

$$\dot{\sigma} = \dot{\sigma}_0^* \left(\frac{R}{R_0^*}\right)^{N_1-2} + \frac{N_1 c\dot{\sigma}_0^*}{b(N_1-2)}\left(1 - \left(\frac{R}{R_0^*}\right)^{N_1-2}\right) \tag{3.22}$$

To determine the trend of $|\dot{\sigma}|$, we check the sign of $\frac{\ddot{\sigma}}{\dot{\sigma}_0^*}$, as

$$\text{sign}\left(\frac{\ddot{\sigma}}{\dot{\sigma}_0^*}\right) = \text{sign}\left((bN_1 - 2b - N_1 c)\left(\frac{R}{R_0^*}\right)^{N_1-3}\left(\frac{\dot{R}}{R_0^*}\right)\right) = -1 \tag{3.23}$$

The last equal sign is because of the conditions given in (3.11). (3.23) shows that $|\dot{\sigma}|$ will decrease from $\dot{\sigma} = \dot{\sigma}_0^*$. It follows that the commanded acceleration reduces after TPN is applied, and thus will not exceed the saturation constraints again.

Following the approach in the proof of Theorem 3.2, when TPN guidance law is used to direct the pursuer, condition (3.12) should hold for an effective interception within the capture area which is given in (3.13). □

Remark 3.3 An important fact emerges from (3.17) that *interception with zero miss distance is not attainable if acceleration saturation persists throughout the entire pursuit.* This is because $\dot{\sigma}$ approaches infinity when R approaches zero near the end of engagement.

3.3 Case III: Non-saturated TPN first, saturation then occurs

Figure 3.3 illustrates this operational mode.

Theorem 3.4 *A TPN based pursuer cannot intercept a manoeuvrable target modelled by (2.5) when operates in an initially unsaturated mode followed by a saturated mode.*

Proof: TPN is attainable during the first portion of the pursuit if $N_1 < b$, i.e., $N_1|\dot{R}_0\dot{\sigma}_0| < b|\dot{R}_0\dot{\sigma}_0| = A_{max}$. The LOS angular rate under the TPN guidance law is given in (3.3). If $c+2 > N_1$, we then have from (3.4) that $\frac{\ddot{\sigma}}{\dot{\sigma}_0} > 0$, which shows $|\dot{\sigma}|$ will monotonically increase.

If $b < \frac{N_1 c}{N_1-2}$, then

$$A_{max} < N_1 \frac{c}{N_1-2}|\dot{R}\dot{\sigma}_0| = N_1|\dot{R}\dot{\sigma}_f| \tag{3.24}$$

Inequality (3.24) suggests that before an interception can occur, acceleration becomes saturated. This is because as $|\dot{\sigma}|$ increases to such a value that $N_1|\dot{R}\dot{\sigma}| = A_{max} < N_1|\dot{R}\dot{\sigma}_f|$, the acceleration commanded by TPN reaches the limit A_{max}.

When A_{max} is applied, the LOS angular rate becomes

$$\dot{\sigma} = \dot{\sigma}^* \left\{ \left(\frac{R}{R^*}\right)^{-2} + \frac{N_1(b-c)}{2b}\left(1 - \left(\frac{R}{R^*}\right)^{-2}\right) \right\} \tag{3.25}$$

where R^* denotes the range at the transition, and $\dot{\sigma}^*$ is the LOS angular rate at the transition. With condition $b < \frac{N_1 c}{N_1 - 2}$, we have $\frac{\ddot{\sigma}}{\dot{\sigma}^*} > 0$. This result indicates that $|\dot{\sigma}|$ increases monotonically after the transition, and will lead to an acceleration saturation for the rest of the pursuit. Consequently, the pursuer cannot intercept the target because $|\dot{\sigma}|$ will approach infinity at the closing moment of the engagement when $R \to 0$. □

3.4 Case IV: Saturation throughout

The condition for saturation to occur throughout the entire engagement is

$$b < \min(N_1, c+2). \tag{3.26}$$

Since $b < N_1$ implies that pursuer acceleration saturates at the beginning, while $b < c+2$ implies that commanded acceleration must increase, hence saturation is maintained throughout the engagement. As we have noted in Remark 3.1 of Theorem 3.3, zero miss distance is not achievable for this case.

4 OBSERVABILITY WITH SATURATION CONSTRAINT

Many guidance systems utilise limited and basic measurements in order to minimise cost as well as jamming interference from an intelligent target. State variables unavailable but needed in the guidance law are often estimated by an extended Kalman filter. The question on whether and how saturation constraints influence the system observability will naturally be raised. We analyse the observability of the guidance systems subject to acceleration saturation constraints in the four modes of operation discussed earlier.

To begin, we use the range observability index $I_A(t_0, t)$ to measure how well the range information can be obtained from a system. The index is defined as

$$I_A(t_0, t) = \int_{t_0}^{t} (t-s)\left[A_{act}(s) - A_T(s)\right] ds \tag{4.1}$$

where A_{act} is the saturation constrained pursuer acceleration. The target acceleration A_T may be obtained from the composition $A_T = \sqrt{A_{TR}^2 + A_{T\sigma}^2}$ applied to (2.5) to give $A_T = c\dot{\sigma}_0\dot{R}$. A_{act} is similarly derived from (2.4).

4.1 Case I: Non-saturated TNP throughout

We investigate the system observability under non-saturated TPN control with non-manoeuvring and manoeuvring targets.

Non-manoeuvring targets with TPN. When the pursuer guidance system engages a non-manoeuvring target, we have $A_T = 0$. The range observability index $I_A(t_0, t)$, with non-saturated acceleration, reduces to

$$\int_{t_0}^{t} (t-s)A_{act}(s)ds. \tag{4.2}$$

Under TPN, when t approaches final interception time t_f, the LOS angular rate $\dot{\sigma}$ approaches zero, and the range rate $\dot{R}$ approaches a constant.

The range observability index is

$$\begin{aligned} I_A(t_0, t_f) &= \int_{t_0}^{t_f} (t_f - s)A_{act}(s)ds \\ &= \int_{t_0}^{t_f} (t_f - s)(N_1\dot{\sigma}\dot{R})ds \\ &\approx \int_{t_0}^{t_f} (t_f - s)(N_1 \cdot 0 \cdot \dot{R}) = 0 \end{aligned} \tag{4.3}$$

As the LOS angular rate $\dot{\sigma}$ is nullified by TPN as intended in the control scheme, the integral vanishes for any value of the navigation constant N_1. I_A approaches zero means that the sufficient unobservability condition shown in Lim and Li (1999) is met. This implies that when the TPN guidance law is used in a non-manoeuvring target engagement, the system is most likely becoming unobservable near the end of engagement. A direct consequence is that when a state estimator such as the extended Kalman filter is used, the TPN based guidance system will suffer from the state estimate divergence problem. With no reliable estimates available to realise the guidance laws, the pursuer will most likely fail to accomplish the interception mission.

Manoeuvring targets with TPN. The manoeuvring target acceleration is modelled as $A_T = c\dot{\sigma}_0\dot{R}$, and under non-saturated acceleration, the range observability index is

$$\begin{aligned} I_A(t_0, t_f) &= \int_{t_0}^{t_f} (t_f - s)[A_{act}(s) - A_T(s)]ds \\ &= \int_{t_0}^{t_f} (t_f - s)(N_1\dot{\sigma}\dot{R} - c\dot{\sigma}_0\dot{R})ds \\ &\approx \frac{c\dot{\sigma}_0\dot{R}_f}{N_1 - 2}(t_f - t_0)^2. \end{aligned} \tag{4.4}$$

Comparing (4.3) with (4.4), the range observability index indicates that for TPN guidance systems using limited and basic measurements such as bearings-angle only, when engaging a manoeuvring target, the range R is more likely to be observable than those engaging a non-manoeuvring target. It can be interpreted as the manoeuvring target may follow a curved trajectory which provides more information about itself than a straight line trajectory of a non-manoeuvring target does.

4.2 Case II: A_{max} is followed by TPN

Under this mode, the range observability index defined in (4.1) becomes

$$I_A(t_0, t_f) = \int_{t_0}^{\tau} (t-s)(A_{max} - A_T)ds + \int_{\tau}^{t_f} (t-s)(N_1\dot{R}\dot{\sigma} - A_T)ds \tag{4.5}$$

where τ denotes the time at transition, and t_0 represents the time when observability is of particular interest. When analysing the system observability near the end of engagement, the range observability index (4.5) reduces to

$$I_A(t_0, t_f) = \int_{t_0}^{t_f} (t-s)(N_1 \dot{R}\dot{\sigma} - A_T)ds. \tag{4.6}$$

This expression is identical to the index for Case I. It follows that the results obtained on system observability towards the end of engagement under a non-saturated TPN command are applicable to Case II, even though saturation occurs initially.

4.3 *Cases III and IV*

For Case III and Case IV in which saturation with A_{max} is taking effect in the final stage of an engagement, the index is

$$I_A(t_0, t_f) = \int_{t_0}^{t_f} (t-s)(A_{max} - A_T)ds. \tag{4.7}$$

Since A_{max} is a constant, the index obtained in (4.7) is not equal to zero when the system is engaging a non-manoeuvering target, i.e., when $A_T = 0$. It follows that when acceleration is operating in the Case III and Case VI regions, an observable system in pursuit of a non-manoeuvering target is still feasible. This result differs from that of Case I. The overriding condition that an interception cannot occur in Case III and Case IV unfortunately diminishes the potential observability advantage offered here.

4.4 *Summary*

As a target interception required non-saturated acceleration towards the end of engagement, it is obvious that saturation constraints do not affect the system's observability significantly so long as intercepting a target is the only aim. Consequently, when investigating the closed loop state observability, the problem can be reformulated as how well a chosen guidance law is capable of enhancing observability. The use of TPN with bearings-only measurements does not fare well in terms of system observability in general, although manoeuvrable target engagements fare better than their non-manoeuvrable counterpart. A better proportional navigation based guidance law is to use the additive observability proportional navigation (AOPN) control law. While retaining the implementation simplicity of TPN, AOPN offers enhanced system observability and improved capturability. A detailed discussion on AOPN can be found in Li and Lim (1999).

5 SIMULATION RESULTS

The following data are used in the simulationst to check the generality of the derived interception conditions.

Velocity of the pursuer,	$V_P = 600m/s$;
Velocity of the target,	$V_T = 300m/s$;
Initial heading angle of the pursuer,	$\theta_0 = 0°$;
Initial heading angle of the target,	$\phi_0 = 30°$;
Initial LOS angle,	$\sigma_0 = 0°$;
Initial relative range,	$R_0 = 1000m$.

The initial conditions $\dot{R}_0$ and $\dot{\sigma}_0$ may be computed from:

$$\dot{\sigma}_0 = \frac{V_T \sin(\phi_0 - \sigma_0) - V_P \sin(\theta_0 - \sigma_0)}{R_0} \tag{5.1}$$

$$\dot{R}_0 = V_T \cos(\phi_0 - \sigma_0) - V_P \cos(\theta_0 - \sigma_0). \tag{5.2}$$

In realising the guidance law (2.4), $N_1 = 4$ which is a typical value of the TPN navigation constant, is used throughout the simulation.

To verify the derived interception conditions for different cases, the trajectories of a manoeuvring target and a pursuer with acceleration constraints under TPN are plotted in Figure 5.1. The target model is given in (2.5) with the manoeuvrer constant $c = 1$. Three different acceleration constraints selected are $21g$, $16g$ and $13g$.

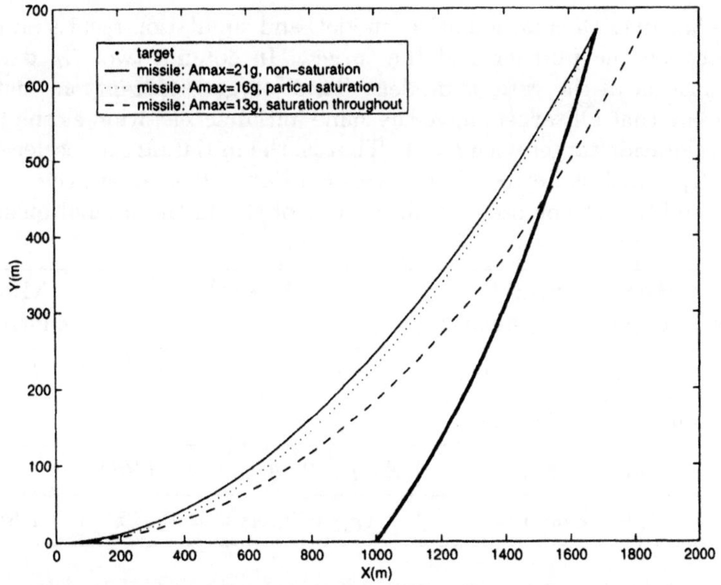

Figure 5.1 Trajectories of a manoeuvring target and a pursuer under TPN with acceleration saturation constraints

When $A_{max} = 21g$, we have, based on inequality (3.1), a scenario representing a non-saturated TPN mode for the entire engagement. As the given system initial values and the navigation constant N_1 satisfy the Case I interception conditions (3.5) and (3.6), an interception can occur. This is confirmed by the trajectory of the pursuer in solid line in Figure 5.1. When $A_{max} = 13g$, a Case IV scenario occurs, according to inequality (3.26). It is observed from Figure 5.1 that the pursuer with $A_{max} = 13g$ is unable to turn sufficiently fast to accomplish the interception task, which confirms the remark in section 3 that an interception is not achievable for Case VI. Interception conditions (3.11)–(3.13) allow us to predict that only when the saturation constraint A_{max} is larger than $15.6g$ for this scenario then can the pursuer successfully hit the target. This predication is confirmed by the ability of the pursuer with $A_{max} = 16g$ to capture the target, in spite of the saturation at the first stage of pursuit. We thus see that the derived interception conditions are practically useful for predicting the occurrence of target interception.

All the interception conditions are derived based on the target model in (2.5), which represents a near constant target acceleration motion with respect to the pursuer. To evaluate how general are the interception conditions when the target model varies, we use four target models in the simulation. They are: near constant acceleration target used by Yuan and Chern (1992a), (i.e., the model used in our analytical study), constant acceleration target of Ohlmeyer (1996), modified smart target of Rao (1997), and TPN based target used by Yang and Yang (1997).

Table 5.1 gives the mathematical models and simulation results on the proximity between the pursuer and the target. In column two, λ_T denotes the target constant of the various models. These target constants are determined by assuming that all targets have the same initial acceleration as the near constant acceleration target with $c = 1$. That is, the initial target acceleration A_{T0} equals $5.2g$, which is derived from $|A_{T0}| = c\dot{R}_0\dot{\sigma}_0$. This allows the same interception conditions to be used for all target models. In the simulation study, the

Target type with respect to pursuer	**Model**	**Miss distance**
near constant acceleration	$A_{TR} = 0;\ A_{T\sigma} = c\dot{\sigma}_0\dot{R}$	$1.2m$
constant acceleration	$A_{TR} = 0;\ A_{T\sigma} = \lambda_T$	$2.5m$
modified smart	$A_{TR} = 0;\ A_{T\sigma} = \lambda_T/(\dot{R}\dot{\sigma})$	$1.5m$
TPN based	$A_{TR} = 0;\ A_{T\sigma} = \lambda_T\dot{R}\dot{\sigma}$	$1.5m$

Table 5.1 Miss distance for different target models under TPN

pursuer is assumed to have $A_{max} = 17g$, which gives a Case II scenario. Based on the Case II interception conditions (3.11)–(3.13), the pursuer is capable of intercepting the near constant acceleration target with $c = 1$.

All the final miss distances shown in Table 5.1 for different target models are well within $5m$, which is considered sufficient in practice to cause damage to the target. The results indicate that the interception conditions (3.11)–(3.13) are sufficiently general to cater for variations in target models.

6 CONCLUSIONS

The effects of acceleration saturation constraints on system performance have been investigated. Analysis of TPN based guidance systems with acceleration saturation constraints demonstrates that saturation constraints do degrade the system's interception performance.

Four different operating modes have been considered when deriving the interception conditions. These conditions are useful to predict the occurrence of target interception.

On observability of saturation constrained TPN based systems, we have shown that for those cases in which target interception can occur, the saturation constraints impose no influence on system observability near the end of pursuit.

Finally, simulations have demonstrated that the derived interception conditions are sufficiently general to cater for significant variations in target models.

References

Murtaugh, S.A. and Criel, H.E. (1966), Fundamentals of proportional navigation. *IEEE Spectrum*, pp. 75-85.

Zarchan, P. (1995), Proportional navigation and weaving targets. *Journal of Guidance, Control and Dynamics*, Vol. 18(5), pp. 969-974.

Ohlmeyer, E.J. (1996), Root-mean-square miss distance of proportional navigation missile against sinusoidal target. *Journal of Guidance, Control and Dynamics*, Vol. 19(3), pp. 563-568.

Hepner, S.A.R. and Geering, H.P. (1990), Observability analysis for target maneuver estimation via bearing-only and bearing-rate-only measurements. *Journal of Guidance, Control and Dynamics*, Vol. 13(6), pp. 977-983.

Yuan, P.J. and Chern, J.S. (1992), Solutions of true proportional navigation for maneuvering and nonmaneuvering targets. *Journal of Guidance, Control and Dynamics*, Vol. 15(1), pp. 267-271.

Yuan, P.J. and Chern, J.S. (1992), Analytic study of biased proportional navigation. *Journal of Guidance, Control and Dynamics*, Vol. 15(1), pp. 185-190.

Yang, C.D. and Yeh, F.B. (1987), Closed-form solution for a class of guidance laws. *Journal of Guidance, Control and Dynamics*, Vol. 10(4), pp. 412-415.

Rao, S.K. (1997), Comments on "True proportional navigation with maneuvering target". *IEEE Transactions on Aerospace and Electronic Systems*, Vol. 33(AES-1), pp. 273-274.

Yang, C.D. and Yang, C.C. (1997), A unified approach to proportional navigation. *IEEE Transactions on Aerospace and Electronic Systems*, Vol. 33(AES-2), pp. 557-567.

Hull, D.G., Speyer, J.L. and Burris, D.B. (1990), Linear-quadratic guidance law for dual control of homing missiles. *Journal of Guidance, Control and Dynamics*, Vol. 13(1), pp. 137-144.

Li, M.Y. and Lim, C.C. (1999), Observability-enhanced proportional navigation guidance with bearings-only measurements. *Journal of Australian Mathematical Society (Series B)*, Vol. 40, pp. 497-512.

Lim, C.C. and Li, M.Y. (1999), Observability analysis of two closed loop guidance systems with bearings-only measurements. *Nonlinear Analysis*, to appear.

5 EVALUATION OF PENALTY FUNCTIONS FOR OPTIMAL CONTROL

Rein Luus, Wichaya Mekarapiruk

Department of Chemical Engineering
University of Toronto
Toronto, Ontario M5S 3E5, Canada

and Colin Storey

Department of Mathematical Sciences
De Montfort University
The Gateway, Leicester LE1 9BH, U.K.

Abstract: To handle final state equality constraints, two types of penalty functions are evaluated and compared in solving three optimal control problems. Both the absolute value penalty function and the quadratic penalty function with shifting terms yielded the optimal control policy in each case. The quadratic penalty function with shifting terms gave the optimum more accurately, and the approach to the optimum is less oscillatory than with the use of absolute value penalty function. In solving for the optimal control in the third example, the use of the absolute value penalty function required a series of runs to get the performance index to within 0.035% of the optimum. A further advantage in using the quadratic penalty function is obtaining sensitivity information with respect to the final state constraints from the shifting terms.

Key words: Penalty functions, iterative dynamic programming, optimal control, state constraints

X. Yang et al. (eds.), Optimization Methods and Applications, 81-103.

1 INTRODUCTION

Optimal control problems involving nonlinear systems where state constraints are specified are generally very difficult to solve. Here we consider the situation where at the final time the desired state must be reached within some given tolerance, while some performance index is minimized. A convenient and effective way to handle such final state constraints is to construct an augmented performance index where the final state constraints are included in a penalty function, and then to minimize the augmented performance index. The problem of choosing an appropriate penalty function such that minimization of the augmented performance index yields the minimum of the originally chosen performance index and also forces the constraints to be satisfied within some tolerance has been recently investigated by Dadebo and McAuley (1995). They concluded that absolute deviation from the desired state is a better choice than the square of deviation in the penalty function, and suggested that each penalty function factor be chosen to be approximately the same size as the performance index. Their work was extended by Mekarapiruk and Luus (1997b) to provide a more systematic way of choosing penalty function factors so that the optimum value of the performance index can be obtained, and so that all the state variables are transferred to the desired values within a specified tolerance.

For steady state optimization Luus and Wyrwicz (1996) showed that the absolute value penalty function works well in handling equality constraints, but recently Luus (1996a) showed that putting shifting terms into a quadratic penalty function enables equality constraints to be handled more readily. The same approach could be used in optimal control where flexible stage lengths are used and the sum of the stage lengths must give the specified final time Luus (1996b). In fact, the quadratic penalty function with shifting terms led to an easy way of dealing with equality constraints Luus and Storey (1997), Mekarapiruk and Luus (1997a). The goal here is to evaluate and compare these two types of penalty functions in the determination of the optimal control policy for three optimal control problems.

2 PROBLEM FORMULATION

Let us consider the system described by the differential equation

$$\frac{d\mathbf{x}}{dt} = \mathbf{f}(\mathbf{x}, \mathbf{u}) \tag{2.1}$$

where the initial state $\mathbf{x}(0)$ is given. The state vector $\mathbf{x}$ is an $(n \times 1)$ vector and $\mathbf{u}$ is an $(m \times 1)$ control vector bounded by

$$\alpha_j \leq u_j \leq \beta_j, \quad j = 1, 2, \cdots, m. \tag{2.2}$$

At the final time t_f, there are k state variables that must be at the desired values, i.e.,

$$h_i = x_i(t_f) - x_i^d = 0, \quad i = 1, 2, \cdots, k \leq n, \tag{2.3}$$

where x_i^d is the desired state of x_i at the final time t_f. The optimal control problem is to find the control policy $\mathbf{u}(t)$ in the time interval $0 \leq t < t_f$ that minimizes the performance index

$$I = \Phi(\mathbf{x}(t_f)) \tag{2.4}$$

where the final time t_f is specified. Numerically the equality constraints can be expressed explicitly as

$$|x_i(t_f) - x_i^d| \leq \epsilon_i, \qquad i = 1, 2, \cdots, k, \tag{2.5}$$

where ϵ_i is the tolerance for the ith constraint and is chosen sufficiently small. Although we could also add state inequality constraints Luus (1991), Mekarapiruk and Luus (1997a), here we wish to consider only the equality constraints imposed on the state variables at the final time. For optimization we use iterative dynamic programming (IDP) as developed by Luus (1989), but use it in a multi-pass fashion Luus (2000). We give the algorithm later. To handle the final state equality constraints we consider two types of penalty functions.

2.1 Absolute Value Penalty Function

Dadebo and McAuley (1995) reported successful results in using in IDP an absolute value penalty function. The augmented performance index is taken in the form

$$J_1 = I + \sum_{i=1}^{k} \theta_i |x_i(t_f) - x_i^d|. \tag{2.6}$$

There are k positive weighting factors θ_i which must be provided by the user, or must be determined during the course of the calculations. One effective method of getting good values for these weighting factors during the course of the calculations is to use the suggestion of Mekarapiruk and Luus (1997b) where the weighting factors are adjusted after every pass of IDP, being increased in magnitude if the tolerance is not met, and decreased in value if the final state constraint is satisfied within the given tolerance. Thus, after starting with some reasonable values for each θ_i, these penalty function factors are changed after every pass by

$$\theta_i^{q+1} = (1 + \delta)\theta_i^q \quad \text{if} \quad |x_i(t_f) - x_i^d| > \epsilon_i \tag{2.7}$$

$$\theta_i^{q+1} = (1 - \delta)\theta_i^q \quad \text{if} \quad |x_i(t_f) - x_i^d| \leq \epsilon_i \tag{2.8}$$

where δ is a small positive factor such as 0.05, q is the pass number, and ϵ_i is the tolerance for the constraint i. To reduce the size of the fluctuations in the control policy, δ is decreased by a small amount such as 1% after every pass. By using this approach, Mekarapiruk and Luus (1997b) were able to solve a very difficult crane optimization problem involving six final state constraints. That problem was first presented by Sakawa and Shindo (1982) and has been considered by several researchers Dadebo and McAuley (1995), Goh and Teo (1988), Teo et al (1991).

2.2 Quadratic Penalty Function with Shifting Terms

Let us now construct an augmented performance index to take into account the k equality constraints by adding to the performance index a quadratic penalty function containing shifting terms

$$J_2 = I + \theta \sum_{i=1}^{k} (h_i - s_i)^2 \tag{2.9}$$

where θ is a positive penalty function factor and the shifting terms s_i are chosen so that computationally (2.3) becomes satisfied and the performance index in (2.4) becomes minimized as the iterative procedure progresses. The use of such shifting terms in the quadratic penalty function was first explored by Luus (1996a) for steady state optimization. It is noted that

$$(h_i - s_i)^2 = h_i^2 - 2s_i h_i + s_i^2 \tag{2.10}$$

so that near the optimum when (2.3) is almost satisfied, the term h_i^2 is negligibly small with respect to the other two terms and the minimization of the augmented performance index in (2.9) becomes equivalent to minimizing

$$J_2 = I - 2\theta \sum_{i=1}^{k} s_i h_i, \tag{2.11}$$

since the term involving the square of the shifting term is simply a constant. Thus, the product $-2\theta s_i$ gives the Lagrange multiplier or sensitivity of the performance index to the change in the ith equality constraint in (2.3). This aspect of the shifting term was pointed out by Luus (1996a) in using penalty functions in dealing with difficult equality constraints in steady state optimization. The use of shifting terms was illustrated with IDP by Luus (1996b), and by Luus and Storey (1997). When prior information on the sensitivity is not available, we choose initially the shifting term s_i equal to zero, and then, after every pass of IDP, update it according to

$$s_i^{q+1} = s_i^q - h_i \tag{2.12}$$

where s_i^q is the value of s_i at pass number q.

It is important to have the shifting terms in the quadratic penalty function, because without such terms the optimum can not be determined very accurately Luus and Rosen (1991).

3 ITERATIVE DYNAMIC PROGRAMMING (IDP)

Let us divide the given time interval $(0, t_f)$ into P subintervals $(0, t_1), (t_1, t_2), \cdots, (t_k, t_{k+1}), \cdots, (t_{P-1}, t_P)$, each of length L, so that

$$L = \frac{t_f}{P}. \tag{3.1}$$

Inside each of these subintervals, the control is either kept constant (piecewise constant control), or we may seek a piecewise linear continuous control policy to minimize the performance index. For piecewise linear continuous control, we calculate the control policy inside the time interval (t_k, t_{k+1}) by the linear expression

$$\mathbf{u}(t) = \mathbf{u}(k) + \frac{\mathbf{u}(k+1) - \mathbf{u}(k)}{L}(t - t_k) \tag{3.2}$$

where $\mathbf{u}(k)$ is the value of $\mathbf{u}$ at the time t_k and $\mathbf{u}(k+1)$ is the value of $\mathbf{u}$ at time t_{k+1}. The optimal control problem is now to find $\mathbf{u}(k), k = 0, 1, 2, \cdots, P$, inside the range specified by (2.2), so that the performance index in (2.4) is minimized. At the last stage we must determine both $\mathbf{u}(P)$ and $\mathbf{u}(P-1)$. This means that for the last stage we must determine $2m$ values for the m control variables. For all the other stages there are only m values to be determined because of the continuity restriction.

For optimization here we use IDP Luus (2000) with a single grid point, i.e., $N = 1$, in a multi-pass fashion, where the region size is contracted by an amount γ after every iteration. After a specified number of iterations making up a pass, the region size is restored to a value η times the region size at the beginning of the pass to give the region size to be used in the subsequent pass. To illustrate this underlying logic in IDP, we present an algorithm to solve the optimal control problem as outlined in (2.1)–(2.4), where it is required to minimize the augmented performance index, either J_1 or J_2, with the use of piecewise linear control over P stages, each of the same length:

1. Divide the time interval $[0, t_f]$ into P time stages, each of length L.
2. Choose the number of test values for $\mathbf{u}$ denoted by R, an initial control policy and the initial region size $\mathbf{r}_{in}$; also choose the region contraction factor γ used after every iteration and the region restoration factor η used after every pass to restore the region sizes at the beginning of each pass.
3. Choose the number of iterations to be used in every pass and the number of passes to be used in the run.
4. Set the pass number index $q = 1$ and the iteration number index to $j = 1$.
5. Set the region size vector $\mathbf{r}^j = \eta^{q-1}\mathbf{r}_{in}$.
6. By using the best control policy (the initial control policy for the first iteration of the first pass), integrate (2.1) from $t = 0$ to t_f to generate the $\mathbf{x}$-trajectory and store the value of $\mathbf{x}$ at the beginning of each time stage, so that $\mathbf{x}(k-1)$ corresponds to the value of $\mathbf{x}$ at the beginning of stage k.
7. Starting at stage P, corresponding to time $t_f - L$, integrate (2.1) from $t_f - L$ to t_f, using as the initial state the stored value $\mathbf{x}(P-1)$ from step 6, once with each of the R allowable values for the control vector calculated from (3.2), using

$$\mathbf{u}(P-1) = \mathbf{u}^{*j}(P-1) + \mathbf{D}_1\mathbf{r}^j \tag{3.3}$$

$$\mathbf{u}(P) = \mathbf{u}^{*j}(P) + \mathbf{D}_2\mathbf{r}^j \tag{3.4}$$

where $\mathbf{u}^{*j}(P-1)$ and $\mathbf{u}^{*j}(P)$ are the best values obtained in the previous iteration and $\mathbf{D}_1$ and $\mathbf{D}_2$ are diagonal matrices of different random numbers between -1 and 1. Clip suitably any of the control variables at α_j or β_j to satisfy (2.2). Out of the R values for the augmented performance index, choose the control values that give the minimum value, and store these values as $\mathbf{u}^*(P-1)$ and $\mathbf{u}^*(P)$. We now have the best control for the last stage.

8. Step back to stage $P-1$, corresponding to time $t_f - 2L$. Choose R values for $\mathbf{u}(P-2)$ as in the previous step, and by taking as the initial state $\mathbf{x}(P-2)$ integrate (2.1) over one stage length by using (3.2) for the control. Continue integration over the last time stage by using the stored value of $\mathbf{u}^*(P-1)$ and $\mathbf{u}^*(P)$ from step 7. Compare the R values of the performance index and store the best control as $\mathbf{u}^*(P-2)$ that gives the minimum value for the augmented performance index.

9. Continue the procedure until stage 1, corresponding to the initial time $t=0$ and the given initial state, is reached. As before, integrate (2.1) and compare the R values of the augmented performance index and store the control $\mathbf{u}(0)$ as $\mathbf{u}^*(0)$ that gives the minimum augmented performance index. Store also the corresponding $\mathbf{x}$-trajectory.

10. Reduce the region size for allowable control

$$\mathbf{r}^{j+1} = \gamma \mathbf{r}^j \tag{3.5}$$

where j is the iteration number index. Use the best control policy from step 9 as the midpoint for the allowable values for the control that we denote by the superscript *.

11. Increment the iteration index j by 1 and go to step 6; continue the procedure for the specified number of iterations to finish a pass.

12. Increment the pass index q by 1 and go to step 5; continue the procedure for the specified number of passes, and interpret the results.

4 NUMERICAL RESULTS

The computations were performed in double precision on personal computers using WATCOM Fortran compiler version 9.5. For integration the standard fourth order Runge–Kutta method was used for Examples 1 and 3. For Example 2 we used the subroutine DVERK Hull et al (1976) with local error tolerance of 10^{-6} to get accurate results.

4.1 Example 1

Let us consider the first example used by Chen and Mills (1981), which was also used by Luus and Storey (1997). The dynamics of the system are described by two state variables, but we introduce the third variable to put the performance index into the standard form for use of IDP:

$$\frac{dx_1}{dt} = x_2 + u_1 \tag{4.1}$$

$$\frac{dx_2}{dt} = -x_1 + u_2 \tag{4.2}$$

$$\frac{dx_3}{dt.} = u_1^2 + u_2^2 \tag{4.3}$$

with $\mathbf{x}(0) = [\,-1\ 1\ 0\,]^T$, and final time $t_f = \pi/2$. There are no constraints on the control variables, but there are two equality constraints to be satisfied at the final time

$$h_1 = x_1(t_f) - 1 = 0 \tag{4.4}$$

and

$$h_2 = x_2(t_f) + 1 = 0. \tag{4.5}$$

It is required to find the unconstrained control policy that minimizes the performance index

$$I = x_3(t_f). \tag{4.6}$$

This is a good example to choose, since it can be solved analytically by using Pontryagin's maximum principle Lapidus and Luus (1967) to minimize the Hamiltonian

$$H = z_1(x_2 + u_1) + z_2(-x_1 + u_2) + u_1^2 + u_2^2 \tag{4.7}$$

to give the minimum performance index of $I = 8/\pi = 2.5464791$.

To obtain the result numerically, we choose the augmented performance indices to be minimized as

$$J_1 = I + \theta_1|h_1| + \theta_2|h_2| \tag{4.8}$$

for the absolute value penalty function, and

$$J_2 = I + \theta[(h_1 - s_1)^2 + (h_2 - s_2)^2] \tag{4.9}$$

for the quadratic penalty function with shifting terms. To determine how closely the equality constraints are satisfied, we consider the sum of absolute values of deviations

$$S = |h_1| + |h_2|. \tag{4.10}$$

For optimization with IDP, we chose $P = 20$ stages, each of equal length and sought to find piecewise linear control policy for each of the control variables. We chose the initial control policy $\mathbf{u} = \mathbf{0}$, the initial region size $\mathbf{r}_{in} = \mathbf{2}$, the reduction factor $\gamma = 0.85$ and the region restoration factor $\eta = 0.90$, $R = 5$, and allowed 100 passes, each consisting of 30 iterations. For integration we used the fourth order Runga–Kutta method with an integration time step of $0.01t_f$. The computation time for 100 passes was 47.4 s on a PentiumII/350.

Absolute Value Penalty Function. In choosing the augmented performance index J_1, we took initially $\theta_1 = \theta_2 = 1$, and $\delta = 0.05$ which was reduced by 1% after every pass. The tolerance ϵ_i was chosen as 10^{-5}. As is shown in Figure 4.1, the sum of absolute values is reduced in a rather oscillatory manner. At pass 98 a minimum value of $I = 2.5471725$ is obtained with $S = 1.564 \times 10^{-9}$; whereas at pass 100 we get $I = 2.5466307$ with the sum of absolute deviations $S = 2.114 \times 10^{-4}$.

Changing the tolerance ϵ_i to 10^{-4} or to 10^{-6} did not improve the results. Although at pass 98 the value of the performance index is within 0.03% of the optimum, we can do much better with the use of the quadratic penalty function with shifting terms.

Quadratic Penalty Function with Shifting Terms. With the use of quadratic penalty function with shifting terms, the convergence, as observed from the values of S in Figure 4.2, is much more regular than with the use of absolute value penalty function. Very similar convergence pattern results for a wide range of the penalty function factor θ.

Table 4.1 Effect of the penalty function factor θ on convergence for Example 1

Penalty function factor θ	Sum of abs. deviations S	Performance index I	Shifting terms s_1	s_2
0.5	5.55×10^{-8}	2.5464792	0.00001	-2.54648
1.0	2.61×10^{-8}	2.5464792	0.00001	-1.27324
10	3.15×10^{-8}	2.5464792	0.00000	-0.12732
100	2.02×10^{-9}	2.5464792	0.00000	-0.01273
500	9.66×10^{-8}	2.5464795	0.00000	-0.00255
1000	6.21×10^{-7}	2.5472788	0.00000	-0.00127

As is seen in Table 4.1, there is a wide range $0.5 \leq \theta \leq 100$ for which convergence to $I = 2.5464792$ occurs. At the optimum, the shifting parameter s_1 is very small, and the factor $-2\theta s_2$ is constant at 2.54648 which is equal to the optimum value of the performance index. It can be shown analytically Luus (2000) that at the final time, the adjoint variables are $z_1 = 0$, and $z_2 = 8/\pi$. The optimal results obtained with 20 stages of piecewise linear control are therefore excellent. The optimal control policy is given in Figure 4.3.

4.2 *Example 2: Two-stage CSTR system*

Let us now consider the two-stage CSTR system used for control by Edgar and Lapidus (1972), Luus (1974a); Luus (1974b), Nishida et al (1976), Rosen

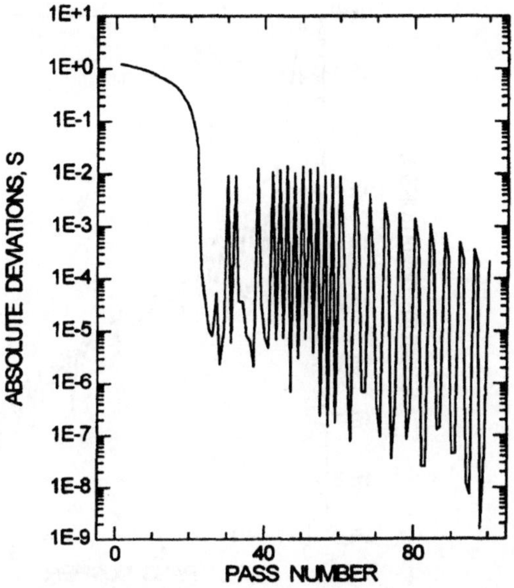

Figure 4.1 Convergence with the use of absolute deviation penalty function for Example 1

and Luus (1991) and Dadebo and McAuley (1995). Instead of choosing the performance index as the sum of squares of the state variables at the final time, we introduce another variable x_5 to reflect the sum of squares of deviations from the desired state and the control effort, and impose the final state constraints at the final time $t_f = 0.325$. The system is described by

$$\frac{dx_1}{dt} = -3x_1 + g_1(\mathbf{x}) \tag{4.11}$$

$$\frac{dx_2}{dt} = -11.1558x_2 + g_1(\mathbf{x}) - 8.1558(x_2 + 0.1592)u_1 \tag{4.12}$$

$$\frac{dx_3}{dt} = 1.5(0.5x_1 - x_3) + g_2(\mathbf{x}) \tag{4.13}$$

$$\frac{dx_4}{dt} = 0.75x_2 - 4.9385x_4 + g_2(\mathbf{x}) - 3.4385(x_4 + 0.122)u_2 \tag{4.14}$$

$$\frac{dx_5}{dt} = x_1^2 + x_2^2 + x_3^2 + x_4^2 + 0.1(u_1^2 + u_2^2) \tag{4.15}$$

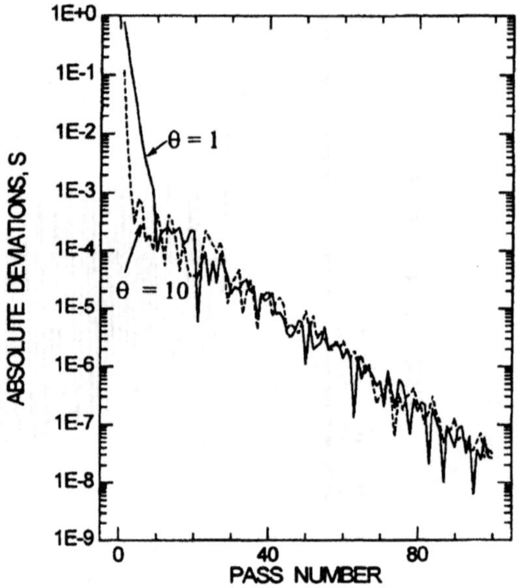

Figure 4.2 Convergence with the use of quadratic penalty function with shifting terms for Example 1

where the reaction term in the first tank is given by

$$\begin{aligned} g_1(\mathbf{x}) &= 1.5 \times 10^7(0.5251 - x_1)\exp(\frac{-10}{x_2 + 0.6932}) \\ &\quad -1.5 \times 10^{10}(0.4748 + x_1)\exp(\frac{-15}{x_2 + 0.6932}) - 1.4280 \end{aligned} \quad (4.16)$$

and the reaction term in the second tank is

$$\begin{aligned} g_2(\mathbf{x}) &= 1.5 \times 10^7(0.4236 - x_2)\exp(\frac{-10}{x_4 + 0.6560}) \\ &\quad -1.5 \times 10^{10}(0.5764 + x_3)\exp(\frac{-15}{x_4 + 0.6560}) - 0.5086. \end{aligned} \quad (4.17)$$

The initial condition is given by

$$\mathbf{x}(0) = [\, 0.1962 \quad -0.0372 \quad 0.0946 \quad 0 \quad 0\,]^T \quad (4.18)$$

and the constraints on the control are

$$-1 \le u_j \le 1, \quad j = 1, 2. \quad (4.19)$$

The state variables x_1 and x_3 are normalized concentration variables in tanks 1 and 2 respectively, and x_2 and x_4 are normalized temperature variables in tanks

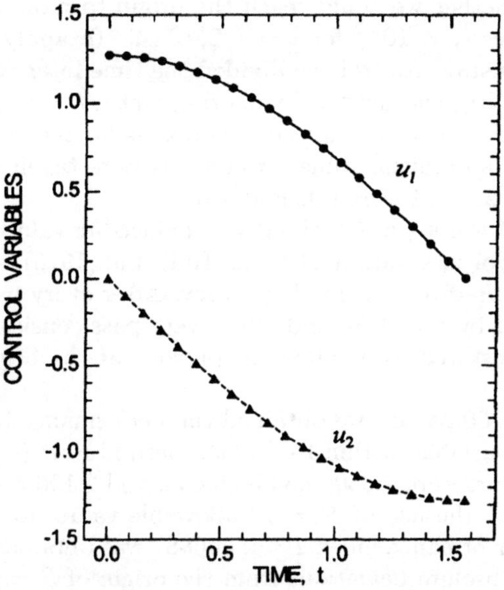

Figure 4.3 Optimal control policy with 20 piecewise linear sections for Example 1, giving $I = 2.5464792$

1 and 2 respectively. It is required to find the control policy that minimizes the performance index

$$I = x_5(t_f) \tag{4.20}$$

where the dimensionless final time $t_f = 0.325$. We impose the final state constraint

$$x_i(t_f) = 0, \quad i = 1, 2, \cdots, 4. \tag{4.21}$$

To measure the closeness of meeting the final state constraints, we use the sum of the absolute deviations

$$S = \sum_{i=1}^{4} |x_i(t_f)|. \tag{4.22}$$

Absolute Value Penalty Function. Incorporating the final state constraints yields the augmented performance index:

$$J_1 = I + \sum_{i=1}^{4} \theta_i |x_i(t_f)|, \tag{4.23}$$

where four penalty function factors θ_i are to be determined in the augmented performance index.

To see whether we could reach the origin to more than six figures, we set the tolerances $\epsilon_i = 10^{-7}$ for $i = 1, 2, \cdots, 4$. To apply IDP to obtain optimal piecewise constant control, we divided the time interval into 20 stages of equal length ($P = 20$), and used a single grid point ($N = 1$) with 15 allowable values for control ($R = 15$). The initial region sizes for the two control variables were chosen as 2, and initial value for controls were taken as 0. The initial values for θ_i, $i = 1, 2, \cdots, 4$ were taken as 1.0.

Initially, δ was set at 0.2 and it was reduced in value by a factor of 0.98 after every pass. For integration, DVERK Hull et al (1976) with local error tolerance of 10^{-6} was used to give good accuracy. After every iteration, the region sizes were reduced by $\gamma = 0.70$, and after every pass consisting of 30 iterations, the region was restored to $\eta = 0.95$ of its value at the beginning of the iterations in that pass.

By using 250 passes, we obtained the performance index $I = 0.046889$ with all the state variables within 10^{-7} of the desired state (origin). The final penalty function factors were 7.650, 0.3459, 5.440, and 2.196 respectively. The run was repeated with the use of $R = 10$ allowable values for control. A marginally improved performance index $I = 0.046887$ was obtained with a lower value for the sum of absolute deviations from the origin of $S = 9.8979 \times 10^{-10}$, so that all the variables are within 10^{-9} of the origin. For this run, the final penalty function factors were $\theta_1 = 7.081$, $\theta_2 = 0.1542$, $\theta_3 = 5.095$, and $\theta_4 = 2.039$.

The convergence profile is very oscillatory, as is seen in Figure 4.4. Although the norm of the final states is oscillatory, the reduction is quite systematic. The computation time for 250 passes, each consisting of 30 iterations, with $R = 10$ was 478 s on a PentiumII/400 personal computer.

Quadratic Penalty Function with Shifting Terms. Now we introduce the augmented performance index

$$J_2 = I + \theta \sum_{i=1}^{4} (x_i(t_f) - s_i)^2 \tag{4.24}$$

where the shifting terms s_i are chosen initially zero and then are updated after every pass according to (2.12). We use IDP to minimize this augmented performance index for several values of the penalty function factor θ.

Except for the number of randomly chosen points R, we used the same parameters as were used by Luus et al (1998) for this problem. We chose $P = 20$ stages of equal length, $N = 1$ grid point, 30 iterations per pass with region contraction factor $\gamma = 0.70$, and allowed 250 passes with region restoration factor $\eta = 0.95$. The initial region sizes were chosen as $r_1^{(0)} = r_2^{(0)} = 1$, and the initial control was chosen zero.

Figure 4.5 shows that the convergence is quite smooth and $R = 10$ gives marginally faster convergence than the use of $R = 15$ when the penalty factor $\theta = 2500$ was used. With the use of $R = 10$, the computation time for 250 passes was 512 s on a PentiumII/400 personal computer.

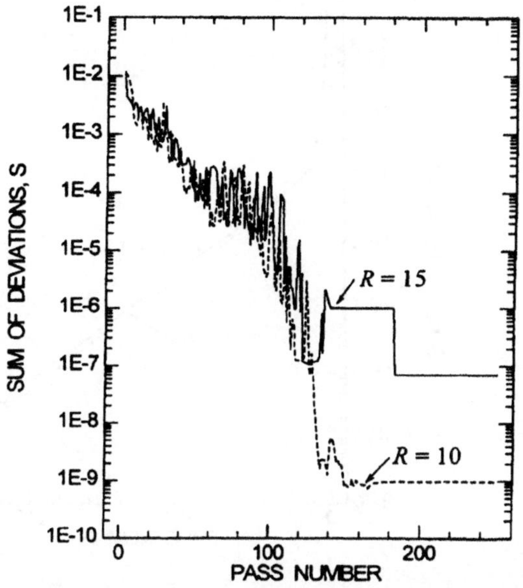

Figure 4.4 Convergence profile for Example 2 with absolute value penalty function, showing the sum of absolute values of deviations as a function of pass number

There is a relatively large range for the penalty function factor $2000 \leq \theta \leq 10000$ for which convergence to the minimum value of the performance index $I = 0.046872$ occurs so that all the final states are within 10^{-8} of the origin. With $\theta = 2500$ the shifting terms are $s_1 = 2.319 \times 10^{-3}$, $s_2 = -4.879 \times 10^{-5}$, $s_3 = -1.681 \times 10^{-3}$, and $s_4 = -6.899 \times 10^{-4}$. With $\theta = 5000$ the shifting terms are $s_1 = 1.159 \times 10^{-3}$, $s_2 = -2.440 \times 10^{-5}$, $s_3 = -8.403 \times 10^{-4}$, and $s_4 = -3.450 \times 10^{-4}$. When θ is too small (less than 1000), the final state constraint is not adequately satisfied, and if θ is too large (larger than 10000), then the performance index is greater than the minimum value. This is shown graphically in Figure 4.6.

The resulting optimal control policy is shown in Figure 4.7. This control policy compares favorably to the policy reported by Luus (1974a), but the control u_1 increases more gradually than the policies reported by Rosen and Luus (1991), and by Bojkov and Luus (1994). However, the optimal control policy is substantially different from the control policies obtained by Edgar and Lapidus (1972), Luus (1974b), and Nishida et al (1976). The trajectories of the first four state variables are given in Figure 4.8.

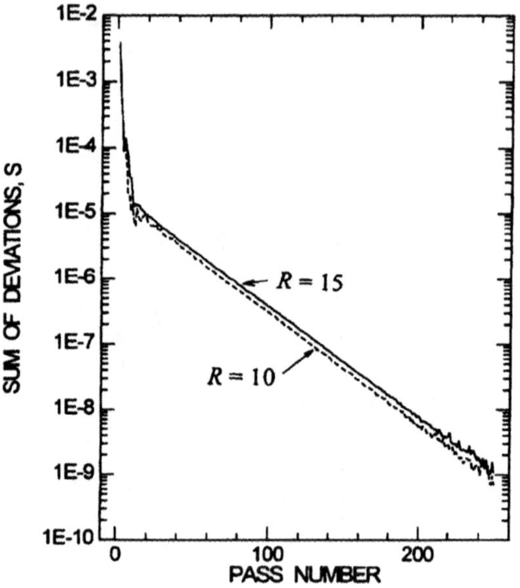

Figure 4.5 Convergence profile for Example 2 with quadratic penalty function, showing the sum of absolute values of deviations as a function of pass number with penalty function factor $\theta = 2500$

4.3 Example 3

Let us now take an optimal control problem involving three control variables, where four final states are specified, as used by Chen and Mills (1981) and by Luus and Storey (1997). The system is described by

$$\frac{dx_1}{dt} = x_2 + \cos 2t + tu_1 + \exp(t^2)u_2 + t^2u_3 \tag{4.25}$$

$$\frac{dx_2}{dt} = -x_1 + \sin 5t + t^2u_2 \tag{4.26}$$

$$\frac{dx_3}{dt} = x_3 + e^t + 2tu_1 + tu_3 \tag{4.27}$$

$$\frac{dx_4}{dt} = 2x_4 + \sin 10t + t^2u_1 + \cos 3tu_2 + 4tu_3 \tag{4.28}$$

$$\begin{aligned}\frac{dx_5}{dt} = & \; [t^4x_1 + \sin tx_2 + (4t_2 - 1)x_3 + 3e^{-t}x_4 - \cos 4t]^2 \\ & +(3+t)u_1^2 + u_1u_2 + 2u_1u_3 + (4+t^2)u_2^2 + (3+t^3)u_3^2\end{aligned} \tag{4.29}$$

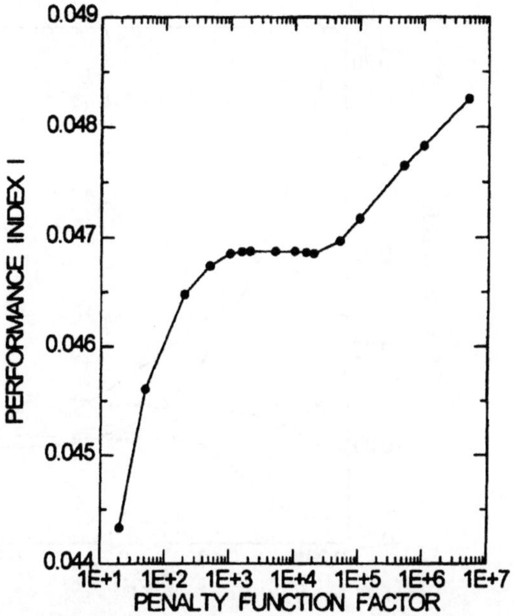

Figure 4.6 Effect of the penalty function factor on the converged value of the performance index I for Example 2

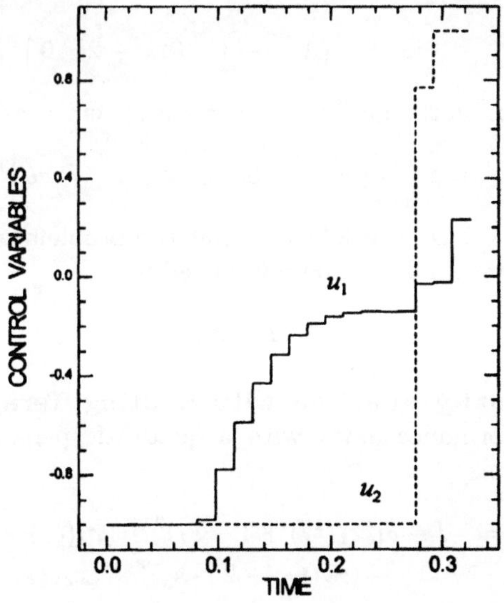

Figure 4.7 Optimal control policy for Example 2

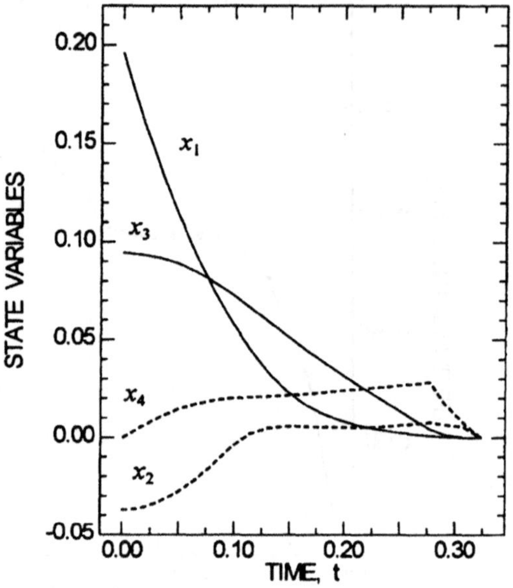

Figure 4.8 Trajectories of the state variables resulting from the optimal control policy for Example 2

with the initial state

$$\mathbf{x}(0) = [\,1 \quad -1 \quad 0 \quad -2 \quad 0\,]^T. \tag{4.30}$$

The desired state at the final time $t_f = 2$ is specified as

$$\mathbf{x}^d = [\,-1 \quad 0.5 \quad -2 \quad 1 \quad free\,]^T. \tag{4.31}$$

The fifth variable has been added to put the problem into the standard form; so, the performance index to be minimized is

$$I = x_5(t_f). \tag{4.32}$$

Quadratic Penalty Function with Shifting Terms. We introduce the augmented performance index with a quadratic penalty function containing four shifting terms

$$\begin{aligned} J_2 \quad = \quad & I + \theta[(x_1(t_f) + 1 - s_1)^2 + (x_2(t_f) - 0.5 - s_2)^2 \\ & +(x_3(t_f) + 2 - s_3)^2 + (x_4(t_f) - 1 - s_4)^2]. \end{aligned} \tag{4.33}$$

Again, we use the same penalty function factor θ for all the equality constraints, and choose zero for the initial values of the shifting terms s_i.

The optimal control problem is then to find the control policy $\mathbf{u}$ in the time interval $0 \leq t < 2$ that minimizes the augmented performance index J. To solve this optimal control problem, Luus and Storey (1997) used initially 50 stages of constant length and piecewise constant control, and then showed that the use of 40 stages of piecewise linear continuous control yielded a lower value of the performance index, namely $I = 114.81$, as opposed to $I = 121.20$. Here we use $P = 25$ stages of piecewise linear control with $N = 1$ and provide detailed computational results.

To minimize the augmented performance index we chose initial control policy $\mathbf{u}^{(0)} = \mathbf{0}$ with initial region sizes equal to 2. For optimization with IDP, we chose 30 iterations per pass with reduction factor $\gamma = 0.90$ and 200 passes with the region restoration factor $\eta = 0.95$. The shifting terms were initially put to zero and updated after every pass according to (2.12). To measure the rate of convergence we examine the sum of absolute values of deviations from the desired states

$$S = |x_1(t_f) + 1| + |x_2(t_f) - 0.5| + |x_3(t_f) + 2| + |x_4(t_f) - 1|. \tag{4.34}$$

It was shown by Luus and Storey (1997) that it makes very little difference whether $R = 3$ or $R = 21$ randomly chosen values for control are chosen. Therefore, here we chose $R = 5$. The penalty factor θ, however, influences the convergence somewhat, as is seen in Figure 4.9. Reliable convergence to the minimum performance index $I = 121.528$ with S less than 1.2×10^{-5} was obtained in the range $0.5 \leq \theta \leq 2$, as is shown in Table 4.2. It is noted that in this range the product θs_i remains constant for each shifting term.

Table 4.2 Effect of the penalty function factor θ on convergence for Example 3

Penalty function factor θ	Sum of deviations S	Performance index I	Shifting terms in J_2 s_1	s_2	s_3	s_4
0.25	2.730×10^{-4}	121.523	-29.4691	24.5376	-44.0952	0.00254
0.5	4.367×10^{-6}	121.528	-14.7357	12.2692	-22.0488	0.00124
1.0	9.032×10^{-6}	121.528	-7.3678	6.1346	-11.0244	0.00062
2.0	1.150×10^{-5}	121.528	-3.6839	3.0673	-5.5122	0.00031
4.0	5.451×10^{-5}	121.529	-1.8420	1.5337	-2.7561	0.00016
8.0	5.770×10^{-5}	121.542	-0.9210	0.7665	-1.3782	0.00009

The optimal control policy is given in Figure 4.10. The control policy for u_3 is smooth for the entire time interval. The control u_1 is smooth up to $t = 1.7$. It is interesting to note, however, the highly oscillatory nature of the control

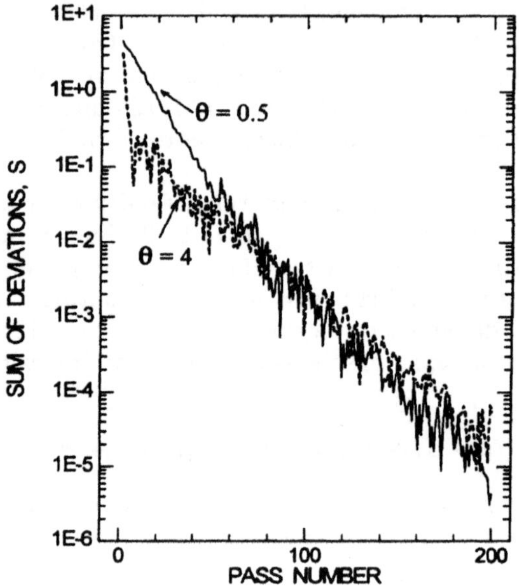

Figure 4.9 Sum of absolute values of deviations from the desired states as a function of the number of passes, using piecewise linear control with $P = 25$ time stages of equal length

variable u_2 after $t = 1.5$. The same type of behavior was observed by Luus and Storey (1997) with the use of 40 stages.

The corresponding state trajectories are shown in Figure 4.11 where a very rapid change in the state variable x_1 is noted near the final time.

Absolute Value Penalty Function. Now let us introduce the augmented performance index with absolute value penalty function containing four penalty function factors

$$\begin{aligned} J_1 &= I + \theta_1 | x_1(t_f) + 1 | + \theta_2 | x_2(t_f) - 0.5 | \\ &\quad + \theta_3 | x_3(t_f) + 2 | + \theta_4 | x_4(t_f) - 1 |. \end{aligned} \tag{4.35}$$

From the starting policy of zero for all control variables, taking the penalty function factors θ_i equal to 1.0 and using the same parameters as in Example 2, we could not get convergence to the vicinity of the optimum. However, taking $\theta_1 = \theta_2 = \theta_3 = 1$ and $\theta_4 = 0.01$, with ϵ_i relaxed to 10^{-3} yielded after 200 passes, each consisting of 30 iterations, $I = 121.695$ with $S = 1.146 \times 10^{-3}$, with $\theta_1 = 14.7518$, $\theta_2 = 12.7591$, $\theta_3 = 22.0294$, and $\theta_4 = 0.1303$. It is noted

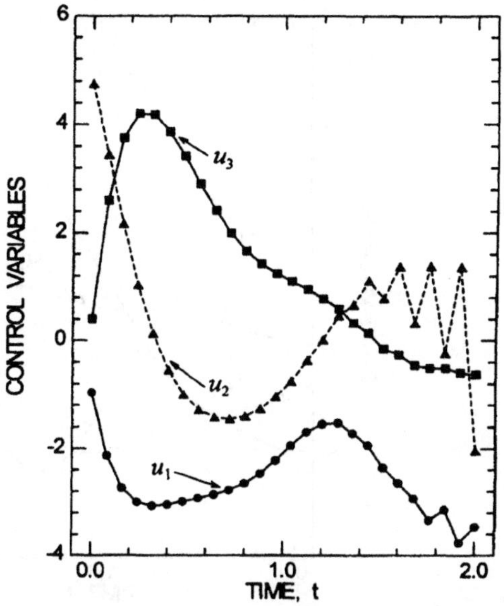

Figure 4.10 Optimal control policy using $P = 25$ piecewise linear sections for Example 3, yielding $I = 121.528$ and $S = 4.37 \times 10^{-6}$

that these penalty function factors, with the exception of θ_4 are very close in magnitude to the Lagrange multipliers ($-2\theta s_i$) in Table 4.2.

A series of 10 further runs were now performed, where the initial value of δ was decreased in value to 0.01, and the tolerance ϵ_i, and initial region size were progressively decreased. At the beginning of each run, the best control policy from the previous run was used as the starting control policy, and the penalty function factors obtained in the previous run were used as the initial values. For the second run we used $\epsilon_i = 10^{-4}$ and initial region size of 10^{-2}. For the third run we reduced ϵ_i to 10^{-5}. In the fourth run we reduced the initial region size to 10^{-3}. For the fifth and remaining runs we took $\epsilon_i = 10^{-6}$ and the initial region size for the control variables 10^{-4}. The results, as shown in Table 4.3, show a very slow approach toward the optimum $I = 121.528$, but the best result is still 0.035% from the optimum. It is interesting to note that the penalty function factors tend to approach the absolute values of the Lagrange multipliers ($-2\theta s_i$) given in Table 4.2. Therefore it appears that close to the optimum the best values for the penalty function factors are absolute values of the Lagrange multipliers. However, these may not be the best values to use far from the optimum.

As is seen in Figure 4.12, the resulting control profiles are not as smooth as the ones obtained with the use of the quadratic penalty function with shifting terms, shown in Figure 4.10.

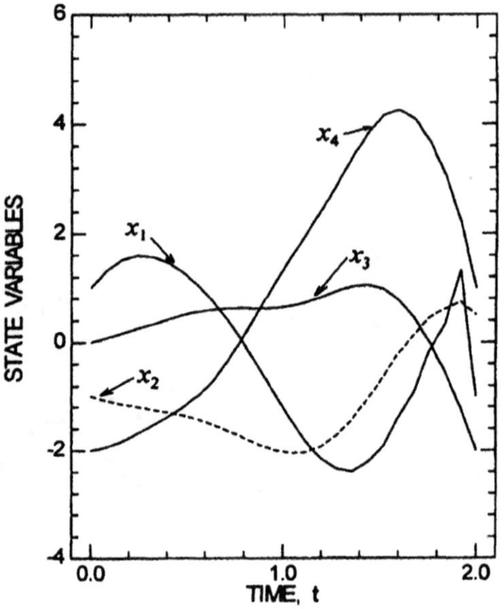

Figure 4.11 State trajectories for Example 3 resulting from the use of the optimal control policy shown in Figure 4.10

Table 4.3 Progress toward the optimum by using a series of runs, each consisting of 200 passes

Run number	Sum of deviations S	Performance index I	Penalty function factors in J_1 θ_1	θ_2	θ_3	θ_4
1	1.146×10^{-3}	121.695	14.7518	12.7591	22.0294	0.1303
2	3.047×10^{-4}	121.600	14.7354	12.2782	22.0122	0.0878
3	1.322×10^{-5}	121.599	14.7451	12.2525	22.0163	0.1241
4	3.817×10^{-5}	121.587	14.7368	12.3436	22.0142	0.0773
5	1.423×10^{-6}	121.588	14.7271	12.3764	22.0331	0.0511
6	3.538×10^{-6}	121.587	14.7226	12.3311	22.0583	0.0346
7	3.543×10^{-6}	121.583	14.7380	12.3589	22.0318	0.0235
8	3.439×10^{-6}	121.579	14.7300	12.3447	22.0698	0.0087
9	2.807×10^{-6}	121.576	14.7337	12.2649	22.0683	0.0061
10	1.833×10^{-6}	121.572	14.7496	12.3124	22.0567	0.0068
11	2.758×10^{-6}	121.571	14.7310	12.2477	22.0491	0.0066

5 CONCLUDING REMARKS

The examples show that penalty functions can be used successfully for handling final state equality constraints. Although for the second example both the

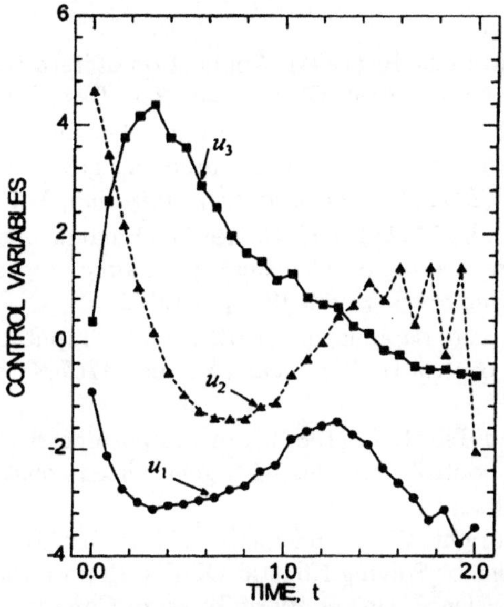

Figure 4.12 Control policy using $P = 25$ piecewise linear sections for Example 3, yielding $I = 121.571$ and $S = 2.758 \times 10^{-6}$, obtained by the use of absolute value penalty function after 11 runs

absolute value penalty function and the quadratic penalty function with shifting terms showed good convergence characteristics, for the other two examples the quadratic penalty function gave the optimum more accurately. Also, the use of quadratic penalty function tended to be more robust, so that starting conditions are not very important.

Although iterative dynamic programming was used here for all the examples, the penalty functions are not restricted to IDP. In fact, Luus and Hennessy (1999) used a quadratic penalty function with a shifting term to handle the volume constraint in two fed-batch reactor models in optimization with the Luus–Jaakola (Luus and Jaakola (1973)) optimization procedure. The most attractive feature of the quadratic penalty function with shifting terms is the requirement of specifying only a single penalty function factor, and there is a wide range over which the penalty function factor can be used to give optimal results.

Acknowledgments

We wish to acknowledge the Natural Sciences and Engineering Research Council of Canada for supporting some of this work.

References

Bojkov, B. and Luus, R. (1994), Application of iterative dynamic programming to time optimal control, *Chem. Eng. Res. Des.*, Vol. 72, pp. 72-80.

Chen, G. and Mills, W. H. (1981), Finite elements and terminal penalization for quadratic cost optimal control problems governed by ordinary differential equations, *SIAM J. Control and Optimization*, Vol. 19, pp. 744-764.

Dadebo, S. A. and McAuley, K. B. (1995), Dynamic optimization of constrained chemical engineering problems using dynamic programming, *Computers and Chemical Engineering*, Vol. 19, pp. 513-525.

Edgar, T. F. and Lapidus, L. (1972), The computation of optimal singular bang-bang control II. Nonlinear systems, *AIChE Journal*, Vol. 18, pp. 780-785.

Goh, C. J. and Teo, K.L. (1988), Control parametrization: a unified approach to optimal control problems with generalized constraints, *Automatica*, Vol. 24, pp. 3-18.

Hull, T. E., Enright, W. D. and Jackson, K. R. (1976), User Guide to DVERK– a Subroutine for Solving Nonstiff ODE's. Report 100 (1976); Department of Computer Science, University of Toronto, Canada.

Lapidus, L. and Luus, R. (1967), *Optimal Control of Engineering Processes*, Blaisdell, Waltham, Mass.

Luus, R. (1974a), Optimal control by direct search on feedback gain matrix, *Chemical Engineering Science*, Vol. 29, pp. 1013-1017.

Luus, R. (1974b), A practical approach to time-optimal control of nonlinear systems, *Industrial and Engineering Chemistry Process Design and Development*, Vol. 13, pp. 405-408.

Luus, R. (1989), Optimal control by dynamic programming using accessible grid points and region reduction, *Hungarian Journal of Industrial Chemistry*, Vol. 17, pp. 523-543.

Luus, R. (1991), Application of iterative dynamic programming to state constrained optimal control problems, *Hungarian Journal of Industrial Chemistry*, Vol. 19, pp. 245-254.

Luus, R. (1996a), Handling difficult equality constraints in direct search optimization, *Hungarian Journal of Industrial Chemistry*, Vol. 24, pp. 285-290.

Luus, R. (1996b), Use of iterative dynamic programming with variable stage lengths and fixed final time, *Hungarian Journal of Industrial Chemistry*, Vol. 24, pp. 279-284.

Luus, R. (2000), *Iterative Dynamic Programming*, Chapman & Hall, CRC, London, UK.

Luus, R. and Hennessy, D. (1999), Optimization of fed-batch reactors by the Luus–Jaakola optimization procedure, *Ind. Eng. Chem. Res.*, Vol. 38, pp. 1948-1955.

Luus, R. and Jaakola, T. H. I. (1973), Optimization by direct search and systematic reduction of the size of search region, *AIChE Journal*, Vol. 19, pp. 760-766.

Luus, R., Mekarapiruk, W. and Storey, C. (1998), Evaluation of penalty functions for optimal control, in: *Proc. of International Conference on Optimization Techniques and Applications (ICOTA '98)*, Perth, Western Australia, July 1-3, 1998, Curtin Printing Services, pp. 724-731.

Luus, R. and Rosen, O. (1991), Application of dynamic programming to final state constrained optimal control problems, *Ind. Eng. Chem. Res.*, Vol. 30, pp. 1525-1530.

Luus, R. and Storey, C. (1997), Optimal control of final state constrained systems, in: *Proc. IASTED International Conference on Modelling, Simulation and Control*, Singapore, Aug.11-14, 1997, pp. 245-249.

Luus, R. and Wyrwicz, R. (1996), Use of penalty functions in direct search optimization, *Hungarian Journal of Industrial Chemistry*, Vol. 24, pp. 273-278.

Mekarapiruk, W. and Luus, R. (1997a), Optimal control of inequality state constrained systems, *Ind. Eng. Chem. Res.*, Vol. 36, pp. 1686-1694.

Mekarapiruk, W. and Luus, R. (1997b), Optimal control of final state constrained systems, *Canadian Journal of Chemical Engineering*, Vol. 75, pp. 806-811.

Nishida, N., Liu, Y. A., Lapidus, L. and Hiratsuka, S. (1976), An effective computational algorithm for suboptimal singular and/or bang-bang control, *AIChE Journal*, Vol. 22, pp. 505-523.

Rosen, O. and Luus, R. (1991), Evaluation of gradients for piecewise constant optimal control, *Computers and Chemical Engineering*, Vol. 15, pp. 273-281.

Sakawa, Y. and Shindo, Y. (1982), Optimal control of container cranes, *Automatica*, Vol. 18, pp. 257-266.

Teo, K. L. Goh, C. J. and Wong, K. H. (1991), *A Unified Computational Approach to Optimal Control Problems*, Longman Scientific & Technical, Wiley, New York.

6 ON THE OPTIMAL CONTROL SYSTEMS WITH MULTIPLE CONTROLLERS

Katsumi Moriwaki

Department of Mechanical Systems Engineering
The University of Shiga Prefecture
Hikone, Shiga 522-8533, Japan

Abstract: The synthesis problem of a control system with multiple controllers is one of interesting design problems because the larger the dimension of the plant becomes the more difficult it becomes to control the plant with a single controller. In this paper we consider the problem of finding a set of internally stabilizing controllers which is globally equivalent to a single internally stabilizing controller such that the state of the plant $x(t)$ is driven to the origin as $t \to \infty$ and a certain performance index on the state is minimized. As the simplest case of control problems with multiple controllers, the control problem for the system with 2 controllers is considered. Under the assumptions of reachability and observability, it is shown that the decoupling problem of a Linear Quadratic Regulator (LQR) can be formulated and the solvability condition is derived for a set of admissible controllers. For the more general system which includes the system does not satisfy the solvability condition of the decoupling problem of a Linear Quadratic Regulator, we will introduce a set of internally stabilizing controller which is almost equivalent to a single internally stabilizing controller via internally balanced state space representation.

Key words: Discrete-time linear quadratic optimal regulator, multiple controllers, reduction of Riccati equation.

X. Yang et al. (eds.), Optimization Methods and Applications, 105-124.

1 INTRODUCTION

The linear quadratic optimal regulator (LQR) is one of the powerful design methods for linear control systems (Anderson et al (1971)), (Anderson et al (1990)). The goal of the LQR problem is to find an optimal control input $u^{\mathrm{opt}}(k)$, if it exists, which minimizes the quadratic performance index J. As is well known, the optimal control input $u^{\mathrm{opt}}(k)$, which minimizes J, uniquely exists under the assumption that the weighting matrices of the quadratic terms on state and control input are both positive definite, and it can be expressed as a linear function of the state $x(k)$, that is, $u^{\mathrm{opt}}(k) = K^{\mathrm{opt}}(k)x(k)$, where $K^{\mathrm{opt}}(k)$ includes the solution to the matrix Riccati equation. We know that it is a tough task to execute the process of obtaining the solution to the matrix Riccati equation.

As the dimension of a dynamic system (plant) to be controlled becomes large, it is often too costly (sometimes practically impossible) to have only one decision maker (or controller) in the system who possesses all available information on the system and make all the decisions for the system, for example, the optimal control input. Hence, much literature has been devoted to control problems of the system with multiple control agents or decentralized dynamic systems (Ho et al (1972)), (Yoshikawa (1978)), (Sandell et al (1978)). The control system with multiple control agents has the advantages that each control agent can treat the corresponding subsystem in which a control input can be obtained by the appropriate computation algorithm with lower dimension than one for the overall system and each control agent can make an appropriate decision on control. In this paper we, therefore, consider the relational problem of model reduction (Davison (1966)), (Moore (1981)), (Enn (1984)), (Glover (1984)) or controller reduction (Anderson et al (1989)), (Mustafa (1991)). Although the main idea underlying the model reduction work is to eliminate any weak subsystem which contributes little to the system response, the main idea underlying the controller decoupling work (or multi-agents control problem), considered here, is to construct the controller with multiple sub-controllers which is globally or nearly equivalent to a single controller.

In this paper, we consider the optimization problem with quadratic criterion for linear multivariable control systems with 2 control agents. The structure of the optimal control system with 2 controllers is derived, which is the globally equivalent to the optimal control system with a single controller. Figure 1.1 illustrates the optimal control system with 2 control agents.

In Section 2 optimal regulator theory and related preliminaries are reviewed. Section 3 contains the problem formulation and the results of LQR problem with 2 controllers which gives the strictly equivalent optimal controller to a single optimal controller. In Section 4 the problem of control decoupling by principal component analysis is considered and the near optimal control input for the overall system is derived. Numerical examples are given in Section 5.

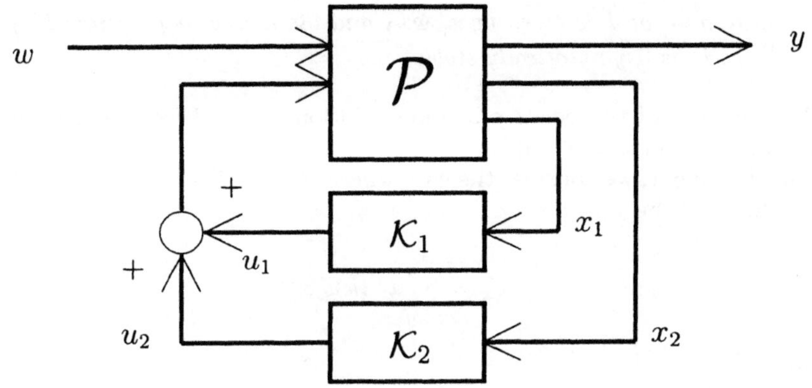

Figure 1.1 The system with a pair of controllers

2 OPTIMAL REGULATOR AND PRELIMINARIES

Consider the linear system:

$$x(k+1) = Ax(k) + Bu(k)\ ,\ \ x(0) = x_0,\ k = 0, 1, 2, \ldots, \tag{2.1}$$

with the performance index to be minimized:

$$J(u) = \sum_{k=0}^{\infty}[x^T(k)Qx(k) + u^T(k)Ru(k)]\ , \tag{2.2}$$

where $x \in \mathbf{R}^n$ is the state of the system to be controlled, $u \in \mathbf{R}^r$ the control input, and $Q \in \mathbf{R}^{n \times n}$ and $R \in \mathbf{R}^{r \times r}$ are positive definite symmetric matrices.

It is assumed that the pair (A, B) is reachable and, for any matrix D such that $D^T D = Q$, the pair (D^T, A) is observable. The latter assumption is sufficient to guarantee the asymptotic stability of the closed-loop system. For this optimal control problem, it is well-known that the solution is given by the following proposition (Anderson et al (1971)), (Anderson et al (1990)).

Proposition 2.1 *The optimal control input $u^{opt}(t)$ with respect to (2.1)–(2.2) is given by the state feedback control:*

$$u^{opt}(k) = K^{opt}x(k)\ , \tag{2.3}$$

where K^{opt} is the optimal time-invariant feedback control gain given by

$$K^{opt} = -(B^T\Phi B + R)^+ B^T\Phi A, \tag{2.4}$$

where Φ is the unique real symmetric solution to the matrix Riccati equation:

$$\Phi = A^T\Phi A - A^T\Phi B(B^T\Phi B + R)^+ B^T\Phi A + Q\ . \tag{2.5}$$

The minimum of J is given by $x_0^T \Phi x_0$ and the closed-loop system $\dot{x}(t) = (A + BK^{\mathrm{opt}})x(t)$ is asymptotically stable.

The existence of the positive definite solution to the Riccati equation (2.5) is guaranteed (Potter (1966)).

In this paper, we consider the case where R in (2.2) is null matrix, thus (2.2) and (2.5) will be

$$J_x = \sum_{k=0}^{\infty} x^T(k)Qx(k) , \tag{2.6}$$

and

$$\Phi = A^T \Phi A - A^T \Phi B(B^T \Phi B)^+ B^T \Phi A + Q , \tag{2.7}$$

respectively. For this optimal control problem formulation, we will consider the problem whether the single optimal control input u^{opt} can be decoupled into two noninteractive optimal control inputs u_1^{opt} and u_2^{opt} where these decoupled optimal control inputs can work as equivalently as the single optimal control input in the control process.

Preparatory to the investigation of such a decoupling problem of optimal control input, some lemmas are needed.

Lemma 2.1 *An arbitrary real symmetric matrix L ($\in \mathbf{R}^{n \times n}$) can be represented by*

$$L = \sum_{i=1}^{n} \lambda_i t_i t_i^T , \tag{2.8}$$

where λ_i ($i \in \underline{n}$) are the real eigenvalues of L and $t_i(\in \mathbf{R}^n, i \in \underline{n})$ are the eigenvectors of L corresponding to λ_i.

Lemma 2.1 is known as the spectral resolution of the real symmetric matrix (Yanai et al (1983)).

Lemma 2.2 *The matrix $t_i t_i^T$ is the projection matrix on $Im(t_i)$, that is ,*

$$t_i t_i^T = P_{Im(t_i)} =: P_{(t_i)}.$$

Proof. From the decomposition of projection matrix, $P_{(t_i)}$ can be written as

$$P_{(t_i)} = t_i(t_i^T t_i)^{-1} t_i^T .$$

Since $t_i^T t_i = 1$, $P_{(t_i)} = t_i t_i^T$.

Applying Lemmas 2.1 and 2.2 to the quadratic term $x^T(k)Qx(k)$ in (2.6), it follows that

$$x^T(k)Qx(k) = \sum_{i=1}^{n} \gamma_i x^T(k) P_{(t_i)} x(k), \tag{2.9}$$

where γ_i $(i = 1, 2, \ldots, n)$ are eigenvalues of Q.

3 LQR PROBLEM WITH 2 CONTROLLERS

3.1 Problem Statement

Now we introduce the following assumption. Given any subspaces $\mathcal{F}_1$ and $\mathcal{F}_2$ $(\subset \mathcal{X} \subset \mathbf{R}^n)$ such that

$$\mathcal{X} = \mathcal{F}_1 \bigoplus \mathcal{F}_2 \text{ and } \mathcal{F}_1^{\perp} = \mathcal{F}_2. \tag{3.1}$$

This decoupling of the state space is usually possible, indeed, since for $i \neq j$, $\mathrm{Im}\{P_{(i)}\} \perp \mathrm{Im}\{P_{(j)}\}$, it is sufficient to set, for example,

$$\mathcal{F}_1 = \mathrm{Im}\{P_{(t_1, t_2, \cdots, t_\pi)}\} \text{ and } \mathcal{F}_2 = \mathrm{Im}\{P_{(t_{\pi+1}, t_{\pi+2}, \cdots, t_n)}\}, \tag{3.2}$$

where π is an integer such that $0 < \pi < n$. According to the given subspaces $\mathcal{F}_1$ and $\mathcal{F}_2$, (2.9) can be decoupled into two terms :

$$\begin{aligned} x^T(k)Qx(k) &= \textstyle\sum_{i=1}^{\pi} \gamma_i x^T(k) P_{(t_i)} x(k) + \sum_{i=\pi+1}^{n} \gamma_i x^T(k) P_{(t_i)} x(k) \\ &=: x^T(k)\Lambda_1 x(k) + x^T(k)\Lambda_2 x(k). \end{aligned} \tag{3.3}$$

such that $x^T(\cdot)\Lambda_j x(\cdot)$ depends only on $\mathcal{F}_j$ $(j = 1, 2)$.

Based on this decoupling of the state space, and introducing some linear maps $H_j : \mathcal{U} \to \mathcal{U}$ $(j = 1, 2)$, the following two optimal control subsystems are considered with the different performance indices to be minimized.

Subsystem j $(j = 1, 2)$:

$$x(k+1) = Ax(k) + BH_j u_j(k), \quad x(0) = x_{j0} = P_{\mathcal{F}_j} x_0 \in \mathcal{F}_j, \tag{3.4}$$

Performance index to be minimized for the subsystem j $(j = 1, 2)$:

$$J_{jx} = \sum_{k=0}^{\infty} \{x^T(k)\Lambda_j x(k)\}. \tag{3.5}$$

Now we assume that, for $j = 1, 2$,

1. The pair $(A,\ BH_j)$ is reachable in $\mathcal{F}_j$.

2. The pair $(\Lambda_j^{\frac{1}{2}},\ A)$ is observable in $\mathcal{F}_j$.

Then, the discrete Riccati equation for the subsystem j $(j = 1, 2)$ is given by

$$\Phi_j = A^T\Phi_j A - A^T\Phi_j BH_j[(BH_j)^T\Phi_j BH_j]^+(BH_j)^T\Phi_j A + \Lambda_j, \tag{3.6}$$

The problem which we are of interest is as follows.

[The Decoupling Problem of Linear Quadratic Regulator(DPLQR)]

Given a linear system (2.1), find (if possible) the linear map $H_j : \mathcal{U} \to \mathcal{U}$ $(j = 1, 2)$ such that the optimal input $u^{\rm opt}$ with respect to (2.1) and (2.6) can be replaced by the linear combination $H_1u_1^{\rm opt} + H_2u_2^{\rm opt}$, where $u_j^{\rm opt}$ $(j = 1, 2)$ are the optimal control inputs with respect to (3.4) and (3.5).

3.2 *Solvability Condition of DPLQR*

In order to solve the decoupling problem of optimal regulator, the supremal element of the class of *A-invariant* subspaces is introduced to consider the optimal control input in a decoupled subspace of the state space. For the subclass of *A-invariant* subspaces contained in a subspace $\mathcal{G}(\subset \mathcal{X})$, denoted by $\underline{\mathcal{I}}(A;\mathcal{G})$. It is known that there exists the supremal element of $\underline{\mathcal{I}}(A;\mathcal{G})$ (Wonham (1979)):

$$\mathcal{V}^* := sup\,\underline{\mathcal{I}}(A;\mathcal{G}). \tag{3.7}$$

For the computation of $\mathcal{V}^*$, we have the following lemma.

Lemma 3.1 *Let $A : \mathcal{X} \to \mathcal{X}$, and $\mathcal{G} \subset \mathcal{X}$. Define the sequence $\mathcal{V}^\mu$ according to*

$$\begin{aligned} \mathcal{V}^0 &= \mathcal{G}, \\ \mathcal{V}^\mu &= \mathcal{G} \cap A^{-1}\mathcal{V}^{\mu-1}, \\ &\qquad (\mu = 1, 2, \ldots, n). \end{aligned} \tag{3.8}$$

Then $\mathcal{V}^\mu \subset \mathcal{V}^{\mu-1}$, and for some $k \le dim(\mathcal{G})$,

$$\mathcal{V}^k = sup\,\underline{\mathcal{I}}(A;\mathcal{G}). \tag{3.9}$$

Proof. We first observe that the sequence $\mathcal{V}^\mu$ is nonincreasing. Clearly $\mathcal{V}^1 \subset \mathcal{V}^0$, and if $\mathcal{V}^\mu \subset \mathcal{V}^{\mu-1}$, then

$$\mathcal{V}^{\mu+1} = \mathcal{G} \cap A^{-1}\mathcal{V}^\mu \subset \mathcal{G} \cap A^{-1}\mathcal{V}^{\mu-1} = \mathcal{V}^\mu.$$

Thus for some $k \le dim(\mathcal{G})$, $\mathcal{V}^\mu = \mathcal{V}^k$ $(\mu \ge k)$. Now $\mathcal{V} \in \underline{\mathcal{I}}(A;\mathcal{G})$ if and only if

$$\mathcal{V} \subset \mathcal{G} \text{ and } \mathcal{V} \subset A^{-1}\mathcal{V}. \tag{3.10}$$

Therefore,

$$\mathcal{V} \subset \mathcal{V}^k \in \underline{\mathcal{I}}(A; \mathcal{G}),$$

and as $\mathcal{V}$ is arbitrary, the result follows.
We also obtain the following theorem.

Theorem 3.1 *If there exists $\mathcal{R}$ such that for an arbitrary $\mathcal{G} \subset \mathcal{X}$,*

$$\langle A | \mathcal{B} \cap \mathcal{R} \rangle \subset \mathcal{G},$$

there exists a linear map $H : \mathcal{U} \to \mathcal{U}$ such that

$$\langle A | Im(BH) \rangle = \langle A | \mathcal{B} \cap \mathcal{R} \rangle. \tag{3.11}$$

Proof. Let $\{b_1, b_2, \cdots, b_m\}$ be a basis for $\mathcal{B} \cap \mathcal{R}$. Then $b_i = Bu_i$ $(u_i \in \mathcal{U})$, where these u_i $(i \in \underline{m})$ are independent. Let $\{u_1, \cdots, u_m, \cdots, u_r\}$ be a basis for $\mathcal{U}$, and define H such that

$$\begin{aligned} Hu_i &= u_i \quad (i = 1, 2, \ldots, m), \\ Hu_i &= 0 \quad (i = m+1, m+2, \ldots, r). \end{aligned}$$

Then we have

$$\text{Im}(BH) = \mathcal{B} \cap \mathcal{R}.$$

When we consider the problem of control input optimal decoupling, the following lemma is useful.

Lemma 3.2 *Let $\mathcal{F}_j$ $(j = 1, 2)$ be any subspaces which satisfy (3.1). For any real symmetric matrix Φ and the matrix S such that $\Phi = S^T S$, there exist the real symmetric matrices Φ_j $(j = 1, 2)$ and the matrices S_j $(j = 1, 2)$ so that the following relations hold :*

$$\begin{aligned} \Phi &= \Phi_1 + \Phi_2, & & \\ \Phi_j &= S_j^T S_j & & (j = 1, 2), \\ S_i^T S_j &= 0 & & (i, j = 1, 2,\ i \neq j), \\ Im\Phi_j &\subset \mathcal{F}_j, & & (j = 1, 2). \end{aligned} \tag{3.12}$$

Proof. The matrix S_j can be constructed by using the projection matrix in Lemma 2.2. The relations of (3.12) are proved straightforward from the property of the decoupled spaces $\mathcal{F}_j$ $(j = 1, 2)$.

In Theorem 3.1, setting $\mathcal{G} = \mathcal{F}_j$ $(j = 1, 2)$ and

$$\mathcal{R} = \mathcal{V}_j^* = sup\, \underline{\mathcal{I}}(A; \mathcal{F}_j), \quad (j = 1, 2),$$

defined by (3.7) and noting that Lemma 3.2, we obtain the following theorem as the solution to the decoupling problem of optimal regulator (Moriwaki et al (1994)), (Moriwaki (1995)).

Theorem 3.2 *Let $\Phi, \Phi_j, (j = 1, 2)$ be the real symmetric (time-invariant) solutions of the Riccati equations with respect to the overall system (2.1) with (2.6) and the subsystems (3.4) with (3.5), respectively. Then, there exist the maps H_j $(j = 1, 2)$ that give a solution for the decoupling problem of linear quadratic regulator (DPLQR) if the linear maps H_j $(j = 1, 2)$ are chosen so that*

$$\langle A | Im(BH_j)\rangle = \langle A|\mathcal{B} \cap \mathcal{V}_j^*\rangle = \mathcal{F}_j, \quad (j = 1, 2), \tag{3.13}$$

where

$$\mathcal{V}_j^* = sup\, \underline{\mathcal{I}}(A; \mathcal{F}_j), \quad (j = 1, 2).$$

Proof. Substituting the matrices S and S_j $(j = 1, 2)$ into Φ and Φ_j $(j = 1, 2)$ of the corresponding Riccati equations (2.7) and (3.6) yields

$$S^T S = A^T S^T SA - A^T S^T SB(B^T S^T SB)^+ B^T S^T SA + Q, \tag{3.14}$$

$$\begin{aligned} S_j^T S_j &= A^T S_j^T S_j A - A^T S_j^T S_j BH_j \\ &\quad \times (H_j^T B^T S_j^T S_j BH_j)^+ H_j^T B^T S_j^T S_j A \\ &\quad + \Lambda_j, \quad (j = 1, 2). \end{aligned} \tag{3.15}$$

DPOR is solvable if

1. (3.14) is equivalent to

$$\begin{aligned} \sum_{j=1}^{2} S_j^T S_j &= \sum_{j=1}^{2} \{A^T S_j^T S_j A - A^T S_j^T S_j BH_j \\ &\quad \times (H_j^T B^T S_j^T S_j BH_j)^+ H_j^T B^T S_j^T S_j A \\ &\quad + \Lambda_j\}. \end{aligned} \tag{3.16}$$

2. $H_1 u_1^{\text{opt}} + H_2 u_2^{\text{opt}}$, which is the linear combination of the optimal control inputs (u_j^{opt} $(j = 1, 2)$) with respect to (3.4) and (3.5) is the optimal control input for the linear system (2.1) with the performance index (2.6).

Obviously, we see from (3.12) of Lemma 3.2 that

$$\Phi = S^T S = (S_1 + S_2)^T (S_1 + S_2) = S_1^T S_1 + S_2^T S_2 = \Phi_1 + \Phi_2. \tag{3.17}$$

For the map H_j $(j = 1, 2)$, it is easily shown that the following relations hold.

$$\begin{array}{rcll} \mathrm{Im}(BH_j) & \subset & \mathcal{F}_j, & (j = 1, 2), \\ \mathrm{Im}(\Phi_j B) & = & \mathrm{Im}(\Phi_j BH_j), & (j = 1, 2), \\ \Phi_i BH_j & = & 0, & (i, j = 1, 2\ i \neq j). \end{array}$$

Then, from some computation, if the equality

$$S^T SB(B^T S^T SB)^+ B^T S^T S = \sum_{j=1}^{2} \{S_j^T S_j BH_j (H_j^T B^T S_j^T S_j BH_j)^+ H_j^T B^T S_j^T S_j\}, \tag{3.18}$$

holds, we can conclude that (3.14) is equivalent to (3.15). Noting that the relations between the orthogonal projector and the Moore-Penrose pseudoinverse, it follows that

$$SB(B^T S^T SB)^+ B^T S^T = P_{\mathrm{Im}SB}, \tag{3.19}$$

and for $j = 1, 2$,

$$S_j BH_j (H_j^T B^T S_j^T S_j BH_j)^+ H_j^T B^T S_j^T = P_{\mathrm{Im}S_j BH_j}, \tag{3.20}$$

$$P_{\mathrm{Im}S_j BH_j} = P_{\mathrm{Im}SBH_j} = P_{\mathrm{Im}S_j B}, \tag{3.21}$$

$$P_{\mathrm{Im}S_1 BH_1} + P_{\mathrm{Im}S_2 BH_2} = P_{\mathrm{Im}SB}, \tag{3.22}$$

$$S_i^T P_{\mathrm{Im}S_j BH_j} S_i = 0,\ (i \neq j), \tag{3.23}$$

$$S_i^T P_{\mathrm{Im}S_j BH_j} S_j = 0,\ (i \neq j). \tag{3.24}$$

From (3.20), (3.23) and (3.24), we obtain

The right-hand side of (3.18)

$$\begin{array}{cl} = & \sum_{j=1}^{2} \{S_j^T P_{\mathrm{Im}S_j BH_j} S_j\} \\ = & (S_1 + S_2)^T \{P_{\mathrm{Im}S_1 BH_1} + P_{\mathrm{Im}S_2 BH_2}\}(S_1 + S_2). \end{array}$$

Noting that (3.21) and (3.23) it follows that

$$
\begin{aligned}
(S_1+S_2)^T\{P_{\mathrm{Im}S_1BH_1}+P_{\mathrm{Im}S_2BH_2}\}(S_1+S_2) & \\
&= S^TP_{\mathrm{Im}SB}S \\
&= \text{the left-hand side of (3.18).}
\end{aligned}
$$

Thus, we have shown that (3.15) is equivalent to (3.14).

Taking account of the form of optimal regulator given by Proposition 2.1, it turns out that

$$H_1u_1^{\mathrm{opt}}(k)+H_2u_2^{\mathrm{opt}}(k)=-\sum_{j=1}^{2}H_j(H_j^TB^TS_j^TS_jBH_j)^+H_j^TB^TS_j^TS_jAx(k). \tag{3.25}$$

Using the following relations (Ben-Israel et al (1974)),

$$(H_j^TB^TS_j^TS_jBH_j)^+H_j^TB^TS_j^T=(S_jBH_j)^+,$$

$$(S_jBH_j)^+S_i=0,\ H_i(S_jBH_j)^+=0,\ (i\neq j),$$

$$(SBH_1)^++(SBH_2)^+=\{SB(H_1+H_2)\}^+,\ ,$$

and noticing that the equality $HH^T(HH^T)^+=I_r$ where $H:=H_1+H_2$ is nonsingular, and

$$H^T(HH^T)^+\{(SB)^TSB\}^+(SB)^T=(SBH)^+,$$

it follows that

$$
\begin{aligned}
&\sum_{j=1}^{2}H_j(H_j^TB^TS_j^TS_jBH_j)^+H_j^TB^TS_j^TS_jA \\
&= H_1(S_1BH_1)^+S_1A+H_2(S_2BH_2)^+S_2A \\
&= (H_1+H_2)(S_1BH_1)^+(S_1+S_2)A+(H_1+H_2)(S_2BH_2)^+(S_1+S_2)A \\
&= (H_1+H_2)\{(S_1BH_1)^++(S_2BH_2)^+\}(S_1+S_2)A \\
&= (H_1+H_2)\langle\{(S_1+S_2)BH_1\}^++\{(S_1+S_2)BH_2\}^+\rangle(S_1+S_2)A \\
&= (H_1+H_2)\{(SBH_1)^++(SBH_2)^+\}SA \\
&= (H_1+H_2)\{SB(H_1+H_2)\}^+SA \\
&= (H_1+H_2)(H_1+H_2)^T\{(H_1+H_2)(H_1+H_2)^T\}^+ \\
&\qquad\times\{(SB)^TSB\}^+(SB)^TSA \\
&= \{(SB)^TSB\}^+(SB)^TSA,
\end{aligned} \tag{3.26}
$$

which shows that $H_1u_1^{\mathrm{opt}}+H_2u_2^{\mathrm{opt}}$ is the optimal control input for the linear system (2.1) with the performance index (2.6).

4 CONTROLLERS DECOUPLING VIA INTERNALLY BALANCED STATE SPACE REPRESENTATIONS

In this section we shall illustrate one way that principal component analysis (Moore (1981)) can be applied to the optimal control decoupling problem for the case where the full model is asymptotically stable.

Assume the system (2.1) with the output equation $y(k) = Cx(k)$ is asymptotically stable and reachable. It is also assumed that the output matrix C is always able to be chosen so that the system becomes to be observable. The reachability grammian and observability grammian are defined as

$$W_c = \sum_{i=0}^{\infty} A^i BB^T (A^T)^i, \tag{4.1}$$

$$W_o = \sum_{i=0}^{\infty} (A^T)^i A^i. \tag{4.2}$$

It can be shown that it is possible to find a linear transformation which makes the two grammians equal to some diagonal positive definite matrix Σ^2 (Moore (1981)). The corresponding state representation is said to be *internally balanced over the interval* $[\, 0, \infty\,)$. The grammians, which are both equal to Σ^2, are given by the unique positive definite solution to the Lyapunov equations,

$$A\Sigma^2 A^T - \Sigma^2 = -BB^T, \tag{4.3}$$

$$A^T \Sigma^2 A - \Sigma^2 = -C^T C. \tag{4.4}$$

Let assume the internally balanced system (2.1) be internally balanced and be partitioned as

$$x(k+1) = \begin{bmatrix} A_{11} & A_{12} \\ A_{21} & A_{22} \end{bmatrix} x(k) + \begin{bmatrix} B_1 \\ B_2 \end{bmatrix} u(k), \tag{4.5}$$

$$y(k) = [C_1 \;\; C_2] x(k), \tag{4.6}$$

and let the grammians be partitioned compatibly as $\Sigma^2 = \mathrm{diag}\{\Sigma_1^2, \Sigma_2^2\}$. The grammians are then given by the unique positive definite solution to the Lyapunov equations.

$$\begin{bmatrix} A_{11} & A_{12} \\ A_{21} & A_{22} \end{bmatrix} \begin{bmatrix} \Sigma_1^2 & 0 \\ 0 & \Sigma_2^2 \end{bmatrix} \begin{bmatrix} A_{11}^T & A_{21}^T \\ A_{12}^T & A_{22}^T \end{bmatrix} - \begin{bmatrix} \Sigma_1^2 & 0 \\ 0 & \Sigma_2^2 \end{bmatrix} = - \begin{bmatrix} B_1 \\ B_2 \end{bmatrix} [B_1^T \; B_2^T], \tag{4.7}$$

$$\begin{bmatrix} A_{11}^T & A_{21}^T \\ A_{12}^T & A_{22}^T \end{bmatrix} \begin{bmatrix} \Sigma_1^2 & 0 \\ 0 & \Sigma_2^2 \end{bmatrix} \begin{bmatrix} A_{11} & A_{12} \\ A_{21} & A_{22} \end{bmatrix} - \begin{bmatrix} \Sigma_1^2 & 0 \\ 0 & \Sigma_2^2 \end{bmatrix} = - \begin{bmatrix} C_1^T \\ C_2^T \end{bmatrix} [C_1 \; C_2]. \tag{4.8}$$

It follows that if the system is internally balanced, then every subsystem is asymptotically stable (Pernebo et al (1982)).

Proposition 4.1 *Assume that a system (4.5) and (4.6) is asymptotically stable and that either the reachability grammian or the observability grammian is nonsingular and diagonal. Then every subsystem is asymptotically stable.*

On the other hand, the reachability and the observability of the each subsystem are not necessarily reserved (Pernebo et al (1982)).

Proposition 4.2 *Assume that the partitioning in (4.5) and (4.6) is made so that $\sigma_{min}(\Sigma_1^2) > \sigma_{max}(\Sigma_2^2)$, in which $\sigma_{min}(\Sigma_1^2)$ and $\sigma_{max}(\Sigma_2^2)$ are the smallest singular value of Σ_1^2 and the largest singular value of Σ_2^2, respectively. Then the subsystem (A_{11}, B_1, C_1) is reachable and observable.*

Proposition 4.2 gives a good set of tools for model reduction problem and controller decoupling problem. The main idea underlying the model reduction work for the internally balanced system is to eliminate any weak subsystem which contributes little to the impulse response matrix (Moore (1981)). Furthermore, the main idea underlying the controller decoupling work is to disregard the correlation between subsystems with a larger singular value set and a small singular value set of the grammians. In the case that we consider the decoupling problem of optimal regulator for the internally balanced system, which satisfies Proposition 4.2, we can choose the subspaces $\tilde{\mathcal{F}}_1$ and $\tilde{\mathcal{F}}_2$ ($\subset \mathcal{X}$) such that

$$\tilde{\mathcal{F}}_1 = \mathrm{Im}\{A_{11}\} \text{ and } \tilde{\mathcal{F}}_2 = \mathrm{Im}\{A_{22}\}, \tag{4.9}$$

note that $\mathcal{X} = \tilde{\mathcal{F}}_1 \bigoplus \tilde{\mathcal{F}}_2$ and $\tilde{\mathcal{F}}_1^{\perp} = \tilde{\mathcal{F}}_2$. This decoupling of the state space is usually possible, indeed, assuming that the real symmetric matrix Q is given and using the same linear transformation by which (2.1) is transformed into (4.5), it is sufficient to set as (3.2) using Lemma 2.1 and Lemma 2.2 where $\pi = \dim(A_{11})$. According to the given subspaces (4.9), the quadratic term $x^T(k)Qx(k)$ in (2.6) can be decoupled into two terms :

$$\begin{aligned} x^T(k)Qx(k) &= \textstyle\sum_{i=1}^{\pi} \gamma_i x^T(k) P_{(t_i)} x(k) + \sum_{i=\pi+1}^{n} \gamma_i x^T(k) P_{(t_i)} x(k) \\ &=: \begin{bmatrix} x_1^T & x_2^T \end{bmatrix} \begin{bmatrix} \Gamma_1 & 0 \\ 0 & \Gamma_2 \end{bmatrix} \begin{bmatrix} x_1 \\ x_2 \end{bmatrix}. \end{aligned} \tag{4.10}$$

Based on this decoupling of the state space, the following two optimal control subsystems are considered with the performance index to be minimized.

Subsystem ℓ $(\ell = 1, 2)$:

$$x_\ell(k+1) = A_{\ell\ell}x_\ell(k) + B_\ell u_\ell(k), \quad x(0) = x_{\ell 0} = P_{\tilde{\mathcal{F}}_\ell} x_0 \in \tilde{\mathcal{F}}_\ell, \tag{4.11}$$

Performance index to be minimized for the subsystem ℓ $(\ell = 1, 2)$:

$$\tilde{J}_{\ell x} = \sum_{k=0}^{\infty} \{x_\ell^T(k)\Gamma_\ell x_\ell(k)\}. \tag{4.12}$$

Proposition 4.1 guarantees both subsystems are asymptotically stable. Proposition 4.2, however, says that subsystem 1 (i.e., (A_{11}, B_1, C_1)) is only guaranteed to be reachable.
Now we assume that, for $\ell = 1, 2$,

1. The pair $(A_{\ell\ell},\ B_\ell)$ is reachable in $\tilde{\mathcal{F}}_\ell$.
2. The pair $(\Gamma_\ell^{\frac{1}{2}},\ A_{\ell\ell})$ is observable in $\tilde{\mathcal{F}}_\ell$.

Then, discrete Riccati equation for the subsystem ℓ $(\ell = 1, 2)$ is given by

$$\Psi_\ell = A_{\ell\ell}^T \Psi_\ell A_{\ell\ell} - A_{\ell\ell}^T \Psi_\ell B_\ell [(B_\ell)^T \Psi_\ell B_\ell]^+ (B_\ell)^T \Psi_\ell A_{\ell\ell} + \Gamma_\ell, \tag{4.13}$$

The optimal controller $u^{\text{opt}}(k)$ for the internally balanced system (2.1) (= (4.5)) can be almost equivalently replaced by the controller composed of two optimal controllers for subsystems $\ell, \ell = 1, 2$ via the following algorithm.
(Step 1) Compute the optimal control $\tilde{u}_\ell^{\text{opt}}(k), \ell = 1, 2$ for the subsystems (4.11) with the performance indeces (4.12) to be minimized.
(Step 2) Replace the optimal control input $u^{\text{opt}}(k)$ for the internally balanced system (2.1) (= (4.5)) with the performance index to be minimized,

$$\tilde{J}_x = \sum_{k=0}^{\infty} \{x^T(k)Qx(k)\}. \tag{4.14}$$

by $\tilde{u}(k)$ where

$$\tilde{u}(k) := \begin{bmatrix} \tilde{u}_1^{\text{opt}}(k) \\ \tilde{u}_2^{\text{opt}}(k) \end{bmatrix}. \tag{4.15}$$

(Remark) The control input $\tilde{u}(k)$ is not optimal for the system (4.5) $(\tilde{u} \neq u^{\text{opt}})$, but $\tilde{u}$ is near optimal control input for the overall system (4.5).

5 NUMERICAL EXAMPLES AND DISCUSSIONS

In Example 1, we consider the system which is treated in the formulation of DPLQR and solved by Theorem 3.2 (see Section 3). Next, in Example 2, we consider the controller decoupling problem for the same system as Example 1 in the problem formulation of controller decoupling via internally balanced state space representation (see Section 4).

(Example 1.)

Consider the linear system

$$x(k+1) = \begin{bmatrix} 0.5 & 0 & 0.4 \\ 0 & 0.25 & 0 \\ 0.4 & 0 & 0.5 \end{bmatrix} x(k) + \begin{bmatrix} 0 & 0 \\ 0 & 1 \\ 1 & 0 \end{bmatrix} u(k), \tag{5.1}$$

with the performance index to be minimized :

$$J_x = \sum_{k=0}^{\infty} x^T(k) Q x(k) = \sum_{k=0}^{\infty} x^T(k) \begin{bmatrix} 1 & 0 & 0 \\ 0 & 2 & 0 \\ 0 & 0 & 1 \end{bmatrix} x(k). \tag{5.2}$$

Using Proposition 2.1, the solution Φ to Riccati equation with respect to the system (5.1)–(5.2) is given by

$$\Phi = \begin{bmatrix} 1.316 & 0 & 0.253 \\ 0 & 2 & 0 \\ 0.253 & 0 & 1.202 \end{bmatrix}. \tag{5.3}$$

Since the eigenvalues of the weighting matrix Q in (5.2) are $\{2, 1, 1\}$, by Lemma 2.1, the matrix Q can be decoupled as follows :

$$\begin{aligned} Q &= \begin{bmatrix} 1 & 0 & 0 \\ 0 & 2 & 0 \\ 0 & 0 & 1 \end{bmatrix} \\ &= \begin{bmatrix} 1 \\ 0 \\ 0 \end{bmatrix} \begin{bmatrix} 1 & 0 & 0 \end{bmatrix} + 2 \begin{bmatrix} 0 \\ 1 \\ 0 \end{bmatrix} \begin{bmatrix} 0 & 1 & 0 \end{bmatrix} + \begin{bmatrix} 0 \\ 0 \\ 1 \end{bmatrix} \begin{bmatrix} 0 & 0 & 1 \end{bmatrix}. \end{aligned} \tag{5.4}$$

The transformation matrix T such that

$$\text{diag}\,\{1, 2, 1\} = T^{-1} Q T,$$

is given by the unit matrix I_3.

Let the subspaces $\mathcal{F}_1$ and $\mathcal{F}_2$ be the decoupled subspaces of the state space $\mathcal{X}$ which satisfy (3.1) :

$$\mathcal{F}_1 = \text{Im}(\begin{bmatrix} 1 & 0 \\ 0 & 0 \\ 0 & 1 \end{bmatrix}), \quad \mathcal{F}_2 = \text{Im}(\begin{bmatrix} 0 \\ 1 \\ 0 \end{bmatrix}). \tag{5.5}$$

Now consider the following two optimal control subsystems.

$$x(k+1) = \begin{bmatrix} 0.5 & 0 & 0.4 \\ 0 & 0.25 & 0 \\ 0.4 & 0 & 0.5 \end{bmatrix} x(k) + \begin{bmatrix} 0 & 0 \\ 0 & 1 \\ 1 & 0 \end{bmatrix} H_j u_j(k), \ (j = 1, 2), \tag{5.6}$$

with the performance index to be minimized :

$$J_{jx} = \sum_{k=0}^{\infty} x^T(k) \Lambda_j x(k), \ (j = 1, 2), \tag{5.7}$$

where

$$\Lambda_1 = \begin{bmatrix} 1 & 0 & 0 \\ 0 & 0 & 0 \\ 0 & 0 & 1 \end{bmatrix}, \quad \Lambda_2 = \begin{bmatrix} 0 & 0 & 0 \\ 0 & 2 & 0 \\ 0 & 0 & 0 \end{bmatrix}. \tag{5.8}$$

Using $\mathcal{V}_j^* = \mathcal{F}_j, (j = 1, 2)$ in Theorem 3.2, we can obtain the maps H_j $(j = 1, 2)$ as

$$H_1 = \begin{bmatrix} \alpha & 0 \\ 0 & 0 \end{bmatrix}, \ H_2 = \begin{bmatrix} 0 & 0 \\ 0 & \beta \end{bmatrix}, \ (\alpha \neq 0, \beta \neq 0). \tag{5.9}$$

The Riccati equations of the optimal control problem for the subsystems (5.6)–(5.8) with (5.9) have the solutions Φ_j $(j = 1, 2)$ as

$$\begin{aligned} \Phi_1 &= \begin{bmatrix} 1.316 & 0 & 0.253 \\ 0 & 0 & 0 \\ 0.253 & 0 & 1.202 \end{bmatrix} \\ &= \begin{bmatrix} 1.147 & 0 & 0 \\ 0 & 0 & 0 \\ 0.221 & 0 & 1.096 \end{bmatrix} \begin{bmatrix} 1.147 & 0 & 0.221 \\ 0 & 0 & 0 \\ 0 & 0 & 1.096 \end{bmatrix} = S_1^T S_1, \end{aligned} \tag{5.10}$$

$$\Phi_2 = \begin{bmatrix} 0 & 0 & 0 \\ 0 & 2 & 0 \\ 0 & 0 & 0 \end{bmatrix} = \begin{bmatrix} 0 & 0 & 0 \\ 0 & \sqrt{2} & 0 \\ 0 & 0 & 0 \end{bmatrix} \begin{bmatrix} 0 & 0 & 0 \\ 0 & \sqrt{2} & 0 \\ 0 & 0 & 0 \end{bmatrix} = S_2^T S_2. \tag{5.11}$$

As is shown, the matrices Φ_j $(j = 1, 2)$ have the decomposition matrices S_j $(j = 1, 2)$ satisfying Lemma 3.2. And the matrix Φ has the decomposition matrix S which is $S_1 + S_2$:

$$\begin{aligned} \Phi = S^T S &= \begin{bmatrix} 1.147 & 0 & 0 \\ 0 & \sqrt{2} & 0 \\ 0.221 & 0 & 1.096 \end{bmatrix} \begin{bmatrix} 1.147 & 0 & 0.221 \\ 0 & \sqrt{2} & 0 \\ 0 & 0 & 1.096 \end{bmatrix} \\ &= (S_1 + S_2)^T (S_1 + S_2). \end{aligned} \tag{5.12}$$

Let $u^{\text{opt}}(k), u_j^{\text{opt}}(k)$ $(j = 1, 2)$ be the optimal control input for the system (5.1)–(5.2) and the subsystem $j (j = 1, 2)$ (5.6)–(5.7), respectively. Then we obtain that

$$\begin{aligned} u^{\text{opt}}(k) &= -(B^T \Phi B)^+ B^T \Phi A x(k), \\ &= \begin{bmatrix} -0.505 & 0 & -0.584 \\ 0 & -0.25 & 0 \end{bmatrix} x(k), \end{aligned} \tag{5.13}$$

$$\begin{aligned} u_1^{\text{opt}}(k) &= -[(BH_1)^T \Phi_1 B H_1]^+ (BH_1)^T \Phi_1 A x(k), \\ &= \frac{1}{\alpha} \begin{bmatrix} -0.505 & 0 & -0.584 \\ 0 & 0 & 0 \end{bmatrix} x(k), \quad (\alpha \neq 0), \end{aligned} \tag{5.14}$$

$$\begin{aligned} u_2^{\text{opt}}(k) &= -[(BH_2)^T \Phi_2 B H_2]^+ (BH_2)^T \Phi_2 A x(k), \\ &= \frac{1}{\beta} \begin{bmatrix} 0 & 0 & 0 \\ 0 & -0.25 & 0 \end{bmatrix} x(k), \quad (\beta \neq 0). \end{aligned} \tag{5.15}$$

Finally, we can easily check that the equation

$$u^{\text{opt}}(k) = H_1 u_1^{\text{opt}}(k) + H_2 u_2^{\text{opt}}(k) \tag{5.16}$$

is satisfied by using (5.9) and (5.13)–(5.15).

(Example 2.)

Consider the system (5.1) with the output

$$y(k) = \begin{bmatrix} 0 & 1 & 0 \\ 0 & 0 & 1 \end{bmatrix} x(k). \tag{5.17}$$

The system (5.1) and (5.17) is asymptotically stable, reachable and observable. We here consider the controller decoupling problem of LQR using internally balanced state space representation, although the system (5.1) can be solved by Theorem 3.2. From the system (5.1) and (5.17), we can obtain the internally balanced representation by some steps of linear transformations as follows.

$$x(k+1) = \begin{bmatrix} 0.25 & 0 & 0 \\ 0 & 0.145 & 0.184 \\ 0 & 0.1836 & 0.855 \end{bmatrix} x(k) + \begin{bmatrix} 0 & 1 \\ -0.520 & 0 \\ 0.854 & 0 \end{bmatrix} u(k), \tag{5.18}$$

$$y(k) = \begin{bmatrix} 1 & 0 & 0 \\ 0 & -0.520 & 0.854 \end{bmatrix}. \tag{5.19}$$

We also consider the LQR problem for the system (5.18) with the performance index (5.2) to be minimized under the condition all state are available. The reachability grammian and observability grammian are both equal to Σ^2,

$$\Sigma^2 = \begin{bmatrix} 1.067 & 0 & 0 \\ 0 & 0.372 & 0 \\ 0 & 0 & 2.765 \end{bmatrix}. \tag{5.20}$$

According to Proposition 4.2, let Σ_1^2 and Σ_2^2 set

$$\Sigma_1^2 = \begin{bmatrix} 1.067 & 0 \\ 0 & 2.765 \end{bmatrix}, \qquad \Sigma_2^2 = \begin{bmatrix} 0.372 \end{bmatrix}. \tag{5.21}$$

Then, changing rows and columns of the state space, we have a partitioned system,

$$x(k+1) = \left[\begin{array}{cc|c} 0.25 & 0 & 0 \\ 0 & 0.855 & 0.184 \\ \hline 0 & 0.184 & 0.145 \end{array}\right] x(k) + \left[\begin{array}{cc} 0 & 1 \\ 0.854 & 0 \\ \hline -0.520 & 0 \end{array}\right] u(k), \tag{5.22}$$

and the corresponding performance index for (5.22) is

$$\tilde{J}_x = \sum_{k=0}^{\infty} x^T(k) \left[\begin{array}{cc|c} 1 & 0 & 0 \\ 0 & 1 & 0 \\ \hline 0 & 0 & 2 \end{array}\right] x(k). \tag{5.23}$$

From (5.22) and (5.23), subsystems and performance indeces are given by

$$\begin{aligned} x_1(k+1) &= \begin{bmatrix} 0.25 & 0 \\ 0 & 0.855 \end{bmatrix} x_1(k) + \begin{bmatrix} 0 & 1 \\ 0.854 & 0 \end{bmatrix} u_1(k), \\ \tilde{J}_{1x} &= \textstyle\sum_{k=0}^{\infty} x_1^T(k) \begin{bmatrix} 1 & 0 \\ 0 & 1 \end{bmatrix} x_1(k), \end{aligned} \tag{5.24}$$

$$\begin{aligned} x_2(k+1) &= [0.145]x_1(k) + [-0.520 \quad 0]u_2(k), \\ \tilde{J}_{2x} &= \textstyle\sum_{k=0}^{\infty} 2x_2^2(k). \end{aligned} \tag{5.25}$$

Thus, for the system (5.22) and derived subsystems with the performance indeces to be minimized, the optimal controllers are obtained as follows.

$$\begin{aligned} u^{\text{opt}}(k) &= \begin{bmatrix} 0 & -0.641 & -0.084 \\ -0.25 & 0 & 0 \end{bmatrix} x(k), \\ \tilde{u}_1^{\text{opt}}(k) &= \begin{bmatrix} 0 & 1.001 \\ 0.25 & 0 \end{bmatrix} x_1(k), \\ \tilde{u}_2^{\text{opt}}(k) &= \begin{bmatrix} -0.278 \\ 0 \end{bmatrix} x_2(k), \end{aligned} \tag{5.26}$$

Although $u^{\text{opt}}(k)$ is seems to be different from $\tilde{u}(k)$ which is composed of $\tilde{u}_\ell^{\text{opt}}(k), \ell = 1, 2$, those closed-loop systems are given by

$$\begin{aligned} x(k) &= \begin{bmatrix} 0 & 0 & 0 \\ 0 & 0.308 & 0.112 \\ 0 & 0.517 & 0.188 \end{bmatrix} x(k), \\ \tilde{x}(k) &= \begin{bmatrix} 0 & 0 & 0 \\ 0 & 0 & 0.184 \\ 0 & 0.184 & 0 \end{bmatrix} \tilde{x}(k), \end{aligned} \tag{5.27}$$

Eqs(5.27) show that both closed-loop systems are driven to the origin by the optimal control scheme.

6 CONCLUSION

It has been considered that the structure of the optimal control system with 2 controllers, whose performance is globally equivalent to the optimal regulator with a single controller. In the problem formulation of Section 3, the dimension of each optimal control subsystem is the same as the dimension of the overall control system. However, each subsystem works in the corresponding orthogonal state subspace. Therefore, we can obtain the solution to the matrix Riccati equation for the subsystem much easier than the one for the overall system.

In the problem formulation of Section 4, the dimension of each optimal control subsystem becomes lower than the dimension of the overall control system according to the dimension of the partitioned state space derived from the internally balanced representation. Although the solvability condition of

Theorem 3.2 is severe, Theorem 3.2 gives the strict solution for the decoupling problem of optimal regulator. On the other hand, the treatment of controllers decoupling via internally balanced representation is able to apply to many systems, but the solution is not optimal. It is near optimal solution.

Acknowledgments

The author wishes to thank Professor K. L. Teo of The Hong Kong Polytechnic University for his encouragement.

References

Anderson, B.D.O. and Moore, J.B. (1971), *Linear Optimal Control*, Prentice-Hall, Englewood Cliffs, New Jersey.

Anderson, B.D.O. and Moore, J.B. (1990), *Linear Optimal Control -Linear Quadratic Methods-*, Prentice-Hall, Englewood Cliffs, New Jersey.

Ho, Y.C. and Chu, K.C. (1972), Team decision theory and information structures in optimal control problems - Part I, *IEEE Trans. Automat. Contr.*, Vol. AC-17, pp. 15-22.

Yoshikawa, T. (1978), Decomposition of dynamic team decision problems, *IEEE Trans. Automat. Contr.*, Vol. AC-23, pp. 627-632.

Sandell, N.R.Jr., et al. (1978), Survey of decentralized control methods for large scale systems, *IEEE Trans. Automat. Contr.*, Vol. AC-23, pp. 108-128.

Davison, E.J. (1966), A method for simplifying dynamic systems, *IEEE Trans. Automat. Contr.*, Vol. AC-11, pp. 93-101.

Moore, B.C. (1981), Principal component analysis in linear systems: controllability, observability, and model reduction, *IEEE Trans. Automat. Contr.*, Vol. AC-26, pp. 17-32.

Enn, D.F. (1984), Model reduction with balanced realizations: an error bound and a frequency weighted generalization, *Proc. 23rd IEEE Conf. on Decis. and Contr.*, pp. 127-132.

Glover, K. (1984), All optimal Hankel-norm approximations of linear multivariable systems and their L^∞-error bounds, *International J. of Contr.*, Vol. 39, pp. 1115-1193.

Anderson, B.D.O. and Liu, Y. (1989), Controller reduction : concepts and approaches, *IEEE Trans. Automat. Contr.*, Vol. AC-34, pp. 802-812.

Mustafa, D. and Glover, K. (1991), Controller reduction by H_∞-balanced truncation, *IEEE Trans. Automat. Contr.*, Vol. AC-36, pp. 668-682.

Potter, J.E. (1966), Matrix quadratic solutions, *SIAM J. Appl. Math.*, Vol. 14, pp. 496-501.

Yanai, H. and Takeuchi, K. (1983), *Projection Matrix, Generalized Inverse Matrix, Singular Value Decomposition*, The University of Tokyo Press, Tokyo (in Japanese).

Wonham, W.M. (1979), *Linear Multivariable Control: A Geometric Approach*, 2nd. ed., Springer-Verlag, New York.

Ben-Israel, A. and Greville, T.N.E. (1974), *Generalized Inverses - Theory and Applications*, Wiley-Interscience, New York.

Moriwaki, K. and Akashi, H. (1994), A design method of the optimal regulator with multiple controllers, *Trans. ISCIE*, Vol. 7, pp. 382-384 (in Japanese).

Moriwaki, K. (1995), *Studies on the Design of Robust Controllers for Linear Multivariable Discrete-Time Systems with Uncertainties*, Ph.D Thesis, Kyoto University, Kyoto.

Pernebo, L. and Silverman, L.M. (1982), Model reduction via balanced state space representations, *IEEE Trans. Automat. Contr.*, Vol. AC-27, pp. 382-387.

7 MULTILEVEL OPTIMIZATION OF OPTIMAL CONTROL PROBLEMS

Volker Rehbock

Department of Mathematics and Statistics
Curtin University of Technology, GPO Box U1987
Perth, WA 6845, Australia

Abstract: A multilevel optimization approach for the numerical solution of optimal control problems is proposed. It is shown that optimal control problems involving ordinary differential equations can be easily cast into a form which allows optimization over two levels. At the upper level, optimization with respect to the state variables is carried out while optimization with respect to the control variables occurs at the lower level. We calculate the gradients of the upper level optimization problems and show how feasibility for the lower level optimization problems can be maintained. Several numerical studies show that the proposed method can result in significant savings of computation time. Furthermore, it is ideally suited to implementation in a parallel programming environment.

Key words: Optimal control, large scale optimization, parallel computing.

X. Yang et al. (eds.), Optimization Methods and Applications, 125-138.

1 INTRODUCTION

Optimal control problems arise in a wide variety of disciplines. Apart from traditional areas such as aerospace engineering, robotics and chemical engineering, optimal control theory has proven itself to be useful in fields as diverse as economics and biomedicine. However, most practical problems are far too difficult to solve analytically and there is a continued need for new computational algorithm to numerically solve increasingly complex problems. Several successful families of algorithms for the numerical solution of optimal control problems involving ordinary differential equations (ODEs) have been devised. These include the sequential gradient restoration algorithms (Miele and Wang (1986)) and algorithms based on the control parametrization approach (Sirisena and Chou (1979), Teo et al. (1991)), amongst others (Sakawa and Shindo (1980)). The control parametrization approach involves a discretization of the original problem, yielding what is, in essence, a standard mathematical programming problem. This mathematical programming problem can then be solved by standard techniques such as sequential quadratic programming (Schittkowski (1985)). This approach is quite satisfactory for solving a many practical optimal control problems. One of its limitations is the accuracy to which a given problem can be solved. The discretization of the original problem gives rise to a large number of parameters to be optimized. Higher accuracy can only be obtained by a finer discretization which in turn leads to more parameters in the discretized problem. While some problems can be solved relatively quickly with a fine partition leading to around 1000 parameters, many practical problems involving nontrivial constraints can require significant computational time with much coarser discretizations. As a guide, the second numerical example we consider in this paper can be solved by the standard optimal control software MISER3 (see Jennings et al. (1991)). Running on a Pentium 166 MHz processor and choosing a parametrization leading to 251 control parameters, 1 hour and 21 minutes of computational time is required to solve this relatively simple problem. Clearly, the accuracy of solutions is limited by the speed of the computer one uses.

Current trends in the computing industry are moving towards multiprocessor parallel machines, since individual processor speeds are beginning to approach their theoretical limits. In the context of optimal control computation, note that most existing algorithms are essentially sequential in nature and unable to fully utilize the computational advantages offered by a multiprocessor environment. In this paper, we propose a new approach to overcome these limitations.

The proposed approach is based on ideas from the area of large scale numerical optimization (see, for example, Wismer (1971)), where a large problem is usually decomposed into a series of subproblems. This decomposition is most effective when a large proportion of the variables in a subproblem occur in (or are local to) only that subproblem, while the small proportion of the remaining variables also occur in other subproblems (*i.e.* are global variables of the overall problem). The solution of the overall problem is then carried out in two levels. At the *upper* level, one defines a global problem which depends only on the global variables. For each set of values of the global variables, it is assumed that the local variables have been chosen optimally by *lower* level optimization of each of the subproblems. The overall solution is then found by optimization at the upper level, each iteration and update of the global variables involving an optimization of the subproblems. Variations of this basic theme include optimizing at more than two levels and exploiting the special structure that particular problems may have. For some recent work, see Coleman and Li (1989).

Optimal control problems can be very naturally cast into such a two level optimization procedure. The resulting lower level problems are identical in structure to the original problem. Hence, while here we implement this idea in the context of control parametrization, it is possible to apply the same approach no matter what method of numerically solving optimal control problems is employed. The basic idea should thus be widely applicable in the area of computational optimal control.

An important aspect of the two level optimization formulation is that each lower level optimization can be carried out independently. Hence the approach is ideally suited to computation in a parallel processing environment with each processor being assigned the task of optimizing one or more of the subproblems.

Ralph (1996) presents this decomposition approach for a general class of unconstrained discrete time optimal control problems. Dynamic programming methods are then used to solve the resulting sequence of problems and the algorithm is successfully implemented in a parallel computing environment. We show that the two level decomposition can be easily applied to continuous optimal control problems involving canonical constraints, thus allowing a variety of optimal control methods to be employed. Furthermore, our proposed algorithm differs from that of Ralph (1996) in its treatment of the continuity of the overall solution across successive subproblems. In particular, we maintain this continuity throughout the optimization by imposing additional terminal state constraints within each subproblem.

Other than Ralph (1996) and some of the references cited therein, little work of this nature has been done until now. The idea of decomposing the optimization problems arising from the discretization of optimal control problems does appear in Wright (1990) and Wright (1991), but the decomposition and the resulting parallelism occur in the underlying linear algebra computations rather than in the original problem. Another related approach is given in Markine et al. (1996), where the problem of controlling vibrations in a particular structure is considered. The construction of the two level optimization in Markine et al. (1996) is based on the physical nature of the problem, though, where as our approach is applied to a general mathematical description of optimal control problems.

2 GENERAL PROBLEM

Consider the following general optimal control problem, to be referred to as Problem (P):

$$\text{Minimize}_{\boldsymbol{u}\in\mathcal{U}} \quad g_0(\boldsymbol{u}) = \Phi_0(\boldsymbol{x}(T)) + \int_0^T \mathcal{L}_0(t, \boldsymbol{x}(t), \boldsymbol{u}(t))dt \tag{2.1}$$

subject to

$$\dot{\boldsymbol{x}}(t) = \boldsymbol{f}(t, \boldsymbol{x}(t), \boldsymbol{u}(t)) \tag{2.2}$$
$$\boldsymbol{x}(0) = \boldsymbol{x}_0 \tag{2.3}$$

and

$$g_i(\boldsymbol{u}) = \Phi_i(\boldsymbol{x}(T)) + \int_0^T \mathcal{L}_i(t, \boldsymbol{x}(t), \boldsymbol{u}(t))dt \begin{smallmatrix} = \\ \geq \end{smallmatrix} 0, \qquad i = 1, \ldots, n_c. \tag{2.4}$$

Here, $\boldsymbol{x} \in \mathbb{R}^n$, $\boldsymbol{u} \in \mathbb{R}^r$ and $\mathcal{U}$ is the set of admissible controls (defined as the set of piecewise continuous and bounded functions on the interval $[0, T]$). We assume the usual continuity and differentiability properties of the functions involved (see Teo et al. (1991)). Note that a variety of constraints can be transformed into the canonical form (2.4).

3 MULTILEVEL APPROACH

Following the approach of Ralph (1996) for discrete time optimal control problems, we show how Problem (P) can be decomposed into a sequence of subproblem which are linked by a set of state values. This decomposition then yields a problem which is equivalent to the original, but the solution of which requires optimization on two levels. At the upper level, we have a master problem, where, at each iteration, a sequence of lower level problems must be solved.

We start with a coarse partitioning the original time horizon $[0, T]$ into N_u elements: $h_u = T/N_u$, $T_j = jh_u$, $j = 0, \ldots, N_u$. On each subinterval of this partition, we can formulate the lower level Problem (P_j), $j = 1, \ldots, N_u$, as follows:

$$\text{Minimize}_{\boldsymbol{u}_j\in\mathcal{U}_j} \qquad g_0^j(\boldsymbol{u}_j) = \int_{T_{j-1}}^{T_j} \mathcal{L}_0(t, \boldsymbol{x}(t), \boldsymbol{u}_j(t))dt \tag{3.1}$$

subject to

$$\dot{\boldsymbol{x}}(t) = \boldsymbol{f}(t, \boldsymbol{x}(t), \boldsymbol{u}_j(t)) \tag{3.2}$$
$$x(T_{j-1}) = \boldsymbol{\xi}_{j-1} \tag{3.3}$$

and the terminal constraint

$$\boldsymbol{x}(T_j) = \boldsymbol{\xi}_j, \qquad i.e. \quad \boldsymbol{g}_{loc}^j(\boldsymbol{u}_j) = \boldsymbol{x}(T_j) - \boldsymbol{\xi}_j = \mathbf{0}. \tag{3.4}$$

Note that $\boldsymbol{g}_{loc}^j(\boldsymbol{u}_j) \in \mathbb{R}^n$ is a vector valued function.

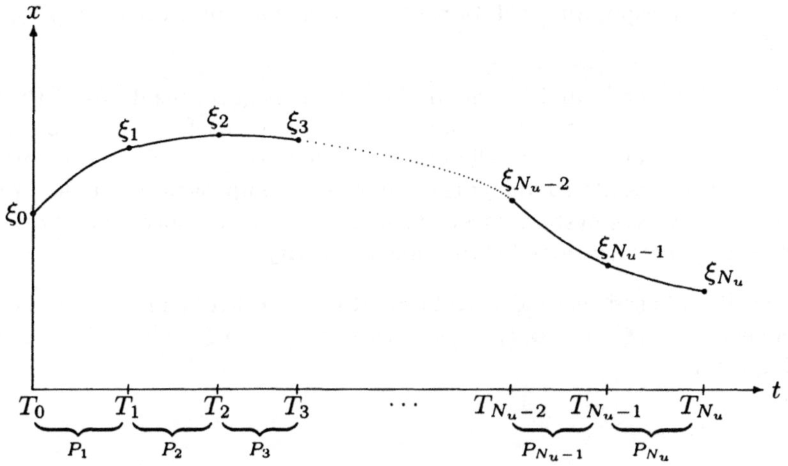

Figure 3.1 Decomposition into subproblems

Before defining the overall master problem, let $\boldsymbol{u}_j^*$ denote the optimal solution of (P_j) corresponding to a given $\boldsymbol{\xi}_{j-1}$ and $\boldsymbol{\xi}_j$. Furthermore, for $j = 1, \ldots, N_u$, let

$$g_i^j(\boldsymbol{u}_j) = \int_{T_{j-1}}^{T_j} \mathcal{L}_i(t, \boldsymbol{x}(t), \boldsymbol{u}_j(t))dt,$$

$i = 1, \ldots, n_c$, and $\boldsymbol{\xi} = [\boldsymbol{\xi}_0^\top, \boldsymbol{\xi}_1^\top, \ldots, \boldsymbol{\xi}_{N_u}^\top]^\top$.

The master problem, denoted as Problem (P_u), is then defined as

$$\text{Minimize}_{\boldsymbol{\xi}} \qquad G_0(\boldsymbol{\xi}) = \Phi_0(\boldsymbol{\xi}_{N_u}) + \sum_{j=1}^{N_u} g_0^j(\boldsymbol{u}_j^*) \tag{3.5}$$

subject to

$$G_i(\boldsymbol{\xi}) = \Phi_i(\boldsymbol{\xi}_{N_u}) + \sum_{j=1}^{N_u} g_i^j(\boldsymbol{u}_j^*) \begin{matrix} = \\ \geq \end{matrix} 0, \quad i = 1, \ldots, n_c. \tag{3.6}$$

Note that the only variables in (P_u) are the state values $\boldsymbol{\xi}$, referred to as the *separator* variables in Ralph (1996). In problems involving interior point constraints or all time inequality constraints on the state of the system, it may be useful to impose upper and/or lower bounds on some of these variables, but for our purposes, we assume that they are unconstrained (except possibly $\boldsymbol{\xi}_0$, if an initial condition for the dynamics is given, and $\boldsymbol{\xi}_{N_u}$, if a terminal state constraint is specified in Problem (P)).

The alternative to solving Problem (P) directly then is to solve Problem (P_u) instead. When this is done numerically, each function evaluation for (P_u)

requires the corresponding solution of the sequence of Problems (P_j), $j = 1, \ldots, N_u$.

With the described partitioning in place, various generalizations of Problem (P) are possible. For example, in some cases it may be useful to have the system dynamics (3.2) described by a different function $\boldsymbol{f}_j$ on each $[T_{j-1}, T_j]$. Furthermore, it is also possible to follow the partitioning approach when the original Problem (P) involves system parameters in addition to control functions. We neglect these generalizations for the sake of brevity.

If a gradient based method is to be used to solve Problem (P_u), we require the gradient of $G_i(\boldsymbol{\xi})$, $i = 0, 1, \ldots, n_c$, with respect to $\boldsymbol{\xi}$. This is given in the following result.

Theorem 3.1 *(Gradients of (P_u)) Let $G_i(\boldsymbol{\xi})$, $i = 0, 1, \ldots, n_c$, be defined as above. For $j = 1, \ldots, N_u$, let*

$$H_i^j(t, \boldsymbol{x}(t|\boldsymbol{u}_j, \boldsymbol{\xi}_{j-1}), \boldsymbol{u}_j(t), \boldsymbol{\xi}_{j-1}, \boldsymbol{\lambda}_i^j(t))$$
$$= \mathcal{L}_i(t, \boldsymbol{x}(t|\boldsymbol{u}_j, \boldsymbol{\xi}_{j-1}), \boldsymbol{u}_j(t)) + \left(\boldsymbol{\lambda}_i^j(t)\right)^{\top} \boldsymbol{f}(t, \boldsymbol{x}(t|\boldsymbol{u}_j, \boldsymbol{\xi}_{j-1}), \boldsymbol{u}_j(t))$$

$$\left(\dot{\boldsymbol{\lambda}}_i^j(t)\right)^{\top} = -\frac{\partial H_i^j(t, \boldsymbol{x}(t|\boldsymbol{u}_j, \boldsymbol{\xi}_{j-1}), \boldsymbol{u}_j(t), \boldsymbol{\xi}_{j-1}, \boldsymbol{\lambda}_i^j(t))}{\partial \boldsymbol{x}}, \quad t \in [T_{j-1}, T_j] \tag{3.7}$$

$$\boldsymbol{\lambda}_i^j(T_j) = \boldsymbol{0}. \tag{3.8}$$

For $j = 1, \ldots, N_u$ and $k = 1, \ldots, n$, let

$$H_{loc,k}^j(t, \boldsymbol{x}(t|\boldsymbol{u}_j, \boldsymbol{\xi}_{j-1}), \boldsymbol{u}_j(t), \boldsymbol{\xi}_{j-1}, \boldsymbol{\lambda}_{loc,k}^j(t))$$
$$= \left(\boldsymbol{\lambda}_{loc,k}^j(t)\right)^{\top} \boldsymbol{f}(t, \boldsymbol{x}(t|\boldsymbol{u}_j, \boldsymbol{\xi}_{j-1}), \boldsymbol{u}_j(t))$$

$$\left(\dot{\boldsymbol{\lambda}}_{loc,k}^j(t)\right)^{\top} = -\frac{\partial H_{loc,k}^j(t, \boldsymbol{x}(t|\boldsymbol{u}_j, \boldsymbol{\xi}_{j-1}), \boldsymbol{u}_j(t), \boldsymbol{\xi}_{j-1}, \boldsymbol{\lambda}_{loc,k}^j(t))}{\partial \boldsymbol{x}},$$
$$t \in [T_{j-1}, T_j], \tag{3.9}$$

$$\boldsymbol{\lambda}_{loc,k}^j(T_j) = \boldsymbol{e}_k, \tag{3.10}$$

where $\boldsymbol{x}(\cdot|\boldsymbol{u}_j, \boldsymbol{\xi}_{j-1})$ denotes the solution of (3.2) and (3.3) corresponding to $\boldsymbol{\xi}_{j-1}$ and $\boldsymbol{u}_j \in \mathcal{U}_j$ and $\boldsymbol{e}_k$ denotes the k-th standard unit basis vector in $\mathbb{R}^n$. Let $\Lambda_{loc}^j = \left[\boldsymbol{\lambda}_{loc,1}^j\ \boldsymbol{\lambda}_{loc,2}^j\ \cdots\ \boldsymbol{\lambda}_{loc,n}^j\right] \in \mathbb{R}^{n \times n}$. Furthermore, let the Lagrangian corresponding to the i-th functional for Problem (P_j) be given by $g_i^j(\boldsymbol{u}_j^*) - (\boldsymbol{\eta}_i^{j,*})^{\top} \boldsymbol{g}_{loc}^j(\boldsymbol{u}_j^*)$, where $\boldsymbol{\eta}_i^{j,*} \in \mathbb{R}^n$ is the optimal Lagrange multiplier vector with respect to the terminal constraint (3.4) of Problem (P_j). Then, for $i =$

$0, 1, \ldots, n_c$,

$$\frac{\partial G_i}{\partial \boldsymbol{\xi}_0} = (\boldsymbol{\lambda}_i^1(0|\boldsymbol{u}_1, \boldsymbol{\xi}_0))^\top - \left(\boldsymbol{\eta}_i^{1,*}\right)^\top \Lambda_{loc}^1(0|\boldsymbol{u}_1, \boldsymbol{\xi}_0), \tag{3.11}$$

$$\frac{\partial G_i}{\partial \boldsymbol{\xi}_j} = \left(\boldsymbol{\lambda}_i^{j+1}(T_j|\boldsymbol{u}_{j+1}, \boldsymbol{\xi}_j)\right)^\top - \left(\boldsymbol{\eta}_i^{j+1,*}\right)^\top \Lambda_{loc}^{j+1}(T_j|\boldsymbol{u}_{j+1}, \boldsymbol{\xi}_j) + \left(\boldsymbol{\eta}_i^{j,*}\right)^\top, \quad j = 1, \ldots, N_u - 1, \tag{3.12}$$

$$\frac{\partial G_i}{\partial \boldsymbol{\xi}_{N_u}} = \frac{\partial \Phi_i(\boldsymbol{\xi}_{N_u})}{\partial \boldsymbol{\xi}_{N_u}} + \left(\boldsymbol{\eta}_i^{N_u,*}\right)^\top, \tag{3.13}$$

where $\boldsymbol{\lambda}_i^j(\cdot|\boldsymbol{u}_j, \boldsymbol{\xi}_{j-1})$ denotes the solution of (3.7) and (3.8) corresponding to $\boldsymbol{\xi}_{j-1}$ and $\boldsymbol{u}_j \in \mathcal{U}_j$ and $\Lambda_{loc}^j(\cdot|\boldsymbol{u}_j, \boldsymbol{\xi}_{j-1})$ denotes the solutions of (3.9) and (3.10), $k = 1, \ldots, n$, corresponding to $\boldsymbol{\xi}_{j-1}$ and $\boldsymbol{u}_j \in \mathcal{U}_j$.

Proof: First, note that each $G_i(\boldsymbol{\xi})$ can be expressed as

$$G_i(\boldsymbol{\xi}) = \sum_{j=1}^{N_u} g_i^j(\boldsymbol{u}_j^*) - (\boldsymbol{\eta}_i^{j,*})^\top \boldsymbol{g}_{loc}^j(\boldsymbol{u}_j^*), \tag{3.14}$$

since, at the upper level, it is assumed that each Problem (P_j) has been optimized and $\boldsymbol{g}_{loc}^j(\boldsymbol{u}_j^*) = \boldsymbol{x}(T_j) - \boldsymbol{\xi}_j = \mathbf{0}$. The gradient (3.11) and the first two terms on the right hand side of (3.12) arise through the dependence of $g_i^j(\boldsymbol{u}_j^*) - (\boldsymbol{\eta}_i^{j,*})^\top \boldsymbol{g}_{loc}^j(\boldsymbol{u}_j^*)$ on the initial condition (3.3) and follow from standard gradient formulae (see, for example, Teo et al. (1991)). The first term on the right hand side of (3.13) is obvious. The second term on the right hand side of (3.13) and the third term on the right hand side of (3.12) can be found by direct differentiation of $g_i^j(\boldsymbol{u}_j^*) - (\boldsymbol{\eta}_i^{j,*})^\top \boldsymbol{g}_{loc}^j(\boldsymbol{u}_j^*)$ with respect to $\boldsymbol{\xi}_j$. □

The question arises as to whether we can always guarantee the existence of feasible solutions to Problems (P_j), $j = 1, \ldots, N_u$. The answer is clearly no and this leads to some difficulties in the numerical implementation of the algorithm, to be addressed in the next section.

4 IMPLEMENTATION

Any numerical implementation of the proposed approach requires the solution of Problems (P_j), which must be discretized by some means. For our numerical examples in the next section, we use a simple Euler scheme to discretize the dynamics. In practice, a more accurate higher order Runge Kutta method would be appropriate. We assume that the control values at each time point in the discretization are variable and hence end up with a large number of control variables. With the cost and constraint functionals calculated via appropriate quadratures, the resulting discretized problem is then a mathematical programming problem. In order to use more efficient gradient based mathematical programming software, the gradients of the discretized cost and control functionals need to be calculated. Furthermore, to ensure good performance from such software, accurate values for the gradients are required. Rather than discretizing gradient formulae derived for the original continuous time problem, which can easily introduce some error if not carried out carefully, we choose to use the technique of automatic differentiation (see Griewank and Corliss (1991)) to calculate the gradients for the discretized problem. Automatic differentiation is a unified methodology for calculating the gradient of a function of many variables and it is particularly well suited to problems which arise from discretizing continuous time optimal control problems.

Since (P_u) and (P_j), $j = 1, \ldots, N_u$, are standard (possibly nonlinear) mathematical programming problems with known gradients, we can use any suitable software, such as `nlpql` (see Schittkowski (1985)) to solve them.

An issue that arises with the use of software such a `nlpql` is how accurate the final solution is desired to be. `nlpql` requires an input which specifies the final desired accuracy (in terms of the Kuhn-Tucker optimality criteria) for the problem. In our use of nlpql for the solution of both (P_u) and (P_j), we need to specify a desired accuracy for each of these. In practice we have found that the Problems (P_j) should be solved quite accurately (the accuracy required should be of the order of 10^{-10}) as their solution contributes to the gradients of Problem (P_u). These gradients, given in the theorem of the previous section, need to be accurate to allow an efficient solution of (P_u). The accuracy to which (P_u) is solved is up to the user. This then provides a stopping criterion for the overall algorithm.

As pointed out above, we cannot guarantee a feasible solution of (P_j) for any arbitrary $\boldsymbol{\xi}$. The following scheme is proposed to deal with this difficulty. We firstly need to assume the continuous dependence of optimal solutions to Problems (P_j), $j = 1, \ldots, N_u$, on $\boldsymbol{\xi}$. Suppose that, at a particular iteration in the optimization of (P_u), the current iterate $\boldsymbol{\xi}^k$ is such that all (P_j), $j = 1, \ldots, N_u$, are feasible. If the next iterate, $\boldsymbol{\xi}^{k+1}$, is such that one or more of the (P_j) become infeasible (*i.e.* there is no $\boldsymbol{u}_j \in \mathcal{U}_j$ such that $\boldsymbol{x}(T_j) = \boldsymbol{\xi}_j^{k+1}$ can be satisfied), we can go back to the previous iterate, artificially introduce upper and lower bounds on each component of $\boldsymbol{\xi}$ so that it is restricted to be within a distance ε of the corresponding component of $\boldsymbol{\xi}^k$, and then continue the optimization. If the next iterate still results in one or more infeasible (P_j), this step can be repeated with a smaller value of ε. The artificial bounds can be removed after the next successful iteration or kept active for a small number of iterations. To implement such an idea, it is necessary that the software used provides some means of reverse communication. Fortunately, `nlpql` does have such a feature. This approach of maintaining feasibility of the lower level problems is quite effective for the numerical examples we considered. It is rather crude, though, and does impact on the performance of the overall algorithm. A more elegant approach would be to try and adjust the line search for each iteration of the upper level optimization in some optimal fashion. Unfortunately, most optimization software, including `nlpql`, does not allow the user to alter the line search.

One difficulty that remains is to find an initial value for $\boldsymbol{\xi}$ which will result in a feasible set of (P_j), $j = 1, \ldots, N_u$. This is easily achieved if we use the idea of control parametrization to first solve a coarse version of the original Problem

(P). The control functions are parametrized as piecewise constant functions consistent with the coarse partition $\{T_0, T_1, T_2, \ldots, T_{N_u-1}, T_{N_u}\}$ (a piecewise linear or higher order parametrization may also be used if desirable). The resulting parametrized problem can then be solved quickly. From its solution, a value of $\boldsymbol{\xi}$ can be extracted, together with a set of control function values that can be used as initial guesses for the lower level Problems (P_j). This method has been implemented for the numerical examples in the next section.

Finally, the question arises as to how the coarse partition defining Problem (P_u) should be chosen, particularly in relation to the partitions used for the lower level problems. We investigate this question with the numerical examples in the next section.

5 NUMERICAL RESULTS

Example 1: This is a simple problem in which the aim is to heat a room in an efficient manner:

Minimize $g_0 = \frac{1}{2}\int_0^T \left[(x(t) - 15)^2 + (u(t))^2\right] dt$

subject to the dynamics $\dot{x}(t) = -x(t) + 2u(t)$, $t \in [0, T]$, $x(0) = 0$ and subject to the terminal state constraint $x(T) = 15$, *i.e.* $g_1 = x(T) - 15 = 0$. We can discretize the time horizon by choosing $h = T/N$, $t_i = ih$, and assume that the control is piecewise constant, *i.e.* $u(t) = u_i$, $t \in [t_i, t_{i+1}]$. The dynamics and cost can be integrated exactly with respect to this partition, yielding the following discretized problem:

Minimize

$$\begin{aligned} g_0 &= \sum_{i=0}^{N-1} hu_i^2 + h(2u_i - 15)^2 - 2e^{-h}(2u_i - 15)(x_i - 2u_i) \\ &\quad +2(2u_i - 15)(x_i - 2u_i) - \frac{(x_i - 2u_i)^2}{2}e^{-2h} + \frac{(x_i - 2u_i)^2}{2} \end{aligned}$$

subject to

$$x_{i+1} = 2u_i + (x_i - 2u_i)e^{-h}, \qquad i = 0, 1, \ldots, N-1, \quad x_0 = 0$$

and

$$g_1 = x_N - 15 = 0.$$

Next, we solve the discretized problem for $T = 3$ and with a variety of values of N. As N increases, the solution becomes a better approximation to the solution of the original continuous problem. As can be seen in Fig. 5.1, though, any additional accuracy comes at the expensive of more computation time. For $N > 1000$, the computation time becomes too excessive, considering the simplicity of the problem. Taking the $N = 1000$ case as a benchmark, we now apply the proposed two level method to the problem. If we have N_u intervals in the coarse partition at the upper level, we require N_l intervals in

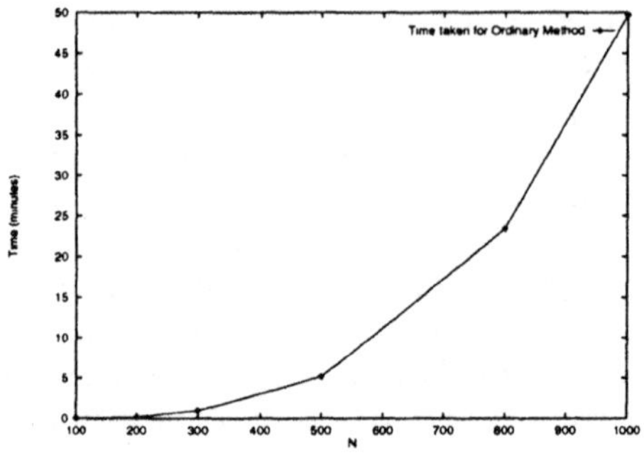

Figure 5.1 Computation times for Example 1 using different N.

the partitions for the lower level problems (different sized subproblems could be constructed without difficulty if so desired), such that $N_u \times N_l = N$. The results and computation times should then be comparable. We solved the two level problem for a variety of N_u and N_l values such that $N_u \times N_l = 1000$ and the final solution in each case was identical to the one obtained previously. Owing to the simplicity of the problem, no difficulties with feasibility of the lower level problems was encountered. Hence, it was not necessary to invoke the backtracking scheme mentioned in Section 4. A plot of the times taken for these computations, as a function of N_u, is shown in Fig. 5.2. Clearly, the two level method has little or no advantage for very small or very large values of N_u. In fact, the method performs worse than the standard approach for $N_u = 500$. This seems reasonable, as the two level method has introduced an additional terminal constraint for each lower level problem and there is the added cost of communication between the upper and lower levels. More importantly, the best computation time achieved by the two level method is 2 minutes and 40 seconds for the case of $N_u = 25$. This compares with a computation time of about 50 minutes for the ordinary approach. It would appear that the two level method performs best when $N_u = N_l \approx \sqrt{N}$ are chosen in this case.

The results in Fig. 5.1 seem to indicate a quadratic or higher order dependency of the required computational time on the size of the problem, N. This is in contrast to the approximately linear dependency of computational time on N shown in studies of similar problems in Ralph (1996) and Wright (1993) where interior point methods with efficient linear algebra were used to solve the underlying optimization problems. Our use of nlpql as a 'black box' routine is clearly not as efficient, but perhaps more representative of commonly used computational techniques in optimal control. It is for such techniques in particular that the proposed multilevel technique can yield very significant savings in computational time.

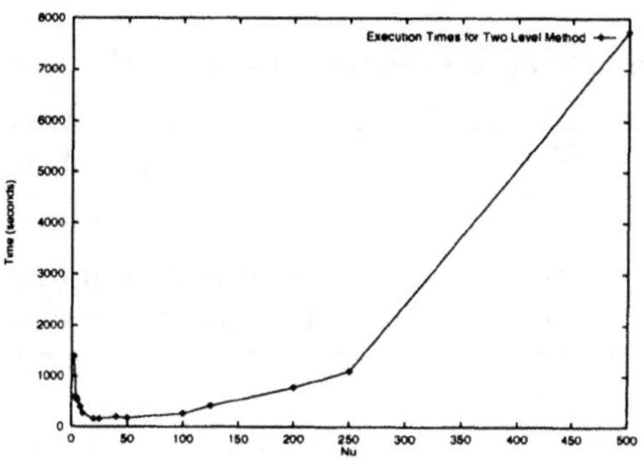

Figure 5.2 Computation times taken using the two level approach for Example 1.

Example 2: This is a well known problem, taken from Teo et al. (1991). It involves a continuous state inequality constraint. The problem is to minimize

$$g_0 = \int_0^1 \left[(x_1(t))^2 + (x_2(t))^2 + 0.005(u(t))^2\right] dt$$

subject to the dynamics

$$\begin{aligned} \dot{x}_1(t) &= x_2(t), & x_1(0) = 0 \\ \dot{x}_2(t) &= -x_2(t) + u(t), & x_2(0) = -1 \end{aligned}$$

and subject to the constraints

$$\begin{aligned} x_1(1) &= 0 \\ x_2(1) &= -1 \\ 8(t-0.5)^2 - 0.5 - x_2(t) &\geq 0 \quad \forall t \in [0,1] \end{aligned}$$

As for Example 1, we partition the time horizon and thereby discretize the problem. The dynamics are integrated by Euler's method and the integral in the cost functional is approximated by the trapezoidal rule. The continuous inequality constraint on the state is handled using a penalty function method described in Rehbock et al. (1992). Using a partition with $N = 900$ intervals, the resulting problem, solved as a standard mathematical programming problem, takes 33 hours, 8 minutes and 32 seconds. Applying the two level approach, we now encounter difficulties with the feasibility of the lower level problems and the backtracking scheme described in the previous section is invoked. It was found that most of the necessary backtracking occurs early in the optimization process. Once near the optimal solution, there is no further need for backtracking. The computation times shown in Table 5.2 were achieved. While not as significant as that achieved in Example 1, there is still a clear improvement in computation time when using the two level method. In this

Table 5.1 Computation times achieved by two level method for Example 2.

N_u	N_l	Computation Time
20	45	8 hours, 18 minutes, 11 seconds
30	30	5 hours, 27 minutes, 28 seconds
36	25	4 hours, 3 minutes, 47 seconds

example, it appears that N_u should be chosen somewhat larger than N_l for better performance of the two level method.

Note that so far we have not made use of the inherent parallelism of the two level method in either of the two examples.

6 PARALLEL IMPLEMENTATION

To test the performance of the proposed method in a parallel computing environment, we ran the method on a 4 processor Silicon Graphics Origin 2000 shared-memory computer. Only minor changes in the code were required to take advantage of the multiprocessor architecture. A lesson learnt early in these simulations was that the task of solving individual subproblems should not be shared across different processors, as the resulting increase in communication between processors is very time consuming. Hence the code was structured in such a way that only 4 subproblems are solved at any one time, one on each processor. Consequently, it seems reasonable to choose N_u to be a multiple of 4. The parallel implementation has one further difficulty in that additional communication between the upper lower levels is required to determine when a set of subproblems has been completed before the next step at the upper level is taken.

We solved Example 1 once more, this time with $N = 800$. Both the sequential and the parallel implementation of the two level method were tested for a range of N_u and N_l values, where $N_u \times N_l = 800$. The results are shown in Fig. 6.1, with the hashed line indicating the results obtained from the parallel implementation. As expected, the parallel implementation is considerably faster than the sequential one, although only by a factor of about 3, rather than 4, as one might expect. Clearly, the additional communication between the upper and lower levels does have some impact.

One would expect these computation times to improve as more processors are used, although Ralph (1996) found in the implementation of his algorithm that, for a large number of processors, the usefulness of parallelism diminishes due the additional communication required.

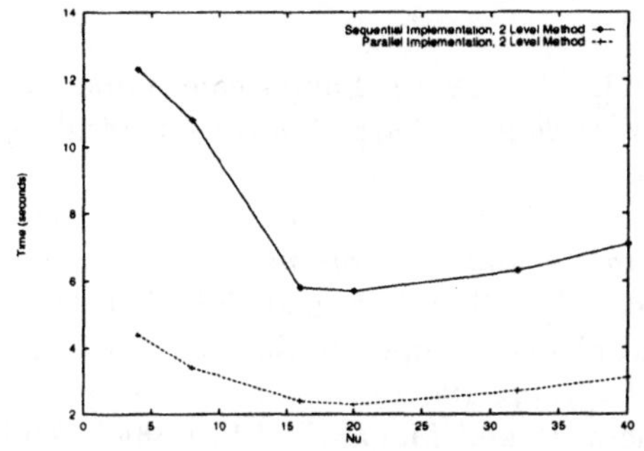

Figure 6.1 Computation times for Example 1 with sequential and parallel implementation.

7 CONCLUSIONS

We have proposed a method to decompose a general continuous time optimal control problem governed by ODEs. The decomposition results in a two level optimization problem where optimization with respect to a set of separator variables takes place at the upper level. Each iteration of the upper level optimization requires the solution of a set of lower level problem, where optimization is performed with respect to a set of control variables. The decomposed version of the problem can then be solved efficiently in a parallel computing environment, since the subproblems can be solved independently.

The proposed method has been successfully implemented in a number of numerical studies. Apart from the advantages through a parallel implementation, significant savings in computation time were also observed with a sequential implementation of the method.

More work is required to establish the convergence properties of the proposed method. A refinement of the upper level optimization to maintain the feasibility of lower level problems is also required to avoid excessive backtracking and improve the performance of the proposed method.

The possibility of decomposing a problem into more than two levels has been looked at, but such a decomposition is not as natural for optimal control problems as the two level decomposition and no performance improvement could be demonstrated.

Acknowledgments

The author wishes to thank Desmond Bylund for his extensive assistance with generating the numerical results in this paper and for his ideas towards the proposed algorithm. Furthermore, the author is grateful for useful comments from Daniel Ralph and an anonymous referee on a draft version of the paper.

References

Coleman, T.F., and Li, Y. (1989), Large-scale numerical optimization, *Proceedings of the Workshop on Large-Scale Numerical Optimization, Cornell University, Ithaca, October 19-20, SIAM.*

Griewank, A. and Corliss, G.F. (1991), *Automatic differentiation of algorithms: theory, implementation and applications*, SIAM, Philadelphia.

Jennings, L.S., Fisher, M.E., Teo, K.L. and Goh, C.J. (1991), MISER3: solving optimal control problems - an update, *Advances in Engineering Software and Workstations*, Vol. 13, pp. 190-196.

Markine, V.L., Meijers, P. and Meijaard, J.P. (1996), Multilevel optimization of the dynamic behaviour of a linear mechanical system with multi-point approximation, *Engineering Optimization*, Vol. 25, pp. 295-307.

Miele, A. and Wang, T.(1986), Primal-dual properties of sequential gradient-restoration algorithms for optimal control problems, part 2, general problem, *Journal of Mathematical Analysis and Applications*, Vol. 119, pp. 21-54.

Ralph, D. (1996), A parallel method for unconstrained discrete - time optimal control problems, *SIAM Journal of Optimization*, Vol. 6(2), pp. 488-512.

Rehbock, V., Teo, K.L. and Jennings, L.S. (1992), An exact penalty function approach to all-time-step constrained discrete-time optimal control problems, *Applied Mathematics and Computation*, Vol. 49, pp. 215-230.

Sakawa, Y., and Shindo, Y. (1980), On global convergence of an algorithm for optimal control, *IEEE Transactions on Automatic Control*, Vol. AC-25, pp. 1149-1153.

Schittkowski, K. (1985), NLPQL: A FORTRAN subroutine solving constrained nonlinear programming problems, *Annals of Operations Research*, Vol. 5, pp. 485-500.

Sirisena, H.R., and F.S. Chou, F.S. (1979), Convergence of the control parametrization Ritz method for nonlinear optimal control problems, *Journal of Optimization Theory and Applications*, Vol. 29, pp. 369-382.

Teo, K.L., Goh, C.J., and Wong, K.H. (1991), *A Unified Computational Approach to Optimal Control Problems*, (Longman Scientific and Technical, Essex, England, 1991).

Wismer, D.A. (1971), *Optimization Methods for Large Scale Systems with Applications*, McGraw-Hill, New York.

Wright, S.J. (1990), Solution of discrete-time optimal control problems on parallel computers, *Parallel Computing*, Vol. 16, pp. 221-238.

Wright, S.J. (1991), Partitioned dynamic programming for optimal control, *SIAM Journal on Optimization*, Vol. 1, pp. 620-642.

Wright, S.J. (1993), Interior point methods for optimal control of discrete time systems, *Journal of Optimization Theory and Applications*, Vol. 77, pp. 161-187.

8 MODIFIED DIRECT GRADIENT DESCENT CONTROL OF NONLINEAR SYSTEMS

K. Shimizu,

Department of System Design Engineering
Keio University, Yokohama, Japan

S. Ito

Institute of Statistical Mathematics
Tokyo, Japan

and K. Otsuka

Department of System Design Engineering
Keio University, Yokohama, Japan

Abstract: This paper concerns with Direct Gradient Descent Control (DGDC) of nonlinear systems. DGDC manipulates control inputs directly so as to decrease a performance function like the squared error, based on the gradient of the performance function and the steepest descent method. The convergence speed or the asymptotical stability of DGDC is improved by considering the time derivative of the performance function. Namely, in order to improve the convergence, we modify the DGDC by decreasing both the performance function and its time derivative. The proposed method can be implemented in a very simple and practical manner. Our simulation results with the modified DGDC showed remarkably good performance for various plants.

Key words: Nonlinear control system, dynamic state feedback control, steepest descent method, performance based stabilizing control.

X. Yang et al. (eds.), Optimization Methods and Applications, 139-149.

1 INTRODUCTION

Studies on nonlinear feedback control have been extensively made in recent years. Needless to say, stabilization and optimization are central concerns. Lyapunov stability theory and Hamilton–Jacobi–Bellman equation for optimal control appear to be main tools to design stabilizing feedback control law.

For affine nonlinear systems, lots of researches have been done based on the feedback linearization (Isidori (1995)), nonlinear optimal regulator (Moylan and Anderson (1973)), the Hamilton–Jacobi–Bellman equation and inverse optimality theory (Freeman and Kokotović (1996)), control Lyapunov function stabilization (Sontag (1990); Kristic et al. (1995)), backstepping technique (Kristic et al. (1995)), nonlinear H^{∞} control and passivity-based control theory (van der Schaft (1996)), etc.

Meanwhile, for general nonlinear systems, receding horizon control (Mayne and Michalska (1990)) is known as one of few studies on on-line nonlinear optimal control. Besides, there are many studies on neuro-controllers based on the error back propagation method (Saerens et al. (1996)), but the stability and generalization ability remain unsolved questions.

The direct gradient descent control (DGDC) was proposed by Shimizu et al. (1996) (see also Shimizu et al. (1998)), which directly manipulates control inputs so as to decrease a performance function such as the squared error from a desired equilibrium state, based on the gradient of the performance function with respect to the control input. The gradient is derived from sensitivity equations. Stability of DGDC can be proved by use of Lyapunov's direct method (Shimizu et al. (1998)) and of the comparison function method (Shimizu et al. (1999)). Similar Studies are found in Fradkov (1996) and in Vincent and Grantham (1997).

In this paper we propose a new control scheme to improve the stability and the convergence of the direct gradient descent control. For that purpose we consider to decrease both the performance function and its time derivative at the same time. This modified direct gradient descent control improves also the Lyapunov asymptotical stability in case of squared error performance function. The related research is found in Shimizu et al. (1999), where the convergence speed of DGDC is improved by introducing a new artificial control input and the second performance function, while the original control input is manipulated to decrease the first one.

The proposed method is of an on-line implementation and can be executed in a very simple and practical manner. It was seen from simulation results that the modified DGDC showed satisfactory improvement of the convergence properties.

2 DIRECT GRADIENT DESCENT CONTROL

Let us consider to control a general nonlinear plant. The aim of control is to manipulate a control input $\boldsymbol{u}(t)$ so that a performance function $F(\boldsymbol{x}(t), \boldsymbol{u}(t))$

decreases. Then the problem is written as follows:

$$\underset{\boldsymbol{u}(t)}{\text{decrease}}\ F\big(\boldsymbol{x}(t), \boldsymbol{u}(t)\big) \tag{2.1a}$$

$$\text{subj. to} \quad \dot{\boldsymbol{x}}(t) = \boldsymbol{f}\big(\boldsymbol{x}(t), \boldsymbol{u}(t)\big), \quad \boldsymbol{x}(0) = \boldsymbol{x}_0 \tag{2.1b}$$

where $\boldsymbol{x}(t) \in R^n$ is a state vector and $\boldsymbol{u}(t) \in R^r$ is a control vector. We assume that plant (2.1b) is locally controllable throughout the paper.

We make the following assumptions for problem (2.1):

Assumption 2.1 *$\boldsymbol{f}$ and F are continuously differentiable in $(\boldsymbol{x}, \boldsymbol{u})$;*

Assumption 2.2 *$\boldsymbol{f}_{\boldsymbol{u}}$, $F_{\boldsymbol{x}}$ and $F_{\boldsymbol{u}}$ are Lipschitz continuous.*

For any continuous $\boldsymbol{u}$: $\boldsymbol{u}(t)$, $t \geq 0$, plant (2.1b) has a unique continuous solution $\boldsymbol{x}$: $\boldsymbol{x}(t)$, $t \geq 0$. We denote the state trajectory $\boldsymbol{x}$ associated with a given $\boldsymbol{u}$ by $\boldsymbol{x}(\boldsymbol{u})$, whose value at t will be denoted by $\boldsymbol{x}(t; \boldsymbol{u})$.

As the class of admissible controls for an arbitrarily fixed t, we consider the space U_t consisting of the r-dimensional-vector-valued functions which are continuous on $[0, t)$, and define the following inner product:

$$\langle \boldsymbol{u}, \boldsymbol{v} \rangle_{U_t} := \int_0^t \boldsymbol{u}(\tau)^T \boldsymbol{v}(\tau) d\tau. \tag{2.2}$$

Let us define a functional $\phi^t : U_t \to R$ by

$$\phi^t[\boldsymbol{u}] := F\big(\boldsymbol{x}(t; \boldsymbol{u}), \boldsymbol{u}(t)\big) \tag{2.3}$$

for the fixed t. Then the following theorem holds.

Theorem 2.1 *Under Assumptions 2.1 and 2.2, the functional $\phi^t : U_t \to R$, defined by (2.3), is Gâteaux differentiable, and the gradient $\nabla\phi^t[\boldsymbol{u}] \in U_t$ at $\boldsymbol{u} \in U_t$ is given, at the final time t, by*

$$\nabla\phi^t[\boldsymbol{u}](t) = \nabla\boldsymbol{x}(t; \boldsymbol{u})(t) F_{\boldsymbol{x}}\big(\boldsymbol{x}(t; \boldsymbol{u}), \boldsymbol{u}(t)\big)^T + F_{\boldsymbol{u}}\big(\boldsymbol{x}(t; \boldsymbol{u}), \boldsymbol{u}(t)\big)^T \tag{2.4}$$

$$\textit{where} \quad \nabla\boldsymbol{x}(t; \boldsymbol{u})(t) = \boldsymbol{f}_{\boldsymbol{u}}\big(\boldsymbol{x}(t; \boldsymbol{u}), \boldsymbol{u}(t)\big)^T. \tag{2.5}$$

Proof. See Shimizu et al. (1996) or Shimizu et al. (1998).

Remark 2.1 *The gradient $\nabla\phi^t[\boldsymbol{u}] \in U_t$ represents the steepest ascent direction of ϕ^t at $u \in U_t$, and $\nabla\phi^t[\boldsymbol{u}](t)$ denotes its value at the right-end time t. See Shimizu et al. (1996) for the details.*

As an on-line control procedure for problem (2.1), we apply the steepest descent method at each time $t \in [0, \infty)$ by using $\nabla\phi^t[\boldsymbol{u}](t)$, i.e., the right-end value of the gradient $\nabla\phi^t[\boldsymbol{u}]$. Namely, $\boldsymbol{u}(t)$ is modified according to

$$\dot{\boldsymbol{u}}(t) = -K\nabla\phi^t[\boldsymbol{u}](t), \quad \boldsymbol{u}(0) = \boldsymbol{u}_0 \tag{2.6}$$

where $K := \mathrm{diag}(k_1, k_2, \ldots, k_r)$, $k_i > 0$, is a proportional coefficient matrix. We call the control law (2.6) the direct gradient descent control (DGDC).

Substituting (2.4) and (2.5) into (2.6) yields

$$\dot{\boldsymbol{u}}(t) = -K\{\boldsymbol{f}_{\boldsymbol{u}}(\boldsymbol{x}(t;\boldsymbol{u}), \boldsymbol{u}(t))^T F_{\boldsymbol{x}}(\boldsymbol{x}(t;\boldsymbol{u}), \boldsymbol{u}(t))^T + F_{\boldsymbol{u}}(\boldsymbol{x}(t;\boldsymbol{u}), \boldsymbol{u}(t))^T\}, \quad \boldsymbol{u}(0) = \boldsymbol{u}_0. \tag{2.7}$$

In order to realize this control, we let F satisfy the following assumption.

Assumption 2.3 *For every $i = 1, 2, \ldots, n$,*

$$F_{x_i}(\boldsymbol{x}, \boldsymbol{u}) \neq 0 \quad \forall (\boldsymbol{x}, \boldsymbol{u}) \neq (\boldsymbol{x}_d, \boldsymbol{u}_d)$$

where $\boldsymbol{x}_d$ is a desired equilibrium state and $\boldsymbol{u}_d$ is the corresponding control.

Let us set the performance index F as a quadratic form. Our purpose of control is to transfer the state $\boldsymbol{x}(t)$ to the desired equilibrium state $\boldsymbol{x}_d$. At the equilibrium state, it must hold that $\boldsymbol{0} = \boldsymbol{f}(\boldsymbol{x}_d, \boldsymbol{u}_d)$. In general, we can arbitrarily specify r of n components of $\boldsymbol{x}_d$, but the remaining $(n-r)$ components and $\boldsymbol{u}_d$ are dependently determined. We thus consider

$$F(\boldsymbol{x}(t), \boldsymbol{u}(t)) = \frac{1}{2}\{(\boldsymbol{x}_d - \boldsymbol{x}(t))^T Q(\boldsymbol{x}_d - \boldsymbol{x}(t)) + (\boldsymbol{u}_d - \boldsymbol{u}(t))^T R(\boldsymbol{u}_d - \boldsymbol{u}(t))\} \tag{2.8}$$

as the performance function to be decreased, where Q and R are positive definite matrices. Then the gradient is written as

$$\nabla\phi^t[\boldsymbol{u}](t) = -\boldsymbol{f}_{\boldsymbol{u}}(\boldsymbol{x}(t;\boldsymbol{u}), \boldsymbol{u}(t))^T Q(\boldsymbol{x}_d - \boldsymbol{x}(t;\boldsymbol{u})) - R(\boldsymbol{u}_d - \boldsymbol{u}(t)) \tag{2.9}$$

and hence the DGDC algorithm (2.6) is given by

$$\dot{\boldsymbol{u}}(t) = K\{\boldsymbol{f}_{\boldsymbol{u}}(\boldsymbol{x}(t;\boldsymbol{u}), \boldsymbol{u}(t))^T Q(\boldsymbol{x}_d - \boldsymbol{x}(t;\boldsymbol{u})) + R(\boldsymbol{u}_d - \boldsymbol{u}(t))\}, \quad \boldsymbol{u}(0) = \boldsymbol{u}_0. \tag{2.10}$$

The stability of direct gradient descent control is proved in Appendix by use of Lyapunov's direct method.

3 MODIFIED DIRECT GRADIENT DESCENT CONTROL

The direct gradient descent control is a scheme for manipulating the control input $\boldsymbol{u}(t)$ so as to decrease a performance function most rapidly based on the steepest descent method. It makes control action so as to decrease $F(\boldsymbol{x}(t), \boldsymbol{u}(t))$ by using the gradient $\nabla\phi^t[\boldsymbol{u}](t)$. Accordingly, it can be said that DGDC is a control achieving the best action at least momentarily and locally.

Now let us consider modifying DGDC to improve convergence properties. In order to improve the convergence, it is considered effective to decrease the time derivative dF/dt of the performance function F also. When one regards the squared error performance function $F(\boldsymbol{x},\boldsymbol{u})$ like (2.8) as a Lyapunov function, decreasing dF/dt is helpful for the improvement of Lyapunov asymptotical stability.

So let us consider the following problem:

$$\underset{\boldsymbol{u}(t)}{\text{decrease}}\ \omega_1 F\big(\boldsymbol{x}(t),\boldsymbol{u}(t)\big) + \omega_2 F_{\boldsymbol{x}}\big(\boldsymbol{x}(t),\boldsymbol{u}(t)\big)\boldsymbol{f}\big(\boldsymbol{x}(t),\boldsymbol{u}(t)\big) \tag{3.1a}$$

$$\text{subj. to}\quad \dot{\boldsymbol{x}}(t) = \boldsymbol{f}\big(\boldsymbol{x}(t),\boldsymbol{u}(t)\big),\quad \boldsymbol{x}(0)=\boldsymbol{x}_0 \tag{3.1b}$$

where $\omega_1, \omega_2 > 0$ are weighting coefficients. Here the second term of (3.1a) can be regarded as dF/dt for a fixed control parameter $\boldsymbol{u}(t)$. We take $F(\boldsymbol{x},\boldsymbol{u}) + F_{\boldsymbol{x}}(\boldsymbol{x},\boldsymbol{u})\boldsymbol{f}(\boldsymbol{x},\boldsymbol{u})$ as a new performance function and apply the theory of DGDC (Theorem 2.1) to problem (3.1). For the first term of (3.1a) we have the expression of its gradient function $\nabla\phi^t[\boldsymbol{u}](t)$, which has been obtained in (2.4) already:

$$\nabla\phi^t[\boldsymbol{u}](t) = \boldsymbol{f}_{\boldsymbol{u}}\big(\boldsymbol{x}(t),\boldsymbol{u}(t)\big)^T F_{\boldsymbol{x}}\big(\boldsymbol{x}(t),\boldsymbol{u}(t)\big)^T + F_{\boldsymbol{u}}\big(\boldsymbol{x}(t),\boldsymbol{u}(t)\big)^T. \tag{3.2}$$

Letting the second term of (3.1a) be

$$\psi^t[\boldsymbol{u}] := F_{\boldsymbol{x}}\big(\boldsymbol{x}(t;\boldsymbol{u}),\boldsymbol{u}(t)\big)\boldsymbol{f}\big(\boldsymbol{x}(t;\boldsymbol{u}),\boldsymbol{u}(t)\big)$$

we calculate its gradient function $\nabla\psi^t[\boldsymbol{u}](t)$. For this purpose, let $\widehat{F}(\boldsymbol{x},\boldsymbol{u}) := F_{\boldsymbol{x}}(\boldsymbol{x},\boldsymbol{u})\boldsymbol{f}(\boldsymbol{x},\boldsymbol{u})$ and calculate

$$\begin{aligned}\widehat{F}_{\boldsymbol{x}}(\boldsymbol{x},\boldsymbol{u}) = \frac{\partial}{\partial \boldsymbol{x}}\big(F_{\boldsymbol{x}}(\boldsymbol{x},\boldsymbol{u})\boldsymbol{f}(\boldsymbol{x},\boldsymbol{u})\big) &= \big(F_{\boldsymbol{x}\boldsymbol{x}}(\boldsymbol{x},\boldsymbol{u})\boldsymbol{f}(\boldsymbol{x},\boldsymbol{u})\big)^T + F_{\boldsymbol{x}}(\boldsymbol{x},\boldsymbol{u})\boldsymbol{f}_{\boldsymbol{x}}(\boldsymbol{x},\boldsymbol{u})\\ &= \boldsymbol{f}(\boldsymbol{x},\boldsymbol{u})^T F_{\boldsymbol{x}\boldsymbol{x}}(\boldsymbol{x},\boldsymbol{u})^T + F_{\boldsymbol{x}}(\boldsymbol{x},\boldsymbol{u})\boldsymbol{f}_{\boldsymbol{x}}(\boldsymbol{x},\boldsymbol{u})\end{aligned}$$

and

$$\begin{aligned}\widehat{F}_{\boldsymbol{u}}(\boldsymbol{x},\boldsymbol{u}) = \frac{\partial}{\partial \boldsymbol{u}}\big(F_{\boldsymbol{x}}(\boldsymbol{x},\boldsymbol{u})\boldsymbol{f}(\boldsymbol{x},\boldsymbol{u})\big) &= \big(F_{\boldsymbol{x}\boldsymbol{u}}(\boldsymbol{x},\boldsymbol{u})\boldsymbol{f}(\boldsymbol{x},\boldsymbol{u})\big)^T + F_{\boldsymbol{x}}(\boldsymbol{x},\boldsymbol{u})\boldsymbol{f}_{\boldsymbol{u}}(\boldsymbol{x},\boldsymbol{u})\\ &= \boldsymbol{f}(\boldsymbol{x},\boldsymbol{u})^T F_{\boldsymbol{x}\boldsymbol{u}}(\boldsymbol{x},\boldsymbol{u})^T + F_{\boldsymbol{x}}(\boldsymbol{x},\boldsymbol{u})\boldsymbol{f}_{\boldsymbol{u}}(\boldsymbol{x},\boldsymbol{u}).\end{aligned}$$

Consequently, we obtain

$$\begin{aligned}\nabla\psi^t[\boldsymbol{u}](t) &= \boldsymbol{f}_{\boldsymbol{u}}\big(\boldsymbol{x}(t),\boldsymbol{u}(t)\big)^T \widehat{F}_{\boldsymbol{x}}\big(\boldsymbol{x}(t),\boldsymbol{u}(t)\big)^T + \widehat{F}_{\boldsymbol{u}}\big(\boldsymbol{x}(t),\boldsymbol{u}(t)\big)^T\\ &= \boldsymbol{f}_{\boldsymbol{u}}\big(\boldsymbol{x}(t),\boldsymbol{u}(t)\big)^T \big\{F_{\boldsymbol{x}\boldsymbol{x}}\big(\boldsymbol{x}(t),\boldsymbol{u}(t)\big)\boldsymbol{f}\big(\boldsymbol{x}(t),\boldsymbol{u}(t)\big)\\ &\quad + \boldsymbol{f}_{\boldsymbol{x}}\big(\boldsymbol{x}(t),\boldsymbol{u}(t)\big)^T F_{\boldsymbol{x}}\big(\boldsymbol{x}(t),\boldsymbol{u}(t)\big)^T\big\}\\ &\quad + F_{\boldsymbol{x}\boldsymbol{u}}\big(\boldsymbol{x}(t),\boldsymbol{u}(t)\big)\boldsymbol{f}\big(\boldsymbol{x}(t),\boldsymbol{u}(t)\big) + \boldsymbol{f}_{\boldsymbol{u}}\big(\boldsymbol{x}(t),\boldsymbol{u}(t)\big)^T F_{\boldsymbol{x}}\big(\boldsymbol{x}(t),\boldsymbol{u}(t)\big)^T.\end{aligned} \tag{3.3}$$

From the above-mentioned, the modified DGDC is obtained as follows:

$$\dot{\boldsymbol{u}}(t) = -K\{\omega_1\nabla\phi^t[\boldsymbol{u}](t) + \omega_2\nabla\psi^t[\boldsymbol{u}](t)\},\quad \boldsymbol{u}(0)=\boldsymbol{u}_0$$

i.e.,

$$\begin{aligned}\dot{\boldsymbol{u}} = & -\omega_1 K\{\boldsymbol{f_u}(\boldsymbol{x},\boldsymbol{u})^T F_{\boldsymbol{x}}(\boldsymbol{x},\boldsymbol{u})^T + F_{\boldsymbol{u}}(\boldsymbol{x},\boldsymbol{u})^T\} \\ & -\omega_2 K\big[\boldsymbol{f_u}(\boldsymbol{x},\boldsymbol{u})^T\{F_{\boldsymbol{xx}}(\boldsymbol{x},\boldsymbol{u})\boldsymbol{f}(\boldsymbol{x},\boldsymbol{u}) + \boldsymbol{f_x}(\boldsymbol{x},\boldsymbol{u})^T F_{\boldsymbol{x}}(\boldsymbol{x},\boldsymbol{u})^T\} \\ & \qquad + F_{\boldsymbol{xu}}(\boldsymbol{x},\boldsymbol{u})\boldsymbol{f}(\boldsymbol{x},\boldsymbol{u}) + \boldsymbol{f_u}(\boldsymbol{x},\boldsymbol{u})^T F_{\boldsymbol{x}}(\boldsymbol{x},\boldsymbol{u})^T\big]\end{aligned} \tag{3.4}$$

where we make an additional assumption:

Assumption 3.1 *$\boldsymbol{f_x}$, $F_{\boldsymbol{xx}}$ and $F_{\boldsymbol{xu}}$ are Lipschitz continuous.*

In particular, when $F(\boldsymbol{x}(t),\boldsymbol{u}(t))$ is of a quadratic form

$$F(\boldsymbol{x}(t),\boldsymbol{u}(t)) = \frac{1}{2}(\boldsymbol{x}_d - \boldsymbol{x}(t))^T Q(\boldsymbol{x}_d - \boldsymbol{x}(t)) + \frac{1}{2}(\boldsymbol{u}_d - \boldsymbol{u}(t))^T R(\boldsymbol{u}_d - \boldsymbol{u}(t)) \tag{3.5}$$

where $Q, R > 0$, (3.4) becomes

$$\begin{aligned}\dot{\boldsymbol{u}} = & -\omega_1 K\{-\boldsymbol{f_u}(\boldsymbol{x},\boldsymbol{u})^T Q(\boldsymbol{x}_d - \boldsymbol{x}) - R(\boldsymbol{u}_d - \boldsymbol{u})\} \\ & -\omega_2 K\big[\boldsymbol{f_u}(\boldsymbol{x},\boldsymbol{u})^T\{Q\boldsymbol{f}(\boldsymbol{x},\boldsymbol{u}) - \boldsymbol{f_x}(\boldsymbol{x},\boldsymbol{u})^T Q(\boldsymbol{x}_d - \boldsymbol{x})\} \\ & \qquad - \boldsymbol{f_u}(\boldsymbol{x},\boldsymbol{u})^T Q(\boldsymbol{x}_d - \boldsymbol{x})\big].\end{aligned} \tag{3.6}$$

4 SIMULATION

Example 1 (Rayleigh). Let us consider the following Rayleigh model:

$$\dot{x}_1 = x_2 \tag{4.1a}$$

$$\dot{x}_2 = -x_1 + (1.4 - 0.14x_2^2)x_2 + 4u. \tag{4.1b}$$

Any equilibrium point (x_{d1}, x_{d2}, u_d) must satisfy $0 = x_{d2}$ and $0 = -x_{d1} + (1.4 - 0.14x_{d2}^2)x_{d2} + 4u_d$, from which we obtain $x_{d1} = 4u_d$ and $x_{d2} = 0$. We executed the modified DGDC with $K = I$, $\omega_1 = 3$, $\omega_2 = 1/2$, $Q = \left(\begin{smallmatrix}3 & 1\\ 1 & 3\end{smallmatrix}\right)$ and $r = 3/2$. The result is shown in Fig. 4.1 for initial values $\boldsymbol{x}(0) = (5,4)^T$, $u(0) = 5$, and desired values $\boldsymbol{x}_d = (0,0)^T$ and $u_d = 0$. For comparison, we show the simulation result by the basic DGDC without improvement ($\omega_2 = 0$) in Fig. 4.2. For different initial states $\boldsymbol{x}(0)$ distributed within the range $(-3,-4) \sim (5,4)$, Fig. 4.3 shows the trajectories in the state space by the same control parameters as described above. Every trajectory starting from an initial state within this range converges to the equilibrium point. This figure also demonstrates an advantage of the modified control scheme.

Example 2 (Single link manipulator). Next we consider a link of length $2l$ and weight m, at one end of which a torque $\boldsymbol{u}(t)$ is added as a control input. Then a single link manipulator is modeled by the following state equation:

$$\dot{x}_1 = x_2 \tag{4.2a}$$

$$\dot{x}_2 = -\frac{D}{I}x_2 + \frac{mlg}{I}\sin x_1 + \frac{1}{I}u \tag{4.2b}$$

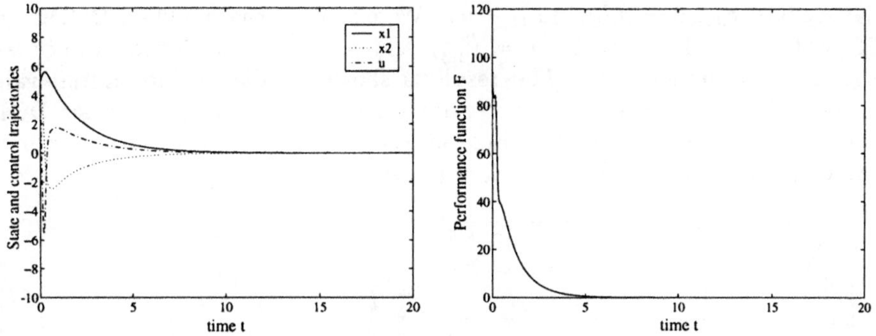

Figure 4.1 Rayleigh model with the modified DGDC

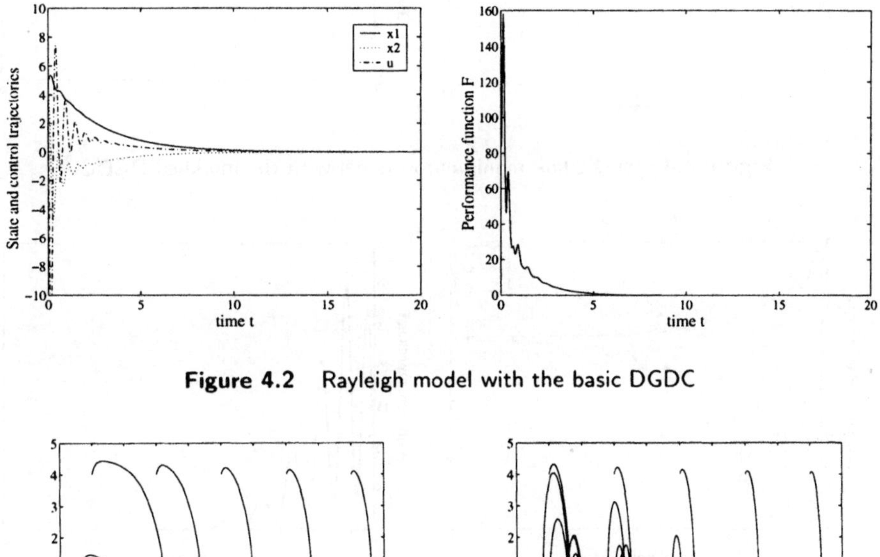

Figure 4.2 Rayleigh model with the basic DGDC

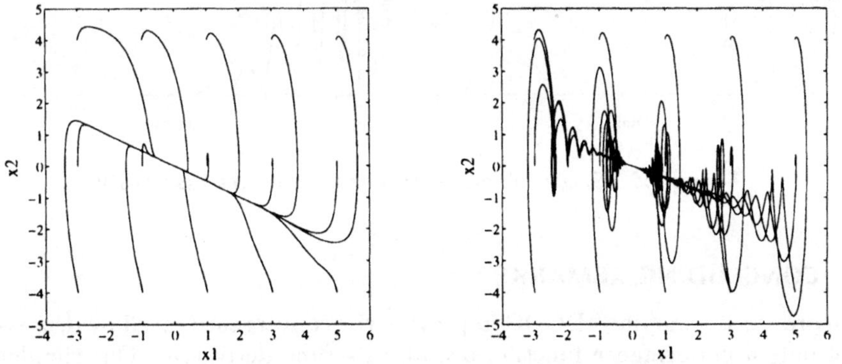

Figure 4.3 State trajectories by the modified (left) and basic (right) DGDC

where $l = 1/2$, $m = 1$, $I = 1/3$, $D = 0.00198$, $g = 9.81$, x_1 is the angle of rotation, I is the moment of inertia of the link, and D is the viscous friction coefficient at the end of the link. Any equilibrium point $(x_{d1},\, x_{d2},\, u_d)$ must satisfy $0 = x_{d2}$, $0 = mlg\sin x_{d1} + u_d$. We executed the modified DGDC with $K = I$, $\omega_1 = 1$, $\omega_2 = 1$, $Q = \begin{pmatrix} 3 & 2 \\ 2 & 3 \end{pmatrix}$ and $r = 1$. The desired values were set as $\boldsymbol{x}_d = \mathbf{0}$, $u_d = 0$. The result is shown in Fig. 4.4 for initial values $\boldsymbol{x}(0) = (3\pi/4, 0)^T$, $u(0) = 0$. For comparison, the result by the basic DGDC ($\omega_2 = 0$) is shown in Fig. 4.5. Fig. 4.6 shows the state trajectories for initial states given in the range $(-\pi, -3) \sim (\pi, 3)$.

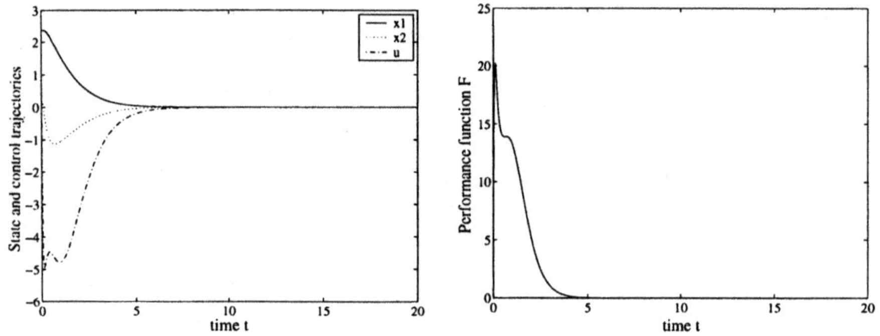

Figure 4.4 Single link manipulator model with the modified DGDC

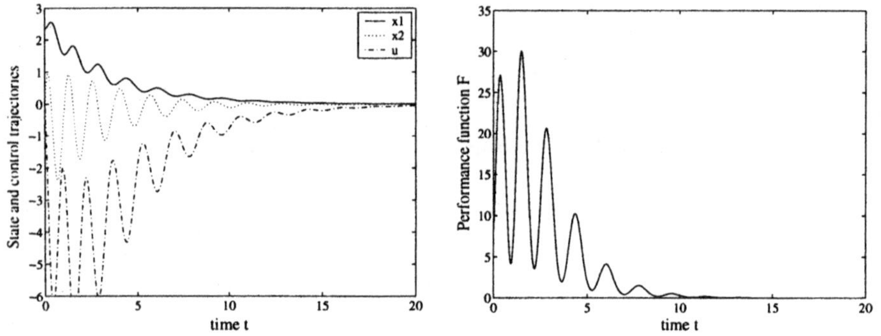

Figure 4.5 Single link manipulator model with the basic DGDC

5 CONCLUDING REMARKS

We proposed a modified DGDC improving the convergence speed by decreasing not only a performance function but also its time derivative. Our simulation results for various plants showed remarkably good performance, and the effectiveness of the modified DGDC was confirmed with many examples.

References

Fradkov, A. L. (1996), Swing control of nonlinear oscillations, *Int. J. Contr.*, Vol. 64, pp. 1189-1202.

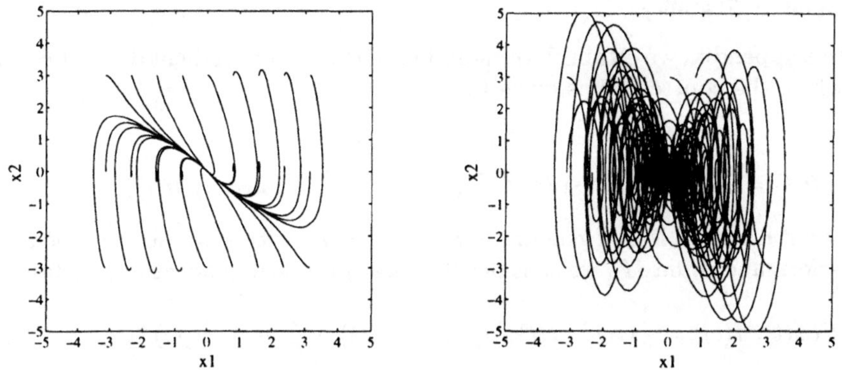

Figure 4.6 State trajectories by the modified (left) and basic (right) DGDC

Freeman, R. A. and Kokotović, P. V. (1996), *Robust Nonlinear Control Design: State-Space and Lyapunov Techniques*, Birkhäuser, Boston.

Isidori, A. (1995), *Nonlinear Control Systems*, 3rd ed., Springer-Verlag, London.

Kristic, M., Kanellakopoulos, I. and Kokotović, P. (1995), *Nonlinear and Adaptive Control Design*, Wiley, New York.

Mayne, D. Q. and Michalska, H. (1990), Receding horizon control of nonlinear systems, *IEEE Trans. Automat. Contr.*, Vol. 35, pp. 814-824.

Moylan, P. J. and Anderson, B. D. O. (1973), Nonlinear regulator theory and an inverse optimal control problem, *IEEE Trans. Automat. Contr.*, Vol. 18, pp. 460-465.

Saerens, M., Renders, J. M. and Bersini, N. (1996), Neuro control based on the backpropagation algorithm, in *Intelligent Control Systems* (Gupta, M. M. and Sinha, N. K., eds.), IEEE Press, New York.

Shimizu, K., Nukumi, H. and Ito, S. (1996), Direct steepest descent control of nonlinear dynamical systems, in *Nonlinear Control Systems Design 1995* (Krener, A. J. and Mayne, D. Q., eds.), Pergamon, Oxford, pp. 801-806.

Shimizu, K., Ito, S. and Suzuki, S. (1998), Tracking control of general nonlinear system by direct gradient descent method, in *Proceedings of the IFAC Nonlinear Control Systems Design Symposium*, Vol. 1, pp. 185-190.

Shimizu, K., Otsuka, K. and Naiborhu, J. (1999), Improved direct gradient descent control of general nonlinear systems, in *Proceedings of the 1999 European Control Conference* (CD-ROM), No. F676.

Sontag, E. D. (1990), *Mathematical Control Theory: Deterministic Finite Dimensional Systems*, Springer-Verlag, New York.

van der Schaft, A. (1996), *L_2-Gain and Passivity Techniques in Nonlinear Control*, Springer-Verlag, London.

Vincent, T. L. and Grantham, W. J. (1997), *Nonlinear and Optimal Control Systems*, Wiley, New York.

Appendix: Stability

In this appendix, we establish the stability of the direct gradient descent control, in which the control law is given by

$$\dot{\boldsymbol{x}}(t) = \boldsymbol{f}(\boldsymbol{x}(t), \boldsymbol{u}(t)) \tag{A.1}$$

$$\dot{\boldsymbol{u}}(t) = -K\{\boldsymbol{f}_{\boldsymbol{u}}(\boldsymbol{x}(t), \boldsymbol{u}(t))^T F_{\boldsymbol{x}}(\boldsymbol{x}(t), \boldsymbol{u}(t))^T + F_{\boldsymbol{u}}(\boldsymbol{x}(t), \boldsymbol{u}(t))^T\}. \tag{A.2}$$

Without essential loss of generality, we set $K = kI$ with $k > 0$ for simplicity. As a performance index F, we consider the most practical quadratic error function

$$F(\boldsymbol{x}(t), \boldsymbol{u}(t)) = \frac{1}{2}(\boldsymbol{x}_d - \boldsymbol{x}(t))^T Q(\boldsymbol{x}_d - \boldsymbol{x}(t)) + \frac{1}{2}(\boldsymbol{u}_d - \boldsymbol{u}(t))^T(\boldsymbol{u}_d - \boldsymbol{u}(t)) \tag{A.3}$$

where $Q > 0$ and $(\boldsymbol{x}_d, \boldsymbol{u}_d)$ is a desired equilibrium point, and assume:

Assumption 0.1 *Plant (A.1) is Lyapunov asymptotically stable for the fixed $\boldsymbol{u}_d$. That is, for a Lyapunov function $V_1(\boldsymbol{x}) = \frac{1}{2}(\boldsymbol{x}_d - \boldsymbol{x})^T Q(\boldsymbol{x}_d - \boldsymbol{x})$, there exists a positive definite function σ such that*

$$V_{1\boldsymbol{x}}(\boldsymbol{x})\boldsymbol{f}(\boldsymbol{x}, \boldsymbol{u}_d) = -(\boldsymbol{x}_d - \boldsymbol{x})^T Q f(\boldsymbol{x}, \boldsymbol{u}_d) \leq -\sigma(\|\boldsymbol{x}_d - \boldsymbol{x}\|).$$

Assumption 0.2 *The function $V_{1\boldsymbol{x}}(\boldsymbol{x})\boldsymbol{f}(\boldsymbol{x}, \boldsymbol{u}) = -(\boldsymbol{x}_d - \boldsymbol{x})^T Q f(\boldsymbol{x}, \boldsymbol{u})$ is convex with respect to $\boldsymbol{u}$. (This always holds for affine nonlinear systems.)*

Assumption 0.1 is a sufficient condition for the internal stability of plant (A.1) when the input is fixed to $\boldsymbol{u}_d$, and this implies that the equilibrium point $\boldsymbol{x}_d$ of the plant $\dot{\boldsymbol{x}}(t) = \boldsymbol{f}(\boldsymbol{x}(t), \boldsymbol{u}_d)$ is asymptotically stable.

Under these assumptions, one can show the asymptotical stability of extended system (A.1) and (A.2) by means of Lyapunov's direct method in a similar way to Fradkov (1996). Let us consider a Lyapunov function candidate

$$V(\boldsymbol{x}, \boldsymbol{u}) = \frac{1}{2}k(\boldsymbol{x}_d - \boldsymbol{x})^T Q(\boldsymbol{x}_d - \boldsymbol{x}) + \frac{1}{2}(\boldsymbol{u}_d - \boldsymbol{u})^T(\boldsymbol{u}_d - \boldsymbol{u}) > 0 \qquad \forall(\boldsymbol{x}, \boldsymbol{u}) \neq (\boldsymbol{x}_d, \boldsymbol{u}_d).$$

For (A.1) and (A.2), the time derivative of $V(\boldsymbol{x}, \boldsymbol{u})$ is given by

$$\begin{aligned}\frac{dV(\boldsymbol{x}, \boldsymbol{u})}{dt} &= V_{\boldsymbol{x}}(\boldsymbol{x}, \boldsymbol{u})\boldsymbol{f}(\boldsymbol{x}, \boldsymbol{u}) - V_{\boldsymbol{u}}(\boldsymbol{x}, \boldsymbol{u})k\{\boldsymbol{f}_{\boldsymbol{u}}(\boldsymbol{x}, \boldsymbol{u})^T F_{\boldsymbol{x}}(\boldsymbol{x}, \boldsymbol{u})^T + F_{\boldsymbol{u}}(\boldsymbol{x}, \boldsymbol{u})^T\} \\ &= -k(\boldsymbol{x}_d - \boldsymbol{x})^T Q\boldsymbol{f}(\boldsymbol{x}, \boldsymbol{u}) - k(\boldsymbol{x}_d - \boldsymbol{x})^T Q\boldsymbol{f}_{\boldsymbol{u}}(\boldsymbol{x}, \boldsymbol{u})(\boldsymbol{u}_d - \boldsymbol{u}) \\ &\qquad - k(\boldsymbol{u}_d - \boldsymbol{u})^T(\boldsymbol{u}_d - \boldsymbol{u}).\end{aligned}$$

On the other hand, since, by Assumption 0.2, $V_{\boldsymbol{x}}(\boldsymbol{x}, \boldsymbol{u})\boldsymbol{f}(\boldsymbol{x}, \boldsymbol{u}) = kV_{1\boldsymbol{x}}(\boldsymbol{x})\boldsymbol{f}(\boldsymbol{x}, \boldsymbol{u}) = -k(\boldsymbol{x}_d - \boldsymbol{x})^T Q f(\boldsymbol{x}, \boldsymbol{u})$ is convex with respect to $\boldsymbol{u}$, we have

$$\begin{aligned}- k(\boldsymbol{x}_d - \boldsymbol{x})^T Q\boldsymbol{f}(\boldsymbol{x}, \boldsymbol{u}_d) &\geq -k(\boldsymbol{x}_d - \boldsymbol{x})^T Q\boldsymbol{f}(\boldsymbol{x}, \boldsymbol{u}) \\ &\qquad - k(\boldsymbol{x}_d - \boldsymbol{x})^T Q\boldsymbol{f}_{\boldsymbol{u}}(\boldsymbol{x}, \boldsymbol{u})(\boldsymbol{u}_d - \boldsymbol{u}).\end{aligned}$$

We thus obtain

$$\frac{dV(\boldsymbol{x},\boldsymbol{u})}{dt} \leq -k(\boldsymbol{x}_d-\boldsymbol{x})^T Q\boldsymbol{f}(\boldsymbol{x},\boldsymbol{u}_d) - k(\boldsymbol{u}_d-\boldsymbol{u})^T(\boldsymbol{u}_d-\boldsymbol{u}).$$

Since the first term of the righthand side is negative definite by Assumption 0.1, we have $dV(\boldsymbol{x},\boldsymbol{u})/dt < 0$ for all $(\boldsymbol{x},\boldsymbol{u}) \neq (\boldsymbol{x}_d,\boldsymbol{u}_d)$. The system (A.1) and (A.2) is hence asymptotically stable by the Lyapunov's theorem, i.e., $\boldsymbol{x}(t) \to \boldsymbol{x}_d$ and $\boldsymbol{u}(t) \to \boldsymbol{u}_d$ as $t \to \infty$.

9 COMPUTATION OF FEEDBACK CONTROL FOR INFINITE TIME OPTIMAL CONTROL PROBLEMS

K.H. Wong and G. Peter

Department of Computational & Applied Mathematics
The University of the Witwatersrand
South Africa

Abstract: In this paper, we use an interactive multivariate interpolation method proposed by T.A. Foley and G.M. Nielson in conjuction with a cardinal product spline approach to compute a suboptimal feedback controller for a class of infinite time optimal control problems. Two examples have been solved to illustrate the efficiency of the method.

Key words: infinite time problem, cubic spline interpolation scheme, finite approximation.

X. Yang et al. (eds.), Optimization Methods and Applications, 151-163.

1 INTRODUCTION

Due to many efficient schemes in the literature for solving optimal control problems, it is possible to efficiently compute open-loop optimal controls for optimal control problems with very general nonlinear dynamics. However, it is well known that open-loop optimal controllers are very sensitive to disturbances, and thus feedback controllers are much more desirable in practice because of their robustness properties.

Although the solution of the Hamilton-Jacobi-Bellman (HJB) equation considered by Ahmed (1988) and Teo, Goh and Wong (1991) provides an optimal feedback control law, the equation is very difficult to solve numerically. The computation of a feedback controller for a class of infinite time nonlinear regulator optimal control problems is investigated by Teo, Fisher and Moore (1993). By assuming a specific feedback structure for the controller, a sequence of finite time optimal control problems is constructed so that each of them is solvable as a constrained optimal parameter selection problem using existing numerical software. A computational algorithm for solving the infinite time problem sequentially is presented. Asymptotic stability of the controlled system is also investigated.

This same class of infinite time optimal control problems has also been considered by Teo, Wong and Yan (1995). A sequence of approximate problems is constructed, where each problem consists of a finite time problem (containing a nonlinear system) combined with an infinite time problem (containing a linear system) in a certain way. The linear system is obtained by linearizing the nonlinear system about the origin. A gradient-flow based algorithm is then used to solve these approximate problems. Asymptotic stability of the system is automatically guaranteed by the gradient-flow method.

A method for approximating the suboptimal feedback controller for a class of fixed terminal time optimal control problems involving general nonlinear systems is proposed by Lee, Teo and Rehbock (1995). Since the value of the optimal control at time t is characterized by the value of the optimal state $x^*(t)$ and time t, the optimal feedback controller $\Psi(.)$ should be a function of the state as well as time, i.e. $\mu^* = \Psi(y)$ where $y^T = \left[[x^*(t)]^T, t\right]$. The suboptimal feedback controller is approximated by first evaluating several open-loop optimal trajectories corresponding to different initial conditions using existing numerical software, and then applying an iterative multivariate interpolation method together with the cardinal product spline approach given by Rehbock, Teo and Jennings (1995) to the scattered optimal trajectory data.

In this paper, we first approximate the infinite time problem by a sequence of finite-time optimal control problem. We then use the computational method proposed by Lee, Teo and Rehbock (1995) to compute the feedback control laws for each of these finite time control problems. Since the structure of our method involves a cubic spline feedback intepolation, our method is much better than that used by Teo, Fisher and Moore (1993), which merely assumes a quadratic feedback. Similar to work done by Teo, Fisher and Moore (1993) we also investigate properties of asymptotic stability of the controlled system.

Two numerical examples are solved to illustrate the stability and robustness of the feedback controller in the presence of disturbances.

2 PROBLEM STATEMENT

Consider the following system of nonlinear differential equations over $[0, \infty]$:

$$\dot{x}(t) = f(x(t), u(t)) \tag{2.1a}$$

$$x(0) = x_0 \tag{2.1b}$$

where $x(t) \in R^n$ and $u(t) \in R^r$ are the state and control vectors respectively; $x_0 \in R^n$ is a given vector.

Let $\mathcal{U}$ be the class of all continuously differentiable functions defined on the domain

$$\{y|y = x(t) \in R^n, \quad t \in [0, \infty)\}$$

The problem, denoted by (P), is stated as follow:
Find a feedback control $u(y) \in \mathcal{U}$ such that the cost functional

$$J(u) = \int_0^\infty f_0(x(t), u(t))dt \tag{2.2}$$

is minimized over $\mathcal{U}$ subject to the system (2.1). Throughout this paper, we assume that the following conditions are satisfied:

Assumption A

(A1) f and f_0 are continuously differentiable functions on their domains.

(A2) f_0 is strictly increasing function with respect to $|x|$ and $|u|$ and $f_0(0,0) = 0$.

3 THE APPROXIMATION PROBLEM $(P(T))$

In order to synthesis the optimal feedback control law, we first need to approximate the control problem (P) by the problem $(P(T))$.

Let $\mathcal{U}[0,T]$ be the class of all continously differentiable functions defined on the co-domain

$$\{y|y = x(t) \in R^n, \quad t \in [0, T]\}$$

The approximate problem, denoted by $(P(T))$ is stated as follow:

Find a feedback control $u(y) \in \mathcal{U}[0,T]$ such that the cost functional

$$J(U) = \int_0^T f_0(x(t), \quad u(t))dt \tag{3.1}$$

is minimized over $\mathcal{U}[0,T]$ subject to the system (2.1).

4 SYNTHESIZING THE OPTIMAL FEEDBACK CONTROL LAW FOR THE APPROXIMATE PROBLEM $(P(T))$

For each approximate problem $(P(T))$, in order to synthesize the optimal control law (i.e. the function $\Psi_T(\cdot)$), we first need to determine the region in the state space over which the control is expected to be active. That is, we need to determine the fitting domain $\mathcal{D}_T$.

The method of determining the fitting domain $\mathcal{D}_T$ is similar to that given by Lee, Teo and Rehbock (1995). Once the fitting domain $\mathcal{D}_T$ has been determined, we then interpolate over $\mathcal{D}_T$ (i.e., we interpolate the data $y_{i,k}$, where $y_{i,k} = x^*(t_{i,k})$ and i, k belong to a pre-specified set) using an iterative multivariate interpolation method together with a cardinal product spline approach, (see Lee, Teo and Rehbock (1995)) in order obtain the function $\Psi_T(\cdot)$.

Thus, it is extremely important that the region of interpolation (i.e. the fitting domain $\mathcal{D}_T$) is determined correctly. The method of determining the fitting domain $\mathcal{D}_T$ is as follows:
For each given T, let $\mathcal{U}[0,T]$ be the class of admissible controls defined on $[0,T]$. Let $x^\mu(t;\xi,t_1)$ denote the solution of equation (1.1a) at time t resulting from the open-loop control $\mu(t) \in \mathcal{U}[0,T]$ which passes through $\xi \in \Re^n$ at time t_1. Let $\mathcal{K}$ be the set of all nonempty compact subsets of $\Re^n$. Let the set-valued mapping $\Xi : \mathcal{K} \times [0,T] \times [0,T] \to \mathcal{K}$ be defined as follows:

$$\Xi(\Gamma, t_1, t) = \left\{ x^{\mu^*}(t;\xi,t_1) | \xi \in \Gamma \text{ and } \mu^* = \arg\left(\min \hat{J}(u|\xi,t_1)\right) \right\} \quad \text{for } t_1 \leq t, \tag{4.1}$$

where

$$\hat{J}(\mu|\xi,t_1) = \int_{t_1}^{T} f_0\left(x^\mu(\tau;\xi,t_1),u\right) d\tau \tag{4.2}$$

Note that, if Γ contains a finite number of initial conditions, then $\Xi(\Gamma,0,t)$ consists of a finite number of points for every time $t \in [0,T]$.

We first use the control parameterization technique to compute a nominal open-loop optimal control $\mu^*_{nominal}(t)$ and the corresponding optimal trajectory $x^*_{nominal}(t)$, for $t \in [0,T]$. Since we are using numerical software to compute the nominal open-loop optimal control and the corresponding optimal trajectory, we can only obtain a finite number of discrete point-pairs along the nominal trajectory. Say that the time axis is partitioned into p partitions, and the resulting knot points are $\{t_0, t_1, t_2, ..., t_p\}$, where

$$0 = t_0 < t_1 < t_2 < \cdots < t_p = T. \tag{4.3}$$

Then the nominal open-loop optimal control and the corresponding optimal trajectory are given by $\mu^*_{nominal}(t_k)$ and $x^*_{nominal}(t_k)$, for $k = 0,1,2,...,p$, respectively. It is clear that,

$$x^*_{nominal}(t_k) = \Xi(x^*_{nominal}(0), 0, t_k), \tag{4.4}$$

and

$$\mu^*_{nominal}(t_k) = \Psi(x^*_{nominal}(t_k)), \quad \text{for } k = 0,1,2,...,p. \tag{4.5}$$

From this partition of the time axis, we choose a much less refined partition,

$$0 < t_{l_1} < t_{l_2} < \cdots < t_{l_q} < T, \tag{4.6}$$

where $l_i \in [0, 1, 2, ..., p]$ for $i = 1, 2, ..., q$ and $q \ll p$. We also choose a set of positive integers $\omega = \{\omega_1, \omega_2, ..., \omega_n\}$, and a set of positive real numbers $\mu = \{\mu_1, \mu_2, ..., \mu_n\}$.

Let $\vartheta_i = [-\omega_i, -\omega_i + 1, ..., -1, 0, 1, ..., \omega_i - 1, \omega_i]$, $\omega_i \in \omega$, $i = 1, 2, ..., n$. Define,

$$\Gamma_0 = \{x_{nominal}(0) + [z_1\mu_1, z_2\mu_2, ..., z_n\mu_n]^T | z_i \in \vartheta_i, \mu_i \in \mu\}, \tag{4.7}$$

and

$$\Gamma_i = \{x_{nominal}(t_{l_i}) + [z_1\mu_1, z_2\mu_2, ..., z_n\mu_n]^T | z_i \in \vartheta_i, \mu_i \in \mu\}, \quad \text{for } i = 1, 2, ..., q. \tag{4.8}$$

We then use the control parameterization technique to obtain $\Xi(\Gamma_0, 0, t), t \in [0, T]$, and $\Xi(\Gamma_i, t_{l_i}, t), t \in [t_{l_i}, T]$ for $i = 1, 2, ..., q$, as well as the corresponding open-loop optimal control for each individual trajectory. Since we are using numerical software to compute these open-loop optimal controls and the corresponding optimal trajectories, these trajectories are obtained in discrete data form. That is

$$\Xi(\Gamma_0, 0, t_{0,k}), \ t_{0,k} \in [0, T], \quad 0 = t_{0,0} < t_{0,1} < \cdots < t_{0,p} = T, \tag{4.9}$$

and

$$\Xi(\Gamma_i, t_{l_i}, t_{i,k}), \ t_{i,k} \in [t_{l_i}, T], \ t_{l_i} = t_{i,0} < t_{i,1} < \cdots < t_{i,p} = T, \ i = 1, 2, ..., q \tag{4.10}$$

Each element in these sets corresponds to an open-loop optimal control value, which are considered as function values of $\Psi^T(\cdot)$. The fitting domain is then defined as $\mathcal{D}_T = C_v(D_T)$, where $C_v(\cdot)$ denotes the convex hull of its argument and D_T is the scattered data set defined by

$$D_T = \left\{ \left\{ \bigcup_{k=0}^{P} \Xi(\Gamma_0, 0, t_{0,k}) \right\} \bigcup \left\{ \bigcup_{i=1}^{q} \bigcup_{k=0}^{p} \Xi(\Gamma_i, t_{l_i}, t_{i,k}) \right\} \right\} \tag{4.11}$$

If the fitting domain $\mathcal{D}_T$ is not wide enough for the fit, then the values of $w_i, i = 1, 2, ..., n$, should be increased.

Remark 4.1 In view of assumption (A2), it is clear that whenever $T \to \infty$, we have

$$\lim_{t \to \infty, t \leq T} x^{\mu^*}(t, \xi, t_1) = 0.$$

Remark 4.2 In view of Remark 4.1, it is clear that there exists $\hat{T} > 0$ such that $\mathcal{D}_T$ is **almost** a constant with respect to the Ecludean norm of R^n for all $T > \hat{T}$.

Remark 4.3 We need to use the property of the fitting domain $\mathcal{D}_T$ given in Remark 4.2 in order to establish some convergence properties of our method.

5 INTERPOLATION USING A CUBIC SPLINE APPROACH ON THE FITTING DOMAIN $\mathcal{D}_{\hat{T}}$

In this section, we give a brief summary of the fitting method given by Lee, Teo and Rehbock (1995) to fit a surface through all the function values on the fitting domain $\mathcal{D}_{\hat{T}}$. Once $\mathcal{D}_{\hat{T}}$ has been determined, we can select a rectangular grid $G_{\hat{T}}$ covering the fitting domain $\mathcal{D}_{\hat{T}}$. Suppose that

$$G_{\hat{T}} = [a_1, b_1] \times [a_2, b_2] \times \cdots \times [a_n, b_n], \tag{5.1}$$

and that each axis of the grid $[a_j, b_j]$ is partitioned into m subintervals of equal size. The associated $(m+7)^n$ knots are defined as

$$\begin{bmatrix} x_1^{-3} \\ \vdots \\ x_n^{-3} \end{bmatrix}, \begin{bmatrix} x_1^{-3} \\ \vdots \\ x_n^{-2} \end{bmatrix}, \cdots \begin{bmatrix} x_1^{m+3} \\ \vdots \\ x_n^{m+3} \end{bmatrix},$$

where

$$x_j^i = a_j + \frac{i(b_j - a_j)}{m}, \qquad i = -3, \ldots, m+3; \qquad j = 1, \ldots, n \tag{5.2}$$

Since the partition of each axis, is equally spaced, a product spline can be used to interpolate the function f. The product spline $S(x)$ defined for $x \in R^n$ is expressed in terms of a linear combination of the basis splines. That is

$$S(x) = \sum_{k \in I} \hat{c}_k \hat{\Omega}_k(x), \tag{5.3}$$

where

$$I = \{k \in Z^n : k(j) \in \{-1, 0, 1, \cdots, m, m+1\} \quad \text{for all} \quad j = 1, \ldots, n\}, \tag{5.4}$$

where the cubic basis spline function $\hat{\Omega}_k(x)$ is defined as

$$\hat{\Omega}_k(x) = \prod_{j=1}^{n} \Omega_{k(j)}(x_j), \tag{5.5}$$

where the one-dimensional spline function Ω is defined as

$$\Omega_{k(j)}(x_j) = \frac{1}{4(b_j - a_j)^3} \Big[(x_j - x_j^{k(j)-2})_+^3 - 4\left(x_j - x_j^{k(j)-1}\right)_+^3$$
$$+6\left(x_j - x_j^{k(j)}\right)_+^3 - 4\left(x_j - x_j^{k(j)+1}\right)^3 + \left(x_j - x_j^{k(j)+2}\right)_+^3 \Big] \tag{5.6}$$

and

$$(y)_+ = \begin{cases} y & \text{if} \quad y \geq 0 \\ 0 & \text{if} \quad y < 0 \end{cases} \tag{5.7}$$

The method of finding $\hat{c}_k$ in (5.3) is given by Lee, Teo and Rehbock (1995).

Remark 5.1 Although this cubic spline interpolation method is computational efficient, it has a major disadvantage in that it requires the function values at the knots of a rectangular grid rather than at scattered points as in the case of the shepherd surface.

Thus, we need to use the iterative multivariate interpolation method proposed by T.A. Foley and G.M. Nielson (see Lee, Teo and Rehbock (1995)) in conjunction with the cardinal product spline approach described above to compute the suboptimal feedback controller. The purpose of the iterative multivariate interpolation method is to obtain the function values at the rectangular grid from those given values at scattered points.

6 CREATING AN APPROXIMATE PROBLEM USING CUBIC SPLINE FEEDBACK STRUCTURE ON THE DOMAIN $\mathcal{D}_{\hat{T}}$

Once the domain $\mathcal{D}_{\hat{T}}$ has been determined, we can create an approximate problem defined as follows:
Let I and $\hat{\Omega}_k(x)$ be as defined in Section 4. Let us approximate the control $u \in \mathcal{U}$ by the following cubic spline feedback structure

$$u_j(t) = \sum_{k \in I} \hat{c}_{k,j} \hat{\Omega}_k(x(t)), \quad j = 1, \dots, r \tag{6.1}$$

where $\hat{c}_{k,j}$ are the spline coefficients of the jth component of the control. Thus, any control having the same structure of (6.1) is completely specified by the spline coefficients $\hat{c}_{k,j}$. For convenience, let these coefficients be referred to collectively as Θ, and let Θ be the set containing all such feedback elements.

For each $k \in I$, let

$$c_k = [\hat{c}_{k,1}, \dots, \hat{c}_{k,r}]^T, \tag{6.2}$$

The feedback system at $x = 0$ can be approximated as follows:

$$\dot{x} = A_L x + B_L \sum_{k \in I} c_k \Omega_k(x) \tag{6.3a}$$

where

$$A_L = \left. \frac{\partial f(x,u)}{\partial x} \right|_{x=0}, B_L = \left. \frac{\partial f(x,u)}{\partial u} \right|_{u=0} \tag{6.3b}$$

Let $\exists$ be the set containing all those feedback elements in Θ such that

$$u(x)\big|_{x=0} = 0$$

and the system at (6.3) is asymptotically stable. The last condition is equivalent to requiring that H_L is stabilizable, where H_L is the matric associated with the linear term of x in (6.3). We may now state our new optimal control problem, denoted by (Q), as follow:

Subject to the nonlinear dynamical system (1), find a feedback element $\theta^* \in \exists$ such that the cost functional

$$J(\theta) = \int_0^\infty f_0(x(t|\theta), u(t|\theta))dt \tag{6.4}$$

is minimized over $\exists$. To continue, we assume that the following additional condition is safised:

Assumption B

(B1) There exists a feedback element $\theta \in \exists$ such that

$$J(\theta) < \infty.$$

The last condition presumes controllability of the system (2.1) for the class of feedback controllers under consideration.

Remark 6.1 In view of assumption (B1), suppose $\theta \in \exists$ is such that $J(\theta) < \infty$, then for any $\varepsilon > 0$, there exists a $T > 0$ such that

$$\int_t^\infty f_0(x(\tau|\theta), u(\tau|\theta))d\tau < \varepsilon \tag{6.5}$$

for all $t > T$.

7 PROPERTIES OF ASYMPTOTIC STABILITY OF THE OPTIMAL CONTROL FEEDBACK CONTROL OF THE PROBLEM (Q)

Let θ^* be the optimal feedback element of the problem (Q). We want to establish certain results concerning local asymptotic stability of the controlled system with respect to the initial condition x_0 under the optimal feedback element θ^*.

Lemma 7.1 Suppose that assymptions (A) and (B) are satisfied. Let θ^* be the optimal feedback element of the problem (Q). Then for any $\varepsilon < 0$, there exists $T_1 > 0$ such that $|x(T_1|\theta^*)| < \varepsilon$ and $|u(T_1|\theta^*)| < \varepsilon$.

Proof: To prove this lemma, we assume the contrary. Then, there exists $\varepsilon_1 > 0$ such that either $|x(t|\theta^*)| > \varepsilon$, or $|u(t|\theta^*)|^2 > \varepsilon_1$

Then, in view of $(A2)$, we have for any $T_0 \in (0, \infty)$,

$$\int_{T_0}^\infty f_0(x(t|\theta^*), u(t|t\theta^*))dt = \infty. \tag{7.1}$$

This is clearly a contradition to Remark 6.1 and the fact that θ^* is the optimal feedback element corresponding to the problem (Q). Hence the proof is complete.

Theorem 7.1 Consider the system (2.1) under assumptions (A) and (B). Then the controlled system under the feedback element θ^* is locally asymptotically stable with respect to the initial condition x_0.

Proof: The proof is similar to that given in Theorem 3.1 of Teo, Fisher and Moore (1993).

8 FINITE TIME APPROXIMATION TO THE PROBLEM (Q)

The problem (Q) is an infinite time problem. We now propose a computation algorithm for solving problem (Q) sequentially as a sequence of finite time problem. To begin, we consider the following finite time optimal control problems for each $T > 0$.

Subject to the dynamical system (2.1), find a feedback element $\theta \in \exists$ such that the functional

$$J(\theta|T) = \int_0^T f_0(x(t|\theta), u(t|\theta))dt \tag{8.1}$$

is minimized over $\exists$. Let this problem be denoted by $(Q(T))$.

Theorem 8.1 Let θ^* and θ_T^* be optimal feedback elements from problems (Q) and $(Q(T))$ respectively. Suppose that Assumptions (A) and (B) are satisfied. Then, for any $\varepsilon > 0$, there exists $\hat{T} > 0$ such that

$$0 \leq J(\theta^*|T) - J(\theta_T^*|T) \leq \varepsilon \tag{8.2}$$

for all $T > \hat{T}$. Moreover, the corresponding controlled system under the feedback element θ_T^* is locally as symptotically stable with respect to the initial condition x_0.

Proof: The proof is similar to that given in Theorem 4.1 of Teo, Fisher and Moore (1993).

Theorem 8.2 Let $\hat{\theta}^T$ be the feedback element obtained by the cubic spline interpolation method described in Section 4 of the finite-time problem $(P(T))$ in section 2. Then there exists $T_2 > 0$ such that θ_T is also an optimal feedback element of the problem $(Q(T))$.

Hence, for all $T > T_2$.

Theorem 8.1 remains valid if θ_T^* is replaced by $\hat{\theta}_T$ in Theorem 7.1.

9 A PRACTICAL EXAMPLE

We shall solve the infinite time optimal control problem given as Example 5.1.

Consider the following nonlinear infinite time optimal control problem based on the Rayleigh equation:

$$\text{Minimize } J(u) = \frac{1}{2}\int_0^\infty \left\{[x_1(t)]^2 + 0.1[x_2(t)]^2 + 0.2[u(t)]^2\right\}dt \tag{9.1}$$

subject to the system:

$$\begin{aligned} \dot{x}_1(t) &= x_2(t) \\ \dot{x}_2(t) &= -x_1(t) + \left\{1.4 - 0.14[x_2(t)]^2\right\}x_2(t) + u(t) \end{aligned}$$

(9.2a)

with initial conditions:

$$x_1(0) = -5 \tag{9.2b}$$

$$x_2(0) = -5 \tag{9.2c}$$

We solve the problem $(Q(T))$ with $T = 15$. The suboptimal feedback control was computed using the following parameter values: $p = 101$, $q = 0$, $\omega_1 = \omega_2 = 2$, $\mu_1 = 1.0$ and $\mu_2 = 0.5$. The number of partitions for the control parameterization method was 10 and the number of iterations of Foley and Nielson's iterative method was 10. The software package MISER3 was used to compute the open-loop optimal control trajectory data. The dynamical system (9.2) was integrated using a fourth order Runge-Kutta method. The nominal open-loop and optimal feedback (closed-loop) trajectories of $|x|$ are shown in Figure 9.1.

In order to test the robustness property of the feedback controller, we then applied the feedback (closed-loop) and open-loop control laws already obtained to the dynamical system (9.2a) with the following disturbed initial conditions;

(a) $(x_1, x_2) = (-5, -4.5)$,

(b) $(x_1, x_2) = (-6, -5)$,

(c) $(x_1, x_2) = (-4.5, -5.25)$.

Table 9.1 shows the cost corresponding to the nominal case as well as the cost corresponding to cases (a), (b) and (c), when the open-loop and feedback control laws were used.

Table 9.1 Cost comparison for the nominal case as well as each hypothetical case.

	cost of applying the feedback control law	cost of applying the open-loop control law	optimal cost when the disturbances are considered
nominal	23.69	22.81	-
case(a)	22.65	59.77	21.83
case(b)	29.97	80.10	29.33
case(c)	21.46	58.70	20.55

Table 9.1 shows that the cost obtained when using the feedback control law is significantly closer to the nominal cost than the cost obtained when the open-loop control law was used.

Figure 9.1 shows the nominal open-loop trajectory of $|x|$ together with the feedback trajectory of $|x|$. These trajectories are very close to each other and are difficult to distinguish. Figures 9.2, 9.3 and 9.4 shows the nominal trajectory of $|x|$ together with the trajectories of $|x|$ resulting from using the open-loop and feedback (closed-loop) control laws for the cases (a), (b) and (c) respectively.

It is obvious that in all three cases that the feedback control law is far superior than the open-loop control law in terms of robustness, as the feedback trajectory of $|x|$ is very close to the nominal trajectory of $|x|$. The results show that, for disturbances in the initial conditions of the dynamical system, the feedback control law produces trajectories which are close to being optimal. In all cases, the open-loop control law is essentially useless when disturbances are present in the dynamical system.

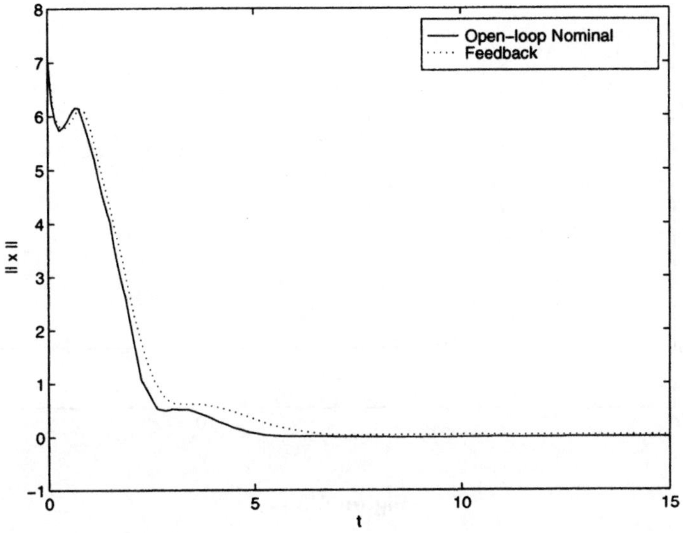

Figure 9.1

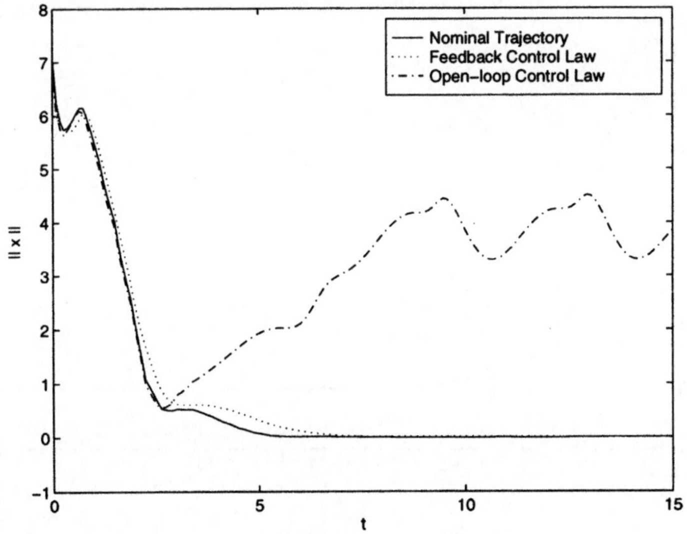

Figure 9.2

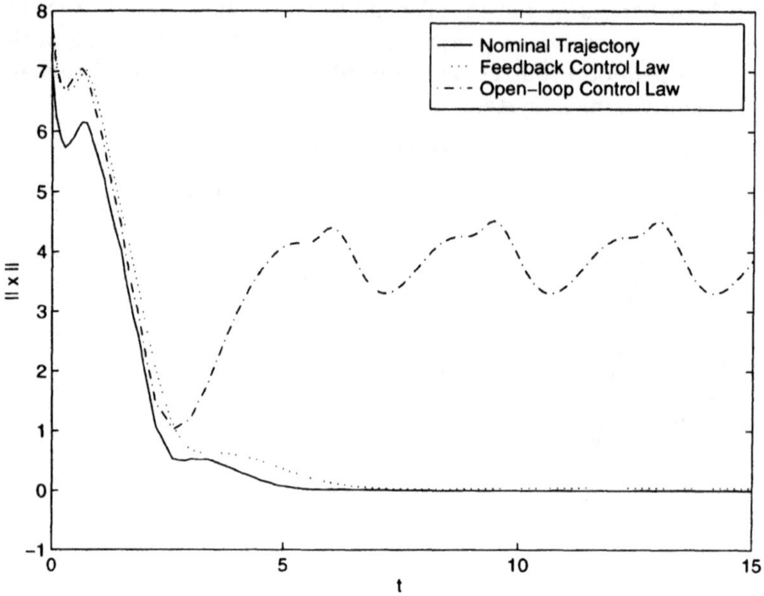

Figure 9.3

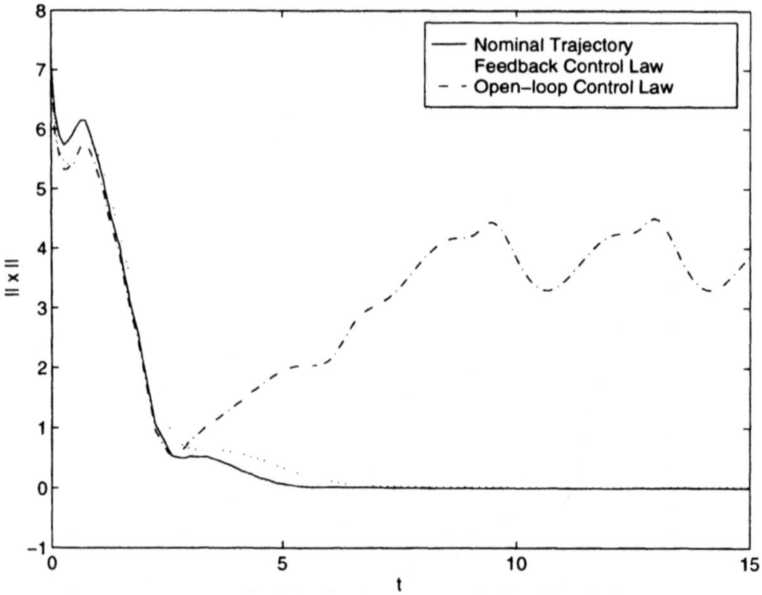

Figure 9.4

References

Ahmed, N.U. (1988), *Elements of Finite-Dimensional Systems and Control Theory,* Pitman Monographs and Surveys in Pure and Applied Mathematics 37, Longman Sciences and Technical.

Bryson, A.E., and Ho, Y.C. (1978), *Applied Optimal Control,* Hemisphere Publishing Corporation.

Edwards, N. (1991), *Synthesising Optimal Feedback Control with Neural Networks,* Honours disertation, The University of Western Australia.

Foley, T.A. (1979), *Smooth Multivariate Interpolation to Scattered Data,* Ph.D. dissertation, Arizona State University.

Foley, T.A., and Nielson, G.M. (1980), *Multivariate Interpolation to Scattered Data Using Delta Iteration,* Approximation Theory III.

Jennings, L.S., Fisher, M.E., Teo, K.L., and Goh, C.J. (1990), *MISER3 Optimal Control Software: Theory and User Manual,* EMCOSS Pty Ltd.

Lee, H.W.J., Teo, K.L., and Rehbock, V. (1995), *Sub-optimal local feedback control for a class of nonlinear control problems,* Dynamics of Continuous, Discrete and impulsive Systems, Vol. 1, pp. 37-51.

Narendra, K.S., and Parthasarathy, K. (1990), *Indentification and control of dynamical systems using neural networks,* IEEEE Trans. on Neural Networks, Vol. 1, No. 1.

Rehbock, V., Teo, K.L., and Jennings, L.S. (1995), *Suboptimal feedback control for a class of nonlinear systems using Ssline interpolation,* Discrete and Continuous Dynamical Systems, Vol. 1, No. 2.

Shepard, D. (1968), *A two-dimensional interpolation function for irregularly-spaced data,* Proceedings 23rd ACM National Conference.

Sirisena, H.R., and Chou, F.S. (1979), *convergence of the control parameterization Ritz method for nonlinear optimal control problems,* Journal of Optimization Theory and Applications, Vol. 29, pp. 369-382.

Teo, K.L., Fisher, M.E., and Moore, J.B. (1993), *A suboptimal feedback stabilizing controller for a class of nonlinear regulator problems,* Applied Mathematics and Computation, Vol. 59, pp. 1-17.

Teo, K.L., Goh, C.J., and Wong, K.H. (1991) *A Unified Computational Approach to Optimal Control Problems,* Longman Sciences and Technical.

Teo, K.L., Wong, K.H., and Yan, W.Y. (1995), *Gradient-flow approach for computing a nonlinear-quadratic optimal-output feedback gain matrix,* Journal of Optimization Theory and Applications, Vol. 85, No. 1 pp. 75-96.

Wong, K.H. (1988), *A control parameterization algorithm for nonlinear time-lag optimal control problems,* Journal of the Operation Research Society of India, Vol 25, No. 3, pp. 177-184.

II OPTIMIZATION METHODS

10 A HOMOGENIZED CUTTING PLANE METHOD TO SOLVE THE CONVEX FEASIBILITY PROBLEM

E. D. Andersen, J. E. Mitchell*, C. Roos and T. Terlaky†

Faculty of ITS, TWI/SSOR
TU Delft, 2628 CD Delft
The Netherlands

Abstract: We present a cutting plane algorithm for the feasibility problem that uses a homogenized self-dual approach to regain an approximate center when adding a cut. The algorithm requires a fully polynomial number of Newton steps. One novelty in the analysis of the algorithm is the use of a powerful proximity measure which is widely used in interior point methods but not previously used in the analysis of cutting plane methods. Moreover, a practical implementation of a variant of the homogenized cutting plane for solution of LPs is presented. Computational results with this implementation show that it is possible to solve a problem having several thousand constraints and about one million variables on a standard PC in a moderate amount of time.

Key words: Interior-point, homogeneous, cutting plane, set-partitioning.

*Permanent address: Math Sciences, RPI, Troy NY 12180, USA. Supported in part by fellowships from NWO and TU Delft
†Current address: Department of Computing and Software, McMaster University, 1280 Main Street West, Hamilton, Ontario, Canada L8S 4L7.

X. Yang et al. (eds.), Optimization Methods and Applications, 167-190.

1 INTRODUCTION

We are trying to find a point y in a convex set $C \subseteq \Re^m$. We assume that if C is nonempty then it contains a ball of radius ϵ, and we assume that $C \subseteq B := \{y \in \Re^m : -e \leq y \leq e\}$, where e denotes a vector of ones. The set C is defined by a *separating oracle:* given a point in $\Re^m$, the oracle will either state that the point is in C, or it will return a hyperplane that separates the point and C.

We will solve the feasibility problem by solving a sequence of linear programming relaxations. The initial relaxation will correspond to the box B, and this relaxation will be extended by adding separating hyperplanes returned by the oracle. Thus, we will always have the primal-dual pair of linear programming problems:

$$\begin{array}{llcl} \min & c^T x & & \\ \text{subject to} & Ax & = & 0 \\ & x & \geq & 0 \end{array} \qquad (P)$$

and

$$\begin{array}{lllll} \max & 0 & & & \\ \text{subject to} & A^T y & + & s = & c \\ & & & s \geq & 0, \end{array} \qquad (D)$$

where A is an $m \times n$ matrix and c, x, y and s are dimensioned appropriately. We assume $n \geq 2m$ and the constraints $A^T y \leq c$ contain all the constraints of the box B. Any point $y \in C$ will satisfy $A^T y \leq c$ in the current relaxation (D). We assume that the feasible region for (D) is defined by an oracle. Thus, we are going to solve this problem using a column generation method, where we add constraints to (D). We will assume that we always know feasible points $x > 0$ and (y, s) with $s > 0$ for the current relaxation — this assumption will be justified later. We will assume throughout, without loss of generality, that every column of A has norm 1.

Recent papers on interior point cutting plane algorithms include Goffin *et al.* (1996), who described a fully polynomial algorithm which added one constraint at a time, right through the current analytic center. This work has been extended in Ye (1997) and Goffin and Vial (1998), who considered adding many cuts simultaneously; in Ye (1997) the analytic center was recovered using a primal approach, while a primal-dual approach requiring the solution of a nonlinear programming problem was proposed in Goffin and Vial (1998). A simpler primal-dual updating approach with multiple cuts was proposed in Ramaswamy and Mitchell (1994), at the cost of weakening the cuts. Nesterov and Vial (1997) have described a homogeneous analytic center cutting plane algorithm with a very good complexity bound; this method is not a primal-dual method. Atkinson and Vaidya (1995) described a polynomial cutting plane algorithm that required that unimportant constraints be dropped; this was extended to a long-step method for optimization problems by Mitchell and Ramaswamy (1993). Other work on interior point cutting plane methods includes that of den Hertog *et al.* (1995).

We propose to solve the original linear programming problem by solving the homogenized self-dual linear programming problem

$$
\begin{array}{llllllll}
\min & & & & & \bar{d}\phi & & \\
\text{subject to} & & Ax & & & & = & 0 \\
 & -A^T y & & -s & +c\tau & & = & 0 \\
 & & -c^T x & & & -\kappa \quad +\bar{d}\phi & = & 0 \\
 & & & -\bar{d}\tau & & & = & -\bar{d} \\
 & & x, s, \tau, \kappa \geq 0, & & & & &
\end{array}
\qquad (HSDF)
$$

where $\bar{d}$ is a positive constant to be specified later. One novel aspect of our analysis is the use of a different proximity measure than those used in other interior point cutting plane algorithms.

In §2, we give some preliminaries on barrier functions and proximity measures. Our algorithm is described in §3. A method that requires $O(1)$ Newton steps to recover the analytic center after the addition of a cutting plane is described in §4. We show that the algorithm is fully polynomial in §5. In §6 we discuss a practical implementation of the homogenized cutting plane method for solution of LPs. In the following §7 we present our computational results for solving some large-scale set-partioning and set-covering problems. Finally, we offer our conclusions in §8.

Notation:

Given an n-vector x, we use X to denote the diagonal matrix with $X_{ii} = x_i$. The matrix S is defined similarly. As stated earlier, we use e to denote a vector of ones of an appropriate dimension.

2 BARRIER FUNCTIONS AND PROXIMITY MEASURES

The primal-dual logarithmic barrier function we use is

$$\Psi(x, s, \tau, \kappa) := x^T s + \tau\kappa - (n+1) - \sum_{i=1}^{n} \ln(x_i s_i) - \ln(\tau\kappa). \qquad (2.1)$$

Note that we will work throughout with a barrier parameter $\mu = 1$, so this parameter is omitted from our definitions of $\Psi(x, s, \tau, \kappa)$ and $\delta(x, s, \tau, \kappa)$.

It can be shown that $\Psi \geq 0$, with equality only at the analytic center. The analytic center is thus the point that minimizes Ψ, and at the analytic center we have $x_i s_i = 1$ for each i and $\tau\kappa = 1$.

One proximity measure is

$$\delta(x, s, \tau, \kappa) := 0.5 \parallel u - u^{-1} \parallel, \qquad (2.2)$$

where $u, u^{-1} \in \Re^{n+1}$, with components indexed from 0 to n, and

$$u_i \;:=\; \begin{cases} \sqrt{\tau\kappa} & \text{if } i = 0 \\ \sqrt{x_i s_i} & \text{otherwise} \end{cases} \qquad (2.3)$$

$$(u^{-1})_i \;:=\; \frac{1}{u_i}. \qquad (2.4)$$

Further, we define the n-vectors $\bar{u}$ and $\bar{u}^{-1} \in \Re^n$:

$$\bar{u}_i = u_i, \quad (\bar{u}^{-1})_i = \frac{1}{u_i} \quad \text{for } i = 1, \ldots, n. \tag{2.5}$$

We will call a feasible point (x, s, τ, κ) *approximately centered* if it satisfies the inequality $\delta(x, s, \tau, \kappa) < \theta$ for some constant θ to be specified later. We will have $0 < \theta < 0.5$.

Define

$$\psi(t) := t - \ln(1 + t) \quad \text{for } t > -1 \tag{2.6}$$
$$\rho(\delta) := \delta + \sqrt{1 + \delta^2}. \tag{2.7}$$

Lemma II.67 in Roos *et al.* (1997) allows us to relate $\Psi(x, s, \tau, \kappa)$ and $\delta(x, s, \tau, \kappa)$:

Lemma 2.1 *Let $\delta := \delta(x, s, \tau, \kappa)$. We have*

$$\psi(-2\delta/\rho(\delta)) \leq \Psi(x, s, \tau, \kappa) \leq \psi(2\delta\rho(\delta)).$$

We also have the following lemma concerning the performance of Newton's method for this algorithm when μ and ϕ are both fixed at one.

Lemma 2.2 *Let $\delta := \delta(x, s, \tau, \kappa)$.*

1. *(Theorem II.49, Roos* et al. *(1997).) If $\delta \leq \frac{1}{\sqrt{2}}$ then the primal-dual Newton step is feasible and $\delta(x^+, s^+, \tau^+, \kappa^+) \leq \delta^2$.*

2. *(Lemma II.70, Roos* et al. *(1997).) If $\delta \geq \frac{1}{\sqrt{2}}$ then the barrier function Ψ is decreased by at least $\frac{1}{6}$ if a damped primal-dual Newton step with a specified step length is taken.*

The global convergence analysis in §5 will use the *dual logarithmic barrier function:*

$$\Psi_D(s) := \sum_{i=1}^{n} \ln(s_i). \tag{2.8}$$

The more usual proximity measure is

$$\| Xs - e \| = \| \bar{U}(\bar{u} - \bar{u}^{-1}) \| .$$

The eigenvalues of $\bar{U}$ are the components of the vector $\bar{u}$; it follows that if $\bar{u}_i < 1 + \zeta$ for each i then $\| Xs - e \| \leq (1 + \zeta) \| \bar{u} - \bar{u}^{-1} \|$.

3 A COLUMN GENERATION METHOD

3.1 *Overview:*

Assume that the initial dual constraints are given by $-e \leq y \leq e$, where e denotes the vector of ones of the appropriate dimension. Thus, initially, $n = 2m$. The vector $\hat{y} = 0$ is then feasible in (D), and the vector $\hat{x} = e$ is feasible in (P). It should be noted that with this assumption, the dual feasible region is bounded, so if κ is nonzero at the optimal complementary pair then there exists a primal ray, and the primal problem has unbounded optimal value since it is feasible.

In order to satisfy the constraints of $(HSDF)$, we will use $\tau^0 = 1$ and $\kappa^0 = 1$ initially; we can then start with $y^0 = 0$ and $x^0 = e$. We can set $\bar{d} := n+1$. Since we are using a column generation method, $\bar{d}$ will change whenever the number of dual constraints changes. Using $s^0 = e$ this gives $c^T x^0 = n = (x^0)^T s^0$. With $\phi = 1$, we then have an analytic center, with $x_i s_i = 1$ for each component i and $\tau\kappa = 1$ also.

The algorithm can be summarized as follows:

1. Initialize with $\phi = 1$.

2. If the problem has too many dual constraints, STOP with infeasibility.

3. Find an approximate center for the current value of ϕ and with $\mu = 1$.

4. Call the separation oracle at the point $\frac{y}{\tau}$.

5. If cuts are found, add them through the current dual iterate, obtain a new feasible iterate with $x > 0$ and $s > 0$, and return to step 2.

6. Otherwise, STOP with feasibility.

Clearly, we need to define what we mean by "too many dual constraints" in Step 2; this will be discussed in §5. Step 5 is the subject of §4. Notice that ϕ is never changed in this algorithm. This is because we are solving the feasibility problem, and so there is no need to decrease the duality gap by decreasing ϕ.

3.2 *A direction for restarting*

Our old iterate (x^k, y^k, s^k) is on the boundary of the feasible region in both the primal and the dual spaces. Let the new primal problem be

$$\begin{array}{llllll} \min & c^T x & + & c_{n+1} x_{n+1} & & \\ \text{subject to} & Ax & + & a_{n+1} x_{n+1} & = & 0 \\ & & & x, x_{n+1} & \geq & 0 \end{array} \qquad (P^+)$$

where a_{n+1} is the new primal column, and x_{n+1} is the new primal scalar variable with objective function coefficient $c_{n+1} := a_{n+1}^T y^k$. Note that we add the new constraint in a central position, that is, right through the current dual iterate y^k. The new dual problem is

$$\begin{array}{lrcrcl} \max & 0 & & & & \\ \text{subject to} & A^T y & + & s & = & c \\ & a_{n+1}^T y & + & s_{n+1} & = & c_{n+1} \\ & & & s, s_{n+1} & \geq & 0, \end{array} \qquad (D^+)$$

where s_{n+1} is the slack for the additional constraint.

Since we add the constraint right through the current iterate, we can get a primal-dual feasible iterate by taking $x_{n+1} = s_{n+1} = 0$. Unfortunately, this is not an interior point, so we need to take a direction to obtain positive iterates. We can use directions d_x and d_y which move us to the points in the corresponding Dikin ellipsoids that maximize the new variables:

$$d_x = -D_P^2 A^T (A D_P^2 A^T)^{-1} a_{n+1} \tag{3.1}$$

$$d_y = -(A D_D^2 A^T)^{-1} a_{n+1}, \tag{3.2}$$

where D_P and D_D are diagonal matrices. We could use, for example, $D_P = X$, $D_D = S^{-1}$, or $D_P = D_D = D_{PD} := X^{0.5} S^{-0.5}$. We now restrict attention to the primal-dual scaling $D_P = D_D = D_{PD}$. This considerably helps the analysis in §4. Moving in these directions is guaranteed to give an interior point, provided we don't take too large a step. So we update for some positive scalars α_P and α_D:

$$x^+ := x + \alpha_P d_x \tag{3.3}$$

$$x_{n+1} := \alpha_P \tag{3.4}$$

$$y^+ := y + \alpha_D d_y \tag{3.5}$$

$$s^+ := s - \alpha_D A^T d_y \tag{3.6}$$

$$s_{n+1} := \alpha_D a_{n+1}^T (A D_{PD}^2 A^T)^{-1} a_{n+1}. \tag{3.7}$$

Using (3.6) and (3.2), we define

$$d_s := -A^T d_y = A^T (A D_{PD}^2 A^T)^{-1} a_{n+1}. \tag{3.8}$$

Using the orthogonality of $(d_x, 1)$ and $(d_s, d_{s_{n+1}})$, the reader easily verifies the following lemma.

Lemma 3.1 *We have* $d_x^T d_s = -a_{n+1}^T (A D_{PD}^2 A^T)^{-1} a_{n+1}$.

Thus, the quantity

$$\Delta_{PD} := \sqrt{a_{n+1}^T (A D_{PD}^2 A^T)^{-1} a_{n+1}} \tag{3.9}$$

will have an effect on the performance of the algorithm, both through affecting $d_x^T d_s$ and through determining the value of s_{n+1}. Methods that back off the additional constraint usually back it off by a quantity proportional to Δ_{PD}, or to the equivalent quantity using a different scaling matrix. (See, for example, Atkinson and Vaidya (1995) and Goffin *et al.* (1993).)

3.3 *Bounds on the step lengths*

The algorithm has found an approximate analytic center for $\phi = 1$, $\mu = 1$ when it adds a constraint. Therefore, the current iterates satisfy

$$0.5 \parallel u - u^{-1} \parallel \leq \theta \tag{3.10}$$

for some positive scalar $\theta < 0.5$, where u and u^{-1} are defined in equation (2.3).

Lemma 3.2 *If the iterates are approximately centered, then*

$$(1 - 2\theta) \leq x_i s_i \leq 1 + 2\theta(2\theta + 1) \tag{3.11}$$

for each component i. *Further, the duality gap is bounded:*

$$n - 2\theta(2\theta + 1) \leq x^T s \leq n + 2\theta.$$

Proof. We have

$$\begin{aligned} 4\theta^2 &\geq (u - u^{-1})^T(u - u^{-1}) \\ &= u^T u + (u^{-1})^T(u^{-1}) - 2(n+1) \\ &= (\sqrt{\tau\kappa} - \frac{1}{\sqrt{\tau\kappa}})^2 + \sum_{i=1}^{n}(x_i s_i + \frac{1}{x_i s_i} - 2). \end{aligned}$$

Since each term in the summand is nonnegative, each term must be less than the left hand side. This implies the following, multiplying by the positive term $x_i s_i$.

$$(x_i s_i)^2 - (2 + 4\theta^2)x_i s_i + 1 \leq 0$$

The roots of this quadratic are

$$1 + 2\theta^2 \pm \sqrt{(1 + 2\theta^2)^2 - 1} = 1 + 2\theta^2 \pm 2\theta\sqrt{\theta^2 + 1},$$

so $x_i s_i$ must lie between these two roots. We can simplify these expressions slightly by observing that $\sqrt{\theta^2 + 1} \leq \theta + 1$. We thus get the bounds:

$$1 - 2\theta \leq x_i s_i \leq 1 + 2\theta(2\theta + 1).$$

To bound the duality gap, notice that $\kappa\tau$ must also satisfy these inequalities. Since $\tau = 1$, we obtain a restriction on κ. From the definition of $(HSDF)$, we obtain

$$c^T x = -\kappa + \bar{d}\phi = n + 1 - \kappa,$$

since $\bar{d} = n + 1$ and $\phi = 1$. The result follows. ∎

If $\alpha_P > 0$ then $x_{n+1} > 0$. To ensure that the other components of x remain positive, it is necessary to keep $x + \alpha_P d_x > 0$. We need positivity of the dual slacks as well, so we need $s + \alpha_D d_s > 0$.

Lemma 3.3 *The following step length choices ensure positive iterates.*

1. *If* $0 < \alpha_P < \frac{\sqrt{1-2\theta}}{\Delta_{PD}}$ *then* $x + \alpha_P d_x > 0$.

2. *If* $0 < \alpha_D < \frac{\sqrt{1-2\theta}}{\Delta_{PD}}$ *then* $s + \alpha_D d_s > 0$.

Proof. To prove part 1, we have

$$\begin{aligned} x + \alpha_P d_x &= Xe - \alpha_P D_{PD}^2 A^T (A D_{PD}^2 A^T)^{-1} a_{n+1} \qquad \text{from (3.1)} \\ &= D_{PD}(X^{0.5} S^{0.5} e - \alpha_P D_{PD} A^T (A D_{PD}^2 A^T)^{-1} a_{n+1}) \\ &\geq D_{PD}(\sqrt{1-2\theta} - \alpha_P \Delta_{PD}) e \end{aligned}$$

where the last line follows from Lemma 3.2 and equation (3.9). Similarly, to show part 2, we have

$$\begin{aligned} s + \alpha_D d_s &= Se + \alpha_D A^T (A D_{PD}^2 A^T)^{-1} a_{n+1} \qquad \text{from (3.8)} \\ &= D_{PD}^{-1}(X^{0.5} S^{0.5} e + \alpha_P D_{PD} A^T (A D_{PD}^2 A^T)^{-1} a_{n+1}) \\ &\geq D_{PD}(\sqrt{1-2\theta} - \alpha_P \Delta_{PD}) e. \end{aligned}$$

The result follows. ∎

3.4 *Adding multiple cuts*

If we add more than one constraint, we can still get a new primal interior point fairly easily as in Mitchell and Todd (1992); Ye (1997):

1. Add together the additional columns to give a new column $\bar{a}$ and a corresponding variable $\bar{x}_{n+1}$.

2. Generate the direction d_x given above with $\bar{a}$ taking the place of a_{n+1} and $\bar{x}_{n+1}$ taking the place of x_{n+1}.

3. Disaggregate this additional variable, and set each of the new variables equal to the value of $\bar{x}_{n+1}$.

Unfortunately, it is not so easy to restart with the dual problem when adding multiple constraints. If only two constraints are added, then we can take d_y to be the average of the directions that would be calculated if only one of the constraints was added. If more than two constraints are added, one way to find a direction is by solving a small linear programming feasibility problem:

> Let $\bar{A}$ consist of the additional columns of A. Find a direction $\eta \geq e$ satisfying $\bar{A}^T (A D_D^2 A^T)^{-1} \bar{A} \eta \geq e$.

Notice that this is an LP in p variables, with p constraints, where p is the number of additional variables. Solving it with an interior point algorithm should result in a well-scaled direction if all the slacks are approximately equal. An appropriate direction is $d_y = -\bar{A}\eta$. Such a method is proposed in Ramaswamy and Mitchell (1994). A related, more complicated approach that requires approximately solving a nonlinear programming problem in p variables is proposed in Goffin and Vial (1998). The approach proposed in Goffin and Vial (1998) allows the recovery of an approximate analytic center for the modified problem in $O(p \ln(p))$ Newton steps.

4 CHOOSING A RESTART POINT

If the step lengths are chosen small enough, then the new x and s will still be approximately centered for the original n components. For the last component, we show that it is possible to bound an appropriate quantity, and that the distance from optimality will then be bounded by a constant. We can keep $\tau^+ = 1$. The constant $\bar{d}$ is increased by one, since the number of dual constraints is increased. We then set

$$\kappa^+ = \bar{d} - c^T x^+ - c_{n+1}x_{n+1} = \bar{d} - (x^+)^T s^+ - x_{n+1}s_{n+1}. \tag{4.1}$$

We will see that this will enable us to regain centrality in $O(1)$ iterations. The proof takes the following form:

- After taking our restart step to an interior point, the proximity measure is bounded by a constant. We prove this in §4.1.
- This implies, from Lemma 2.1 that the primal-dual logarithmic barrier function is also bounded by a constant.
- From Lemma 2.2, the logarithmic barrier function decreases by a constant as long as δ is large. Once δ is small, it will converge quadratically to zero. Therefore, a new approximate center can be found in $O(1)$ iterations. We discuss this further in §4.2.

4.1 *Updating the measure of centrality*

We had an approximately centered iterate before adding the constraint, so the proximity measure $\delta(x, s, \tau, \kappa) \leq \theta$, say. We define

$$\bar{\delta} :=\| \bar{u} - \bar{u}^{-1} \| \tag{4.2}$$

for this approximate center. Note that

$$\delta(x, s, \tau, \kappa)^2 = 0.5 \left[\bar{\delta}^2 + (\sqrt{\tau\kappa} - \frac{1}{\sqrt{\tau\kappa}})^2\right]. \tag{4.3}$$

After updating, the measure of centrality becomes

$$\begin{aligned}(\delta^+)^2 &:= (\delta((x^+, x_{n+1}), (s^+, s_{n+1}), \tau^+, \kappa^+))^2 \\ &= 0.5 \left[\| \bar{u}^+ - (\bar{u}^+)^{-1} \|^2 + (x_{n+1}s_{n+1} + \frac{1}{x_{n+1}s_{n+1}} - 2)\right. \\ &\qquad \left. + (\kappa^+ + \frac{1}{\kappa^+} - 2)\right],\end{aligned} \tag{4.4}$$

where $\bar{u}^+$ and $(\bar{u}^+)^{-1}$ are defined in the natural way using equations (2.3–2.5).

It will prove useful to use the same step length in the primal and dual problems, particularly in Lemma 4.2. Therefore, we take

$$\alpha_P = \alpha_D = \alpha = \frac{\beta}{\Delta_{PD}} \tag{4.5}$$

for an absolute constant β to be specified later, with $0 < \beta < \sqrt{1-2\theta}$.

Lemma 4.1 *After taking the specified step, the term in equation (4.4) corresponding to the new dual constraint satisfies*

$$x_{n+1}s_{n+1} + \frac{1}{x_{n+1}s_{n+1}} - 2 = \beta^2 + \frac{1}{\beta^2} - 2.$$

Proof. From equations (3.4), (3.7), and (4.5), we have

$$x_{n+1}s_{n+1} = \alpha_P \alpha_D \Delta_{PD}^2 = \beta^2.$$

The result follows. ∎

Let us define the vector

$$v := D_{PD}A^T(AD_{PD}^2A^T)^{-1}a_{n+1}. \tag{4.6}$$

Notice that

$$v^Tv = \Delta_{PD}^2 \tag{4.7}$$

and that

$$d_x = -D_{PD}v \text{ and } d_s = D_{PD}^{-1}v. \tag{4.8}$$

This observation enables us to simplify the expression for $\| \bar{u}^+ - (\bar{u}^+)^{-1} \|$ considerably, as shown in the next two lemmas.

Lemma 4.2 *The ith component of the complementary slackness is reduced to* $x_i^+ s_i^+ = x_i s_i - \alpha^2 v_i^2$ *for* $i = 1, \ldots, n$.

Proof. We have

$$\begin{aligned} x_i^+ s_i^+ &= (x_i + \alpha(d_x)_i)(s_i + \alpha(d_s)_i) \\ &= x_i s_i + \alpha x_i^{0.5} s_i^{0.5} v_i - \alpha x_i^{0.5} s_i^{0.5} v_i - \alpha^2 v_i^2 \\ &= x_i s_i - \alpha^2 v_i^2. \end{aligned}$$

∎

Lemma 4.3 *We have the following bound provided* $0 < \beta < \sqrt{1-2\theta}$*:*

$$\| \bar{u}^+ - (\bar{u}^+)^{-1} \|^2 \leq \bar{\delta}^2 + \gamma,$$

where

$$\gamma := \beta^2 \left(\frac{1}{(1-2\theta)(1-2\theta-\beta^2)} - 1 \right).$$

Proof. We have

$$\begin{aligned}
\| \bar{u}^+ - (\bar{u}^+)^{-1} \|^2 &= \sum_{i=1}^{n} \left[\sqrt{x_i s_i - \alpha^2 v_i^2} - \frac{1}{\sqrt{x_i s_i - \alpha^2 v_i^2}} \right]^2 \\
&= \sum_{i=1}^{n} \left[x_i s_i - \alpha^2 v_i^2 + \frac{1}{x_i s_i - \alpha^2 v_i^2} - 2 \right] \\
&= \sum_{i=1}^{n} (x_i s_i + \frac{1}{x_i s_i} - 2) - \alpha^2 \| v \|^2 \\
&\quad + \sum_{i=1}^{n} \left[\frac{1}{x_i s_i - \alpha^2 v_i^2} - \frac{1}{x_i s_i} \right] \\
&= \bar{\delta}^2 - \alpha^2 \Delta_{PD} + \sum_{i=1}^{n} \frac{\alpha^2 v_i^2}{(x_i s_i - \alpha^2 v_i^2)(x_i s_i)} \qquad \text{by (4.7)} \\
&\leq \bar{\delta}^2 - \beta^2 + \frac{\beta^2}{(1-2\theta)(1-2\theta-\alpha^2\Delta_{PD})} \\
&\qquad \text{by (4.5), (4.7) and Lemma 3.2.}
\end{aligned}$$

The result follows from a further application of (4.5). ∎

We can also bound the term in the proximity measure due to $\tau^+ = 1$ and κ^+.

Lemma 4.4 *If* $0 < \beta < \sqrt{1 - 2\theta}$ *then*

$$\kappa^+ + \frac{1}{\kappa^+} - 2 \leq 1 + (\kappa + \frac{1}{\kappa} - 2).$$

Proof. Now,

$$\begin{aligned}
\kappa^+ &= n + 2 - (x^+)^T s^+ - x_{n+1} s_{n+1} && \text{from (4.1)} \\
&= n + 2 - (x^T s - \beta^2) - \beta^2 && \text{from Lemmas 4.2 and 4.1} \\
&= n + 2 - x^T s \\
&= 1 + \kappa && \text{from } (HSDF) \text{ without the new constraint.}
\end{aligned}$$

Therefore,

$$\kappa^+ + \frac{1}{\kappa^+} = 1 + \kappa + \frac{1}{1+\kappa} < 1 + \kappa + \frac{1}{\kappa}.$$

The result follows. ∎

It is now possible to bound δ^+.

Theorem 4.1 *If* $\beta = \sqrt{0.5(1-2\theta)}$ *then* $(\delta^+)^2 \leq \theta^2 + 0.5(-1 + \frac{3}{1-2\theta})$.

Proof. From equation (4.4) and Lemmas 4.1, 4.3, and 4.4, we have

$$\begin{aligned}(\delta^+)^2 &= 0.5\left[\| \bar{u}^+ - (\bar{u}^+)^{-1} \|^2 + (x_{n+1}s_{n+1} + \frac{1}{x_{n+1}s_{n+1}} - 2) + (\kappa^+ + \frac{1}{\kappa^+} - 2)\right] \\ &\leq 0.5\left[\bar{\delta}^2 + \gamma + \beta^2 + \frac{1}{\beta^2} - 2 + 1 + \kappa + \frac{1}{\kappa} - 2\right] \\ &= (\delta(x, s, \tau, \kappa))^2 + 0.5\left[\gamma + \beta^2 + \frac{1}{\beta^2} - 1\right] \qquad \text{from (4.3).}\end{aligned}$$

Now,

$$\begin{aligned}\gamma + \beta^2 + \frac{1}{\beta^2} &= \beta^2\left[\frac{1}{(1-2\theta)(1-2\theta-\beta^2)} - 1\right] + \beta^2 + \frac{1}{\beta^2} \\ &= \frac{\beta^2}{(1-2\theta)(1-2\theta-\beta^2)} + \frac{1}{\beta^2} \\ &= \frac{3}{1-2\theta}\end{aligned}$$

with the specified choice of β. The result follows. ■

4.2 *Obtaining a new approximate analytic center*

From Theorem 4.1 and Lemma 2.1, the logarithmic barrier function value at the restart point is bounded above by a constant:

$$\Psi^+ := \Psi((x^+, x_{n+1}), (s^+, s_{n+1}), \tau^+, \kappa^+) \leq \psi(2\delta^+\rho(\delta^+)).$$

Notice that if we have an *exact* analytic center with $\theta = 0$ then the proximity measure for the new iterate is no larger than 1, and $\Psi^+ \leq 3.066$. The bound on the proximity measure is achieved if just one component of v is nonzero, in which case the new iterate has $x_{n+1}s_{n+1} = 0.5$, $x_i s_i = 0.5$ for exactly one other component, and $x_i s_i = 1$ for the remaining components. One possible choice for an approximate analytic center is $\theta = 0.25$, which gives $\delta^+ \leq 1.601$ and $\Psi^+ \leq 8.672$.

By using damped Newton steps while $\delta > \frac{1}{\sqrt{2}}$ and full Newton steps thereafter, we can then retrieve an approximate center in $O(1)$ steps.

5 GLOBAL CONVERGENCE

The analysis of Goffin *et al.* (1996) can be used directly to analyze the performance of this algorithm. In Goffin *et al.* (1996), upper and lower bounds are placed on the value of the dual logarithmic barrier function evaluated at the exact analytic center. After a fully polynomial number of iterations, the bounds cross. The lower bound uses the assumption that C contains a ball of radius ϵ. Thus, if the bounds cross then this assumption must be violated, so the algorithm can terminate with infeasibility.

From Lemma 3.2 and the discussion in §2, if an iterate is approximately centered for an appropriate θ in our framework, then it is also approximately centered for an appropriate centering parameter with the standard proximity measure. We add cuts in exactly the same manner as in Goffin *et al.* (1996), so the dual analytic center will behave identically. Thus, we require the same fully polynomial number of cutting planes as in Goffin *et al.* (1996). This gives the following theorem:

Theorem 5.1 *The upper bound on the number of constraints that need to be added before the algorithm can be terminated with infeasibility is* $O(\frac{m^2}{\epsilon^2})$, *and the number of Newton steps required has the same complexity.*

6 AN IMPLEMENTATION

The purpose of this section is to present a practical implementation of the homogenized cutting plane method. The practical implementation differs from the theoretical in several ways, because it has been adapted to solve linear programs directly and not only the feasibility problem. Hence, the problem of study is the primal LP

$$\begin{array}{lllll} \min & c^T x & & \\ \text{subject to} & Ax & = & b, \\ & x & \geq & 0 \end{array} \tag{6.1}$$

and the dual problem

$$\begin{array}{ll} \max & b^T y \\ \text{subject to} & A^T y \leq c, \end{array} \tag{6.2}$$

where A is an $m \times n$ matrix. We assume that c, b, x, and y are appropriately dimensioned.

In general we will assume that $n \gg m$ which for example is the case for LP relaxations of set-partitioning and set-covering problems occurring in practice, see Bixby *et al.* (1992) for an example. In that case it is impractical to solve (6.2) directly or may be impossible due to memory requirements.

The homogeneous and self-dual linear programming formulation (see Roos *et al.* (1997)) of the linear program (6.1) is

$$\begin{array}{lllllll} \min & & & & \bar{g}\phi & & \\ \text{subject to} & & Ax & -b\tau & -\bar{b}\phi & = & 0, \\ & -A^T y & & +c\tau & +\bar{c}\phi & \geq & 0, \\ & b^T y & -c^T x & & +\bar{d}\phi & \geq & 0, \\ & \bar{b}^T y & -\bar{c}^T x & -\bar{d}\tau & & = & -\bar{g}, \\ & & x \geq 0, & \tau \geq 0, & & & \end{array} \tag{6.3}$$

where

$$\begin{array}{rcl}\bar{b} & := & Ax^0 - b\tau^0, \\ \bar{c} & := & -c\tau^0 + A^T y^0 + s^0, \\ \bar{d} & := & c^T x^0 - b^T y^0 + \kappa^0, \\ \bar{g} & := & (x^0)^T s^0 + \tau^0 \kappa^0.\end{array}$$

One obvious choice is

$$(y^0, x^0, \tau^0, s^0, \kappa^0, \phi^0) = (0, e, 1, e, 1, 1), \tag{6.4}$$

where e is a vector of appropriate dimension containing all ones and subsequently it is the choice we will use. We have the following lemma:

Lemma 6.1 *The problem (6.3) always has an optimal solution with zero objective value. Let $(y^*, x^*, \tau^*, \phi^*)$ be any maximally complementary solution of (6.3), then $\tau^* > 0$ if and only if* (6.1) *has an optimal solution. Moreover, x^*/τ^* is an optimal solution to* (6.1).

The central path for the homogeneous model (6.3) is defined by

$$\begin{array}{ccccccl} & Ax & -b\tau & -\bar{b}\phi & & = & 0, \\ -A^T y & & +c\tau & +\bar{c}\phi & -s & = & 0, \\ b^T y & -c^T x & & +\bar{d}\phi & -\kappa & = & 0, \\ \bar{b}^T y & -\bar{c}^T x & -\bar{d}\tau & & & = & -\bar{g}, \\ & & & & Xs & = & \rho\frac{\bar{g}}{n+1}e, \\ & & & & \tau\kappa & = & \rho\frac{\bar{g}}{n+1}, \\ & x \geq 0, & \tau \geq 0, & & s, \kappa \geq 0, & & \end{array} \tag{6.5}$$

where ρ is an additional parameter. Note we have added the slack variables s and κ. It can be proved that this system of linear and nonlinear equations has an unique solution for all positive values of ρ. Moreover, the solution to this system converges to a maximally complementary solution of (6.3) for ρ converging to zero.

Now let $(y, x, \tau, s, \kappa, \phi)$ be any solution to (6.5) for a given ρ, then we have that

$$\begin{array}{rcl} 0 & = & y^T(Ax - b\tau - \bar{b}\phi) + x^T(-A^T y + c\tau + \bar{c}\phi - s) \\ & & +\tau(b^T y - c^T x + \bar{d}\phi - \kappa) + \phi(\bar{b}^T y - \bar{c}^T x - \bar{d}\tau + \bar{g}) \\ & = & -x^T s - \tau\kappa + \phi\bar{g} \\ & = & -e^T X s - \tau\kappa + \phi\bar{g} \\ & = & -\rho\bar{g} + \phi\bar{g} \\ & = & (\phi - \rho)\bar{g} \end{array} \tag{6.6}$$

which implies that $\phi = \rho$ for all solutions on the central path. Moreover, it is easy to verify that

$$-x^T s - \tau\kappa = \phi(\bar{b}^T y - \bar{c}^T x - \bar{d}\tau) \tag{6.7}$$

which implies

$$\begin{aligned} \bar{b}^T y - \bar{c}^T x - \bar{d}\tau &= -\tfrac{1}{\phi}(x^T s + \tau\kappa) \\ &= -\tfrac{\rho\bar{g}}{\phi} \\ &= -\bar{g}. \end{aligned} \tag{6.8}$$

Hence, for any solution on the central path, the fourth equation in (6.5) is redundant. Therefore, the central path equations (6.5) can be reduced to

$$\begin{array}{rrrrcl} & Ax & -b\tau & & = & \phi\bar{b}, \\ -A^T y & & +c\tau & -s & = & -\phi\bar{c}, \\ b^T y & -c^T x & & -\kappa & = & -\phi\bar{d}, \\ & & & Xs & = & \phi\frac{\bar{g}}{n+1}e, \\ & & & \tau\kappa & = & \phi\frac{\bar{g}}{n+1}, \\ & x \geq 0, & \tau \geq 0, & s, \kappa \geq 0, & & \end{array} \tag{6.9}$$

where ϕ is now seen as a parameter which should be reduced to zero. It should be noted that the elimination of the redundant equation in the central path definition also leads to more efficient computation of the search direction to be presented later.

Given our assumptions it will be costly to solve (6.9) using the usual path following method, because n is large. Therefore, a cutting plane (or a column generation) method is devised. Now let

$$\mathcal{J} \subseteq \{1, \ldots, n\}$$

and define the complementary set

$$\bar{\mathcal{J}} := \{1, \ldots, n\} \setminus \mathcal{J}. \tag{6.10}$$

If all the variables in $\bar{\mathcal{J}}$ are fixed i.e.

$$x_{\bar{\mathcal{J}}} = \phi \frac{\bar{g}}{n+1} \tag{6.11}$$

and hence the dual variables $s_{\bar{\mathcal{J}}}$ are free variables, then a reduced central path problem

$$\begin{array}{rrrrcl} & A_{\mathcal{J}} x_{\mathcal{J}} & -b\tau & & = & \phi(A_{\mathcal{J}} x^0_{\mathcal{J}} - b\tau^0), \\ -A^T_{\mathcal{J}} y & & +c_{\mathcal{J}}\tau & -s_{\mathcal{J}} & = & -\phi(-c_{\mathcal{J}}\tau^0 + A^T_{\mathcal{J}} y^0 + s^0_{\mathcal{J}}), \\ b^T y & -(c_{\mathcal{J}})^T x_{\mathcal{J}} & & -\kappa & = & -\phi(b^T y^0 - c^T_{\mathcal{J}} x^0_{\mathcal{J}} + \kappa^0), \\ & & & X_{\mathcal{J}} s_{\mathcal{J}} & = & \phi\frac{\bar{g}}{n+1}e, \\ & & & \tau\kappa & = & \phi\frac{\bar{g}}{n+1}e, \\ & x_{\mathcal{J}} \geq 0, & \tau \geq 0, & s_{\mathcal{J}}, \kappa \geq 0, & & \end{array} \tag{6.12}$$

is obtained. Note that this problem only requires information about the variables in $\mathcal{J}$. Moreover, given that $|\mathcal{J}|$ is not too large, then an approximate solution to this problem can be computed cheaply. By an approximate solution we mean a solution $(y, x_{\mathcal{J}}, \tau, s_{\mathcal{J}}, \kappa)$ that satisfies

$$\begin{array}{rrrrcl} & A_{\mathcal{J}} x_{\mathcal{J}} & -b\tau & & = & \phi(A_{\mathcal{J}} x^0_{\mathcal{J}} - b\tau^0), \\ -A^T_{\mathcal{J}} y & & +c_{\mathcal{J}}\tau & -s_{\mathcal{J}} & = & -\phi(-c_{\mathcal{J}}\tau^0 + A^T_{\mathcal{J}} y^0 + s^0_{\mathcal{J}}), \\ b^T y & -(c_{\mathcal{J}})^T x_{\mathcal{J}} & & -\kappa & = & -\phi(b^T y^0 - c^T_{\mathcal{J}} x^0_{\mathcal{J}} + \kappa^0), \\ & x_{\mathcal{J}} \geq 0, & \tau \geq 0, & s_{\mathcal{J}}, \kappa \geq 0, & & \end{array} \tag{6.13}$$

and belongs to the set

$$\mathcal{N}(\mathcal{J}, \phi, \theta) := \left\{ (y, x_{\mathcal{J}}, \tau, s_{\mathcal{J}}, \kappa) : \begin{array}{l} \min_{j \in \mathcal{J}} (x_j s_j, \tau\kappa) \geq \theta \phi \frac{\bar{g}}{n+1}, \\ x^T_{\mathcal{J}} s_{\mathcal{J}} + \tau\kappa = \phi \frac{\bar{g}}{n+1} |\mathcal{J}| \end{array} \right\}$$

for some $\theta \in (0, 1)$. We call such a solution an approximate analytic center solution. Note, we do not use the proximity measure suggested in Section 2, but rather a relaxed variant. Hence, the practical algorithm does not follow the central path as closely as the theoretical one.

If $(\hat{y}, \hat{x}_{\mathcal{J}}, \hat{\tau}, \hat{s}_{\mathcal{J}})$ is an approximate analytic center solution, then it can be extended to a solution of the full problem by defining $\hat{x}_{\bar{\mathcal{J}}}$ by (6.11) and

$$\hat{s}_{\bar{\mathcal{J}}} := c_{\bar{\mathcal{J}}} \hat{\tau} - A^T \hat{y} + \phi(-c_{\bar{\mathcal{J}}} \tau^0 + A^T_{\bar{\mathcal{J}}} y^0 + s^0_{\bar{\mathcal{J}}}).$$

Clearly, nothing guarantees that $\hat{s}_{\bar{\mathcal{J}}} \geq 0$ i.e. the set

$$\bar{\mathcal{J}}^- := \{ j \in \mathcal{J} : \hat{s}_j < 0 \}$$

is nonempty. However, if $\bar{\mathcal{J}}^-$ is nonempty, then some of the variables in $\bar{\mathcal{J}}^-$ should be moved to $\mathcal{J}$. This corresponds to generating variables in the primal and cutting planes in the dual. On the other hand if $\bar{\mathcal{J}}$ is empty, then ϕ is reduced. In both cases a new approximate analytic center solution is computed to the new relaxations. This process is continued until ϕ is reduced sufficiently, because then an approximate solution to (6.5) has been computed. This method stated formally leads to the algorithm:

Algorithm 6.1

> ***Input:*** $\mathcal{J}^0 \subseteq \{1, \ldots, n\}$, $\beta \in (0, 1)$, $\varepsilon > 0$
> $\phi^0 := 1$, $(x^0, \tau^0, y^0, s^0, \kappa^0) := (e, 1, 0, e, 1)$
> $k := 0$
> ***while*** $\phi^k > \beta\varepsilon$
> $(x^{k+1}, \tau^{k+1}, y^{k+1}, s^{k+1}, \kappa^{k+1}) :=$ ***center***$(\mathcal{J}^k, \phi^k, (x^k, \tau^k, y^k, s^k, \kappa^k))$.
> $\bar{\mathcal{J}}^- := \{ j \notin \mathcal{J}^k : s^{k+1}_j < 0 \}$
>   ***if*** $\bar{\mathcal{J}}^- \neq \emptyset$
>   ***Choose*** $\mathcal{J}^{k+1}$ ***such that*** $\mathcal{J}^k \subset \mathcal{J}^{k+1} \subseteq \mathcal{J}^k \cup \bar{\mathcal{J}}^-$
> $\phi^{k+1} := \phi^k$
>   ***else***
> $\mathcal{J}^{k+1} := \mathcal{J}^k$
> $\phi^{k+1} := \beta\phi^k$
>   ***end if***
> $k := k + 1$
> ***end while***

The procedure "center" used in step 5 means that the approximate analytic center is computed to the reduced center problem defined by $\mathcal{J}^k$. Also note that whenever ϕ is reduced in step 12, then the value of the "inactive" variables is reduced.

Algorithm 6.1 converges in a number iterations bounded by $O(n\frac{\ln(\epsilon)}{\ln(\beta)})$ because in each iteration either ϕ is reduced by the factor β or the set $\mathcal{J}$ is extended. Moreover, the size of $\mathcal{J}$ is bounded by n and ϕ is bounded below by the termination tolerance ε.

6.1 Recomputing the approximate analytic center

One important issue not addressed in Algorithm 6.1 is how to compute the approximate center solution in step 5. Clearly, whenever the approximate analytic center is recomputed after ϕ has been reduced or $\mathcal{J}$ has been extended, then the previous approximate analytic center should serve as a warm-start.

In both cases we are seeking an approximate analytic center solution satisfying

$$\begin{array}{rrrrcl} & A_{\mathcal{J}}x_{\mathcal{J}} & -b\tau & & = & \phi(A_{\mathcal{J}}x^0_{\mathcal{J}} - b\tau^0), \\ -A^T_{\mathcal{J}}y & & +c_{\mathcal{J}}\tau & -s_{\mathcal{J}} & = & \phi(-c_{\mathcal{J}}\tau^0 + A^T_{\mathcal{J}}y^0 + s^0_{\mathcal{J}}), \\ b^Ty & -(c_{\mathcal{J}})^Tx_{\mathcal{J}} & & -\kappa & = & \phi(b^Ty^0 - c^T_{\mathcal{J}}x^0_{\mathcal{J}} + \kappa^0). \end{array} \quad (6.14)$$

Moreover, the complementarity gap should satisfy

$$(x_{\mathcal{J}})^Ts_{\mathcal{J}} + \tau\kappa = \phi((x^0_{\mathcal{J}})^Ts^0_{\mathcal{J}} + \tau^0\kappa^0). \quad (6.15)$$

and the new iterate should be close to the central path i.e.

$$\min_{j\in\mathcal{J}}(x_js_j, \tau\kappa) \geq \theta\phi\frac{\bar{g}}{n+1} \quad (6.16)$$

for $\phi = \phi^{k+1}$ and the given $\theta \in (0, 1)$. The equations (6.14) and (6.15) define a target for the residuals and the complementarity gap respectively, whereas the equation (6.16) is the centering condition.

In the case ϕ is reduced, then a solution satisfying (6.14) and (6.16) is known for $\phi = \phi^k$. Let this solution be denoted as $(y', x'_{\mathcal{J}}, \tau', s'_{\mathcal{J}}, \kappa')$ and let ϕ' denote the corresponding ϕ.

Hence, we want to compute a new approximate analytic center for a smaller ϕ and for that purpose we use the search direction

$$\begin{array}{rrrrcl} & A_{\mathcal{J}}d_{x_{\mathcal{J}}} & -bd_\tau & & = & \eta\phi'(A_{\mathcal{J}}x^0_{\mathcal{J}} - b\tau^0), \\ -A^T_{\mathcal{J}}d_y & & +c_{\mathcal{J}}d_\tau & -d_{s_{\mathcal{J}}} & = & -\eta\phi'(-c_{\mathcal{J}}\tau^0 + A^T_{\mathcal{J}}y^0 + s^0_{\mathcal{J}}), \\ b^Td_y & -(c_{\mathcal{J}})^Td_{x_{\mathcal{J}}} & & -d_\kappa & = & -\eta\phi'(b^Ty^0 - c^T_{\mathcal{J}}x^0_{\mathcal{J}} + \kappa^0), \\ & S'_{\mathcal{J}}d_{x_{\mathcal{J}}} & & +X'_{\mathcal{J}}d_{s_{\mathcal{J}}} & = & -X'_{\mathcal{J}}s'_{\mathcal{J}} + \gamma\phi'\frac{\bar{g}}{n+1}e, \\ & & \kappa d_\tau & +\tau d_\kappa & = & -\tau'\kappa' + \gamma\phi'\frac{\bar{g}}{n+1}e, \end{array}$$

(6.17)

suggested in Xu *et al* (1996). Here, $\gamma, \eta \in [0,1]$ are two parameters to be chosen later. After the search direction has been computed, then a step-size α is chosen and the current solution is updated as follows

$$\begin{bmatrix} x_{\mathcal{J}}^+ \\ \tau^+ \\ y^+ \\ s_{\mathcal{J}}^+ \\ \kappa^+ \end{bmatrix} = \begin{bmatrix} x'_{\mathcal{J}} \\ \tau' \\ y' \\ s'_{\mathcal{J}} \\ \kappa' \end{bmatrix} + \alpha \begin{bmatrix} d_{x_{\mathcal{J}}} \\ d_\tau \\ d_y \\ d_{s_{\mathcal{J}}} \\ d_\kappa \end{bmatrix}. \tag{6.18}$$

It is easy to verify that

$$\begin{array}{rrrrcl} & A_{\mathcal{J}} x_{\mathcal{J}}^+ & -b\tau^+ & & = & (1-\alpha\eta)\phi'(A_{\mathcal{J}} x_{\mathcal{J}}^0 - b\tau^0), \\ -A_{\mathcal{J}}^T y^+ & & +c_{\mathcal{J}}\tau^+ & -s_{\mathcal{J}}^+ & = & -(1-\alpha\eta)\phi'(-c_{\mathcal{J}}\tau^0 + A_{\mathcal{J}}^T y^0 + s_{\mathcal{J}}^0), \\ b^T y^+ & -(c_{\mathcal{J}})^T x_{\mathcal{J}}^+ & & -\kappa^+ & = & -(1-\alpha\eta)\phi'(b^T y^0 - c_{\mathcal{J}}^T x_{\mathcal{J}}^0 + \kappa^0), \end{array}$$

and

$$(x_{\mathcal{J}}^+)^T s_{\mathcal{J}}^+ + \tau^+\kappa^+ = (1 - \alpha(1-\gamma) + \alpha^2\eta(1-\eta-\gamma))\phi'((x_{\mathcal{J}}^0)^T s_{\mathcal{J}}^0 + \tau^0\kappa^0).$$

Given $\alpha > 0$ and the choice $\eta = 1-\gamma > 0$, then $(x_{\mathcal{J}}^+, \tau^+, y, s_{\mathcal{J}}^+, \kappa^+)$ is a solution to (6.14) and (6.15) for

$$\phi = (1-\alpha(1-\gamma))\phi' = (1-\alpha\eta)\phi' =< \phi'.$$

This demonstrates that the updated solution is closer to the target solution. In general it might not be possible to reach the target in one step, because the step size α should be chosen such that $(y^+, x^+, \tau^+, s^+, \kappa^+)$ is not too far from the central path. In our implementation we use a fairly small γ and let intermediate iterates move far from the central path. Therefore, whenever a solution satisfying (6.14) for the target ϕ has been computed, then several (centering) steps using $\gamma = 1$ and $\eta = 0$ may be required.

Note that the search direction given by (6.17) is a highly flexible direction. For example if a solution is known satisfying (6.14), then by choosing $\eta = 0$ and γ either smaller or greater than 1, then the complementarity gap can be reduced or increased respectively. Moreover, the new solution keeps on satisfying the three first equations in (6.14).

In summary, whenever ϕ is reduced in Algorithm 6.1, then it is easy to perform an efficient warmstart. However, in the case the set $\mathcal{J}$ is extended then the situation is more difficult, because the current solution is not feasible i.e. some of the elements in $s_{\mathcal{J}}$ are negative. In that case we define an intermediate solution

$$\begin{array}{rcl} \phi' & := & \phi, \\ y' & := & y, \\ x'_{\mathcal{J}} & := & x_{\mathcal{J}}, \\ x'_{\mathcal{J}^-} & := & \phi' x_{\mathcal{J}^-}^0, \\ \tau' & := & \tau, \\ s'_{\mathcal{J}'} & := & s_{\mathcal{J}}, \\ s'_{\mathcal{J}^-} & := & \frac{\bar{g}}{n+1}(X_{\mathcal{J}^-}^0)^{-1} e, \\ \kappa' & := & \kappa, \end{array} \tag{6.19}$$

where $(y, x_{\mathcal{J}}, \tau, s_{\mathcal{J}}, \kappa)$ was the previous approximate analytic center and the search direction

$$\begin{array}{rrrrcl} A_{\mathcal{J}}d_{x_{\mathcal{J}}} & -bd_{\tau} & & & = & \phi^k(A_{\mathcal{J}}x^0_{\mathcal{J}} - b\tau^0) \\ & & & & & -(A_{\mathcal{J}}x'_{\mathcal{J}} - b\tau'), \\ -A^T_{\mathcal{J}}d_y & & +c_{\mathcal{J}}d_{\tau} & -d_{s_{\mathcal{J}}} & = & \phi^k(-c_{\mathcal{J}}\tau^0 + A^T_{\mathcal{J}}y^0 + s^0_{\mathcal{J}}) \\ & & & & & -(-c_{\mathcal{J}}\tau' + A^T_{\mathcal{J}}y' + s'_{\mathcal{J}}), \\ b^Td_y & -(c_{\mathcal{J}})^Td_{x_{\mathcal{J}}} & & -d_{\kappa} & = & \phi^k(b^Ty^0 - c^T_{\mathcal{J}}x^0_{\mathcal{J}} + \kappa^0) \\ & & & & & -(b^Ty' - c^T_{\mathcal{J}}x'_{\mathcal{J}} + \kappa'), \\ S'_{\mathcal{J}}d_{x_{\mathcal{J}}} & & & +X'_{\mathcal{J}}d_{s_{\mathcal{J}}} & = & -X'_{\mathcal{J}}s'_{\mathcal{J}} + \phi'\frac{g^0}{n+1}e, \\ & \kappa'd_{\tau} & & +\tau'd_{\kappa} & = & -\tau'\kappa' + \phi'\frac{g^0}{n+1}e, \end{array} \tag{6.20}$$

and perform the update (6.18). In this case it can be verified that

$$\begin{array}{rrrrcl} A_{\mathcal{J}}x^+_{\mathcal{J}} & -b\tau^+ & & & = & (1-\alpha)(Ax'_{\mathcal{J}} - b\tau') \\ & & & & & +\alpha\phi'(A_{\mathcal{J}}x^0_{\mathcal{J}} - b\tau^0), \\ -A^T_{\mathcal{J}}y^+ & & +c_{\mathcal{J}}\tau^+ & -s^+_{\mathcal{J}} & = & (1-\alpha)(-c_{\mathcal{J}}\tau' + A^T_{\mathcal{J}}y' + s'_{\mathcal{J}}) \\ & & & & & +\alpha\phi'(-c_{\mathcal{J}}\tau^0 + A^T_{\mathcal{J}}y^0 + s^0_{\mathcal{J}}), \\ b^Ty^+ & -(c_{\mathcal{J}})^Tx^+_{\mathcal{J}} & & -\kappa^+ & = & (1-\alpha)(b^Ty' - c^T_{\mathcal{J}}x'_{\mathcal{J}} + \kappa') \\ & & & & & +\alpha\phi'(b^Ty^0 - c^T_{\mathcal{J}}x^0_{\mathcal{J}} + \kappa^0), \end{array}$$

Note if $\alpha = 1$, then the updated solution satisfies (6.14) exactly. Moreover, for any positive step-size α, then the solution is moved closer to satisfying (6.14). Unfortunately, there is no control over the complementary gap and it may decrease or increase. Therefore, α is chosen such that the complementarity gap is only allowed to be within the range $[0.5, 2.0]$ times the target gap (6.15). If the step-size has to be cutback, because the complementarity gap was moving outside the range, then the search direction (6.17) is used repeatedly with $\eta = 0$ and $\gamma > 1$ ($\gamma < 1$) to increase (decrease) the complementarity gap until the target value of the complementarity gap is reach. Ultimately this procedure generates a solution satisfying (6.14). However, when this procedure is terminated then the complementarity gap might be too small or too large and the solution might not be centered. However, this can be corrected using the search direction (6.17) with $\eta = 0$ and γ chosen appropriately.

7 COMPUTATIONAL RESULTS

In this section we report computational results for the implementation discussed in the previous section. The test problems we are using are LP relaxations of real world set-partitioning and set-covering problems obtained from the OR-Library, see `http://mscmga.ms.ic.ac.uk/info.html`. The characteristic of these problems is that $n \gg m$, all elements in the matrix A are either zero or one, and all coefficients in the right-hand side are one. All constraints for the set-covering type problems are of the greater-than-equal type, whereas all the constraints for the set-partitioning type problems are of the equality type. Therefore, we add a full set of artificial variables, i.e. slack variables with lower and upper bounds zero, to the set-partitioning problems. Hence, the initial set $\mathcal{J}^0$ consists of slack variables which corresponds to the identity matrix. In our implementation we use $\beta = 0.1$ and $\theta = 0.1$, meaning that ϕ is reduced fairly fast and the central path is followed loosely. In each iteration we add multiple variables as follows. Assume that the variables in the set $\mathcal{J}^-$ has been sorted in increasing order after index, then each kth variable is chosen, where k is determined such that at most $m/4$ variables are chosen. The procedure for computing the search direction is a slight modification of those presented in Andersen and Andersen (1997). Finally, the algorithm is terminated when a primal and dual solution has been computed to the original problem that does not violate any inequality by more than 10^{-8} and 8 figures in the corresponding primal and dual objective value are identical.

Name	Con-straints	Variables full	Variables relaxation	Iterations outer	Iterations inner	Time	Primal objective	Relative gap
sppaa01	823	8904	2774	42	121	77.17	5.553544e+004	1.48e-010
sppaa02	531	5198	1592	29	71	15.27	3.049400e+004	6.42e-011
sppaa03	825	8627	2485	37	83	46.69	4.961636e+004	1.13e-010
sppaa04	426	7195	1486	36	93	12.66	2.587761e+004	1.45e-010
sppaa05	801	8308	2443	38	97	54.69	5.373593e+004	9.94e-011
sppaa06	646	7292	2087	33	86	29.51	2.697719e+004	1.74e-011
sppkl01	55	7479	350	21	48	0.48	1.084000e+003	2.19e-010
sppkl02	71	36699	506	26	61	1.15	2.152500e+002	3.69e-010
sppnw01	135	51975	586	38	76	2.55	1.148520e+005	3.50e-010
sppnw02	145	87879	742	39	80	3.91	1.054440e+005	9.24e-010
sppnw03	59	43749	392	28	58	1.31	2.444700e+004	4.55e-010
sppnw04	36	87482	397	28	67	1.94	1.631067e+004	-9.62e-012
sppnw05	71	288507	522	35	77	6.76	1.328780e+005	3.04e-012
sppnw06	50	6774	274	22	37	0.42	7.640000e+003	8.78e-010
sppnw07	36	5172	221	20	34	0.28	5.476000e+003	-2.49e-010
sppnw08	24	434	109	20	35	0.11	3.589400e+004	-2.21e-010
sppnw09	40	3103	226	23	45	0.25	6.776000e+004	1.62e-010
sppnw10	24	853	133	20	37	0.13	6.827100e+004	1.37e-013
sppnw11	39	8820	259	22	45	0.35	1.162545e+005	2.39e-011
sppnw12	27	626	134	20	34	0.12	1.411800e+004	1.07e-010
sppnw13	51	16043	354	27	57	0.61	5.013200e+004	9.07e-011
sppnw14	73	123409	514	35	78	3.35	6.184400e+004	5.79e-011
sppnw15	31	467	150	21	36	0.18	6.774300e+004	1.72e-011
sppnw16	139	148633	625	29	55	4.41	1.181590e+006	1.54e-011
sppnw17	61	118607	414	30	71	2.96	1.087575e+004	2.85e-010
sppnw18	124	10757	584	32	70	1.37	3.388643e+005	2.00e-011
sppnw19	40	2879	207	20	35	0.23	1.089800e+004	-5.31e-010
sppnw20	22	685	113	19	34	0.10	1.662600e+004	-2.09e-010
sppnw21	25	577	108	17	26	0.09	7.380000e+003	2.74e-011
sppnw22	23	619	123	18	32	0.10	6.942000e+003	-6.78e-011
sppnw23	19	711	127	22	38	0.11	1.231700e+004	1.41e-010
sppnw24	19	1366	140	18	27	0.09	5.843000e+003	-4.27e-010
sppnw25	20	1217	141	19	32	0.11	5.852000e+003	5.62e-011
sppnw26	23	771	165	20	33	0.11	6.743000e+003	1.54e-011
sppnw27	22	1355	165	20	36	0.13	9.877500e+003	3.72e-010
sppnw28	18	1210	111	20	33	0.11	8.169000e+003	1.97e-010
sppnw29	18	2540	194	22	43	0.15	4.185333e+003	3.02e-010
sppnw30	26	2653	168	20	34	0.16	3.726800e+003	-4.14e-010
sppnw31	26	2662	173	19	32	0.15	7.980000e+003	2.22e-010
sppnw32	19	294	107	20	38	0.08	1.457000e+004	1.60e-010
sppnw33	23	3068	162	19	30	0.15	6.484000e+003	3.50e-011
sppnw34	20	899	125	18	24	0.08	1.045350e+004	5.43e-011
sppnw35	23	1709	139	18	28	0.11	7.206000e+003	3.98e-010
sppnw36	20	1783	158	21	38	0.13	7.260000e+003	-7.80e-011
sppnw37	19	770	105	18	28	0.09	9.961500e+003	-8.63e-011
sppnw38	23	1220	162	19	42	0.16	5.552000e+003	-6.17e-011
sppnw39	25	677	143	18	28	0.11	9.868500e+003	7.15e-010
sppnw40	19	404	90	19	30	0.07	1.065825e+004	-3.27e-011
sppnw41	17	197	80	18	28	0.06	1.097250e+004	-3.87e-010
sppnw42	23	1079	141	18	31	0.11	7.485000e+003	-4.26e-011
sppnw43	18	1072	151	20	36	0.10	8.897000e+003	-7.00e-011
sppus01	145	1053137	768	39	90	40.33	9.963067e+003	1.30e-010
sppus03	77	85552	405	26	52	2.95	5.338000e+003	5.16e-011
sppus04	163	28016	609	29	54	2.54	1.773167e+004	4.98e-011
rail2536	2536	1081841	11100	42	183	2346.76	6.883992e+002	3.09e-010
rail2586	2586	920683	12663	42	223	1217.80	9.359218e+002	1.54e-010
rail4284	4284	1092610	18206	43	201	14529.23	1.054054e+003	2.48e-010
rail507	507	63009	2489	35	144	26.08	1.721456e+002	4.23e-010
rail516	516	47311	1963	29	100	14.10	1.820000e+002	1.86e-010
rail582	582	55515	2525	35	146	27.93	2.097122e+002	1.54e-010
Sum			74585	1551	3661			

Table 7.1: Computation using the homogenized cutting plane algorithm.

The computational test is performed on a 300mhz Pentium II PC using Windows NT and having 128 megabytes of memory.

In Table 7.1 the test problems and the computational results are presented. All problems having a name starting with "spp" are of the set-partitioning type and the remaining problems are of the set-covering type.

Beyond the name of the test problems the table first shows the number of constraints (m). Next the number of variables in the full problem (n) and the final relaxation is shown.

The column with the heading Iterations shows the number of times ϕ is reduced and the number of times the search direction is computed. It can be observed that for most of the problems few outer iterations are required. Furthermore, approximately 2 to 3 inner iterations are needed per outer iteration. Although for the large rail problems 4 to 5 inner iterations are needed per outer iteration. This indicates that our warm start technique works well. It should also be noted that relatively few variables out of the total number of variables in the full problem are included in the final relaxation. One exception is the set of problems with names starting with "sppa". However, these problems have relatively few variables compared to the number of constraints and in fact the final relaxation contains a number of variables which is only a small multiple of the number of constraints. In computational experiments not reported here it was observed that if the number of variables added per iterations was limited to about 50, then the number of outer iterations grew by around a factor of 2 for a lot of the problems. Moreover, contrary to expectations, the total number of generated variables was not reduced significantly.

The column with the heading Time shows the number of CPU seconds spent to solve the problems. Except for the rail problems these numbers are small. The reason the solution time is large for the rail problems is that for those problems it is expensive to compute the search direction which is caused by the relatively large number of constraints and fill-in in the Cholesky factorization, see Andersen and Andersen (1997) for details about the latter. Finally, it should be noted that the solution time for rail4284 is exceptionally large due to memory swapping caused by needing more than 128 megabytes of memory to solve the problem.

The last two columns shows the optimal primal objective value and the relative gap. In all cases a solution satisfying the termination tolerances was computed.

One obvious question is how the homogenized cutting plane algorithm compares to solving the full problem directly. To answer that question we have made Table 7.2 which reports results for solving the problems having more than 50000 variables using MOSEK v1.0b interior-point optimizer (Andersen and Andersen (1997)). Note that most of the linear algebra routines are shared by MOSEK and the homogenized cutting plane algorithm so the comparison should be fair. Moreover, the largest problems are excluded from Table 7.2, because those problems could not be solved using only 128 megabytes of memory.

Table 7.2 Results for solving some of the problems using MOSEK v1.0b.

Name	Iterations	Time full	Time cut	Primal objective
sppnw01	15	23.2	2.55	1.148520e+005
sppnw02	15	40.2	3.91	1.054441e+005
sppnw04	20	28.1	1.94	1.631067e+004
sppnw05	29	170.5	6.76	1.328780e+005
sppnw14	23	60.2	3.35	6.184400e+004
sppnw16	23	106.1	4.41	1.181590e+006
sppnw17	24	58.7	2.96	1.087575e+004
sppus03	15	31.3	2.95	5.338000e+003
rail507	23	44.7	26.86	1.721456e+002
rail516	15	25.8	14.10	1.820000e+002
rail582	22	43.8	27.93	2.097122e+002

Table 7.2 presents the number of interior-point iterations and the time spent solving the problems for solving the full problem and using MOSEK (full) and the cutting plane approach (cut). Finally, the optimal primal objective value reported by MOSEK is shown. A comparison of the timing results shows that for the chosen problems the homogenized cutting-plane algorithm is significantly better than the full approach with respect to running time. Indeed for the problems sppnw05 and sppnw16 the cutting plane approach is approximately 20 to 25 times faster than the regular approach. Hence, we can conclude at least for LP relaxations of set-partitioning and set-covering problems having many more variables than constraints, then the homogenized cutting plane approach is an attractive alternative to solving the full problem. Moreover, it should be emphasized that the problems with about one million variables can only be solved using the cutting plane approach due to the large memory consumption of the full approach.

In summary, the suggested implementation is capable of solving large LP relaxations of set-covering and set-partitioning problems in a low number of outer and inner iterations. Moreover, for problems having many more variables than constraints, then the approach is much more efficient than solving the full problem directly. However, it seems to be important for efficiency reasons that possibly many variables in each iteration can be generated.

Finally, it should be mentioned that several possibilities exist for improving the efficiency of the implementation. For example a more sophisticated choice of the initial set $\mathcal{J}^0$ is likely to reduce the number of inner and outer iterations.

8 CONCLUSIONS

We have presented a cutting plane algorithm using a self-dual homogenized formulation. The analysis of the algorithm uses a different proximity measure than that in, for example, Goffin *et al.* (1996). The algorithm has the same complexity as that in Goffin *et al.* (1996).

Furthermore, we present an implementation of the homogenized cutting plane algorithm adapted to solve LP problems having many more variables than constraints. The implementation is applied to the LP relaxation of real-world set-covering and set-partitioning problems. The computational results show that the discussed implementation solves the test problems in a low number of inner and outer iterations. Moreover, the results show that for problems having many more variables than constraints, the cutting plane approach saves a lot of computation time and space compared to solving the problems directly.

References

E. D. Andersen and K. D. Andersen. The MOSEK interior point optimizer for linear programming: an implementation of the homogeneous algorithm. In J. B. G. Frenk, C. Roos, T. Terlaky, and S. Zhang, editors, *High Performance Optimization Techniques, Proceedings of the HPOPT-II conference*, 1997, 197-232.

D. S. Atkinson and P. M. Vaidya. A cutting plane algorithm for convex programming that uses analytic centers. *Mathematical Programming*, 69:1–43, 1995.

R. E. Bixby, J. W. Gregory, I. J. Lustig, R. E. Marsten, and D. F. Shanno. Very large-scale linear programming: A case study in combining interior point and simplex methods. *Oper. Res.*, 40(5):885–897, 1992.

J.-L. Goffin, Z.-Q. Luo, and Y. Ye. On the complexity of a column generation algorithm for convex or quasiconvex problems. In *Large Scale Optimization: The State of the Art.* Kluwer Academic Publishers, 1993.

J.-L. Goffin, Z.-Q. Luo, and Y. Ye. Complexity analysis of an interior cutting plane method for convex feasibility problems. *SIAM Journal on Optimization*, 6:638–652, 1996.

J.-L. Goffin and J.-P. Vial. Multiple cuts in the analytic center cutting plane method. Technical Report Logilab Technical Report 98.10, Logilab, Management Studies, University of Geneva, Geneva, Switzerland, June 1998. Accepted for publication in *Mathematical Programming*.

D. den Hertog, J. Kaliski, C. Roos, and T. Terlaky. A logarithmic barrier cutting plane method for convex programming problems. *Annals of Operations Research*, 58:69–98, 1995.

J. E. Mitchell and S. Ramaswamy. A long-step, cutting plane algorithm for linear and convex programming. Technical Report 37–93–387, DSES, Rensselaer Polytechnic Institute, Troy, NY 12180–3590, August 1993. Accepted for publication in *Annals of Operations Research.*

J. E. Mitchell and M. J. Todd. Solving combinatorial optimization problems using Karmarkar's algorithm. *Mathematical Programming*, 56: 245–284, 1992.

Y. E. Nesterov and J. Ph. Vial. Homogeneous analytic center cutting plane methods for convex problems and variational inequalities. *SIAM Journal on Optimization*, 9(3): 707-728, 1999.

S. Ramaswamy and J. E. Mitchell. On updating the analytic center after the addition of multiple cuts. Technical Report 37–94–423, DSES, Rensselaer Polytechnic Institute, Troy, NY 12180, October 1994. Substantially revised: August, 1998.

C. Roos, T. Terlaky, and J.-Ph. Vial. *Theory and Algorithms for Linear Optimization: An Interior Point Approach.* John Wiley, Chichester, 1997.

X. Xu, P. -F. Hung, and Y. Ye. A simplified homogeneous and self-dual linear programming algorithm and its implementation. *Annals of Operations Research*, 62:151–171, 1996.

Y. Ye. Complexity analysis of the analytic center cutting plane method that uses multiple cuts. *Mathematical Programming*, 78:85–104, 1997.

11 ALGORITHMS FOR SOME HARD KNAPSACK PROBLEMS

Louis Caccetta and Araya Kulanoot

School of Mathematics and Statistics
Curtin University of Technology
GPO Box U 1987,
Perth, W.A, 6845 Western Australia.

Abstract: Knapsack problems are typically concerned with selecting from a set of n given items, each with a specified weight and value, a subset of items whose weight sum does not exceed a prescribed capacity and whose value is maximum. This NP-hard problem arises in many applications and has been the focus of considerable research over the past two decades. A number of exact algorithms have been developed for the classical 0-1 Knapsack Problems and its variants. In this paper, exact algorithms are presented for the following variants and data types: the Subset Sum Problem; the Strongly Correlated 0-1 Knapsack Problem; the Inverse Strongly Correlated 0-1 Knapsack Problem; and the corresponding Bounded Strongly Correlated Knapsack Problem and Bounded Subset Sum Problem. All our algorithms consist of three stages: the first stage generates an initial solution by a greedy procedure ; the second stage refines the approximate solution; and the final stage applies a partial lexicographic search procedure to generate an optimal solution. Extensive computational experiments show that our algorithms are able to solve large problems of size up to one million variables in less than 7 seconds CPU time on a Silicon Graphic Workstation (R 5000) running at a clock speed of 150 MHz. A comparative analysis with some recent effective algorithms is given.

Key words: Branch and bound, dynamic programming, knapsack problem, combinatorial optimization.

X. Yang et al. (eds.), Optimization Methods and Applications, 191-217.

1 INTRODUCTION

Let N = {1,2, ... ,n} be a given set of items with item i having a weight of w_i and a value (or profit) of p_i , and c the capacity of the given knapsack. The classical **0-1 Knapsack Problem (0-1KP)** can be stated as :

$$\begin{aligned} \text{maximize} \quad & z = \sum_{i=1}^{n} p_i x_i \\ \text{subject to} \quad & \sum_{i=1}^{n} w_i x_i \leq c, \\ \text{and} \quad & x_i \in \{0,1\}, \qquad \text{for } i = 1,2,\ldots,n. \end{aligned}$$

The $w_i's$ and $p_i's$ are assumed to be positive integers.

KP's have been extensively studied by many researchers particularly over the past two decades. The excellent book of Martello amd Toth (1990a) and the recent survey by Pisinger and Toth (1998) provides a comprehensive account of the theory and applications of the problem.

Knapsack Problems arise in many business and industrial applications including : financial management (Cord (1964); Kaplan (1966); Hansmann (1961)) ; cargo loading (Bellman and Dreyfus (1962)); cutting stock (Gilmore and Gomory (1963; 1966)); fault tolerance (Sinha and Zottners (1979)) ; and cryptography (Diffe and Hellman (1976)).

Theoretical interest in the Knapsack Problems is two fold. Firstly, there are many applications where a general integer programming problem can be transformed into a Knapsack Problem (Salkin (1975); Syslo et al (1983)). Secondly, in the solution of many integer programming problems, the Knapsack Problem arises as a subproblem. This is the case, for example, in the set-partitioning problem and in generating cutting planes for some large-scale 0-1 integer programming.

Data instances which have been used by many researchers in computational experiments concerning the 0-1**KP** are classified into four types depending on the correlation between profit and weight:

Uncorrelated:	there is no correlation between the profit and the weight of an item (p_i and w_i uniformly random in [1,R] where R is the data range).
Weakly Correlated:	the profit is (highly)correlated with the weight (w_i uniformly random in [1,R] and p_i uniformly random in [w_i - R/a, w_i + R/a] where a is an integer).
Strongly Correlated:	profit is linearly related to the weight of an item (w_i uniformly random in [1,R] and $p_i = w_i + \alpha$ where α is positive integer.) .

Inverse Strongly Correlated:	profit is linearly related to the weight of an item (w_i uniformly random in [1,R] and $p_i = w_i + \alpha$ where α is negative integer).
Subset Sum:	profit and weight of each item are equal w_i uniformly random in [1,R] and $p_i = w_i$).

A number of variations to the classical Knapsack Problem have also been studied (Martello amd Toth (1990a)) including :

Bounded Knapsack Problem (BKP):	here a bounded amount of each item is allowed.
Multiple - Choice Knapsack Problem:	here items are to be chosen from disjoint subsets(classes).
Multiple - Knapsack Problem:	here several knapsacks are to be filled simultaneously.
Multi - Constrained Knapsack Problem:	basically a general integer programming problem.

Mathematically, the BKP can be formulated as :

$$\text{maximize} \quad z = \sum_{i=1}^{n} p_i x_i$$

$$\text{subject to} \quad \sum_{i=1}^{n} w_i x_i \leq c,$$

$$\text{and} \quad x_i \in \{0, 1, \ldots, b_i\}, \text{ for } i = 1, 2, \ldots, n \quad \text{and } b_i \text{ is a positive integer.}$$

We have the classical 0-1 Knapsack Problem when $b_i = 1$ for each i.

Bellman (1954; 1957) tackled the 0-1 KP using dynamic programming. Dantzig (1957) proposed the well-known upper bound determined by non-increasing values of the profit and weight ratio. Since this early work, most of the research effort on the 0-1 KP has been on developing the methods of branch-and-bound and dynamic programming.

In brief, Branch-and-bound methods for the 0-1KP have been developed over the past three decades with the early contributions made by Gilmore and Gomory (1963); Kolesar (1967); Greenberg and Hegerich (1970); Horowitz and Sahni (1974); Nauss (1976) . This early work was followed by several well-known hybrid algorithms which aimed to improve performance. A reduction procedure was presented by Ingargiola and Korsh (1973). A number of authors: Martello and Toth (1977); Martello and Toth (1988); Martello and Toth (1997); Müller -Merbach (1978); Dudzinski and Walukiewicz (1987) modified the upper bound improvement of Dantzig's. In the recent paper of Martello and Toth (1997), 0-1KP's involving all data types with n up to 10000 are solved .

Several effective algorithms for the 0-1KP are hybrids of the dynamic programming method. Toth (1980) presented a dynamic programming algorithm based on modified reduction procedures and Balas and Zemel (1980) solved large 0-1KP's by considering the "core problem". The latter approach has given rise to a number of efficient algorithms developed by Fayard and Plateau (1982); Martello and Toth (1988); Pisinger(1995a; 1995b; 1997). Recently, Martello and Toth (1997) proposed an efficient algorithm to solve the general 0-1KP with n up to 10000 in less than 0.12 second. Their algorithm, called COMBO, utilizes dynamic programming combined with maximum cardinality of an optimal solution.

The first approximation scheme for the 0-1KP was proposed by Sahni (1975) and a fully polynomial-time approximation scheme was obtained by Ibarra and Kim (1975). A number of authors have examined the implementation of sequential algorithms for the 0-1KP based on both branch-and-bound and dynamic programming. Work on the branch-and-bound was investigated by Lai and Sahni (1984) and further developed by Janakiram et al (1988); Kindervater and Trienekens (1988). More recently the paper of Loots and Smith (1992) details a method that performs both sorting and branch-and-bound steps in parallel. Their algorithm solves 120 problems with n up to 20000 for all data types except the strongly correlated type. Their computational results report average speedups in the range from 1.35 to 1.99, 1.52 to 3.9 and 0.95 to 7.9 for 2, 4, and 8 processors, respectively. The corresponding analysis for dynamic programming using a hypercube topology was presented by Lee et al (1988). The recent developments by Lin and Storer (1991) have provided the Connection Machine (CM) technique for parallel algorithms to solve large problems with n up to 800 by using 4096 processors. The CM running time as a function of the number of processors and also as a function of the knapsack capacity were presented graphically. This running time for fixed n increases logarithmically in c, the knapsack capacity. A comparative analysis with the algorithm of Lee et al (1988) demonstrated the superiority of the CM technique.

In this paper we present specialized algorithms for the 0-1KP and the BKP. We consider only two data types : subset sum type and strongly correlated type for 0-1KP. For the bounded case we refer to the corresponding problems as Bounded Subset Sum Problem (BSSP) and Bounded Strongly Correlated Knapsack Problem (BSCKP). Only these data types are considered since uncorrelated and weakly correlated instances are easy to solve by general algorithms. For the subset sum and the strongly correlated data instances, good approximate solutions can be obtained through the implementation of a greedy algorithm. We present new greedy algorithms which we use in our solution strategy specialized for each type of problem. In brief, our solution strategy involves three stages:

First stage :	initialize a solution by a Greedy Algorithm.
Second stage:	refine the approximate solution.
Final stage:	apply a procedure for optimality.

All our specialized algorithms solve large-sized problems efficiently as established by an extensive computational analysis. We consider the Subset Sum Problem (SSP), the Strongly Correlated 0-1KP (SCKP), the Inverse Strongly

Correlated 0-1KP and the BKP (BSSP and BSCKP) in Sections 3, 4, 5 and 6 respectively. For all cases, the final stage when required (if an optimal solution is not obtained in the previous stages) involves a partial enumeration. Our procedure "Partial Lexicographic Search" for this enumeration is common to all problem types and is described in next section.

2 PRELIMINARIES

The most effective exact algorithms for solving the 0-1KP utilize the concepts of branch-and-bound or dynamic programming. Usually, the enumeration is curtailed by the use of reduction procedures and upper bound tests. Martello and Toth (1997); Martello et al (1997) propose upper bounds by utilizing additional cardinality constraints. Their upper bounds are determined by solving a relaxed problem. The upper bound test proposed by Pisinger(1995b; 1997) is used to delete unpromising states after each iteration in a dynamic programming algorithm (MINKNAP). This method may lead to a complete enumeration for the Subset Sum Problem (SSP) and the Strongly Correlated 0-1 Knapsack Problem (SCKP).

As mentioned in the introduction our algorithms involve three stages. The first two stages produce an approximate solution which often is optimal. If an optimal solution is not obtained in the first two stages the final stage searches for an optimal solution using a partial lexicographic search (PLS) procedure. After termination of the second stage, the set of items will be partitioned into a "$1 - \textit{variable set}$" and a "$0 - \textit{variable set}$" yielding residual capacity r. If $r > 0$, the PLS procedure is applied to generate subsets of the 1-variable set and subsets of the 0-variable set in order to exchange so as to obtain an improved utilization of the knapsack capacity. More specifically, the procedure is described below:

The Partial Lexicographic Search Procedure (PLS)

Let I^1	:	$\{ i_j := 1;\ w_{i_j} \le w_{i_{j+1}},\ j = 1, 2, \ldots, h \}$ where h is an integer $0 < h < n$,
I_u^1	:	$\{i_j\!: 1 \le j \le u\}$ where $2 \le u \le h$,
$\Im_r$	:	$\bigcup_{s=2}^{r}$ {s-ordered subset of I^1 with the first element in I_u^1},
P_s	:	s-ordered subsets of I^1 with the first element in I_u^1,
$\Im_r$	:	$\bigcup_{s=2}^{r} \mathbf{P}_s$,
P_{js}	:	j^{th} element of P_s,
I^0	:	$\{i_j : x_{i_j} = 0\ ;\ w_{i_j} \le w_{i_{j+1}},\ j = 1, 2, \ldots, , n - h\}$,
I_v^0	:	$\{i_j : 1 \le j \le v\}$ where $2 \le v \le n - h$,
Q_t	:	t-ordered subsets of I^0 with the first element in I_v^0,
$\Re_r$	:	$\bigcup_{t=2}^{r} Q_t$,
q_{mt}	:	m^{th} element of Q_t,
z	:	optimal value,
sum	:	total weight sum(or profit sum)of 1-variables in second stage,
res	:	residual capacity in second stage.

Note that $|\Im_r| = \sum_{s=2}^{r}(h-u)\binom{u}{s-1} + \binom{u}{s}$,

$|P_s| = (h-u)\binom{u}{s-1} + \binom{u}{s}$,

$|R_r| = \sum_{t=2}^{r}(n-h-v)\binom{v}{t-1} + \binom{v}{t}$, and

$|Q_t| = (h-v)\binom{v}{t-1} + \binom{v}{t}$.

Let $J = |P_s|$ and $M = |Q_t|$.

PLS Procedure

input : *sum*, *res* ;

BEGIN

$kk = 2$;

WHILE $\ell \le r$; $kk \le k^*$ DO

$u = kk$; $v = kk$;

WHILE $kk < \ell$ DO

GENERATE all elements of $P_s = \{p_{js},\ j = 1, 2, ..., J\}$;

FIND $VR_j = \sum_{p_{js}} w$ + res , for all j ;

FOR $m = 1$ to M DO

GENERATE q_{mt} (m^{th} element of Q_t)

FIND $V_{mt} = \sum_{q_{mt}} w$;

FOR $j = 1$ to J DO $D_{jt} = VR_j - V_{mt}$;

IF $D_{jt} = 0$ THEN

$x_a = 0$, for all $a \in p_{js}$;

$x_b = 1$, for all $b \in q_{mt}$;

$z = c$;

STOP ;

ELSE

FIND $MD_{kk} = min_j(D_{jt})$;

$kk = kk + 1$; $\ell = \ell + 1$;

FIND $min\text{MD}_{kk} = min_{kk}(\text{MD}_{kk})$;

BACKTRACK TO FIND the corresponding ordered subset p_{js} and q_{mt} of $min\text{MD}_{kk}$;

$x_a = 0$, for all $a \in p_{js}$;

$x_b = 1$, for all $b \in q_{mt}$;

$z = sum + V_{mt} - VR_j$; STOP ;

END.

Experimentally, providing n is not too small, an optimal solution is always obtained by exchanging at most a few items around the item $k = max_j\{j : \sum_{i=1}^{j} w_i \leq c\}$. Therefore, in our implementation we restrict the value of two parameters k^* and r to 4. The indices of 1-variables are started counting from item k in descending order until $|I_u^1| = 4$ and similarly the indices of 0-variables are started counting from item $k+1$ in ascending order until $|I_v^0| = 4$.

3 SUBSET SUM PROBLEM (SSP)

Given a knapsack with a capacity c and a set of n items with item i having a weight w_i , the **Subset Sum Problem (SSP)** is to select a subset of items to fill the knapsack such that the total sum of the weights is maximized without exceeding the capacity c. Without loss of generality, it is assumed that w_i and c are positive integers satisfying $max_i\{w_i\} < c$ and $\sum_{i=1}^{n} w_i > c$.

SSP is called the *Value-Independent Knapsack Problem* (Balas and Zemel (1980)) or Stickstacking Problem (Aherns and Finke (1975)). The most important applications are cargo loading (load a ship with the maximum cargo), cutting stock (find the best way of cutting a portion of space into pieces of different size) and two-processor scheduling (assign n tasks to two processors so as to minimize the finishing time of the entire operation; (Martello and Toth (1984)). Before describing our algorithm for the SSP, we briefly outline the literature work on this problem. The method of dynamic programming has been extensively used.

The first dynamic programming algorithm for SSP was proposed by Faaland (1973). Aherns and Finke (1975) proposed a more effective approach with the average time and space reduction to solve the SSP. Martello and Toth (1984) presented a specialized exact algorithm for SSP which is a mixture of dynamic programming and tree search. This method efficiently solves the SSP with weight ordering. Chvatal (1980) identified SSP as a hard problem when w_i is a randomly chosen integer between 1 and $10^{n/2}$ and $c = \lfloor \frac{1}{2} \sum_{i=1}^{n} w_i \rfloor$ since every recursive algorithm (combination of branch-and-bound, dynamic programming and rudimentary arguments) has an exponential time complexity. Pisinger (1995b) showed that any optimal solution to SSP may be reached through a series of balanced operations applied to dynamic programming with time and space bound of $O(n^2r)$ where r is the weight bound. The recent exact algorithm, COMBO for general 0-1KP's proposed by Martello et al (1997) is also able to solve SSP efficiently.

As an alternative to the above exact dynamic programming based algorithms, fully polynomial-time approximation schemes for SSP have been proposed (Martello amd Toth (1990a); Ibarra and Kim (1975); Lawler (1979)) .

With these schemes any prefixed worst-case performance ratio can be obtained in polynomial time; the time and space complexities grow polynomially in the inverse of the relative error. The schemes are based on the idea of partitioning the items into "*large*" and "*small*" items and solving the problem associated with large items before completing the solution in a greedy way with the small items. Martello and Toth (1984) detailed a worst-case analysis of this polynomial-time approximation scheme for SSP and proved that its worst-case performance is better than that of Johnson's well-known scheme. Fischetti (1990) presented a polynomial-time approximation scheme for SSP requiring linear space. The scheme is based on a local-search heuristic. The complementary knapsack is defined to collect all items which are not inserted into the knapsack with capacity as the difference of total weight and the capacity of SSP. The algorithm is based on the idea that minimizing the excess of overfilling the complementary knapsack is equivalent to optimally filling, without exceeding the knapsack capacity. Soma et al (1995) presented a variant of the polynomial approximation scheme of Martello and Toth (1984) for SSP. The algorithm has a better experimental error behavior and comparable running time with the same error. The most recent approximation algorithms for k-items SSP (kSSP) which is a special case of SSP with the additional constraint $\sum_{j \in N} x_j \leq k$; $N = \{1, 2, \ldots, n\}$, $1 \leq k \leq n$ have been presented by Capara et al (1998).

A number of other approaches have been studied. Probabilistic results for the SSP were first presented by D' Atri and Puech (1982). Assuming that the weights are independently drawn from a uniform distribution over $\{1, 2, \ldots, b\}$ where b is an upper bound on the weight value, they proved that for $b < n$ the simple greedy algorithm can solve the SSP with probability tending to 1 as n goes to infinity. Lagaries and Odlyzko (1983) considered the SSP with equality constraint by assuming that the weights are independently drawn from a uniform distribution over $\{1, 2, \ldots, 2^{bn^2}\}$ and the knapsack capacity is the sum of the weights in a randomly chosen subset. The algorithms solve almost all instances for the case $b > 1$. The recent paper of Pferchy (1998) investigated the expected values of some parameter illustrating the behavior of the same greedy algorithms by using some of the technical tools of D' Atri and Puech (1982).

We now describe our algorithm for SSP which involves three stages. The first stage is based on the same idea of the standard greedy approach of Johnson (1974) and no sorting of the weights is needed. The initial approximate solution is obtained as follows. Items 1, 2,..., k where $k = max\{j : \sum_{i=1}^{j} w_i \leq c\}$ are placed in the knapsack. The second stage attempts to improve the solution. Since in the first stage, the item sequence is randomly ordered, some items from $i = k + 1$ to n may have a weight less than or equal to the residual capacity $r = c - \sum_{i=1}^{k} w_i$ and so an improved solution can be obtained by adding items to the knapsack. After termination of the second stage, a good approximate solution will be obtained. To search for an optimal solution, the PLS procedure is applied in the final stage of the algorithm. The proposed algorithm can be described as follows :

Algorithm CK1

First stage:

```
BEGIN
    sum = 0;
    FOR i = 1 TO n DO
        sum = sum + w_i ;
        x_i = 1;
        IF sum = c THEN STOP;
        ELSE IF sum > c THEN sum = sum - w_i;
                             x_i = 0 ;
                             k = i - 1 ;
                             BREAK;
```

Second stage :

```
    r = c - sum ;
    j = k + 1 ;
    WHILE r > 0 and j ≤ n DO
        IF w_j ≤ r THEN r = r - w_j ;
                        x_j = 1;
                        sum = sum + w_j ;
                        j = j + 1;
        ELSE x_j = 0 ;
    IF r = 0 THEN STOP.
```

Final stage :

```
    ELSE IF r > 0 THEN call PLS procedure ( sum, r) ;
END.
```

We implemented our algorithm CK1 in ANSI C on a Silicon Graphic Workstation (SGI R5000) running at clock speed of 150 MHz. To provide a measure of the quality of our algorithm we also ran our test problems on the available code that implements the COMBO algorithm of Martello et al (1997). A comparative analysis based on 900 random problems is carried out. The integer weights $w_i's$ are randomly generated from a uniform distribution in $[1,R]$ where $R = 1000$ and $R = 10000$. The number of item n varied from 100 to 1000000. We use nine values of n : 100, 300, 1000, 3000, 10000, 30000, 100000,

Table 3.1 Average CPU time (in seconds) for CK1 and COMBO, Silicon Graphic R5000 Workstation.

n	$R = 1000$		$R = 10000$	
	CK1	COMBO	CK1	COMBO
1×10^2	0.000727	0.002433	0.009326	0.026341
3×10^2	0.001230	0.002809	0.004910	0.035066
1×10^3	0.002236	0.002924	0.007666	0.030336
3×10^3	0.001807	0.003451	0.015716	0.025546
1×10^4	0.001321	0.012222	0.022484	0.034772
3×10^4	0.004669	0.040582	0.027919	0.039076
1×10^5	0.015765	0.154293	0.023029	0.154645
3×10^5	0.048148	0.494771	0.049174	0.492164
1×10^6	0.158000	1.669074	0.160124	-

Table 3.2 Performance of first two stages for solving the SSP.

n	$\frac{z^*-z}{z^*}$		%z optimal	
	R=1000	R=10000	R=1000	R=10000
1×10^2	0.000761	0.000747	2	0
3×10^2	0.000081	0.000095	10	2
1×10^3	0.000007	0.000006	42	4
3×10^3	0.000000	0.000001	86	18
1×10^4	0.000000	0.000000	100	56
3×10^4	0.000000	0.000000	100	84
$\geq 10^5$	0.000000	0.000000	100	100

300000, 1000000. For each n and R we generated 50 problems. The capacity c is selected as $c = 0.5 \sum_{i=1}^{n} w_i$.

Table 3.1 shows the average computing time, in seconds, of CK1 and COMBO for each set of 50 problems. All 900 generated problems are solved to optimality within an average time of less than 0.17 second by CK1. Clearly, for all cases CK1 is competitive to COMBO. In the table a " – " means that there is not enough memory to run the program on the computer.

Our computational experiments suggested that the solution obtained at the end of Stage 2 is either optimal or close to optimal. Table 3.2 gives the average value of the statistic $\frac{z^*-z}{z^*}$, where z^* is the optimal value and z is the value of the feasible solution to SSP at the end of stage 2. It also gives the percentage of instances when z is optimal.

4 STRONGLY CORRELATED PROBLEM (SCKP)

The 0-1KP with strongly correlated data in which profit and weight are linearly related ($p_i = w_i + \alpha$) has been classified as a hard Knapsack Problem (Martello and Toth (1988)). The problem can be formulated as:

$$\begin{aligned} \text{maximize} \quad & z = \sum_{i=1}^{n} w_i x_i + \alpha \sum_{i=1}^{n} x_i \\ \text{subject to} \quad & \sum_{i=1}^{n} w_i x_i \leq c, \\ & x_i \in \{0,1\}, \qquad i \in N = \{1,2,\ldots,n\}, \end{aligned}$$

where the weight w_i, $i \in N$, α and c are positive integers. In order to avoid trivial cases we assume that $max_i \{w_i\} < c$, $\sum_{i=1}^{n} w_i > c$.

As algorithms for general 0-1KP's do not efficiently solve the SCKP, a number of specialized algorithms have recently been developed. Pandit and Ravi (1993) proposed a lexicographic search for SCKP based on the concept of determining a lower bound and an upper bound for the capacity when the weights are in non-decreasing order. This specialized algorithm is able to solve the SCKP with n up to 500 and shows a much better performance than MT2 the well-known algorithm of Martello and Toth (1988). However, it has been observed that the algorithm becomes ineffective when $\alpha = 0$ and n is large. For this case a superior algorithm (MTH) is given by Martello and Toth (1997) based on branch and bound combined with a partial dynamic programming recursion.

Pisinger proposed, in order to improve performance for hard KP's, the minimal algorithm (MINKNAP)(Pisinger (1997)) and a fast algorithm for SCKP's (Pisinger (1998)). The MINKNAP algorithm is based on a dynamic programming procedure approached by enumerating a minimal core. The algorithm was extensively tested on all four data types and performed well except for the SCKP. For the SCKP, the algorithm needs to enumerate a large core, in which millions of states may need to be examined. Martello et al (1997) proposed the COMBO algorithm which incorporates strong upper bound on the number of possible items in an optimal solution in the dynamic programming process. The dynamic programming part is essentially a refinement of MINKNAP and MTH. Strong upper bounds are obtained through the use of a maximum cardinality constraint. The algorithm was extensively tested and found to perform very well on all four data types. Recently, a specialized effective algorithm for SCKP (Pisinger (1998)) was presented. In this approach an additional constraint, derived from an extended cover, is added and then through the use of surrogate relaxation with optimal multipliers the problem is transformed into an equivalent SSP. For pre-processing, a two-optimal heuristic is applied. The algorithm can solve such SSP problems in milliseconds.

We note that in the paper of Pisinger (1998), a two-optimal heuristic is used based on the concept of adding the constraint: $\sum_{i=1}^{n} x_i = b - 1$ where $b = min\{j : \sum_{i=1}^{j} w_i > c\}$ when $w_i \leq w_{i+1}$ for $i = 1, 2, \ldots, n-1$. This constraint is utilized to obtain a strong lower bound. The author showed a proof that " an optimal solution x^* to SSP is also an optimal solution to SCKP provided that $\sum_{i=1}^{n} x_i^* = b - 1$". The solution procedure described uses the additional constraint $\sum_{i-1}^{n} x_i = \beta$ with descending values of $\beta < b$. The SCKP is formulated as an equivalent SSP using the model of SSP for optimality at the final stage.

We propose a three stage specialized algorithm to solve the SCKP. In the first stage, the items are ordered according to nondecreasing weights. An initial solution is obtained by the same way as the algorithm CK1. This solution has the maximum number of items. In the second stage, the solution is refined by exchanging between items $x_i = 1,\ i \leq k$ and item $x_j = 0,\ j > k$. The exchange is based on the criterion that for each iteration we choose the pair (i, j) which gives the minimum value of $w_i + r - w_j$, where r is a current residual capacity. This means that in each iteration the objective value will be increased. The iteration is terminated when $w_j > w_i + r$. After termination of the second stage, if the residual capacity r is positive, the PLS procedure will be applied to search for an optimal solution. The algorithm can be sketched as:

Algorithm CK2

First stage:

BEGIN

$sumw = 0$;

SORT ITEMS IN NON-DECREASING ORDER OF WEIGHTS

FOR $i = 1$ to n DO

$sumw = sumw + w_i$; $x_i = 1$;

IF $sumw = c$ THEN STOP; (An optimal solution is obtained)

ELSE IF $sumw > c$ THEN $sumw = sumw - w_i$;

$x_i = 0$;

$k = i - 1$; BREAK ;

Second stage:

$r = c - sumw$;

$maxw = w_k + r$;

$sump = sumw + \alpha k$; (comment: $p_i = w_i + \alpha$)

IF $w_{k+1} > maxw$ THEN call PLS procedure($r, sump$);

$jj = k$; $s = 1$;

DO

FOR $j = k + 1$ to n THEN DO

```
            IF x_j = 0 and w_j < maxw THEN DO
                  FOR i = 1 to k DO
                        IF x_i = 1 and w_i = w_j - r THEN
                                    x_i = 0; x_j = 1;
                                    sump = sump + w_j - w_i ;
                                    STOP ;
                        ELSE IF x_i = 1 and w_i > w_j - r THEN
                                    FIND diff_s = w_i + r - w_j ;
            s = s + 1;
      FIND mindiff = min_s(diff_s)
        BACKTRACK TO FIND THE CORRESPONDING
        x_m, m ∈ I_1 and x_t, t ∈ I_0 OF mindiff
        x_m = 0; x_t = 1 ;
        r = r - w_t + w_m ;
        sump = sump + w_t - w_m ;
        IF r = 0 THEN STOP ;
        maxw = w_t + r ;
        jj = t ;
```

Final stage:

```
        WHILE r > 0 DO call PLS procedure(r, sump);
END.
```

We implemented CK2 in ANSI C on a Silicon Graphic Workstation (SGI R5000) and the following tables present the computational results on the 900 random problems described earlier. We restricted the running time for each problem with $n < 300$ to 1 second. Again we compare our results with COMBO algorithm.

Table 4.1 shows the average computing time, in seconds, of CK2 for each set of 50 problems. Of the 900 random problems, 891 are solved optimally within the specified time of 1 second for each problem with $n < 300$. In the case of $n = 100$, CK2 with the maximum value of k^* and r (in PLS procedure) set at 4, cannot solve, in the specified time of 1 second per problem, 1 and 8 problems when $R = 1000$ and 10000 , respectively. We observe that when n is greater than 1000, the average computing time of our CK2 does not depend on the value of R since for each n, the computing times between the two values of R are very close. The value in bracket shows the average CPU time for the 50 problems solved by COMBO for the case in which CK2 cannot solve

Table 4.1 Average CPU time (in seconds) for CK2 and COMBO, Silicon Graphic R5000 Workstation.

n	$R = 1000$		$R = 10000$		Average
	CK2	COMBO	CK2	COMBO	sorting time CK2
1×10^2	0.000952[a]	0.014473[a] (0.015158)*	0.019675[b] (0.134450)*	0.087283[b]	0.00015
3×10^2	0.000706	0.018304	0.001205	0.073089	0.00051
1×10^3	0.002348	0.024245	0.003129	0.051271	0.00205
3×10^3	0.008289	0.037707	0.008851	0.037001	0.00703
1×10^4	0.030574	0.091637	0.031527	0.089466	0.02721
3×10^4	0.104236	0.224132	0.105222	0.284032	0.09427
1×10^5	0.394602	0.797120	0.397086	1.068551	0.35806
3×10^5	1.369840	2.769327	1.305274	3.274306	1.20310
1×10^6	4.858109	8.974759	4.820616	-	4.48000

a : The average CPU time for the 49 problems solved optimally.
b : The average CPU time for the 42 problems solved optimally.
* : The average CPU time for all 50 problems.

Table 4.2 Performance of first two stages for solving the SCKP.

n	$\frac{z^*-z}{z^*}$		%z optimal	
	R=1000	R=10000	R=1000	R=10000
1×10^2	0.000005	0.000000	86	20
3×10^2	0.000000	0.000000	100	96
$\geq 10^3$	0.000000	0.000000	100	100

all 50 problems. We found that an optimal solution for the problems of size $n \leq 100000$ can be obtained within 0.4 second. The bulk of the CPU time in our algorithm is due to sorting.

Table 4.2 gives the average value of the statistic $\frac{z^*-z}{z^*}$, where z^* is the optimal value and z is the value of the feasible solution to SCKP at the end of stage 2. It also gives the percentage of instances when z is optimal.

5 INVERSE STRONGLY CORRELATED KNAPSACK PROBLEM (ISCKP)

The SCKP may be generalized to the Inverse Strongly Correlated Knapsack Problem (ISCKP) which is defined as :

$$\begin{aligned} \text{maximize} \quad & z = \sum_{i=1}^{n} w_i x_i - \alpha \sum_{i=1}^{n} x_i \\ \text{subjectto} \quad & \sum_{i=1}^{n} w_i x_i \leq c, \\ & x_i \in \{0,1\}, \qquad i \in N = \{1,2,\ldots,n\}, \end{aligned}$$

where the weight w_i , $i \in N$, α and c are positive integers and the weight $w_i > \alpha$.

A specialized algorithm based on lexicographic search was proposed by Pandit and Ravi (1993) for solving the *Strongly Correlated Knapsack Problem* including both SCKP and ISCKP. The algorithm efficiently solves such problems with n up to 500. Recently Martello et al (1997) presented a comparison of MT2, MINKNAP, MTH and COMBO algorithms for ISCKP and concluded that the COMBO algorithm is superior.

We propose a two stage specialized algorithm to solve ISCKP. In the first stage, the items are ordered according to nonincreasing weights. An initial solution is obtained by placing items $1, 2, \ldots, k$ where $k = max_j\{j : \sum_{i=1}^{j} w_i \leq c\}$ in the knapsack. This solution has the minimum number of items. If this stage fails to make full capacity, it means that an optimal solution must contain at least k items. Observe that items i and j with $i \leq k$ and $j > k$ cannot be exchanged since $w_j < w_i$ for all j, i . In addition, since we modified PLS procedure to search for an optimal solution by setting $u = k - 1$ and $v = k$ in the WHILE loop, an optimal solution is obtained whenever the residual capacity $r \leq \alpha$. The algorithm can be sketched as:

Algorithm CK3

First stage :

BEGIN

$sumw = 0$;

SORT ITEMS IN NON-INCREASING ORDER OF WEIGHTS

FOR $i = 1$ to n DO

$sumw = sumw + w_i$; $x_i = 1$;

IF $sumw = c$ THEN STOP; (An optimal solution is obtained)

ELSE IF $sumw > c$ THEN $sumw = sumw - w_i$;

$x_i = 0;$

$k = i - 1$; BREAK ;

Second stage:

$r = c - sumw;$

IF $r \leq \alpha$ THEN STOP; (An optimal solution is obtained)

ELSE IF $r > \alpha$ THEN call PLS procedure (sum, r) ;

(by setting $u = k - 1$ and $v = k$)

END

We implemented CK3 in ANSI C on a Silicon Graphic Workstation (SGI R5000) and the following tables present the computational results on the 900 random problems described earlier. As before, we restrict the maximum value of k^* and r to 4 and the running time for each problem with $n < 300$ to 1 second. A comparative analysis with algorithm COMBO is also presented.

Table 5.1 Average CPU time (in seconds) for CK3 and COMBO, Silicon Graphic R500 Workstation.

n	$R = 1000$		$R = 10000$	
	CK3	COMBO	CK3	COMBO
1×10^2	0.0006973[c]	0.015787[c] (0.015133)*	0.003395[d]	0.091319[d] (0.110414)*
3×10^2	0.001255	0.023907	0.003035	0.067612
1×10^3	0.003201	0.045295	0.008654[e]	0.058169[e] (0.801798)*
3×10^3	0.010553	0.034925	0.012395	0.075980
1×10^4	0.039616	0.085456	0.041611	0.111668
3×10^4	0.141803	0.252560	0.143916	0.291328
1×10^5	0.513364	0.857881	0.518675	1.031159
3×10^5	1.705176	2.891316	1.709795	3.519782
1×10^6	6.067827	9.202884	5.900765	-

c : The average CPU time for the 47 problems solved optimally.
d : The average CPU time for the 45 problems solved optimally.
e : The average CPU time for the 48 problems solved optimally.
* : The average CPU time for all 50 problems.

Table 5.1 shows the average computing time, in seconds, of CK3 for each set of 50 problems. Of the 900 random problems, 890 are solved optimally. The 10 that are not solved within 1 second by CK3 include : 3, 5 and 2 problems with $(n, R) = (100, 1000)$, $(100, 10000)$ and $(100, 10000)$, respectively. We observe

Table 5.2 Performance of first stage for solving the ISCKP.

n	$\frac{z^*-z}{z^*}$	
	R=1000	R = 10000
1×10^2	0.014542	0.011837
3×10^2	0.005212	0.004416
1×10^3	0.001452	0.001425
3×10^3	0.000508	0.000525
1×10^4	0.000156	0.000150
3×10^4	0.000045	0.000048
1×10^5	0.000014	0.000015
3×10^5	0.000004	0.000004
1×10^6	0.000001	0.000002

that when n is greater than 10000, the average computing time of CK3 does not depend on the value of R since for each n, the computing times between the two values of R are very close. We found that an optimal solution for problems of size $n \leq 100000$ can be obtained within 0.52 second. The value in brackets gives the average CPU time for 50 problems solved by algorithm COMBO for the case which CK3 cannot solve all 50 problems.

Table 5.2 gives the average value of the statistic $\frac{z^*-z}{z^*}$, where z^* is the optimal value and z is the value of the feasible solution to ISCKP at the end of Stage 1.

6 BOUNDED KNAPSACK PROBLEM (BKP)

A generalization of the 0-1KP in which up to b_i items of type i can be selected to fill in a given knapsack capacity c is called the **Bounded Knapsack Problem**. This problem is of course $NP-hard$. Mathematically, BKP can be formulated as:

$$\text{maximize} \quad z = \sum_{i=1}^{n} p_i x_i$$

$$\text{subject to} \quad \sum_{i=1}^{n} w_i x_i \leq c,$$

$x_i \in \{0, 1, 2, ..., b_i\}$ and b_i is a positive integer, for $i = 1, 2, \ldots, n$. We have the classical 0-1KP when $b_i = 1$ for each i.

BKP has been studied intensively by only a few researchers. Ingargiola and Korsh (1973) presented a branch-search algorithm combined with a reduction algorithm to solve uncorrelated data type with common upper bound of items. The algorithm is able to solve problems with n up to 500. Martello and Toth (1977) proposed a branch and bound procedure to solve uncorrelated and weakly correlated data types with up to $n = 100$.

Martello amd Toth (1990a) tackled the BKP by transforming it into a 0-1KP using binary expansion. Each item type i in BKP is expanded to $\tilde{n} = \lceil log_2(b_i + 1) \rceil$ binary variables in the equivalent 0-1KP. The profits and weights are, respectively, $2^{k-1}p_i$, $2^{k-1}w_i$ for $k < \tilde{n}$ and $(b_i - \sum_{j-1}^{\tilde{n}-1} 2^{j-1})p_i$, $(b_i - \sum_{j-1}^{\tilde{n}-1} 2^{j-1})w_i$ for k $=\tilde{n}$. Now any algorithm for 0-1KP can be applied to the transformed problem. The authors suggest applying the MT2 algorithm. This procedure is effective in solving uncorrelated and weakly correlated types with n up to 50000. However, because of the large number of variables produced by the transformation, computer memory restriction limits the size of the BKP that can be solved. As an alternative, the authors also propose the application of the MT2 heuristic (MT2 is halted at the end of step 2) to the transformed problem. This approximation algorithm can solve all types of data but not efficiently.

Recently, Pisinger (1995b) modified the MINKNAP algorithm to solve BKP, this modified algorithm is called BOUKNAP. With $R = 1000$ and b_i uniformly chosen from the interval [5,10], the algorithm can solve almost all types of data with n up to 100000 in less than a second except for the strongly correlated type. This type requires up to 450 seconds computing time for $n = 100000$.

Martello et al (1997) classify the Bounded Strongly Correlated Problem as a class of hard problems. The authors showed that BSCKP can be solved by the algorithm COMBO when the problems are transformed to 0-1KP using the technique described in Martello amd Toth (1990a). The BSCKP are generated with w_j uniformly random in [1,1000] , $p_j = w_j + 100$ and the bounds b_j uniformly random in [1,10]. COMBO is able to solve such problems with n up to 10000 in average CPU time 9.99 seconds. The authors suggest that specialized algorithms for the BSCKP should be developed since the algorithm COMBO is able to solve the similar KP 100 times faster than 0-1KP transformed from the BSCKP.

We modify algorithm CK4 for the BSSP and algorithm CK5 for the BSCKP. The first two stages of the modified algorithm are slightly different from algorithm CK1 and CK2 respectively. If $w_i \leq r$ the item type i will be placed in the knapsack until $w_i > r$ (a current residual capacity) and $x_i < b_i$. The algorithm CK4 and CK5 can be described as :

Algorithm CK4

First stage:

BEGIN

$sum = 0$; $sumb = 0$;

FOR $i = 1$ TO n DO

$sum = sum + w_i b_i$;

$x_i = b_i$;

IF $sum = c$ THEN STOP ; (An optimal solution is obtained)

ELSE IF $sum > c$ THEN $sum = sum - w_i b_i$;

$x_i = 0$;

$k = i$;

BREAK;

Second stage :

$r = c - sum$;

FOR $i = k$ TO n DO

$x_i = 0$;

IF $w_i = r$ THEN STOP (An optimal solution is obtained)

ELSE IF $w_i < r$ THEN WHILE $w_i \leq r$ and $x_i < b_i$ DO

$r = r - w_i$; $x_i = x_i + 1$;

$sum = sum + w_i$;

IF $r = 0$ THEN STOP ; (An optimal solution is obtained)

Final stage :

ELSE IF $r > 0$ THEN call PLS procedure (sum, r) ;

END.

Algorithm CK5

First stage:

BEGIN

$sumw = 0$; $sumb = 0$; $sump = 0$;

SORT ITEMS IN NON-DECREASING ORDER OF WEIGHTS

FOR $i = 1$ to n DO

```
        sumw = sumw + w_i b_i ; x_i = b_i ;
        IF sumw = c THEN STOP; (An optimal solution is obtained)
        ELSE IF sumw > c THEN sumw = sumw − w_i b_i ;
                          x_i = 0 ;
                          k = i − 1 ; kk = k;
                          r = c − sumw ;
                          IF w_i < r THEN WHILE w_i ≤ r DO
                                  r = r − w_i ;
                                  x_i = x_i + 1 ;
                                  sumw = sumw + w_{k+1} x_{k+1};
                                  sumb = sumb + x_{k+1} ;
                                  kk = k + 1;
sump = sumw + α sumb ; (comment: p_i = w_i + α ) BREAK;
IF sumw = c THEN STOP; ( An optimal solution is obtained )
```

Second stage:

```
maxw = w_{kk} + r ;
IF w_{kk+1} > maxw THEN call PLS procedure(r, sump);
jj = kk ; s = 1 ;
DO
FOR j = kk + 1 to n THEN DO
IF w_j = maxw THEN x_j = x_j + 1 ; x_{kk} = x_{kk} − 1 ;
                   sump = sump − w_{kk} + w_j ;
                   STOP ; ( An optimal solution is obtained)
IF x_j = 0 and w_j < maxw THEN DO
              FOR i = 1 to kk + 1 DO
                   IF x_i > 0 and w_i = w_j − r THEN
                           x_i = x_i − 1;   x_j = x_j + 1;
                           sump = sump + w_j − w_i ;
                           STOP ;
                   ELSE IF x_i > 0 and w_i > w_j − r THEN
                           FIND diff_s = w_i + r − w_j ;
```

$s = s + 1;$

FIND $mindiff = min_s(diff_s)$

BACKTRACK TO FIND THE CORRESPONDING x_m, $m \in I_1$ and x_t, $t \in I_0$ OF $mindiff$

$x_m = x_m - 1;\ x_t = x_t + 1$;

$r = r - w_t + w_m$;

$sump = sump + w_t - w_m$;

IF $r = 0$ THEN STOP ;

$maxw = w_t + r;$

$jj = t$;

Final stage:

WHILE $r > 0$ DO call PLS procedure(r, $sump$);

END.

CK4 and CK5 were implemented in ANSI C on a Silicon Graphic Workstation (SGI R5000) and the following tables present the computational results on the 1350 random problems. The integer weight $w_i's$ are randomly generated from a uniform distribution in [1,R] where R = 100, R = 1000, and R = 10000. The bounds b_i is randomly distributed in [1,50] and the capacity $c = 0.5 \sum_{i=1}^{n} w_i b_i$. We also use the nine values of n as described earlier. As before, we restrict the maximum value of k^* and r to 4. Further for $n \leq 300$ the running time for each problem restricted to 1 second. To provide a measure of the quality of our algorithms we also ran our test problems on the available code that implements the BOUKNAP algorithm of Pisinger (1995b).

Table 6.1 shows the average computing time, in seconds, of CK4 and BOUKNAP each set of 50 problems. All 1350 generated instances are solved to optimality within an average time of less than 0.27 second by CK4.

Table 6.3 shows the average computing time, in seconds, of CK5 and BOUKNAP for each set of 50 problems. Of 1350 random problems, 1313 are solved optimally. The 37 that are not solved within 1 second by CK5 include : 4, 25, 3 and 5 problems with (n, R) = (100,1000), (100,10000), (300,1000), and (300, 10000), respectively. The value in bracket shows the average CPU time for 50 problems solved by the algorithm BOUKNAP for the case in which CK5 cannot solve all 50 problems. We found that solving by CK5 an optimal solution of the problem size $n \leq 100000$ can be obtained within 0.42 second.

Our computational experiments suggested that the solution obtained at the end of the stage 2 is either optimal or close to optimal. Table 6.2 and Table 6.4 give the average value of the statistic $\frac{z^*-z}{z^*}$, where z^* is the optimal value and z is the value of the feasible solution to BSSP and BSCKP, respectively, at the end of stage 2. They also give the percentage of problems when z is optimal.

Table 6.1 Average CPU time (in seconds) for CK4 and BOUKNAP, Silicon Graphic R5000 Workstation.

n	$R = 100$		$R = 1000$		$R = 10000$	
	CK4	BNAP*	CK4	BNAP*	CK4	BNAP*
1×10^2	.00017	.00864	.00099	.13393	.01231	1.4468
3×10^2	.00017	.00999	.00092	.14035	.00417	1.2030
1×10^3	.00017	.00857	.00126	.11390	.00869	1.2935
3×10^3	.00051	.01053	.00132	.13755	.01085	1.3211
1×10^4	.00194	.02242	.01893	.12630	.01396	1.3947
3×10^4	.00615	.05592	.00061	.15865	.01131	-
1×10^5	.02493	.01555	.02617	.24837	.02589	-
3×10^5	.07633	.48930	.07768	-	.07856	-
1×10^6	.26009	-	.26688	-	.26072	-

* here BNAP is BOUKNAP

Table 6.2 Performance of first two stages for solving the BSSP

n	$\frac{z^*-z}{z^*}$			%z optimal		
	R=100	R=1000	R=10000	R=100	R=1000	R=10000
1×10^2	0.000007	0.000016	0.000019	78	26	4
3×10^2	0.000001	0.000002	0.000002	90	34	4
1×10^3	0.000000	0.000000	0.000000	100	70	12
3×10^3	0.000000	0.000000	0.000000	100	92	34
1×10^4	0.000000	0.000000	0.000000	100	100	70
3×10^4	0.000000	0.000000	0.000000	100	100	96
$\geq 10^5$	0.000000	0.000000	0.000000	100	100	100

Table 6.3 Average CPU time (in seconds) for CK5 and BOUKNAP, Silicon Graphic R5000 Workstation.

n	R=100		R=1000		R=10000	
	CK5	BNAP	CK5	BNAP	CK5	BNAP
1×10^2	.00018	.02131	$.00037^f$	$.29090^f$ (.27979)*	$.10431^g$	3.6325^g (2.8616)*
3×10^2	.00052	.09058	$.00058^h$	1.2395^h (1.1652)*	$.00656^i$	12.3345^i (11.1186)*
1×10^3	.00205	.38327	.00202	4.0606	.002562	-
3×10^3	.00733	1.5059	.00662	17.720	.00770	-
1×10^4	.05290	7.0201	.02860	60.8290	.02922	-
3×10^4	.10940	-	.10903	-	.11119	-
1×10^5	.41368	-	.40919	-	.41380	-
3×10^5	1.3678	-	1.3620	-	1.3610	-
1×10^6	4.9880	-	4.9413	-	4.8900	-

(here BNAP is BOUKNAP)
f : The average CPU time for the 46 problems solved optimally.
g : The average CPU time for the 25 problems solved optimally.
h : The average CPU time for the 47 problems solved optimally.
i : The average CPU time for the 45 problems solved optimally.
* : The average CPU time for all 50 problems.

Table 6.4 Performance of first two stages for solving the BSCKP.

n	$\frac{z^8 - z}{z^*}$			%z optimal		
	R=100	R=1000	R=10000	R=100	R=1000	R=10000
1×10^2	0.000000	0.000000	0.000000	98	84	16
3×10^2	0.000000	0.000000	0.000000	100	94	76
1×10^3	0.000000	0.000000	0.000000	98	100	100
$\geq 3 \times 10^3$	0.000000	0.000000	0.000000	100	100	100

7 CONCLUSIONS

This paper has focused on the hard 0-1Knapsack Problem of strongly correlated, subset sum and inverse strongly correlated data instances and the Bounded Knapsack Problem of strongly correlated and subset sum data instances. All our algorithms utilize the PLS Procedure in the search strategy for the optimal solution. From our computational experiences, the optimal solutions to most of the problems considered in this paper are obtained when the knapsacks are fully filled. However, there are some instances of strongly correlated data (SCKP, ISCKP and BSCKP) that cannot be solved in a reasonable time. These instances may need more enumerating by PLS procedure since their optimal solution has the total weight less than the knapsack capacity. We note that the small problems for this data type are difficult to solve due to increasing the range of random data.

In comparison to current efficient literature algorithms for solving general 0-1KP and BKP, CK1-5 all perform very well for all large-sized problems. Especially, CK4 and CK5 can solve problems with up to 1 million variables within 0.3 second for BSSP and less than 5 seconds for BSCKP while BOUKNAP can solve only n up to 300000 for BSSP and n up to 10000 for BSCKP.

Acknowledgments The authors are grateful to Professor D. Pisinger for providing the source code of the algorithm COMBO and BOURKNAP.

References

Aherns J.H., and Finke G. (1975), Merging and sorting applied to the 0-1 knapsack problem, *Operations Research*, Vol.23, pp. 1099-1109.

D' Atri G. and Puech C. (1982), Probabilistic analysis of the subset-sum problem, *Discrete Applied Mathematics*, Vol. 4, pp. 329-334.

Balas E. and Zemel E. (1980), An algorithm for large zero-one knapsack problems, *Operations Research*, Vol. 28, pp. 1130-1154.

Bellman, R. (1954), Some applications of the theory of dynamic programming-a o*review*, *Operations Research*, vol. 2, pp.275-288.

Bellman R.(1957), *Dynamic programming*, Princeton NJ.

Bellman R. and Dreyfus, S.E. (1962), *Applied Dynamic Programming*, Princeton University Press, Princeton, NJ.

Capara A., Kellerer H., Pferschy U. and Pisinger D. (1998), *Approximation algorithms for knapsack problems with cardinality constraints,* Technical report, DIKU, University of Copenhagen.

Chvatal V. (1980), Hard knapsack problem, *Operations Research*, Vol. 28, pp. 402-411.

Cord, J. (1964), A Method for Allocating Funds to Investment Projects when Returns Are Subject to Uncertainty, *Management Science*, vol. 10, no. 2, pp. 335-341.

Dantzig G.B.(1957), Discrete variables extremum problems, *Operations Research*, Vol.5 pp. 266-277.

Diffe, W. and Hellman, M.E. (1976), New directions in cryptography, *IEEE Transactions Information Theory*, IT-36, pp.644-654.

Dudzinski K., and Walukiewicz S. (1987), Exact methods for the knapsack problem and its generalizations, *European Journal of Operational Research*, Vol. 28, pp.3-21.

Faaland B. (1973), Solution of the value-independent knapsack problem by partitioning, *Operations Research*, Vol. 21, pp.332-337.

Fayard D. and Plateau G. (1982), An algorithm for the solution of the 0-1 knapsack problem, *Computing*, Vol.28, pp. 269-287.

Fischetti M.(1990), A new linear storage, polynomial-time approximation scheme for the subset-sum problem, *Discrete Applied Mathematics*, Vol. 26, pp. 61-77.

Gilmore P.C and Gomory R.E. (1963), A Linear Programming Approach to the Cutting Stock Problem II, *Operations Research*, vol. 11, pp.863-888.

Gilmore P.C and Gomory R.E.(1966), The theory and computation of knapsack functions, *Operations Research*, Vol.14, pp. 1045-1074.

Greenberg H., and Hegerich R.L.(1970), A branch search algorithm for the knapsack problem, *Management Science*, 16 (1970) 327-332.

Hansmann, F. (1961), Operations Research in the National Planning of Underdeveloped Countries, *Operations Research*, vol. 9, pp. 203-248.

Horowitz E. and Sahni S.(1974), Computing partitions with applications to the knapsack problem, *Journal of ACM*, Vol.21, pp. 277-292.

Ibarra O.H. and Kim C.E. (1975), Fast approximation algorithms for the knapsack and sum of subset problems, *Journal of ACM*, Vol. 22, pp. 463-468.

Ingargiola G.P. and Korsh J.F.(1973), A reduction algorithm for zero-one single knapsack problems, *Management Science*, Vol. 20, pp. 460-463.

Janakiram V.K., Gehringer E.F., Agrawal D.P., and Mehrotra R. (1988), A Randomized Parallel Branch-and-Bound Algorithm, *International Journal of Parallel Programming*, vol. 17, pp.277-300.

Johnson D.S.(1974), Approximation algorithms for combinatorial problems, *Journal of Computer and System Science*, Vol.9, pp. 256-278.

Kaplan S. (1966), Solution of the Lorie-Savage and Similar Integer Programming Problems by the Generalized Lagrange Multiplier Method, *Operations Research*, Vol.14, no. 6, pp.1130-1136.

Kindervater G.A.P. and Trienekens H.W.J.M. (1988), Experiments with Parallel Algorithms for Combinatorial Problems, *European Journal of Operational research*, vol. 33, pp.65-81.

Kolesar P.J.(1967), A branch and bound algorithm for the knapsack problem, *Management Science*, Vol.13, pp. 723-735.

Lagaries J.C. and Odlyzko. (1983), A.M., Solving low-density subset sum problem, *Proceeding 24th Annual IEEE Symposium Foundations of Computer Science*, pp. 1-10.

Lai T.C. and Sahni S.(1984), Anomalies in parallel branch-and-bound algorithms, *Communication of ACM*, Vol. 27, pp. 594-602.

Lawler E.L. (1979), Fast approximation algorithms for knapsack problems, *Mathematics of Operations Research* , Vol. 4, pp. 339-356.

Lee J., Shragowitz E. and Sahni S. (1988), A hypercube algorithm for the 0/1 knapsack problem, *Parallel and Distributed Computing*, Vol. 5, pp. 438-456.

Lin J. and Storer J.A.(1991), Processor-efficient hypercube algorithms for the knapsack problem, *Journal of Parallel and Distributed Computing*, Vol.11, pp.332-337.

Loots W. and Smith T.H.C.(1992), A parallel algorithm for the 0-1 knapsack problem, *International Journal of Parallel Programming*, Vol. 21, pp. 349-362.

Martello S. , Pisinger D., and Toth P. (1997), Dynamic programming and tight bounds for the 0-1 knapsack problem, *working paper* DIKU, University of Copenhagen, July 29, 1997.

Martello S.and Toth P.(1977), An upper bound for the zero-one knapsack problem and a branch and bound algorithm, *European Journal of Operational Research*, Vol. 1 pp. 169-175.

Martello S. and Toth P.(1984), A mixture of dynamic programming and branch-and-bound for the subset-sum problem, *Management Science*, Vol. 30, pp. 765-771.

Martello S. and Toth P.(1988), A new algorithm for the 0-1 knapsack problem, *Management Science*, Vol. 34, pp. 633-644.

Martello S. and Toth P.(1990a), *Knapsack Problems Agorithms and Computer Implementations*, John Wiley and Sons, Chichester.

Martello S. and Toth P.(1990b), An exact algorithm for large unbounded knapsack problems, *Operations Research Letters*, vol. 9, pp. 15-20.

Martello S. and Toth P.(1997), Upper bounds and algorithms for hard 0-1 knapsack problems, *Operations Research*, Vol. 45, pp. 768-778.

Müller -Merbach H. (1978), An improved upper bound for the zero-one knapsack problem : a note on the paper by Martello and Toth, *European Journal of Operational Research*, Vol. 2, pp. 212-213.

Nauss R.M.(1976), An efficient algorithm for the 0-1 knapsack problem, *Management Science*, Vol. 23, pp. 27-31.

Pandit S.N.N. and Ravi Kumar M. (1993), A lexicographic search for strongly correlated 0-1 knapsack problems, *Opsearch*, Vol. 30 pp. 97-116.

Pferchy U.(1998), Stochastic analysis of greedy algorithms for the subset sum problem, *working paper*, Universitat Graz, Institut fur Statistik und Operations Research, Austria.

Pisinger D.(1995a), An expanding-core algorithm for the exact 0-1 knapsack problem, *European Journal of Operational Research*, Vol. 87 pp. 175-187.

Pisinger D.(1995b) , *Algorithms for knapsack problem*, Ph.D. thesis.

Pisinger D.(1997), A minimal algorithm for the 0-1 knapsack problem, *Operations Research*, Vol. 45, pp. 758-767.

Pisinger D.(1998), A fast algorithm for strongly correlated knapsack problems, *Discrete Applied Mathematics*, Vol. 89 pp. 197-212.

Pisinger D. and Toth P.(1998), Knapsack Problems , in D.Z.Du, P. Paralos(eds.) *Handbook of Combinatorial Optimization*, Kluwer, Norwell.

Sahni S. (1975), Approximate algorithms for the 0-1 knapsack problem, *Journal of ACM*, Vol. 22, pp. 115-124.

Salkin, H. M. (1975), *Integer Programming*, Addison-Wesley Publishing Company, Inc.

Sinha, A. and Zottners, A.A. (1979), The multiple-choice knapsack problem, *Operations Research*, 27, pp. 503-515.

Soma N.Y. and Zinober A.S.I., Yanasse H.H. and P.J.Harley P.J. (1995), A polynomial approximation scheme for the subset sum problem, *Discrete Applied Mathematics*, Vol. 57, pp. 243-253.

Syslo, M., Deo, N. and Kowalik, J. S. (1983), *Discrete Optimization Algorithms with PASCAL programs*, Prentice-Hall, Inc., Englewood Cliffs, New Jersey.

Toth P. (1980), Dynamic programming algorithms for the zero-one knapsack problem, *Computing*, Vol. 25, pp. 29-45.

12 NON-STATIC NETWORK OPTIMIZATION PROBLEMS: A SURVEY

X. Cai, D. Sha

Department of System Engineering and Engineering Management
The Chinese University of Hong Kong
Shatin, N.T., Hong Kong

and C. K. Wong

Department of Computer Science and Engineering
The Chinese University of Hong Kong
Shatin, N.T., Hong Kong

Abstract: In this paper we provide a survey of studies on non-static network optimization problems. As contrast to static network optimization which considers networks that are assumed to be time-independent, non-static (or time-varying) network optimization addresses situations where the structure and/or parameters of the network under consideration may change over time. These non-static problems are more adequate and realistic in modelling real-world situations and have thus attracted considerable attention in recent years. We will review the relevant models, analyses, and algorithms reported in this important branch of network optimization.

Key words: Time-varying network optimization, Network flow, Survey

X. Yang et al. (eds.), Optimization Methods and Applications, 219-246.

1 INTRODUCTION

Network flow optimization is a problem domain that concerns with optimization problems over networks, which may arise in a wide variety of important fields. Most network optimization problems that have been studied in the literature are static in nature, in the sense that they assume that it takes zero time to traverse any arc in a network and all attributes of the network, for example, the cost to send flow on an arc and the capacity of an arc, are constant that do not change at all over time. These kinds of *static network flow problems* have been studied extensively in the literature; see Ahuja, Magnanti and Orlin (1991,1993), Adel'son-Vel'ski, Dinic and Karzanov (1975), Bazaraa, Jarvis and Sherali (1990), Ford and Fulkerson (1962), Gupta (1985), Iri (1969), Jensen and Barnes (1980), Lawler (1976), and Minieka (1978).

In the real world, networks are, however, time-varying in essence. Therefore the static model is only an approximation of a dynamic network. In a dynamic network, any flow must take a period of time to traverse an arc and the parameters (such as arc capacities) of the network may also change over time. A flow optimization problem based on this type of networks is called *non-static network flow optimization problem*. In such a problem, how to control the flow becomes very important, since waiting at a node, or traveling along a particular arc, may allow one to catch the best timing along his path, and therefore achieve his overall objective, such as a minimum overall cost or a minimum travel time from the origin to the destination. There are plenty of decision making problems in practice that should be formulated by optimization problems on dynamic networks. These include those occurring in production-distribution systems, economic planning, traffic systems, energy systems, communication systems, manpower system, and many others.

The purpose of this article is to provide a state-of-the-art survey of studies on non-static network flow optimization problems. Relevant models, analyses, and algorithms reported in the literature will be summarized and discussed. The article is organized as follows: Section 2 introduces a basic model of non-static network optimization problems. Some related issues are also discussed. Then, Sections 3-7 are devoted to, respectively, the review of the following problems: Shortest path; Maximum flow; Minimum cost flow; Vehicle routing including the travelling salesman problem and the Chinese postman problem; and other network problems. Section 8 gives some concluding remarks.

2 THE NON-STATIC NETWORK FLOW MODEL

Traditionally, a network is modelled as follows: let $N = (V, A, r, c, l)$, where V is a vertex set, A is an arc set, and $(x, y) \in A$ is an arc that connects a pair of vertices x and y. Associated with each vertex $x \in V$, there is a supply $r(x)$, which represents the amount of flow that can be supplied by the vertex. Thus, $r(x)$ is positive if x is the source vertex; negative if x is the sink vertex, and zero if x is an intermediate vertex. Associated with each arc $(x, y) \in A$, there are two numbers $c(x, y)$ and $l(x, y)$, which represent the cost for a flow to traverse the arc and the capacity limit of the arc, respectively.

As long as the objective is specified, an optimization problem based on the network model above can be formulated. For example, if the objective is to minimize the total cost for sending a given amount of flow from the source vertex to the sink vertex, then the problem is the so-called *minimum cost network flow problem*, which can be formulated as follows, where $f(x, y)$ denotes the flow value on arc (x, y):

$$\min \quad \sum_{(x,y)\in A} c(x,y)f(x,y) \tag{2.1}$$

$$s.t. \quad \sum_{(x,y)\in A} f(x,y) - \sum_{(y,x)\in A} f(y,x) = r(x), \quad \forall x \in V \tag{2.2}$$

$$0 \le f(x,y) \le l(x,y), \quad \forall (x,y) \in A \tag{2.3}$$

Formula (2.2) is called the *conservation conditions.*

The model stated above is static, in the sense that no element of *time* is involved. It is an inherent assumption of a static network that it takes zero time for any flow to traverse any arc. To address more realistic situations that occur in the real world, an important direction in studies of network optimization is to relax the static model to a dynamic one in which a *transit time* is needed to travel on an arc (x, y). Moreover, since the network under consideration (for example, the capacity of some of its arcs) is often varying over time, it is desirable to consider such a time-varying feature. A model to formulate these dynamic problems can be stated as follows:

Let $N = (V, A, b, c, l)$ be a network, where V is the vertex set and A is the arc set (throughout this article, we let $m = |A|$ and $n = |V|$). Associated with each arc $(x, y) \in A$, there are three numbers: $b(x, y, t)$, $c(x, y, t)$, and $l(x, y, t)$, which represent, respectively, the transit time (the amount of time needed to traverse the arc), the transit cost (the cost incurred by traversing the arc), and the capacity of the arc, which are all functions of time t. Waiting at an intermediate vertex x is allowed and the waiting time is a decision variable. Accordingly, a waiting cost $c(x, t)$ may be caused and also there may be a vertex capacity $l(x, t)$ that limits the maximum amount of flow that can wait at the vertex. For each vertex $x \in V$, its supply $r(x, t)$, capacity $l(x, t)$, and waiting cost $c(x, t)$ are, in general, also functions of time t. To illustrate, a typical non-static network is shown in Figure 2.1.

Again, as long as the objective to be optimized is specified, an optimization problem based on the network model above can be formulated. For example, if the objective is to minimize the total cost for sending a given amount of flow from the source vertex to the sink vertex, then we have the non-static minimum cost network flow problem as follows, where $f(x, y, t)$ denotes the flow value transmitting on arc (x, y) during the time period $[t, t + b(x, y, t)]$, and $f(x, t)$

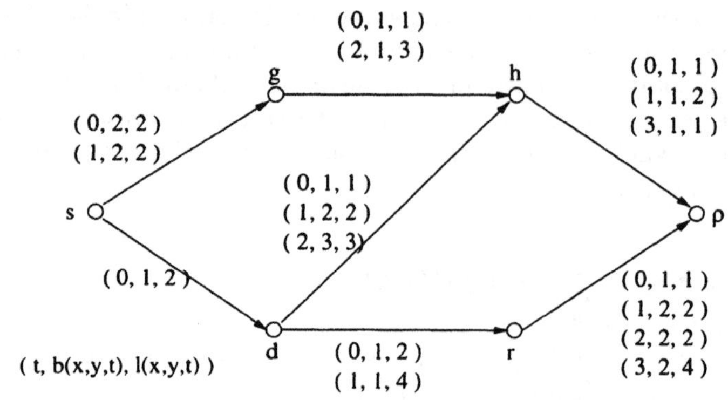

Figure 2.1 In the figure, s is the source vertex and p is the sink vertex. The three numbers inside each pair of brackets associated with an arc are t, $b(x,y,t)$ and $c(x,y,t)$ respectively. For instance, $(0,1,2)$ near the arc (s,d) means at time 0, transit time $b(s,d,0)$ is 1 and capacity limit $c(s,d,0)$ is 2, and at other times, $b(s,d,t) = +\infty$ and $c(s,d,t) = 0$. We assume, in this example, that no waiting time at any internal vertex is allowed, so $l(x,y,t) = 0$ for all x and t.

denotes the flow value waiting at vertex x during time period $[t, t+1)$:

$$\begin{aligned}
\min \quad & \sum_{(x,y)\in A} \sum_{t} c(x,y,t) f(x,y,t) + \sum_{(x)\in V} \sum_{t} c(x,t) f(x,t) \\
s.t. \quad & \sum_{(x,y)\in A} f(x,y,t) - \sum_{(y,x)\in A} \sum_{\{u|u+b(y,x,u)=t\}} f(y,x,u) + f(x,t) = r(x,t) \\
& \qquad \forall x \in V, t = 0,1,...,T \\
& 0 \le f(x,y,t) \le l(x,y,t), \quad \forall (x,y) \in A, t = 0,1,...,T \\
& 0 \le f(x,t) \le l(x,t), \quad \forall x \in V, t = 0,1,...,T
\end{aligned}$$

Obviously, if $b = 0$, and if flow can be sent from any vertex at time zero only, then the model above reduces to the traditional static network model. If $b > 0$ and all r, l and c are time-independent, then the network is usually called a *dynamic network*. Otherwise, if all r, l and c are time-dependent, then the network is called a *time-varying network*. The network model is said to be *discrete* if the time horizon $[0, T]$ can be divided into some time periods of equal width and all numbers associated with the network remain constants within a same time period. Otherwise, the network model is said to be *continuous*. If $T = +\infty$, it is an *infinite model*; otherwise, it is a *finite model*. In general, we call the network problem involving nonzero transit time (that is, $b > 0$) a *non-static network model*.

Remark 1. The non-static network model can describe not only situations in which the parameters of the network considered change over time, but also sce-

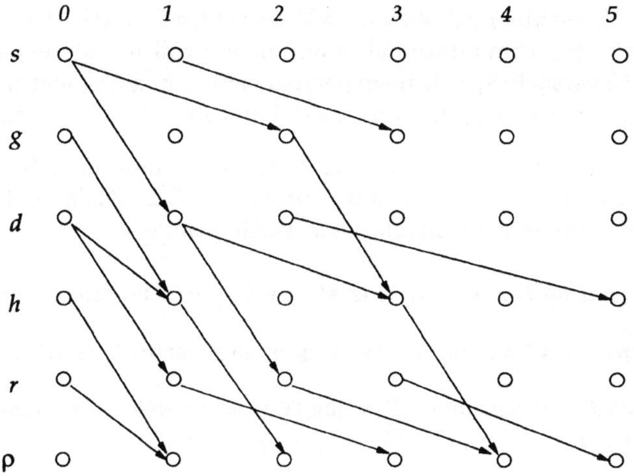

Figure 2.2 The time-expended network generated for the network of Figure 2.1.

narios in which the structure of the network changes at some particular times. Particularly, to describe some cases in which an arc (x, y) is broken at time t, one can let $c(x, y, t) = 0$ or $b(x, y, t) = \infty$. Sancho (1994), Sexton and Choi (1984), and Solomon (1986,1987) discuss a type of dynamic network optimization problems with *time windows*, where a vertex can be visited within a given time period only. This constraint can be absorbed in the non-static network above by such an approach: Split the vertex into two vertices, create an arc to connect them and let the capacity of this arc to be zero for all times outside the the given time window.

Remark 2. Sometimes a discrete non-static network model may be converted to an equivalent static network (see Ford and Fulkerson (1958)). The following describes such a conversion. Let $V(t)$ be the set of vertices in period of time t, $t = 1, 2, ..., T$. A vertex is represented by the two-tuple $(x, t) \in V(t)$. Let $A(t)$ be the set of arcs that have their origin vertices $(x, t) \in V(t)$. An arc is represented by two-tuple of vertices $((x, t), (y, t')) \in A(t)$, where $t' = t + b(x, y, t)$. An arc $((x, t), (x, t+1))$ represents the flow waiting at vertex x during time $[t, t + 1]$. Combine all $V(t)$ to obtain the new vertex set V, and combine all $A(t)$ to obtain the new arc set A. Attach c_t, c_t, l and r to the new vertex and the new arc correspondingly and get a new static network without transit times. This type of static network is usually called a *time-expended network*. Figure 2.2 shows a simple example that converts the network of Figure 2.1 into a time-expended network.

One should note that the size of the time-expended network is substantially larger than the original one. For example, for a time-varying network with time limit T, its time-expanded network will have $O(nT)$ vertices and $O(mT)$ arcs. Consequently, the computational requirements will usually expand rapidly as T grows. Nevertheless, a time-expended network can sometimes be used in analysis of a time-varying network model and its solution algorithms.

As we mentioned earlier, waiting time at an intermediate vertex is now a decision variable in the non-static problem. Cai, Kloks and Wong (1997) consider three different constraints for waiting times:

(i) *Zero waiting times* – Waiting at any vertex x is strictly prohibited;

(ii) *Arbitrary waiting times* – Waiting at any vertex x is arbitrarily allowed;

(iii) *Bounded waiting times* – Waiting time at a vertex x is bounded by a given number u_x.

Although Cai, Kloks and Wong deal with the above constraints for the time-varying shortest path problem, these constraints may exist in other non-static network flow problems. We note that the first two constraints above can be absorbed into a time-expanded network, but the third one can only be dealed with as a side constraint of the time-expanded network if there is a capacity limit for a vertex.

The basic non-static network model as discussed above can lead to various variations of non-static network optimization problems, depending on the objectives to be optimized and the specific constraints to be considered. In the following sections, we will focus our discussions on some important versions of these problems that have been addressed in the literature. These include: the shortest path problem, the maximum flow problem, the minimum cost flow problem, the vehicle routing problem, the minimum spanning tree problem, and the quickest path problem.

3 NON-STATIC SHORTEST PATH PROBLEMS

A static shortest path problem can be stated as: *Given a network $N = (V, A, c)$ where $c(x, y)$ is a cost (length) associated with each arc (x, y), and a source vertex s, find a path from s to $x \in V \backslash \{s\}$ such that the cost (length) of the path is minimum.*

The static shortest path problem can be regarded as a special case of the minimum cost flow problem by letting $r(s) = 1$, $r(x) = -1$ and $r(y) = 0$ for all other vertices. The static shortest path problem and its variants have been widely studied in the literature; see Dreyfus (1969), Papadimitrion and Steiglitz (1982), and Deo and Pang (1984). As also shown below, the shortest path problem plays an important role in the research of network optimization problems. Numerous algorithms for network optimization problems, such as the maximum flow problem, the minimum cost flow problem, and the quickest path problem, use the solution for the shortest path problem (see, e.g., Ahuja, Magnanti and Orlin (1993)).

A non-static shortest path problem can be stated as: *Given a non-static network $N = (V, A, b, c)$, where V is a vertex set, A a arc set, b the transit time, c the cost, and T the time limit, find a path from s to $x \in V\backslash\{s\}$ such that the cost (length) of the path is minimum subject to the constraint that the path can be traversed within the time limit T.*

It is well-known that the static shortest path problem is solvable in polynomial time (see, e.g., Papadimitrion and Steiglitz (1982)). However, the non-static shortest path problem is NP-complete; see Ahuja, Magnanti and Orlin (1993).

Non-static shortest path problems can be grouped into two classes. The first class consider transit times $b(x, y, t)$, transit costs $c(x, y, t)$, and waiting costs $c(x, t)$, which depend upon the starting time t at vertex x (Cooke and Halsey (1966), Orda and Rom (1990,1991), Philpott and Mees (1992), Psaraftis and Tsitsiklis (1993), and Cai, Kloks and Waong (1997)). The second class of problems address the so-called *time-window constraints*, which specify that a certain arc $(x, y) \in A$ can only be traversed within a given time period, and/or a certain vertex $x \in V$ can only be visited within a given time period. These time periods are prespecified and called time windows (Desrsier and Soumis (1988a,1988b), Sancho (1992,1994), and Loachim, Gelinas, Soumis and Desrosiers (1998)).

(1) **On the first class of problems**

Cooke and Halsey (1966) consider a discrete model in which the transit time $b(x, y, t)$ varies as a function of the departure time t at vertex x, the transit cost $c(x, y, t) = b(x, y, t)$, and waiting at vertex is strictly prohibited. The function $b(x, y, t)$ is defined as a positive integer-valued function of $t \in \{t_0, t_0 + 1, ...\}$. For any vertex $x \in V$, the problem is to find the path from s to x with starting time $t = t_0$, so that the total cost (i.e., the total travel time) is minimum. They establish an optimality and develop a dynamic programming algorithm, which can solve this problem in a finite number of iterations.

Orda and Rom (1990) generalize Cooke and Halsey's model to allow for the following additional features: (a) the function of the transit time $b(x, y, t)$ is arbitrary, and (b) waiting at vertices may occur as in the following three cases:

(i) Waiting at any vertex is unrestricted; namely, unlimited waiting is allowed everywhere along the path through the network.

(ii) Waiting is forbidden; namely, waiting is disallowed everywhere along the path through the network.

(iii) Only waiting at the source vertex is allowed; namely, waiting is disallowed everywhere along the path except at the source vertex which permits unlimited waiting.

They consider an infinite time horizon (i.e., $T = \infty$). For the cases (i) and (ii) above, they provide several polynomial labeling algorithms to find the optimal solutions. They also investigate properties of the optimal paths derived and show that for case (iii) where waiting at the source vertex is arbitrary allowed, a shortest path can be found that is simple (that is, each vertex appears in the path at most once) and that can achieve a cost as cheap as the most unrestricted path (that is, the optimal solution for case (i)). Orda and Rom (1991) continue to study an infinite continuous model in which both transit time $b(x, y, t)$ and transit cost $c(x, y, t)$ are continuous functions of the departure time t at vertex x while b is strictly positive and c is nonnegative. Waiting at vertices is allowed, and a waiting cost is introduced. They set $\pi_x(t)$ as waiting cost density where t is the departure time at x, and the function of waiting cost is defined by $P_x(\alpha, t) = \int_\alpha^t \pi_x(\theta)d\theta$, where α is the arrival time at vertex x and $0 \leq \alpha \leq t \leq \infty$. They assume that, for each arc $(x, y) \in A$, there is a countable union of open and non-overlapping intervals of time, during which this arc is unavailable. Similarly, there is a countable union of open and non-overlapping intervals during which waiting is prohibited. This assumption can be regarded as an extension of the time window constraint. They point out that, the problem always has a solution no mater whether the path is finite or infinite. For the former case, an algorithm is provided that can optimally solve the problem.

Philpott and Mees (1992) examine a finite continuous shortest path model for a vehicle traveling problem, in which the transit times, parking costs and restarting costs are all time-varying but the stopping costs (waiting costs) for each unit of time is fixed. They present an algorithm and derive conditions under which the algorithm converges to an optimal solution. Psaraftis and Tsitsiklis (1993) assume that the transit costs of arcs are known functions of certain environment variables at the vertices. Each of these variables evolves according to an independent Markov process. Nachtigall (1995) study a railway model in which the transit cost $c(x, y, t)$ depends on the time t when the passenger enters the vertex x. Suppose a passenger can depart from the original station at time τ, A *transit function* $f(\tau)$ gives the earliest possible arrival time at the destination for the passenger. A label correcting method is used to calculate the desired transit function for all starting times with one path search procedure.

Cai, Kloks and Wong (1997) study a discrete constrained time-varying shortest path problem, which is to find a shortest path, in a time-varying network, that can start from the source vertex at time zero and arrive at the destination no later than a given time limit T. They consider three types of waiting constraints, namely, zero waiting time, arbitrary waiting time and bounded waiting time. In addition, they also consider the possibility that the transit times $b(x, y, t)$ on some arcs can be equal to zero (this in fact increases the difficulty of finding an optimal solution if the solution approach to be used is dynamic programming). Three dynamic programming algorithms are presented, which can find, respectively, the optimal solutions to the three versions of the problem (corresponding to the three types of waiting constraints). They also show that all the three algorithms are efficient, in the sense that only pseudopolynomial running times are required, if some proper data structure is adopted in the solution procedure.

(2) **On the second class of problems**

The shortest path problem with time windows is firstly formulated as a sub-problem in the construction of school bus routes (Dosrosiers, Soumis and Desrochers (1988a)), where the tasks must be carried out according to a specified time schedule and the total cost should be minimized. Dosrosiers, Soumis and Desrochers (1988a) propose a model in which transit time $b(x, y)$ is a positive number, transit cost $c(x, y)$ is an arbitrary number, which are however all time-independent. They develop a column generation method to construct routes covering the set of tasks. At each iteration of their algorithm, the current solution is improved by inserting into the basis the least marginal cost route. This route includes a subset of tasks respecting the time window constraints and is obtained by solving the shortest path problem with time windows. Late, Desrsier and Soumis (1988b) present a labeling method to solve the problem with up to 2500 vertices and 250,000 arcs.

Sancho (1992) investigates the problem where both arcs and vertices are associated with time windows. The situation where one passes the vertex x (without waiting at x) to travel on arc (x, y) within the time window of the arc is considered feasible even though the vertex x is not visited within its time window. Sancho (1994) also considers the problem where arrival at the vertex before its time window is permitted if one is willing to wait at the vertex until the time window is opened. However, arrival at vertex after its time window is not permitted, even if no waiting is incurred at the vertex. A dynamic programming approach is developed to solve this problem.

Loachim, Gelinas, Soumis and Desrosiers (1998) propose a continuous model in which there is a vertex cost, which can be regarded as the waiting cost, at the vertex as a linear function of the service start time within the vertex time window. A dynamic programming algorithm is proposed for finding the optimal solution. A related problem is the *shortest routing problem with time windows*; refer to Section 6 below for more discussions.

4 NON-STATIC MAXIMUM FLOW PROBLEMS

The maximum flow problem is one of the fundamental network flow optimization problems. Suppose there is a network $N = (V, A, l)$, where V is a vertex set, A is an arc set, and $l(x, y)$ is the capacity of the arc $(x, y) \in A$. Let $f(x, y)$ be the flow on arc (x, y) and v be the total flow value to be sent from s to ρ, where s is the source vertex and ρ is the sink vertex. The static maximum network flow problem can be stated as:

$$\begin{aligned} \max \quad & v \\ s.t. \quad & \sum_{(s,x)\in A} (f(s,x) - \sum_{(x,s)\in A} f(x,s) - v = 0 & (4.1) \\ & \sum_{y\in V} (f(x,y) - f(y,x)) = 0, \quad \forall x \in V\backslash\{s,\rho\} & (4.2) \\ & \sum_{(x,\rho)\in A} f(\rho,x) - \sum_{(x,\rho)\in A} f(x,\rho) + v = 0 & (4.3) \\ & 0 \le f(x,y) \le l(x,y), \quad \forall (x,y) \in A & (4.4) \end{aligned}$$

This problem is polynomially solvable. Numerous powerful algorithms have been proposed to solve this problem and its variants. Goldberg (1998) gives a survey of the recent studies on this problem.

A non-static maximum flow problem can be stated as: *Given a non-static network $N = (V, A, b, l)$, where V is a vertex set and A is an arc set, $b(x, y, t)$ is the transit time of the arc (x, y), $l(x, y, t)$ is the capacity of the arc (x, y), $l(x, t)$ is the waiting capacity at the vertex x, and T is a pre-specified time limit, find a schedule to send the maximum flow from s to ρ within the time duration T.* As we mentioned in the previous section, waiting at vertices may be considered in the model.

Although the static maximum flow problem can be solved in polynomial time, Sha, Cai and Wong (2000) show that the non-static maximum flow problem is NP-complete.

Generally, non-static maximum flow problems can be classified into three classes. The first class of problems are the so-called *dynamic maximum flow problems*, in which the transit times and the capacities of the arcs are time-independent; see Bellmore and Vemuganti (1971), Ford and Fulkerson (1958,1962), Gale (1957,1959), Minieka (1973), Orlin (1983), Philpott (1990), Anderson and philpott (1994). The second class of problems consider time-varying transit times, time-varying arc capacities, or time-varying vertex capacities; see Gale (1959), Minieka (1974), Xue, Sun and Rose (1998), Cai, Sha and Wong (2000b), Sha, Cai and Wong (2000) and Orda and Rom (1995). The third class of problems address the situations with continuous time; see Anderson, Nash and Philpott (1982), and Blum (1990, 1993).

(1) **On the first class of problems: Dynamic maximum flows**

Ford and Fulkerson (1958,1962) introduce the concept of *dynamic flows* in a network and propose the *maximal dynamic network flow problem.* In their model, $b(x, y, t) = b(x, y)$ and $l(x, y, t) = l(x, y)$ for any arc $(x, y) \in A$. In other words, the transit time and the arc capacity are all time-independent. No vertex capacity is considered. Although this model is not time-varying, it is widely regarded as the fundamental work on the time-varying maximum flow problem. They propose an algorithm to find an optimal solution to their model, which solves successively a series of static min-cost flow problems. Although waiting at the intermediate vertex is allowed, they claim that there always exists a dynamic maximum flow that avoids hole-overs at these vertices. Orlin (1983) considers the problem with an infinite time horizon and the flow is to be sent through the network in each period of time so as to satisfy the upper and lower bounds. The objective is to maximize the throughput, the net flow circulating in the network in a given period. He formulates the problem as an infinite integer program.

A dynamic maximum flow is said to be *universal* if, when the time limit T is truncated to a smaller t and all subflows which arrive at ρ later than t are removed, the remaining flow is still a dynamic maximum flow within the time duration t. The dynamic maximum flow, however, generated by Ford and Fulkerson's algorithm is not necessarily universal. Gale (1957,1959) proves the existence of a universal dynamic maximum flow, not only for the problem of Ford and Fulkerson, but also for a more general model in which the transit times and the arc capacities are all time-varying. Furthermore, he conjectures that there might exist a mild modification of the Ford and Fulkerson algorithm that could generate a universal dynamic maximum flow. Minieka (1973) resolves this conjecture by devising a modification of the Ford and Fulkerson algorithm that can obtain a universal solution.

The key of Ford and Fulkerson's method for the dynamic maximum flow problem is to find the maximal temporally repeated dynamic flow. But this strategy is hard to be applied to the multi-commodity maximum dynamic flow problem. Bellmore and Vemuganti (1971) examine the multi-commodity maximum dynamic flow problem and claim that for sufficiently large number of time periods, the difference between the multi-commodity maximum dynamic flow and the temporally repeated multi-commodity flow is bounded by a constant. They give a method to calculate this bound.

Philpott (1990), Anderson and philpott (1994) consider a continuous dynamic maximum flow model in which the arc capacities vary as Lebesgue-measurable functions of times, transit time of each arc is constant, and waiting at intermediate vertex is allowed. They generalize the max-flow min-cut theorem to their model. The concept of augmenting path is also generalized and an algorithm is presented.

(2) On the second class of problems – Time-varying maximum flow problems

After Gale (1959) propose his model with time-varying arc capacities, Minieka (1974) discusses a special case of Gale's model, in which each arc capacity has two possible values, a normal value and a zero value (corresponding to, respectively, the situation where the arc is usable and unavailable). He suggests a modification of Fold and Fulkerson's method to solve this problem and indicates that it is computationally more efficient to apply the standard maximal flow algorithm to the time-expanded network when the number of arc changes is excessive. Halpern (1979) examines the problem in which the vertex capacities are also time-varying. Xue, Sun and Rose (1998) study the fast data transmission problem by formulating it as a dynamic maximum flow problem.

Orda and Rom (1995) investigate another version of the time-varying maximum flow problem, in which all transit times, arc capacities and vertex capacities are time-dependent, and waiting at the inmediate vertices is allowed. They establish a generalized max-flow min-cut theorem for their model. However, no algorithm has been given to solve the problem. Cai, Sha and Wong (2000b) consider the problem with three waiting constraints, namely, zero waiting times, arbitrary waiting times and bounded waiting times. A set of labeling algorithms are proposed which can solve the problems optimally in psudopolynomial time. They generalize the concepts of augmenting path and residual network for their time-varying model, and their algorithms use the strategy of finding repeatedly dynamic augmenting paths. They show that the optimal flow produced by their algorithm is universal. For the problem allowing arbitrary waiting, Sha, Cai and Wong (2000) prove that it is NP-complete, even though the underlying graph of N is a planar graph.

(3) **On the third class of problems: Continuous maximum flow**

The continuous maximum flow problem is another kind of non-static maximum flow problems. Even though the transit time may be absent in the network, the flow is to be sent dynamically. In this sense, the problem is considered to involve dynamic flow. In some models reported in the literature, time-varying arc capacities and vertex capacities have also been considered.

Anderson, Nash and Philpott (1982) examine a continuous-time problem in which the transit time of each arc is constant, and both arcs and vertices are subject to capacity limits. The problem is formulated as an infinite linear program, and a continuous-time version of the the well-known labelling algorithm is proposed to solve this problem.

Blum (1990, 1993) examines the issues of approximating a continuous max-flow problem by a sequence of static max-flow models in finite networks. He introduces a definition of a continuous flow and proves that the approximation sequence of network flows generated by his algorithm has a weakly convergent subsequence which converges to a maximal continuous flow. Jacobs and Seiffert (1983) address similar problems.

5 NON-STATIC MINIMUM COST FLOW PROBLEMS

The minimum cost flow model describes a wide variety of problems in practice, including those in transportation, traffic assignment, multi-period decision, scheduling, etc. See Ahuja, Magnanti and Orlin (1991), and Current and Marsh (1993) for reviews on these problems.

Recall the statement of a non-static minimum cost flow problem: *Give a network $N = (V, A, b, c, l)$, where V is the vertex set, A is the arc set, $b(x, y, t)$ is the transit time of the arc (x, y), $c(x, y, t)$ is the transit cost of the arc (x, y), $c(x, t)$ is the waiting cost of the vertex x, $l(x, y, t)$ is the capacity of the arc (x, y) and $l(x, t)$ is the capacity of the vertex x. All parameters b, c and l are functions of time t, where $0 \leq t \leq T$ and T is a given number. The problem is to send a given flow from s to p within the time limit T so that the total cost is minimum.*

The static minimum cost flow problem is polynomially solvable, but the non-static problem is NP-complete. This is because the constrained shortest path problem, a well-known NP-complete problem (Ahuja, Magnanti and Orlin (1993)), is a special case of the non-static minimum cost flow problem.

Non-static minimum cost flow problems can be divided into two classes. The first class of problems consider the situations where the supply and the demand are discrete functions of time. These problems have been studied by Aronson, Morton and Thompson (1984a,1984b,1985a,1985b), Stanley (1987), Aronson and Chen (1988), Orlin (1984), Carey (1987), Carey and Srivivasan (1988,1994), Merchant and Nemhauser (1978a,1978b), Zawach and Thompson (1987), Murthy (1993), and Cai, Sha and Wong (2000a). The second class of problems deal with continuous flow. Some problems in this category have been examined by Anderson and Philpott (1984,1994), and Grossmann, Guariso, Hitz and Werthner (1995).

(1) **On the first class of problems: Discrete models**

For multi-period optimization problems within a time horizon T, an approach called *forward algorithms* is usually used. Its basic idea is to solve successively subproblems for a time horizon $t \leq T$, with t gradually increased. The iteration continues until a stopping rule is satisfied, or until t reaches T. Forward algorithms have been suggested in the literature to deal with multi-period minimum cost flow problems; see Aronson and Thompson (1984b). In general, these methods consider the dynamic (multi-period) structure of minimum cost flow problems, but the parameters involved are usually time-independent (namely, not changing over time). It has been pointed out that the implementation scheme for these forward algorithms is critical for their success in terms of computational efficiency.

Aronson, Morton and Thompson (1984a) use the forward method to find planning horizons for infinite discrete horizon version of the deterministic production smoothing problem without stock outs and inventories. Their model comes from the systems for highly obsolescent or perishable commodities, such as newspapers and fresh products. The computational results they obtained show that the algorithm is linear in problem length while linear programming is at least quadratic.

Aronson, Morton and Thompson (1985a) present a forward simplex method to solve the multi-period, personnel planning, staircase linear programming model. Later, they apply their method to the case in which the objective function is linear (Aronson, Morton and Thompson (1985b)). Stanley (1987) generalize this method to the case in which the objective function is convex. Aronson and Chen (1988) continue their study on a general multi-period minimum cost flow model. They develop a primary/secondary memory implementation of the forward network simplex method, and report their computational results show this implementation scheme is more efficient than others designed before.

Orlin (1984) investigates a dynamic network flow problem, in which the transit time of an arc can be a negative integer, and the transit cost is a function depending on the flow on the arc. The network is finite but the time horizon is infinite. To solve this problem, Orlin first obtains an optimal continuous-valued flow which repeats in every period. The fractional variables of this solution may be rounded in such a way that an integral flow is obtained which repeats in every q periods, where q is the least common denominator of the fractional parts of the continuous flow. This integral flow is proven to be optimal.

Carey (1987) considers another dynamic network flow problem, in which the transit time of an arc depends on the flow rate on the arc. The model arises from some traffic problems. He also extends the model to handle problems with multiple destinations and multiple commodities, although not all of these extensions lead to convex programming. Convex nonlinear programming is suggested to tackle the problems. Carey and Srivivasan (1988) examine a model with multiple origins, single destination, and time-varying demands. Later, they (1994) extend this model to consider the possibility that resources, such as storage capacities, can be shared among neighbouring vertices. They formulate the problem as a multi-period optimization model with an objective to minimize the sum of the travel costs and delay costs. According to their computational experiments, their algorithms are 13 to 42 times faster than the approach of linear programming.

Merchant and Nemhauser (1978a) develop an algorithm to solve a dynamic traffic assignment problem, in which the time is discrete and each vertex has an input value depending on the time t, where $0 \leq t \leq T$. Their problem is, in general, nonlinear and non-convex. A special case, the piecewise linear cost version, is solved by using a one-pass simplex algorithm. An optimality condition is given for this type of problems; see Merchant and Nemhauser (1978b).

Zawach and Thompson (1987) discuss a city traffic congestion model in which input and output are time-varying, capacity changes over time, and cost is equal to transit time. Murthy (1993) studies a multi-period assignment problem with start-up costs, where there are a total of T periods of time. The cost of assigning job i to machine j can vary from one period of time to another period of time. Further, a start-up cost is incurred whenever job i is assigned to machine j in period t, but not so in period $t-1$. The problem is to determine the assignment for each time period so as to minimize the sum of assignment and start-up costs. They develop a dual ascent approximation procedure to solve this problem.

Cai, Sha and Wong (2000a) examine the minimum cost flow problem in which transit times, arc capacities, vertex capacities, waiting costs and transit costs are all time-varying. They consider three types of waiting constraints, namely, waiting at any vertex is prohibited, waiting at any vertex is arbitrarily allowed, and the waiting time at a vertex x is bounded above by a given number u_x. They develop several dynamic programming algorithms, based on repeated computations of shortest dynamic augmenting paths in dynamic residual networks. They prove that their algorithms can optimally solve the problem with the different waiting constraints in pseudopolynomial time.

(2) **The second class of problems: Continuous models**

Anderson and Philpott (1984) investigate a class of continuous minimum cost models. They establish the duality theory for this type of problems and provide an algorithm which can converge to an optimal solution. They (1994) also examine a model in which the transit times of arcs are constant, and flow can be stored at a vertex to be transmitted at a later time.

Grossmann, Guariso, Hitz and Werthner (1995) consider a model in which each arc or vertex capacity can change over time, within an upper and lower bound. The transit time is, however, not considered. During each period of time $0 \leq t \leq T$, there is a given flow value $F^{(t)}$ to be sent from the source vertex s. An objective function is given based on utility functions which are assumed to be piecewise linear and concave. They suggest to solve this problem by converting it to a minimum cost flow problem which can be tackled by existing algorithms.

6 NON-STATIC VEHICLE ROUTING PROBLEMS

A non-static vehicle routing problem can be stated as: *Let $N = (V, A, b, c)$ be a network, where v is a vertex set, A is an arc set, $b(x, y, t)$ is the transit time of the arc (x, y), $c(x, y, t)$ is the transit cost of the arc (x, y), and $c(x, t)$ is the waiting cost at the vertex x. Given a time limit T and a set of vertices and/or arcs which must be serviced by a fleet of vehicles, the problem is to find a set of feasible routes, one for each vehicle, with no restrictions on the order in which these vertices and/or arcs must be serviced, so that each route can be traversed within the time limit T and the total cost is minimized.*

This model includes the following problems:

(a) the traveling salesman problem;

(b) the Chinese postman problem;

(c) the m-traveling salesman problem;

(d) the single depot, multiple vehicle, vertex routing problem;

(e) the multiple depot, multiple vehicle, vertex routing problem;

(f) the capacited Chinese postman problem.

The static versions of the problems have been studied extensively in the literature; see Bodin, Golden, Assad and Ball (1983) for a survey. In the static situations, some problems above are polynomially solvable; e.g. the Chinese postman problem (Eiselt, Gendreau and Laporte (1995a, 1995b)), but some problems above are already strongly NP-hard, such as the traveling salesman problem.

Non-static vehicle routing problems can be divided into two classes. The first class of problems include the so-called *vehicle routing problems with time windows*, in which a vertex and/or arc must be serviced within a given time window; see Baker (1983), Dumas, Desrosiers, Gelinas and Solomon (1995), Sexton and Choi (1986), Solomon (1983,1987), Baker and Schaffer (1986), Desrosiers, Dumas and Soumis (1984), Desrochers, Desrosiers and Solomon (1992), Dell (1996), Fisher, Jornsten and Madsen (1997), Mingozzi, Bianco and Ricciardelli (1997), and Salomon, Solomon, Wassenhove, Dumas and Danzère-Pérès (1997). The second class of problems are so-called *time-varying vehicle routing problems*, in which the transit times, transit costs and waiting costs are functions of the time; see Picard and Quueyranne (1978), Wiel and Sahinidis (1995,1996), Brideau and Cavalier (1994), and Erkut and Zhang (1996).

(1) **On the first class of problems**

The *traveling salesman problem with time windows* is an important variation of the non-static vehicle routing problems. A salesman based at city 1 must visit each of the remaining $n-1$ cities exactly once before returning to city 1. Moreover, each city can only be visited within a given time window. Baker (1983) investigates such a problem in which the salesman can wait at a city, if necessary, for a time window to open. The dual of the problem is shown to be a disjunctive graph. A longest path algorithm is used to derive the bounding information for subproblems in a branch-and-bound solution procedure. Dumas, Desrosiers, Gelinas and Solomon (1995) present an optimal algorithm. This algorithm adopts an elimination test which can greatly enhance the performance of dynamic programming. Based on this algorithm, Salomon, Solomon, Wassenhove, Dumas and Danzère-Pérès (1997) develop an exact algorithm to solve the discrete lot-sizing and scheduling problem with sequence dependent set-up costs and set-up times by transforming it into a traveling salesman problem with time window constraints. Carlton and Barnes (1996) use tabu search (see Glover (1990) and Glover and Laguna (1997)) to solve the problem. Mingozzi, Biance and Ricciardelli (1997) investigate the problem with precedence constraints, i.e., before visiting vertex i, the salesman must visit every vertex of a given set of vertices $\pi_i \subset V\backslash\{1\}$. They present an exact algorithm, based on dynamic programming, to solve the problem.

Sexton and Choi (1986) examine a model called *Pickup and delivery of partial loads with "soft" time windows*, in which the transit times of the arcs are time-independent, and some vertices must be visited within a pre-specified time window. Visiting a vertex at a time later than the time window is permitted, which however will incur a lateness penalty. The problem is to find a sequence of pickups and deliveries which satisfies the capacity restriction of the vehicle as well as precedence requirements so as to minimize the total vehicle operating time and the total customer penalty. Obviously, the constraint of soft time window can be regarded as a generalization of the constraint of "hard" time window, since if the lateness penalty is set to infinite, the former is equivalent to the later. Sexton and Choi provide a heuristic approach to solve the problem.

The *vehicle routing and scheduling problem with time window constraints* is another variation of the vehicle routing problems, in which each of the customers to be serviced may have one or more time windows during which the service must be scheduled. Some heuristics are reported in the literature; see Solomon (1983). Baker and Schaffer (1986) present a modified branch exchange method to improve the heuristic solutions of a vehicle routing problem with time windows. Solomon (1987) conducts an extensive computational study of the performance of a variety of heuristics. They find that several heuristics performed well in different problem environments. In particular, an insertion-type heuristic consistently delivered very good solutions. Desrochers, Desrosiers and Solomon (1992) present a new method for the vehicle routing problem with time windows and vehicle capacity constraints. A linear programming relaxation of the set partitioning formulation of the problem is derived and solved by column generation. Feasible columns are added as needed by solving a shortest path problem with time windows and capacity constraints using dynamic programming. The linear programming solution obtained generally provides an excellent lower bound that is used in a branch-and-bound algorithm to solve the integer set partitioning problem.

Desrosiers, Dumas and Soumis (1984) develop a dynamic programming method to solve a large-scale single-vehicle dial-a-ride problem with time windows. In this problem, a vehicle picks up riders at their origins and then takes them to their destinations, subject to the vehicle capacity and the time window constraints at both the origins and the destinations. Bruggen, Lenstra and Schuur (1993) investigate a single-vehicle pickup and delivery problem with time windows, in which there is one depot, one vehicle, and a set of customers with known demands, each of which must be picked up and delivered at specified locations and within given time windows. The problem is to find the minimum route duration, i.e., the difference between the arrival time and the departure time at the depot. A heuristic approach is proposed to solve the problem. Nanry and Barnes (2000) consider an extension of this problem with a group of vehicles (a homogeneous vehicle fleet) and propose a heuristic algorithm to find a solution.

Dell (1996) investigates a multiple vehicle routing model with time windows and equity constraints over a multi-day horizon. The need to maintain equity in delivery times comes from the need to preserve business from large-value customers. An optimal column generation procedure and several heuristics are proposed. Later, Fisher, Jornsten and Madsen (1997) present two optimization approaches to solve this problem.

(2) **On the second class of problems**

Time-varying vehicle routing problems have been addressed by many researchers. Picard and Queyranne (1978) study a time-varying traveling salesman problem with application to the tardiness problem in one-machine scheduling. They consider a model in which the cost of transition depends not only on the two respective locations involved, but also on their positions (time) in the sequence that defines the tour. The algorithm proposed combines finding shortest paths in an associated multipartite network with subgradient optimization and some branch-and-bound enumeration. Wiel and Sahinidis (1995) present a mixed-integer linear programming formulation for this problem. A directed multipartite graph representation of this model facilitates the development of a local search heuristic which extends the well known Lin-Kernighan tour improvement heuristic to the problem (refer to Lin and Kernighan (1973)). A second upper bound is derived by applying a Benders-decomposition-based heuristic to the mixed-integer linear programming formulation. Later, they (1996) present an exact solution approach for this problem which is a network-flow-based method for finding Pareto-optimal dual solutions of a highly degenerate subproblem. Preliminary computational experience demonstrates that the use of these Pareto-optimal solutions has a large impact on the performance of the algorithm.

Brideau and Cavalier (1994) introduce a related problem called the maximum collection problem with time-dependent rewards. In this problem the transit time is fixed and rewards at vertex are time-varying. They describe an integer programming formulation and a heuristic algorithm for this problem. Erkut and Zhang (1996) suggest an alternative heuristic, which is a penalty-based greedy algorithm. An implicit enumeration algorithm is also developed to solve this problem

7 OTHER NON-STATIC NETWORK OPTIMIZATION PROBLEMS

In this section, we discuss some other non-static network optimization problems, which do not fall in any categories of problems that have been reviewed in the above sections.

The *minimum spanning. tree problem* in a static network is a well-known problem in network optimization, which requires to find a tree that spans a given network, such that the sum of the arc costs in the spanning tree is minimum. This problem in its static setting, can be easily solved by some well-known algorithms (see Kruskal (1956) and Prim (1957)). Solomom (1986) investigates the problem with time constraints. In his model, each arc (x, y) is associated with a transit time and a cost, and each vertex x is associated with a time window during which a visit to the vertex is allowed. He proves that this problem is NP-hard, and presents two heuristics, a greedy algorithm, and an insertion algorithm, to solve the problem. Cai, Sha and Wong (1998) discuss a time-varying minimum spanning tree problem. The problem is formulated as follows: Given a non-static network N and a time limit T, find the minimum spanning tree such that for each vertex $x \in V\backslash\{s\}$, there exists a path from s to x which can be traversed within the given time duration T. Similar to their time-varying minimum cost flow problem (Cai, Sha and Wong (2000a)), they consider three waiting constraints. They show that, even thought the classical static minimum spanning tree problem is polynomially solvable, their time-varying problem is NP-complete even if the underlying graph is a tree. They present several heuristics to deal with the different versions of their problem.

Chen and Chin (1989) propose a new shortest path model, called the *quickest path problem*, which can be described as follows: Given a network $N = (V, A, l, u)$, where $l(x, y)$ and $u(x, y)$ are, respectively, the capacity and the lead time of the arc $(x, y) \in A$, the problem is to find a path connecting s and ρ so as to send a given amount σ of flow from s to ρ in quickest time. Chen and Chin (1989) derive an algorithm, which can find an optimal solution for the problem in polynomial time. Rosen, Sun and Xue (1991) suggest another algorithm, which has the same time complexity as Chen and Chin's algorithm, but requires less space. Furthermore, they develop an algorithm to find the first k quickest paths. Chen and Hung (1994) develop a new algorithm to solve the k-quickest path problem. Their results show an improvement as compared to those by Rosen, Sun and Xue. Lee and Papadopoulou (1993) consider the all-pairs quickest path problem, which is to find the quickest oath to transit a flow σ between any pair of vertices x and y. A polynomial algorithm has been developed to optimally solve the problem.

Cai, Sha and Wong (1998) propose a new problem, called the *time-varying maximum capacity path problem*. Give a network $N = (V, A, b, l)$, where $b(x, y, t)$ and $c(x, y, t)$ are the transit time and the capacity of the arc $(x, y) \in A$, which are all the function of the time t, the problem is to find the path with the maximum capacity between two particular vertices, subject to the constraint that the total traversal time along the whole path is at most T, where T is a given number. Waiting at any vertex is prohibited. They show that the problem is NP-complete and develop a dynamic programming algorithm that can optimally solve the problem in pseudopolynomial time. In addition, they investigate the *time-varying quickest path problem*, in which both the arc capacity and the lead time change over time. They point out that, the time-varying quickest path problem can be solved by finding a set of time-varying maximum capacity paths.

Sherail, Ozbay and Subramanian (1998) discuss a *time-dependent shortest pair of disjoint paths problem*, which is to find a pair of edge-disjoint paths from the source s to the sink ρ so that the total transit time of the two paths is minimum, where the network $N = (V, A, b)$ is time-varying with time-dependent transit times $b(x, y, t)$. They show that this problem is NP-hard, and develop a 0-1 linear programming model to formulate the problem.

8 CONCLUSIONS

Noting that in realistic situations, the parameters and structure of a network under consideration are, in general, changing over time, there have been more and more studies focused on non-static network optimization problems. We have provided a survey of the relevant studies published in the literature on several important types of these problems, including shortest path problems, maximum flow problems, minimum cost flow problems, vehicle routing problems, and some other non-static network optimization problems. As we have observed from the survey, although its static counterparts may be polynomially solvable, a non-static network optimization often becomes NP-hard. Nevertheless, the discrete-time version of a non-static problem is generally solvable by some pseudopolynomial time algorithms, while its continuous-time version usually tends to be very difficult to solve. We have also noticed that, the time complexities of most algorithms for discrete-time problems are proportional to, or a non-linear polynomial function of, the time horizon T. This is usually the main cause why the time complexities of these algorithms are pseudopolynomial. This occurs because these algorithms consider all discrete times t up to T, in order to find an exact optimal solution. An interesting issue for further studies is whether certain points (instead of all) in the interval $[0, T]$ could be considered so that an approximate solution could be found while the time complexity of such an approximate algorithm could become fully polynomial.

In a non-static network, it is very important to control the timing to stay in a vertex or to travel on an arc. We have noted that in many time-varying models, the waiting times at vertices have been treated as decision variables the values of which can be optimally determined. Nevertheless, the transit times on arcs are, usually, assumed to be fixed. Even though a transit time is considered as a function of time in some models, it is usually not a decision variable that the decision maker can change in order to achieve an overall optimal solution by optimally controlling the timing to traverse an arc or to arrive at the end vertex of the arc. Models that allow for the change of transit times, probably at some speedup costs, are another interesting topic for further studies.

Acknowledgments

This research was partially supported by the Research Grants Council of Hong Kong under Earmarked Grant No. CUHK 4135/97E.

References

Adel'son-Vel'ski, G.M., Dinic, E.A. and Karzanov, E.V. (1975), *Flow Algorithms*, Science, Moscow.

Ahuja, R.K. and James, B. Orlin, J.B. (1991), Distance-directed augmenting path algorithms for maximum flow and parametric maximum flow problems, *Naval Research Logistics*, Vol. 38, pp. 413-430.

Ahuja, R.K., Magnanti, T.L. and Orlin, J.B. (1991), Some recent advances in network flows. *SIAM Review*, Vol. 33, pp. 175-219.

Ahuja, R.K., Magnanti, T.L. and Orlin, J.B. (1993), *Network Flows: Theory, Algorithms and Applications*, Prentice Hall, Englewood Cliff, New Jersey.

Anderson, E.J., Nash, P. and Philpott, A.B. (1982), A class of continuous network flow problems, *Mathematics of Operations Research*, Vol. 7, No. 4, pp. 501-514.

Anderson, E.J. and Philpott, A.B. (1984), Duality and algorithm for a class of continuous transportation problems, *Mathematics of Operations Research*, Vol. 9, No. 2, pp. 222-231.

Anderson, E.J. and Philpott, A.B. (1994), *Optimization of flows in networks over time, Probability, Statistics and Optimization (Edited by F. P. Kelly)*, John Wiley & Sons Ltd..

Aronson, J.E. (1986), The multiperiod assignment problem: a multicommodity network flow model and specialized branch and bound algorithm, *European Journal of Operational Research*, Vol. 23, pp. 367-381.

Aronson, J.E. (1989), A survey of dynamic network flows, *Annals of Operations Research*, Vol. 20, No. 1/4, pp. 1-66.

Aronson, J.E. and Chen B.D. (1986), A forward network simplex algorithm for solving multiperiod network flow problems, *Naval Research Logistics Quarterly*, Vol. 33, pp. 445-467.

Aronson, J.E. and Chen, B.D. (1989), A primary/secondary memory implementation of a forward network simplex algorithm for multiperiod network flow problems, *Computers and Operations Research*, Vol. 16, No. 4, pp. 379-391.

Aronson, J.E., Morton, T.E. and Thompson, G.L. (1984a), A forward algorithm and planning horizon procedure for the production smoothing problem without inventory, *European Journal of Operations Research*, Vol. 15, pp. 348-365.

Aronson, J.E., Morton, T.E. and Thompson, G.L. (1985a), A forward simplex method for staircase linear programs, *Management Science*, Vol. 31, No. 6, pp. 664-679.

Aronson, J.E. and Thompson, G.L. (1984b), A survey on forward methods in mathematical programming, *Large Scale Systems*, Vol. 7, pp. 1-16.

Aronson, J.E. and Thompson, G.L. (1985b), The solution of multiperiod personnel planning problems by the forward simplex method, *Large Scale Systems*, Vol 9, pp. 129-139.

Baker, E.K. (1983), An exact algorithm for the time-constrained traveling salesman problem, *Operations Research*, Vol. 31, No. 5, pp. 938-945.

Baker, E.K. and Schaffer, J.R. (1986), Solution improvement heuristics for the vehicle routing and scheduling problem with time window constraints, *American Journal of Mathematical and Management Sciences*, Vol. 6, No. 3 & 4, pp. 261-300.

Bazaraa, M.S., Jarvis, J.J. and Sherali, H.D. (1990), *Linear Programming and Network Flows (2nd ed.)*, Wiley, New York.

Bean, J.C., Birge, J.R. and Smith, R.L. (1987), Aggregation in dynamic programming, *Operations Research*, Vol. 35, No. 2, pp. 215-220.

Bean, J.C., Lohmann, J.R. and Smith, R.L. (1984), A dynamic infinite horizon replacement economy decision model, *The Engineering Economist*, Vol. 30, No. 2, pp. 99-120.

Bellmore, M. and Ramakrishna, R. Vemuganti, R.R. (1973), On multi-commodity maximal dynamic flows, *Operations Research*, Vol. 21, No. 1, pp. 10-21.

Blum, W. (1990), A measure-theoretical max-flow-min-cut problem, *Math. Z.*, Vol. 205, pp. 451-470.

Blum, W. (1993), An approximation for a continuous max-flow problem, *Mathematics of Operations Research*, Vol. 18, No. 1, pp. 98-115.

Bodin, L. Golden, B., Assad, A. and Ball, M. (1983), Routing and scheduling of vehicles and crews: the state of the art, *Computers and Operations Research*, Vol. 10, No. 2, pp63-211.

Brideau, R.J., III, and Cavalier, T.M. (1994), The maximum collection problem with time dependent rewards, presented at the *TIMS International Conference*, Alaska, June.

Cai, X., Kloks, T. and Wong, C.K. (1997), Time-varying shortest path problems with constraints, *Networks*, Vol. 29, No. 3, pp. 141-149.

Cai, X., Sha, D. and Wong, C.K. (1998), The Time-varying minimum spanning tree problem with waiting times, *Advances in Operations Research and Systems Engineering (Edited by Jifa Gu, et al.)*, Global-link Publishing Co., Hong Kong, pp. 1-8.

Cai, X., Sha, D. and Wong, C.K. (2000a), Time-varying minimum cost flow problems, to appear in *European Journal of Operational Research*.

Cai, X., Sha, D. and Wong, C.K. (2000b), Time-varying universal maximum flow problems, to appear in *Mathematical and Computer Modelling.*

Carey, M. (1987), Optimal time-varying flows on congested networks, *Operations Research*, Vol. 35, No. 1, pp. 58-69.

Carey, M. and Srinivasan, A. (1988), Congested network flows: time-varying demands and start-time policies, *European Journal of Operational Research*, Vol. 36, pp. 227-240.

Carey, M. and Srinivasan, A. (1994), Solving a class of network models for dynamic flow control, *European Journal of Operational Research*, Vol. 75, pp. 151-170.

Carlton, W.B. and Barnes J.W. (1996), Solving the traveling-salesman problem with time windows using tabu search, *IIE Transactions*, Vol. 28, pp. 617-629.

Cattrysse, D.G. and Wassenhove, L.N.V. (1992), A survey of algorithms for the generalized assignment problem, *European Journal of Operational Research*, Vol. 60, pp. 260-272.

Cattrysse, L.G., Francis, R.L. and Saunders, P.B. (1982), Network models for building evacuation, *Management Science*, Vol. 28, No. 1, pp. 86-105.

Chang, H. and Chung, K. (1992), Fault-tolerant routing in unique-path multistage omega network, *Information Processing Letters*, Vol. 44, pp. 201-204.

Chardaire, P., Sutter, A. and Costa, M. (1996), Solving the dynamic facility location problem, *Networks*, Vol. 28, pp. 117-124.

Chari, K. and Dutta, A. (1993a), Design of private backbone networks-I: time varying traffic, *European Journal of Operational Research*, Vol. 67, pp. 428-442.

Chari, K. and Dutta, A. (1993b), Design of private backbone networks-II: time varying grouped traffic, *European Journal of Operational Research*, Vol. 67, pp. 443-452.

Chen, H. and Yao, D.D. (1993), Dynamic scheduling of a multiclass fluid network, *Operations Research*, Vol. 41, No. 6, pp. 1104-1115.

Cheung, R.K. and Powell, W.B. (1996), An algorithm for multistage dynamic networks with random arc capacities, with an application to dynamic fleet management, *Operations Research*, Vol. 44, No. 6, pp. 951-963.

Chlebus, B.S. and Diks, K. (1996), Broadcasting in synchronous networks with dynamic faults, *Networks*, Vol. 27, pp309-318.

Choi, W., Hamacher, H.W. and Tufekci, S. (1988), Modeling of building evacuation problems by network flows with side constraints, *European Journal of Operational Research*, Vol. 35, pp. 98-110.

Cooks, K.L. and Halsey, E. (1966), The shortest route through a network with time-dependent internodal transit times, *Journal of Mathematical Analysis and Applications*, Vol. 14, pp. 493-498.

Current, J. and Marsh, M. (1993), Multiobjective transportation network design and routing problems: taxonomy and annotation, *European Journal of Operational Research*, Vol. 65, pp. 4-19.

Dell, R.F., Batta, R. and Karwan, M.H. (1996), The multiple vehicle TSP with time windows and equity constraints over a multiple day horizon, *Transportation Science*, Vol. 30, No. 2, pp. 120-133.

Denardo, E.V. and Fox, B.L. (1979), Shortest-route methods: 1. reaching, pruning, and buckets, *Operations Research*, Vol. 27, No. 1, pp. 161-186.

Deo, N. and Pang, C. (1984), Shortest-path algorithms: taxononny and annotation, *Networks*, Vol. 14, pp. 275-323.

Desrochers, M., Desrosiers, J. and Solomon, M. (1992), A new optimization algorithm for the vehicle routing problem with time windows, *Operations Research*, Vol. 40, No. 2, pp. 342-354.

Desrochers, M. and Soumis, F. (1988a), A reoptimization algorithm for the shortest path problem with time windows, *European Journal of Operational Research*, Vol. 35, pp. 242-254.

Desrochers, M. and Soumis, F. (1988b), A generalized permanent labelling algorithm for the shortest path problem with time windows, *INFOR*, Vol. 26, No. 3, pp. 191-212.

Desrosiers, J., Dumas, Y. and Soumis, F. (1986), A dynamic programming solution of the large-scale single-vehicle dial-a-ride problem with time windows, *American Journal of Mathematical and Management Sciences*, Vol. 6, No. 3 & 4, pp. 301-325.

Dreyfus, S.E. (1969), An appraisal of some shortest-path algorithms, *Operations Research*, Vol. 17, pp. 395-412.

Dumas, Y., Desrosiers, J., Gelinas, E. and Solomon, M.M. (1995), An optimal algorithm for the traveling salesman problem with time windows, *Operations Research*, Vol. 43, No. 2, pp. 367-371.

Eiselt, H.A., Gendreau, M. and Laporte, G. (1995a), Arc routing problems, part I: the Chinese postman problem, *Operations Research*, Vol. 43, No. 2, pp. 231-242.

Eiselt, H.A., Gendreau, M. and Laporte, G. (1995b), Arc routing problems, part I: the rural postman problem, *Operations Research*, Vol. 43, No. 3, pp. 399-414.

Erkut, E. and Zhang, J. (1996), The maximum collection problem with time-dependent rewards, *Naval Research Logistics*, Vol. 43, pp. 749-763.

Federgruen, A. and Groenevelt, H. (1986), Preemptive scheduling of uniform machines by ordinary network flow techniques, *Management Science*, Vol. 32, No. 3, pp. 341-349.

Fisher, M.L. (1994), Optimal solution of vehicle routing problems using minimum k-trees, *Operations Research*, Vol. 42, No. 4, pp. 626-642.

Fisher, M.L., Jornsten, K.O. and Madsen, O.B.G. (1997), Vehicle routing with time windows: two optimization algorithms, *Operations Research*, Vol. 45, No. 3, pp. 488-492.

Ford, L.R., Jr. and Fulkerson, D.R. (1956), Maximal flow through a network, *Can. Journal Mathematics*, Vol. 8, pp. 399-404.

Ford, L.R., Jr. and Fulkerson, D.R. (1958), Constructing maximal dynamic flows from static flows, *Operations Research*, Vol. 6, pp. 419-433.

Ford, L.R., Jr. and Fulkerson, D.R. (1962), *Flows in Networks*, Princeton University Press, Princeton, NJ.

Franca, P.M., Gendreau, M., Laporte, G. and Muller, F.M. (1995), The *m*-traveling salesman problem with minimax objective, *Transportation Science*, Vol. 29, No. 3, pp. 267-275.

Fukuda, K. and Matsul, T. (1992), Finding all minimum-cost perfect matchings in bipartite graphs, *Networks*, Vol. 22, pp. 461-468.

Gaimon, C. (1986), Optimal inventory, backlogging and machine loading in a serial multi-stage, multi-period production environment, *Int. J. Prod. Res.*, Vol. 24, No. 3, pp. 647-662.

Gale, D. (1957), A theorem on flows in networks, *Pacific Journal of Mathematics*, Vol. 7, pp. 1073-1086.

Gale, D. (1959), Transient flows in network, *Michigan Mathematical Journal*, Vol. 6, pp. 59-63.

Glover, F. (1990), Tabu search: a tutorial, *Interfaces*, Vol. 20, pp. 74-94.

Glover, F. and Laguna, M. (1997), *Tabu Search*, Kluwer Academic Publishers, Norwell, MA.

Goczyla, K. and Cielatkowski, J. (1995), Optimal routing in a transportation network, *European Journal of Operations Research*, Vol. 87, pp. 214-222.

Goldberg, A.V. (1998), Recent developments in maximum flow algorithms, *proceedings of Sixth Scandinavian Workshop on Algorithm.*

Gondran, M. and Minoux, M. (1984), *Graphs and Algorithms*, Wiley-Interscience, New York.

Grossmann, W., Guariso, G., Hitz, M. and Werthner, H. (1995), A min cost flow solution for dynamic assignment problems in Networks with storage devices, *Management Science*, Vol. 41, No. 1, pp. 83-93.

Gupta, S.K. (1985), *Linear Programming and Network Models.* Affiliated East-West Press, New Delhi, India.

Halpern, I. (1979), Generalized dynamic flows, *Networks*, Vol. 9, pp. 133-167.

Hamacher, H.W. and Tufekci, S. (1987), On the use of lexicographic min cost flows in evacuation modeling, *Naval Research Logistics*, Vol. 34, pp. 487-503.

Imase, M. and Waxman, B.M. (1991), Dynamic steiner tree problem, *SIAM J. Discrete Mathematics*, Vol. 4, No. 3, pp. 369-384.

Ioachim, I., Gelinas, S., Francois, and Desrosiers, J. (1998), A dynamic programming algorithm for the shortest path problem with time windows and linear node costs, *Networks*, Vol. 31, pp. 193-204.

Iri, M. (1969), *Network Flow, Transportation and Scheduling*, Academic Press, New York.

Jacobs, K. and Seiffert, G. (1983), A measure-theoretical max-flow problem, Part I, *Bulletin Inst. Mathe. Acad. Sinica*, Vol 11, No. 2, pp. 261-280.

Jaillet, P. (1992), Shortest path problems with node failures, *Networks*, Vol. 22, pp. 589-605.

Janson, B.N. (1991), Dynamic traffic assignment for urban road networks, *Transportation Research Part B*, Vol. 25, No. 2/3, pp. 143-161.

Jarvis, J.J. and Ratliff, H.D. (1982), Some equivalent objectives for dynamic network flow problems, *Management Science*, Vol. 28, No. 1, pp. 106-109.

Jensen, P.A. and Barnes, W. (1980), *Network Flow Programming*, Wiley, New York.

Kogan, K., Shtub, A. and Levit, V.E. (1997), DGAP-the dynamic generalized assignment problem, *Annals of Operations Research*, Vol. 69, pp. 227-239.

Lee, D.T. and Papadopoulou, E. (1993), The all-pairs quickest path problem, *Information Processing Letters*, Vol. 45, pp. 261-267.

Lee, H. and Pulat, P.S. (1991), Bicriteria network flow problems: continuous case, *European Journal of Operational Research*, Vol. 51, pp. 119-126.

Levner, E.V. and Nemirovsky, A.S. (1994), A network flow algorithm for just-in-time project scheduling, *European Journal of Operational Research*, Vol. 79, pp. 167-175.

Lawler, E.L. (1976), *Combinational Optimization: Networks and Matroids*, Holt Rinehart and Winston, New York.

Mcquillan, J.M., Richer, I. and Rosen, E.C. (1980), The new routing algorithm for the ARPANET, *IEEE Transactions on Communications*, Vol. 28, No. 5, pp. 711-719.

Merchant, D.K. and Nemhauser, G.L. (1978a), A model and an algorithm for the dynamic traffic assignment problems, *Transportation Science*, Vol. 12, No. 3, pp. 183-199.

Merchant, D.K. and Nemhauser, G.L. (1978b), Optimality conditions for a dynamic traffic assignment model, *Transportation Science*, Vol. 12, No. 3, pp. 200-207.

Mingozzi, A., Bianco, L. and Ricciardelli, S. (1997), Dynamic programming strategies for the traveling salesman problem with time window and precedence constraints, *Operations Research*, Vol. 45, No. 3, pp. 365-377.

Minieka, E. (1973), Maximal, lexicographic, and dynamic network flows, *Operations Research*, Vol. 12, No. 2, pp. 517-527.

Minieka, E. (1978), *Optimization Algorithms for Networks and Graphs*, Marcel Dekker, New York.

Murthy, I. (1993), Solving the multiperiod assignment problem with start-up cost using dual ascent, *Naval Research Logistics*, Vol. 40, pp. 325-344.

Nachtigall, K. (1995), Time depending shortest-path problems with applications to railway networks, *European Journal of Operational Research*, Vol. 83, pp. 154-166.

Nanry, W.P. and Barnes, J.W. (2000), Solving the pickup and delivery problem with time windows using reactive tabu search, *Transportation Research Part B*, Vol. 34, pp. 107-121.

Nozawa, R. (1994), Examples of max-flow and min-cut problems with duality gaps in continuous networks, *Mathematical Programming*, Vol. 63, pp. 213-234.

Orda, A. and Rom, R. (1990), Shortest-path and minimum-delay algorithms in networks with time-dependent edge-length, *Journal of Association for Computing Machinery*, Vol. 37, No. 3, pp. 607-625.

Orda, A. and Rom, R. (1991), Minimum weight paths in time-dependent networks, *Networks*, Vol. 21, pp. 295-319.

Orda, A. and Rom, R. (1995), On continuous network flows, *Operations Research Letters*, Vol. 17, pp. 27-36.

Orlin, J.B. (1983), Maximum-throughput dynamic network flows, *Mathematical programming*, Vol. 27, No. 2, pp. 214-231.

Orlin, J.B. (1984), Minimum convex cost dynamic network flows, *Mathematics of Operations Research*, Vol. 9, No. 2, pp. 190-207.

Papadimitrion, C. and Steiglitz, K. (1982), *Combinatorial Optimization: Algorithms and Complexity*, Prentice-Hall, Endlewood Cliff, NJ.

Papageorgiou, M. (1990), Dynamic modeling, assignment, and route guidance in traffic networks, *Transportation Research Part B*, Vol. 24, No. 6, pp. 471-495.

Philpott, A.B. (1990), Continuous-time flows in networks, *Mathematics of Operations Research*, Vol. 15, No. 4, pp. 640-661.

Philportt, A.B. and Mees, A.I. (1992), Continuous-time shortest path problems with stopping and starting costs, *Appl. Mathe. Lett.*, Vol. 5, No. 5, pp. 63-66.

Picard, J. and Queyranne M. (1978), The time-dependent traveling salesman problem and its application to the tardiness problem in one-machine scheduling, *Operations Research*, Vol. 26, No. 1, pp. 86-110.

Polak, G.G. (1992), On a parametric shortest path problem from primal-dual multicommodity network optimization, *Network*, Vol. 22, pp. 283-295.

Posner, M.E. and Szwarc, W. (1983), A transportation type aggregate production model with backordering, *Management Science*, Vol. 29, No. 2, pp. 188-199.

Psaraftis, H.N., Solomon, M.M., Magnanti T.L., Kim T. (1990), Routing and scheduling on a shoreline with release times, *Management Science*, Vol. 36, No. 2, pp. 212-223.

Psaraftis, H.N. and Tsitsiklis, J.N. (1993), Dynamic shortest paths in acyclic networks with markovian arc costs, *Operations Research*, Vol. 41, No. 1, pp. 91-101.

Rego, C. and Roucairol, C. (1995), Using tabu search for solving a dynamic multi-terminal truck dispatching problem, *European Journal of Operational Research*, Vol. 83, pp. 411-429.

Salomon, M., Solomom, M.M., Wassenhove, L.N.V., Dumas, Y. and Dauzere-Peres, S. (1997), Solving the discrete lotszing and scheduling problem with sequence dependent set-up costs and set-up times using the travelling salesman problem with time windows, *European Journal of Operational Research*, Vol. 100, pp. 494-513.

Sancho, N.G.F. (1992), A dynamic programming solution of a shortest path problem with time constraints on movement and parking, *J. Math. Anal. Appl.*, Vol. 166, pp. 192-198.

Sancho, N.G.F. (1994), Shortest path problem with time windows on nodes and arcs, *Journal of Mathematical Analysis and Applications*, Vol. 186, pp. 643-648.

Saniee, I. (1995), An efficient algorithm for the multiperiod capacity expansion of one location in telecommunications, *Operations Research*, Vol. 43, No. 1, pp. 187-190.

Sexton, T.R. and Choi Y. (1986), Pickup and delivery of partial loads with "soft" time windows, *American Journal of Mathematical and Management Sciences*, Vol. 6, No. 3 & 4, pp. 369-398.

Sha, D., Cai, X. and Wong, C.K. (2000), The maximum flow in a time-varying network, *Optimization*, Lecture Notes in Economics and Mathematical Systems, 481, Edited by Nguyen, V. H., Strodiot, J. -J. and Tossings, P., Springer-Verlag Berlin Heidelberg, pp. 437-456.

Sherali, H.D., Ozbay, K. and Subramanian, S. (1998), The time-dependent shortest pair of disjoint paths problem: complexity, models, and algorithms, *Networks*, Vol. 31, pp. 259-272.

Solomom, M.M. (1986), The minimum spanning tree problem with time window constraints, *America Journal of Mathematical and Management Sciences*, Vol. 6, No. 3 & 4, pp. 399-421.

Solomon, M.M. (1987), Algorithms for the vehicle routing and scheduling problems with time window constraints, *Operations Research*, Vol. 35, No. 2, pp. 254-273.

Stanley, J.D. (1987), A forward convex-simplex method, *European Journal of Operational Research*, Vol.29, pp. 328-335.

Steinberg, E. and Napier, H.A. (1980), Optimal multi-level lot sizing for requirements planning systems, *Management Science*, Vol. 26, No. 12, pp. 1258-1271.

Thangiah, S.R., Potvin, J. and Sun, T. (1996), Heuristic approaches to vehicle routing with backhauls and time windows, *Computers Operations Research*, Vol. 23, No. 11, pp. 1043-1057.

Van Der Bruggen, L.J.J., Lenstra, J.K. and Schuur, P.C. (1993), Variable-depth search for the single-vehicle pickup and delivery problem with time windows, *Transportation Science*, Vol. 27, No. 3, pp. 298-311.

Vemuganti, R.R., Oblak, M. and Aggarrwal, A. (1989), Network models for fleet management, *Decision Science*, Vol. 20, pp. 182-197.

Vythoulkas, P.C. (1990), A dynamic stochastic assignment model for the analysis of general networks, *Transpn. Res. -B*, Vol. 24B, No. 6, pp. 453-469.

Wie, B., Tobin, R.L., Friesz, T.L. and Bernstein, D. (1995), A discrete time, nested cost operator approach to the dynamic network user equilibrium problem, *Transportation Science*, Vol. 29, No. 1, pp. 79-92.

Wiel, R.J.V. and Sahinidis, N.V. (1995), Heuristic bounds and test problem generation for the time-dependent traveling salesman problem, *Transportation Science*, Vol. 29, No. 2, pp. 167-183.

Wiel, R.J.V. and Sahinidis, N.V. (1996), An exact solution approach for the time-dependent traveling-salesman problem, *Naval Research Logistics*, Vol. 43, pp. 797-820.

Wilkinson, W.L. (1971), An algorithm for universal maximal dynamic flows in a network, *Operations Research*, Vol. 19, No. 7, pp. 1602-1612.

Xue, G., Sun, S. and Rosen, J.B. (1998), Fast data transmission and maximal dynamic flow, *Information Processing Letters*, Vol. 66, pp. 127-132.

Zahorik, A., Thomas, L.J. and Trigeiro, W.W. (1984), Network programming models for production scheduling in multi-stage, multi-item capacitated systems, *Management Science*, Vol. 30, No. 3, pp. 308-325.

Zawack, D.J. and Thompson, G.L. (1987), A dynamic space-time network flow model for city traffic congestion, *Transportation Science*, Vol. 21, No. 3, pp. 153-162.

13 ASYMPTOTIC RATES OF CONVERGENCE OF SQP-TYPE METHODS OF FEASIBLE DIRECTIONS

Michael M. Kostreva and Xibin Chen

Department of Mathematical Sciences
Clemson University
Clemson, SC 29634-1907
USA

Abstract: A modified SQP-type MFD was presented by Chen and Kostreva and its global convergence under rather mild assumptions has been proved. The numerical results showed that this modified MFD converges faster than Pironneau - Polak's MFD and Cawood - Kostreva's norm-relaxed MFD. However, the rate of convergence and especially the relationship between this rate and the choice of the introduced speed-up factors have not been investigated yet. This paper studies the asymptotic rates of convergence of SQP-type MFD based on the modified MFD. Our analysis is completely focused on a direction finding subproblem (DFS) which is a QP problem like that of the modified MFD. Based on the modified MFD, we have analyzed the rate of convergence of SQP-type MFD and its bound. This paper shows that SQP-type MFD is at least linearly convergent and induced a better bound on convergence ratio than the Cawood - Kostreva's analysis. In addition, for the case that all constraints are linear, we show that the modified MFD is superlinearly convergent under the assumptions.

Key words: Nonlinear programming(NLP), sequential quadratic programming (SQP), method of feasible directions (MFD), direction finding subproblem (DFS), asymptotic rate of convergence.

X. Yang et al. (eds.), Optimization Methods and Applications, 247-265.

1 INTRODUCTION

The general problem of nonlinear programming is to minimize or maximize an arbitrary objective function subject to nonlinear equality and inequality constraints. We consider the following formulation which minimizes a function with inequality constraints only:

$$\min \{f^0(x) \mid f^j(x) \leq 0;\ j = 1, 2, \cdots, m\} \tag{1.1}$$

where $f^j : \Re^n \to \Re^1$ $(j = 0, 1, 2, \cdots, m)$ are smooth functions.

The method of feasible directions (MFD) is an important method for solving problem (1.1). It is well-known that MFD is now established in the engineering community where it is used in structural optimization (Belegundu et al (1995), Korycki et al (1996b), Vanderplaats (1984a) and Vanderplaats (1984b)). There are successful commercial codes which use MFD to support the engineering design process Vanderplaats (1993). Even though faster algorithms for solving NLP exist, MFD has never been abandoned because of several important advantages such as the descent property, feasibility of all iterations, conceptual simplicity and overall computational efficiency. The basic steps of methods of feasible directions involve solving a direction finding subproblem (DFS) to find the direction vector and then finding the step-length along this direction by performing a constrained line search. After updating the current point, the above steps are repeated until a termination criterion is satisfied. A good DFS should be solved in a finite number of iterations. Usually, DFS is a linear programming problem (LP), quadratic programming problem (QP) or some variant or a combination of LP and QP. The main differences among various algorithms of MFD occur in the direction finding subproblem (DFS). Most of recent research on MFD has been focused on the improvement and the modification of the direction finding subproblem.

According to the type of DFS, MFD can be classified into three main types: SLP-type, SQP-type, and other type. SLP-type uses a linear programming problem as its DFS at each iteration. This type MFD includes Zoutendijk's MFD Zoutendijk (1960) and Topkis - Veinott's MFD Topkis et al (1967). Even though the computational cost to solve DFS at each iteration is cheap, these two methods are not linearly convergent, but only sublinearly convergent Pironneau et al (1972). In addition, Zoutendijk's MFD can not assure global convergence. SQP-type uses a quadratic programming problem as its DFS. This type MFD mainly includes Pironneau - Polak's MFD Pironneau et al (1973), Cawood - Kostreva's norm-relaxed MFD Cawood et al (1994) and Chen - Kostreva's modified MFD Chen et al (1999a). This type MFD converges globally and at least linearly even the computational cost to solve DFS at each iteration is more expensive than SLP-type MFD. A third type of MFD includes all MFD which do not belong to the previous two types. This type of MFD has Polak's GQP MFD Polak (1971), Wiest - Polak's MFD Wiest et al (1992), Polak - Mayne - Trahan's MFD Polak et al (1979) and Panier -Tits' FSQP Panier et al (1993). The DFS of this type MFD is neither a LP nor a QP problem. In this type MFD, the computational cost at each iteration may be higher than the previous two type MFD since they generate feasible directions by solving a LP or QP problem plus one projection operation or solving more than one LP or QP problem. Some of this type MFD are linearly convergent methods such as Wiest - Polak's MFD and Polak - Mayne - Trahan's MFD. While FSQP is locally superlinearly convergent, the computational cost at each iteration is pretty high, since each iteration is required to solve three subproblems. Hence, SQP-type MFD still has overall advantages over other two types of MFD.

A modified SQP-type MFD was presented by Chen and Kostreva in Chen et al (1999a). The global convergence of the modified method under rather mild assumptions has been proved there. The numerical results showed that this modified MFD converges faster than Pironneau - Polak's MFD Pironneau et al (1973) and Cawood - Kostreva's norm-relaxed MFD Cawood et al (1994). However, the rate of convergence and especially the relationship between this rate and the choice of the introduced speed-up factors have not been investigated yet. The convergence rate of Pironneau - Polak's MFD was investigated by Pironneau - Polak (Pironneau et al (1972) and Pironneau et al (1973)) and Chaney (Chaney (1976a) and Chaney (1976b)). They showed that their method is linearly convergent even though Pironneau - Polak and Chaney assumed different conditions. The convergence rate of Cawood - Kostreva's norm-relaxed MFD was discussed by Korycki - Kostreva (Korycki et al (1996a)). Under the same assumptions as Pironneau - Polak's, they showed that Cawood - Kostreva's norm-relaxed MFD is at least linearly convergent, with a better bound on the convergence ratio.

This paper considers the asymptotic rate of convergence of SQP-type MFD based on the modified MFD. In this research, we are not concerned with issues of almost active constraints or line-search techniques. Our analysis is completely focused on the direction finding subproblem (DFS) which contains the novelty of the modified MFD. The direction finding subproblem of the modified MFD is a generalization of the subproblems of Pironneau - Polak's MFD and Cawood - Kostreva's norm-relaxed MFD. The convergence rates of Pironneau - Polak's MFD and Cawood - Kostreva's norm-relaxed MFD can be considered as special cases of that of this modified MFD. The analysis in this paper is under the following assumptions:

(i) $f^j \in C^2, j = 0, 1, \cdots, m$.

(ii) f^j $(j = 1, \cdots, m)$ are convex and f^0 is strictly convex.

(iii) There exists $x_0 \in \Gamma \stackrel{\text{def}}{=} \{x \mid f^j(x) \leq 0, j = 1, \cdots, m\}$ such that the set

$$X \stackrel{\text{def}}{=} \{x \mid f^0(x) \leq f^0(x_0), f^j(x) \leq 0, j = 1, \cdots, m\}$$

is compact and has an interior.

Under the assumptions (i) - (iii), NLP (1.1) has a unique solution $\hat{x}$.
For discussion, the algorithm of the modified MFD is described as follows:

Algorithm A:

Step 1: Give or find an initial feasible point x_0 as in assumption (iii). Let H be some $n \times n$ symmetric, positive definite matrix. Set $i = 0$ and go to Step 2.

Step 2: Compute $h^0(x_i)$ and $h(x_i)$ solving the quadratic direction finding subproblem (DFS):

$$\min\{h^0 + \frac{1}{2}h^T Hh\} \tag{1.2}$$

$$\text{s.t.} \nabla f^0(x)^T h \leq \gamma_0 h^0 \tag{1.3}$$

$$f^j(x) + \nabla f^j(x)^T h \leq \gamma_j h^0, \; j = 1, 2, \cdots, m \tag{1.4}$$

where H a positive definite, symmetric matrix and $\gamma_j > 0$ $(j = 0, 1, 2, \cdots, m)$. If $h^0(x_i) = 0$, then stop and x_i is a solution. Otherwise go to Step 3.

Step 3: Perform the line search in direction $h(x_i)$:

$$\delta_h(x_i) \stackrel{\text{def}}{=} \min_\alpha\{f^0(x_i + \alpha h(x_i)) - f^0(x_i) \mid x_i + \alpha h(x_i) \in \Gamma\} \tag{1.5}$$

Step 4: Set $x_{i+1} = x_i + \alpha_i h(x_i)$, where α_i is the argmin of (1.5). Go to Step 2.

This paper consists of five sections. In Section 2, some notations and definitions are introduced and several propositions and corollaries are presented. Section 3 investigates the asymptotic convergence rate of SQP-type MFD by using a series of lemmas and theorems. Besides the linear convergence rate, a better bound on convergence ratio is given by the introduction of the parameters γ_j. A theoretical relationship of the bound with the previous bounds presented by Pironneau - Polak, Chaney and Korycki - Kostreva is discussed in Section 4. In addition, a better convergence rate is obtained for the case that all constraints are linear under our assumptions. The final conclusions are presented in Section 5.

2 DEFINITIONS AND PROPOSITIONS

2.1 *Notations and Definitions*

To analyze conveniently, we need to introduce the following definitions and notations:

- Define two sets:

$$\Lambda \stackrel{\text{def}}{=} \{\lambda \in \Re^{m+1} \mid \textstyle\sum_{j=0}^m \lambda^j \nabla f^j(\hat{x}) = 0, \sum_{j=0}^m \gamma_j \lambda^j = 1,$$

$$\lambda^j \geq 0, \lambda^j f^j(\hat{x}) = 0, j = 1, \cdots, m\}$$

$$\Omega \stackrel{\text{def}}{=} \{\lambda \in \Re^{m+1} \mid \sum_{j=0}^{m} \gamma_j \lambda^j = 1,\ \lambda^j \geq 0, \lambda^j f^j(\hat{x}) = 0,\ j = 1, \cdots, m\}$$

where $\gamma_j > 0$ ($j = 0, 1, \cdots, m$) are fixed numbers. Λ is a set of all Fritz John multipliers associated with the solution $\hat{x}$ of NLP (1.1) and Ω is a superset of Λ containing all those nonnegative multipliers satisfying the complementary slackness condition. Define a scalar:

$$\hat{\lambda}^0 \stackrel{\text{def}}{=} \min_{\lambda \in \Lambda} \lambda^0$$

- Define a generalized Rayleigh quotient of an $n \times n$ matrix A as the following:

$$GR_A(x) \stackrel{\text{def}}{=} \frac{x^T A x}{x^T H x}$$

where H is an $n \times n$ symmetric, positive definite matrix and $x \in \Re^n \backslash \{0\}$. The Rayleigh quotient of an $n \times n$ matrix A is

$$R_A(x) \stackrel{\text{def}}{=} \frac{x^T A x}{x^T x}$$

where $x \in \Re^n \backslash \{0\}$.

- Define the following scalars:

$$m(A) \stackrel{\text{def}}{=} \inf_{x \neq 0} \{GR_A(x)\}$$

$$M(A) \stackrel{\text{def}}{=} \sup_{x \neq 0} \{GR_A(x)\}$$

where A is an $n \times n$ symmetric, positive semidefinite matrix.

- Define scalars m^j and M^j for each $j = 0, 1, \cdots, m$:

$$m_\epsilon^j \stackrel{\text{def}}{=} \min_{x \in B(\hat{x}, \epsilon) \cap X} m^j(x)$$

$$M_\epsilon^j \stackrel{\text{def}}{=} \max_{x \in B(\hat{x}, \epsilon) \cap X} M^j(x)$$

where for $j = 0, 1, 2, \cdots, m$:

$$m^j(x) \stackrel{\text{def}}{=} m(\nabla^2 f^j(x))$$

$$M^j(x) \stackrel{\text{def}}{=} M(\nabla^2 f^j(x))$$

- Define the scalars m_ϵ and M_ϵ:

$$m_\epsilon \stackrel{\text{def}}{=} \min\{1, \min_{\lambda \in \Lambda} \sum_{j=0}^{m} m_\epsilon^j \lambda^j\}$$

$$M_\epsilon \stackrel{\text{def}}{=} \max\{1, \max_{\lambda\in\Omega}\sum_{j=0}^{m} M_\epsilon^j\lambda^j\}$$

where m_ϵ and M_ϵ are well defined since the sets Λ and Ω are compact and $\sum_{j=0}^{m} m_\epsilon^j\lambda^j$ and $\sum_{j=0}^{m} M_\epsilon^j\lambda^j$ are continuous functions of λ.

- Define two scalars $\bar{m}$ and $\bar{M}$:

$$\bar{m} \stackrel{\text{def}}{=} \min\{1, \min_{\lambda\in\Lambda}\sum_{j=0}^{m} m^j(\hat{x})\lambda^j\}$$

$$\bar{M} \stackrel{\text{def}}{=} \max\{1, \max_{j\in J(\hat{x})\cup\{0\}} M^j(\hat{x})\}$$

2.2 Propositions and Corollaries

The following are some important propositions and corollaries to be used in the analysis of rate of convergence.

Proposition 2.1 $\hat{\lambda}^0 > 0$ *under the assumptions (i)-(iii).*

Proof: By compactness of Λ, there exists $\overline{\lambda} \in \Lambda$ such that $\hat{\lambda}^0 = \overline{\lambda}^0$. Since all functions f^j are convex and continuously differentiable, we have

$$f^j(x) \geq f^j(\hat{x}) + \nabla f^j(\hat{x})^T(x - \hat{x}), \forall x \in \Re^n, j = 0, 1, \cdots, m. \quad (2.1)$$

Multiplying (2.1) by $\overline{\lambda}^j$ for $j = 0, 1, \cdots, m$, and adding the results, we obtain

$$L(x, \overline{\lambda}) \geq L(\hat{x}, \overline{\lambda}) + \nabla L(\hat{x}, \overline{\lambda})^T(x - \hat{x}).$$

Since $\overline{\lambda} \in \Lambda$, this is equivalent to:

$$\sum_{j=0}^{m} \overline{\lambda}^j f^j(x) \geq \hat{\lambda}^0 f^0(\hat{x}), \forall x \in \Re^n.$$

If $\hat{\lambda}^0 = 0$, then $\sum_{j=1}^{m} \overline{\lambda}^j f^j(x) \geq 0$, which is a contradiction with the assumption (iii). Hence, $\hat{\lambda}^0 > 0$. □

Proposition 2.2 *Let H be a symmetric, positive definite matrix and A is a symmetric, positive semidefinite matrix. Let $H = GG^T$ where G denotes the Cholesky factor of H and is a real lower triangular matrix with positive diagonal elements. Then*

$$GR_A(x) = R_B(y)$$

where $B = G^{-1}A(G^{-1})^T$ and $y = G^T x$.

Proof: From the definition of $GR_A(x)$, we have

$$GR_A(x) = \frac{x^T Ax}{x^T Hx} = \frac{x^T Ax}{x^T GG^T x} = \frac{y^T G^{-1}A(G^{-1})^T y}{y^T y} = R_B(y)$$

□

Proposition 2.3 *Let $\eta_1(B)$, $\eta_2(B)$, $\cdots$, $\eta_n(B)$ be the eigenvalues of $B = G^{-1}A(G^{-1})^T$ with order:*

$$0 \leq \eta_{min}(B) = \eta_1(B) \leq \eta_2(B) \leq \cdots \leq \eta_{n-1}(B) \leq \eta_n(B) = \eta_{max}(B)$$

Then $m(A) = \eta_{min}(B)$ and $M(A) = \eta_{max}(B)$.

Proof: Since $m(A) = \inf_{x \neq 0}\{GR_A(x)\} = \inf_{y \neq 0}\{R_B(y)\}$, $m(A) = \eta_{min}(B)$ follows Rayleigh-Ritz Theorem (Theorem 1.3-1 of Ciarlet (1989)). Similarly, $M(A) = \eta_{max}(B)$. □

Corollary 2.1 *If H is an identity matrix I, then $m(A)$ is the smallest eigenvalue of A and $M(A)$ is the largest eigenvalue of A.*

Proof: Follows directly from Proposition 2.3. □

Corollary 2.2 *$m(A)$ and $M(A)$ have the following properties:*

(i) *$m(A)$ and $M(A)$ are continuous functions of A.*

(ii) $0 \leq m(A) \leq M(A) < \infty$,

(iii) $M(A) = 0$ *if and only if* $A = 0$,

(iv) $A \in$ PD *if and only if* $m(A) > 0$.

(v) $A \in$ PSD $\setminus$ PD *if and only if* $m(A) = 0$,

Proof: (i) By Proposition 2.3, $m(A)$ and $M(A)$ are the smallest and the largest eigenvalues of B, i.e., the zeroes of a polynomial $p(\eta) = \det(B - \eta I) = \det(G^{-1}(A - \eta H)(G^{-1})^T)$ with coefficients continuously dependent on components of matrix A. But zeroes of a polynomial depend continuously on its coefficients and thus $m(A)$ and $M(A)$ depend continuously on A. □

The proofs of (ii), (iii) and (iv) are straightforward, and thus omitted.

Proposition 2.4 *$m_\epsilon^j, M_\epsilon^j (j = 0, 1, \cdots, m)$, m_ϵ and M_ϵ satisfy the following properties:*

(i) *$m_\epsilon^j, M_\epsilon^j, (j = 0, 1, \cdots, m)$ are well defined and depend continuously on ϵ.*

(ii) *m_ϵ and M_ϵ are continuous with respect to ϵ.*

(iii) *$M_\epsilon = \max\{1, \max_{j \in J(\hat{x}) \cup \{0\}} M_\epsilon^j\}$, where $J(\hat{x}) = \{j \mid f^j(\hat{x}) = 0\}$ is the set of active constraints at $\hat{x}$.*

(iv) $m_\epsilon > 0, \forall \epsilon \geq 0$.

(v) *$\lim_{\epsilon \to 0} m_\epsilon = \bar{m}$ and $\lim_{\epsilon \to 0} M_\epsilon = \bar{M}$.*

Proof: The proof of this proposition is straightforward, and thus omitted. □

3 ASYMPTOTIC RATE OF CONVERGENCE

This section will show that our modified MFD has a linear convergence rate and its upper bound will be also derived, which is similar to Korycki - Kostreva's analysis Korycki et al (1996a) under the assumptions (i) - (iii). Under the additional assumption, we will induce a tighter upper bound for general cases and a better convergence rate for the case that all constraints are linear. Let μ^0 be a Lagrange multiplier to the constraint (1.3) and μ^j ($j = 1, \cdots, m$) be the multipliers to the constraints (1.4). We use μ to substitute λ to distinguish Lagrange multipliers associated with the direction finding subproblem from the multipliers associated with NLP (1.1).

Lemma 3.1 *(Properties of DFS). Let $(h^0(x), h(x))$ be the solution of DFS (1.2)-(1.4) at $x \in \Gamma$ and $\mu \in \Re^{m+1}$ be its dual solution as above. Then:*

(i) $\sum_{j=0}^{m} \gamma_j \mu^j = 1, \mu \geq 0,$

(ii) $h(x) = -H^{-1}(\sum_{j=0}^{m} \mu^j \nabla f^j(x)),$

(iii) $(h^0(x), h(x)) = (0,0)$ *if and only if x is a Fritz John point of NLP (1.1),*

(iv) $(h^0(x), h(x))$ *is a continuous function of x for $x \in X$,*

(v) $h^0(x) + \frac{1}{2}\|h(x)\|_H^2$ *is a continuous, non-positive function of $x \in X$ and*

$$h^0(x) + \frac{1}{2}\|h(x)\|_H^2 = \sum_{j=1}^{m} \mu^j f^j(x) - \frac{1}{2}\|\sum_{j=0}^{m} \mu^j \nabla f^j(x)\|_{H^{-1}}^2 \tag{3.1}$$

where $\|h\|_H^2 = h^T H h.$

Proof: (i) follows KKT conditions of optimality for (1.2) - (1.4). (ii) - (v) follow directly from Lemmas 3.3, 3.4 and 4.3 of Chen et al (1999a).

Lemma 3.2 *For any $\epsilon > 0$ there exists $r > 0$ such that if $x \in B(\hat{x}, r) \cap X$ then $x + \alpha h(x) \in B(\hat{x}, \epsilon), \forall \alpha \in [0, 1]$.*

Proof: Since H is a symmetric, positive definite matrix, there exists a positive constant C_0 such that $\|x\|^2 \leq C_0^2 \|x\|_H^2, \forall x \in \Re^n$. Thus $\|h\|^2 \leq C_0^2 \|h\|_H^2$, i.e, $\|h\| \leq C_0 \|h\|_H$.

But $\|h\|_H$ is continuous, nonnegative and takes value 0 at $\hat{x}$. Therefore $\forall \epsilon > 0$, $\exists\, r$ such that $\|h(x)\| \leq \frac{\epsilon}{2}$ if $x \in B(\hat{x}, r) \cap X$. Let $\bar{r} = \min\{\frac{\epsilon}{2}, r\}$. Then $\|h(x)\| \leq \frac{\epsilon}{2}$ for $x \in B(\hat{x}, r) \cap X \subset B(\hat{x}, \frac{\epsilon}{2}) \cap X$. Thus $x + \alpha h(x) \in B(\hat{x}, \epsilon)$, for all $\alpha \in [0, 1]$.

Lemma 3.3 *(Limit Property of Dual Solution of the Line Search Problem). Let $\{x_i\}$ be an infinite sequence of points feasible to NLP (1.1) converging to*

$\hat{x}$. For each i select a normalized dual solution μ_i for the line search problem (1.5) at x_i as defined above. If $\hat{\mu}$ is an accumulation point of $\{\mu_i\}$, then $\hat{\mu} \in \Omega$.

Proof: In the line search procedure, we must solve the problem (1.5) which is the minimization problem with a single variable α and m constraints. Let $\bar{\mu}^j$ be the Lagrange multiplier to each constraint $f^j(x+\alpha h(x)) \leq 0\,(j = 1, \cdots, m)$. Define $\mu = (\mu^0, \mu^1, \cdots, \mu^m)$ by setting $\gamma_0 \bar{\mu}^0 = 1$ and $\mu = \bar{\mu}/\sum_{j=0}^m \gamma_j \bar{\mu}^j$.

For each i, the solution α_i and the normalized dual solution μ_i of (1.5) at x_i satisfy the complementary slackness condition:

$$\mu_i^j f^j(x_i + \alpha_i h(x_i)) = 0, \; j = 1, \cdots, m$$

This means that $\mu_i^j f^j(x_{i+1}) = 0$ $(j = 1, \cdots, m)$. Let $\{\mu_{i_k}\}$ be the subsequence of $\{\mu_i\}$ converging to $\hat{\mu}$. Letting $k \to \infty$ we obtain:

$$\hat{\mu}^j f^j(\hat{x}) = 0, \; j = 1, \cdots, m$$

Since $\mu_i \geq 0$ and $\sum_{j=0}^m \gamma_j \mu_i^j = 1$ for each i, then, by taking the limit, it follows that $\hat{\mu} \geq 0$ and $\sum_{j=0}^m \gamma_j \hat{\mu}^j = 1$. So, $\hat{\mu} \in \Omega$.

Lemma 3.4 *Let $\epsilon > 0$ and $\{x_i\}$ be an infinite sequence of points feasible to NLP (1.1) converging to $\hat{x}$. Let β be an arbitrary constant in $(0, 1)$. Then there exists i_1 such that for $i \geq i_1$:*

$$\delta_h(x_i) \leq \frac{h^0(x_i) + \frac{1}{2}\|h(x_i)\|_H^2}{M_\epsilon(1+\beta)} \tag{3.2}$$

Proof: By duality

$$\begin{aligned} \delta_h(x_i) = \;& \max_{\mu \in \Re^m, \mu \geq 0}\{\inf_{\alpha \in \Re^1}\{f^0(x_i + \alpha h(x_i)) \\ - \;& f^0(x_i) + \textstyle\sum_{j=1}^m \mu^j f^j(x_i + \alpha h(x_i))\}\} \end{aligned} \tag{3.3}$$

So,

$$\delta_h(x_i) \leq \inf_{\alpha \in [0,1]} \{f^0(x_i + \alpha h(x_i)) - f^0(x_i) + \sum_{j=1}^m \bar{\mu}_i^j f^j(x_i + \alpha h(x_i))\} \tag{3.4}$$

where $\bar{\mu}_i$ is the arg max in (3.3).

For $\epsilon > 0$, set r according to Lemma 3.2 such that $x \in B(\hat{x}, r) \cap X$ implies that $x + \alpha h(x) \in B(\hat{x}, \epsilon)$, $\forall \alpha \in [0, 1]$. Set $\bar{r} = \min\{r, \epsilon\}$ and i_1' such that $x_i \in B(\hat{x}, \bar{r}) \cap X$ for $i \geq i_1'$. Hence, for $i \geq i_1'$, $x_i \in B(\hat{x}, \epsilon)$ and $x_i + \alpha h(x_i) \in B(\hat{x}, \epsilon)$, where $\alpha \in [0, 1]$ and we may use the following majorization:

$$h(x_i)^T \nabla^2 f^j(\xi) h(x_i) \leq M_\epsilon^j \|h(x_i)\|_H^2, \quad \forall \xi \in [x_i, x_i + \alpha h(x_i)] \tag{3.5}$$

with arbitrary $\alpha \in [0, 1]$ and for any $j = 0, 1, \cdots, m$.

By expanding $f^j(x_i + \alpha h(x_i))$ $(j = 0, 1, \cdots, m)$ in (3.4) in Taylor's series and using majorization (3.5), we have

$$\begin{array}{rl} \delta_h(x_i) \leq & \inf_{\alpha\in[0,1]}\{\sum_{j=1}^m \bar{\mu}_i^j f^j(x_i) + \alpha \sum_{j=0}^m \bar{\mu}_i^j \nabla f^j(x_i)^T h(x_i) \\ + & \frac{1}{2}\alpha^2(\sum_{j=0}^m \bar{\mu}_i^j M_\epsilon^j)\|h(x_i)\|_H^2\} \end{array} \quad (3.6)$$

Set $\mu = \bar{\mu}/S$, where $\gamma_0\bar{\mu}_i^0 = 1$ and $S = \sum_{j=0}^m \gamma_j\bar{\mu}_i^j$.

By Lemma 3.3, every accumulation point of $\{\mu_i\}$ belongs to Ω. Because the function $\sum_{j=0}^m \mu^j M_\epsilon^j$ is continuous with respect to μ, from the definition of M_ϵ, we have

$$\limsup_{k\to\infty} \sum_{j=0}^m \mu_k^j M_\epsilon^j \leq M_\epsilon$$

Thus for any $\beta \in (0, 1)$, $\exists i_1''$ such that for $i \geq i_1''$:

$$\sum_{j=0}^m \mu_i^j M_\epsilon^j \leq (1+\beta) \limsup_{k\to\infty} \sum_{j=0}^m \mu_k^j M_\epsilon^j \leq (1+\beta) M_\epsilon \quad (3.7)$$

Let $i_1 = \max\{i_1', i_1''\}$. By the definition of μ and using (3.7) in (3.6), for $i \geq i_1$, the following holds:

$$\begin{array}{rl} \delta_h(x_i) \leq & \inf_{\alpha\in[0,1]}\{S\sum_{j=1}^m \mu_i^j f^j(x_i) + \alpha S \sum_{j=0}^m \mu_i^j \nabla f^j(x_i)^T h(x_i) \\ + & \frac{1}{2}\alpha^2 S M_\epsilon(1+\beta)\|h(x_i)\|_H^2\} \end{array} \quad (3.8)$$

Let $\alpha = \frac{1}{SM_\epsilon(1+\beta)}$ in (3.8). So, $\alpha \in (0, 1)$ since $S \geq 1, M_\epsilon \geq 1$ and $\beta > 0$. From (3.8), we can obtain

$$\begin{array}{rl} \delta_h(x_i) \leq & S\sum_{j=1}^m \mu_i^j f^j(x_i) + \frac{1}{M_\epsilon(1+\beta)} \sum_{j=0}^m \mu_i^j \nabla f^j(x_i)^T h(x_i) \\ + & \frac{1}{2SM_\epsilon(1+\beta)}\|h(x_i)\|_H^2 \end{array} \quad (3.9)$$

Since $\sum_{j=1}^m \mu_i^j f^j(x_i) \leq 0$, then $S\sum_{j=1}^m \mu_i^j f^j(x_i) \leq \frac{1}{M_\epsilon(1+\beta)} \sum_{j=1}^m \mu_i^j f^j(x_i)$. From (3.9), we have

$$\begin{array}{rl} \delta_h(x_i) \leq & \frac{1}{M_\epsilon(1+\beta)}(\sum_{j=1}^m \mu_i^j f^j(x_i) + \sum_{j=0}^m \mu_i^j \nabla f^j(x_i)^T h(x_i) \\ + & \frac{1}{2}\|h(x_i)\|_H^2) \end{array} \quad (3.10)$$

Because $(h^0(x_i), h(x_i))$ is the solution of DFS (1.2) - (1.4) at x_i, $\nabla f^0(x_i)^T h(x_i) \leq \gamma_0 h^0(x_i)$, $f^j(x_i) + \nabla f^j(x_i)^T h(x_i) \leq \gamma_j h^0(x_i)$ $(j = 1, \cdots, m)$ and $\sum_{j=0}^m \gamma_j \mu_i^j = 1$. So,

$$\sum_{j=1}^m \mu_i^j f^j(x_i) + \sum_{j=0}^m \mu_i^j \nabla f^j(x_i)^T h(x_i) \leq (\sum_{j=0}^m \gamma_j \mu_i^j) h^0(x_i) = h^0(x_i) \quad (3.11)$$

By substituting (3.11) into (3.10), then for $i \geq i_1$, (3.2) holds.

Lemma 3.5 and Lemma 3.6 are connected with the following subproblem from the method of centers Huard (1967). Let $x_i \in \Gamma$. Define the following quantity $\delta(x_i)$:

$$\delta(x_i) \stackrel{\text{def}}{=} \min_{x\in\Gamma} \max\{f^0(x) - f^0(x_i), f^j(x),\ j = 1, \cdots, m\} \tag{3.12}$$

In some sense, this subproblem finds the center of the set $\{x \mid f^0(x) \leq f^0(x_i), f^j(x) \leq 0,\ j = 1, \cdots, m\}$. The equivalent subproblem to (3.12) is:

$$\delta(x_i) = \min_{x\in\Gamma,\delta} \{\delta \mid f^0(x) - f^0(x_i) \leq \gamma_0\delta, f^j(x) \leq \gamma_j\delta,\ j = 1, \cdots, m\} \tag{3.13}$$

where $\delta(x_i) \leq 0$ for $x_i \in \Gamma$ and $\delta(\hat{x}) = 0$. $\delta(x_i)$ is introduced technically for purposes of the analysis of the rate of convergence. We do not use any elements of the method of centers in the algorithm of MFD. Let μ_i^0 be the Lagrange multiplier to the first constraint in (3.13) and μ_i^j $(j = 1, \cdots, m)$ be the multipliers to the rest of the constraints of (3.13).

Lemma 3.5 *Let $\epsilon > 0$ and $\{x_i\}$ be an infinite sequence of points feasible to NLP (1.1) converging to $\hat{x}$. Let $\beta \in (0,1)$ be some arbitrary constant. Then there exists i_2 such that for $i \geq i_2$:*

$$h^0(x_i) + \frac{1}{2}\|h(x_i)\|_H^2 \leq m_\epsilon(1-\beta)\delta(x_i) \tag{3.14}$$

Proof: The proof is similar to that of Lemma 3.6 in Korycki et al (1996a).

Lemma 3.6 *Let $\{x_i\}$ be an infinite sequence of points feasible to NLP (1.1) converging to $\hat{x}$. Given arbitrary constant $\beta \in (0,1)$, there exists i_3 such that for $i \geq i_3$:*

$$\delta(x_i) \leq \hat{\lambda}^0(1-\beta)(f^0(\hat{x}) - f^0(x_i))$$

Proof. Let $\mu_i \in \Re^{m+1}$ denote the dual solution of (3.13) at x_i. By duality:

$$\delta(x_i) = \inf_{x\in\Re^n} \{\sum_{j=1}^{m} \mu_i^j f^j(x) + \mu_i^0(f^0(x) - f^0(x_i))\}. \tag{3.15}$$

Replace x by $\hat{x}$ in (3.15):

$$\delta(x_i) \leq \sum_{j=1}^{m} \mu_i^j f^j(\hat{x}) + \mu_i^0(f^0(\hat{x}) - f^0(x_i)) \leq \mu_i^0(f^0(\hat{x}) - f^0(x_i)). \tag{3.16}$$

We want to show that if $\{x_i\}$ converges to $\hat{x}$ and $\hat{\mu}$ is an accumulation point of the corresponding sequence $\{\mu_i\}$, then $\hat{\mu} \in \Lambda$. The KKT conditions of optimality for (3.13) satisfied at each iteration i imply the following:

$$\mu_i \geq 0 \tag{3.17}$$

$$\sum_{j=0}^{m} \gamma_j \mu_i^j = 1 \tag{3.18}$$

$$\sum_{j=0}^{m} \mu_i^j \nabla f^j(x_i) = 0 \tag{3.19}$$

$$\mu_i^j (f^j(x_i) - \gamma_j \delta_i) = 0, j = 1, \cdots, m \tag{3.20}$$

where $\delta_i = \max\{f^0(x_{i+1}) - f^0(x_i); f^j(x_{i+1}), j = 1, \cdots, m\}$. Consider the subsequence $\{\mu_{i_k}\}$ convergent to $\hat{\mu}$. By continuous differentiability of f^j, (3.17) - (3.19) yield in the limit: $\hat{\mu} \geq 0, \sum_{j=0}^{m} \gamma_j \hat{\mu}^j = 1$ and $\nabla L(\hat{x}, \hat{\mu}) = 0$. Also, since $\lim_{k \to \infty}(f^0(x_{i_k+1}) - f^0(x_{i_k})) = 0$, then $\lim_{k \to \infty} \delta_{i_k} = 0$. Thus (3.20) by taking the limit as $k \to \infty$ yields $\hat{\mu}^j f^j(\hat{x}) = 0 (j = 1, \cdots, m)$. Hence, $\hat{\mu} \in \Lambda$.

We will now use this limit result to obtain a bound on μ_i^0. Since all accumulation points of $\{\mu_i\}$ are in Λ, then from the definition of $\hat{\lambda}^0$, for given a constant $\beta \in (0, 1)$, there exists i_3 such that for $i \geq i_3$, we have

$$\mu_i^0 \geq \hat{\lambda}^0 (1 - \beta). \tag{3.21}$$

Substituting (3.21) into (3.16) completes the proof.

Lemma 3.7 *Let assumptions (i)-(iii) about NLP (1.1) hold. Then either Algorithm A stops after a finite number of steps with the last iterate equal to* $\hat{x}$ *or an infinite sequence* $\{x_i\}$ *is generated such that:* $\lim_{i \to \infty} x_i = \hat{x}$.

Proof: By Theorem 4.1 in Chen et al (1999a), Algorithm A either stops at the Fritz John point of NLP (1.1) or an infinite sequence $\{x_i\}$ is produced such that every accumulation point is a Fritz John point of NLP (1.1). Under the assumptions (i) - (iii), $\hat{x}$ is the only Fritz John point in the set X. Hence, this lemma holds.

Theorem 3.1 *(Linear Convergence). Let* $\epsilon > 0$ *and* $\beta \in (0, 1)$, *arbitrary. Let* $\{x_i\}$ *be an infinite sequence generated by Algotithm A. Then there exists* i_0 *such that for* $i \geq i_0$:

$$f^0(x_{i+1}) - f^0(\hat{x}) \leq (1 - \frac{m_\epsilon}{M_\epsilon} \hat{\lambda}^0 (1 - \beta))(f^0(x_i) - f^0(\hat{x})) \tag{3.22}$$

Proof: The proof is similar to that of Theorem 3.1 of Korycki et al (1996a). Let β_1 satisfy $1 - \beta = (1 - \beta_1)^2/(1 + \beta_1)$. From Step 3 of the algorithm of the new MFD and the definition of $\delta_h(x_i)$, we can see that for each i,

$$f^0(x_{i+1}) - f^0(x_i) = \delta_h(x_i) \tag{3.23}$$

By Lemma 3.7, we have $x_i \to \hat{x}$.

Since $x_i \to \hat{x}$, by Lemma 3.4, there exists i_1 such that for $i \geq i_1$:

$$\delta_h(x_i) \leq \frac{1}{M_\epsilon(1+\beta_1)}(h^0(x_i) + \frac{1}{2}\|h(x_i)\|_H^2) \tag{3.24}$$

Since $x_i \to \hat{x}$, by Lemma 3.5 there exists i_2 such that for $i \geq i_2$:

$$h^0(x_i) + \frac{1}{2}\|h(x_i)\|_H^2 \leq m_\epsilon(1-\beta_1)\delta(x_i) \tag{3.25}$$

Since $x_i \to \hat{x}$, by Lemma 3.6 there exists i_3 such that for $i \geq i_3$:

$$\delta(x_i) \leq \hat{\lambda}^0(1-\beta_1)(f^0(\hat{x}) - f^0(x_i)) \tag{3.26}$$

Let $i_0 = \max\{i_1, i_2, i_3\}$. From (3.23), (3.24), (3.25) and (3.26), we can obtain that for $i \geq i_0$:

$$f^0(x_{i+1}) - f^0(x_i) \leq \frac{m_\epsilon}{M_\epsilon}\hat{\lambda}^0\frac{(1-\beta_1)^2}{1+\beta_1}(f^0(\hat{x}) - f^0(x_i)) \tag{3.27}$$

After (3.27) is rearranged, (3.22) holds.

Theorem 3.2 *(Bound of Convergence Ratio). Let $\{x_i\}$ be a sequence generated by Algorithm A. Under the assumptions (i) - (iii), either the algorithm stops after finite number of steps at the solution $\hat{x}$ or the infinite sequence $\{f^0(x_i)\}$ converges at least linearly to $f^0(\hat{x})$ with convergence ratio which is no more than a bound $\theta \in [0, 1)$, where*

$$\theta = 1 - \frac{\bar{m}}{\bar{M}}\hat{\lambda}^0 \tag{3.28}$$

Proof: If an infinite sequence $\{x_i\}$ is generated, by Theorem 3.1, there exists i_0 such that for $i \geq i_0$ and arbitrary $\epsilon > 0$ and $\beta \in (0, 1)$, we have

$$\frac{| f^0(x_{i+1}) - f^0(\hat{x}) |}{| f^0(x_i) - f^0(\hat{x}) |} \leq 1 - \frac{m_\epsilon}{M_\epsilon}\hat{\lambda}^0(1-\beta)$$

Thus the following holds for all $\epsilon > 0$ and $\beta \in (0, 1)$:

$$q = \limsup_{i\to\infty} \frac{| f^0(x_{i+1}) - f^0(\hat{x}) |}{| f^0(x_i) - f^0(\hat{x}) |} \leq 1 - \frac{m_\epsilon}{M_\epsilon}\hat{\lambda}^0(1-\beta) \tag{3.29}$$

But q does not depend on ϵ and β. Moreover, according to Proposition 2.4(v):

$$\lim_{\epsilon\to 0}\frac{m_\epsilon}{M_\epsilon} = \frac{\bar{m}}{\bar{M}} \in (0, 1)$$

Thus taking the limit of the right hand side of (3.29) as $\epsilon \to 0$ and $\beta \to 0$ we obtain (3.28).

The bound θ depends directly on three parameters: $\hat{\lambda}^0$, $\bar{m}$ and $\bar{M}$. The constants $\bar{m}$ and $\bar{M}$ depend on Hessians of the objective function and constraint functions active at $\hat{x}$. $\hat{\lambda}^0$ and $\bar{m}$ depend on the dual solution of NLP (1.1). We can see that the choice of the weight vector γ has effect on the dual solution of DFS. A small $\hat{\lambda}^0$ and a small $\bar{m}$ result in slower convergence of the MFD. To achieve a small θ, we must increase $\bar{m}$ and $\hat{\lambda}^0$ when $\bar{M}$ is fixed. For attaining some stronger results, we require the following assumption (iv).

(iv) $\hat{x}$ is a regular point and strict complementary slackness holds at $\hat{x}$, i.e. a dual solution λ satisfies $\lambda^j > 0$ if $j \in J(\hat{x})$.

Lemma 3.8 *(Limit Property of Dual Solution of DFS). Let $\{x_i\}$ be an infinite sequence of points feasible to NLP (1.1) converging to $\hat{x}$. For each i select a dual solution μ_i for DFS (1.2) - (1.4) at x_i. If $\hat{\mu}$ is an accumulation point of $\{\mu_i\}$, then $\hat{\mu} \in \Lambda$.*

Proof: Let $\{\mu_{i_k}\}$ be a subsequence of $\{\mu_i\}$ converging to $\hat{\mu}$. Without loss of generality, we can identify the subsequence with the sequence $\{\mu_i\}$ itself. By Lemma 3.1 (ii), we have

$$h(x_i) = -H^{-1} \sum_{j=0}^{m} \mu_i^j \nabla f^j(x_i) \tag{3.30}$$

Also, complementary slackness must hold for DFS (1.2)-(1.4) at x_i:

$$\mu_i^j (f^j(x_i) + \nabla f^j(x_i)^T h(x_i) - \gamma_j h^0(x_i)) = 0, \qquad j = 1, \cdots, m \tag{3.31}$$

By descent character of MFD, all $x_i \in X$. By Lemma 3.1 (iii) and 3.1 (iv), $(h^0(\cdot), h(\cdot))$ is a continuous function in X and $h^0(\hat{x}) = 0, h(\hat{x}) = 0$. Since $\{x_i\}$ converges to $\hat{x}$, then $\lim_{i \to \infty} h^0(x_i) = 0$ and $\lim_{i \to \infty} h(x_i) = 0$.

By taking the limit as $i \to \infty$ in (3.30), $H^{-1} \sum_{j=0}^{m} \hat{\mu}^j \nabla f^j(\hat{x}) = 0$. But since H^{-1} is nonsingular, then

$$\sum_{j=0}^{m} \hat{\mu}^j \nabla f^j(\hat{x}) = 0$$

Also, since $\mu_i \geq 0$ for each i and $\sum_{j=0}^{m} \gamma_j \mu_i^j = 1$, then

$$\hat{\mu} \geq 0, \qquad \sum_{j=0}^{m} \gamma_j \hat{\mu}^j = 1$$

By taking the limit as $i \to \infty$ in (3.31), we have

$$\hat{\mu}^j f^j(\hat{x}) = 0, \; j = 1, \cdots, m$$

Therefore, $\hat{\mu} \in \Lambda$.

Lemma 3.9 *Let $\{x_i\}$ be an infinite sequence of points feasible to NLP (1.1) convergent to $\hat{x}$. Under the assumptions (i)-(iv), there exists i_0 such that for $i \geq i_0$ the j-th constraint (1.4) in the direction finding subproblem is active if and only if the corresponding j-th constraint in NLP (1.1) is active.*

Proof: Since $\hat{x}$ is a regular point, then $\Lambda = \{\hat{\lambda}\}$. Let μ_i be a dual solution of the direction finding subproblem (1.2)-(1.4) at x_i. Since Λ contains only a single point $\hat{\lambda}$ and the set $\{\hat{\lambda} \mid \hat{\lambda} \geq 0, \sum_{j=0}^{m} \gamma_j \hat{\lambda}^j = 1\}$ is compact, then, by Lemma 3.8, $\lim_{i \to \infty} \mu_i = \hat{\lambda}$.

By the complementary slackness condition, the following holds at each iteration i:

$$\mu_i^j (f^j(x_i) + \nabla f^j(x_i)^T h(x_i) - \gamma_j h^0(x_i)) = 0 \tag{3.32}$$

If $j \in J(\hat{x})$, then there exists i_j such that $\mu_i^j > 0$ for $i \geq i_j$ by strict complementary slackness $\lambda^j > 0$. From (3.32), we know that the j-th constraint (1.4) is active for $i \geq i_j$. If $j \notin J(\hat{x})$, then $f^j(\hat{x}) < 0$. Since $x_i \to \hat{x}, h(x_i) \to 0$ and $h^0(x_i) \to 0$, by (3.32), there exists i_j such that for $i \geq i_j$ the j-th constraint (1.4) is inactive. Let $i_0 = \max\{i_j \mid j = 1, \cdots, m\}$, the proof is completed.

By Lemma 3.9, under the assumptions (i) - (iv), there exists i_0 such that for all $i \geq i_0$, $\mu_i^j = 0$ if and only if $\hat{\lambda^j} = 0$ for $j = 1, 2, \cdots, m$.

Lemma 3.10 *Under the assumptions (i) - (iv), $\forall\, \epsilon > 0$, there exists i_0 such that for all $i \geq i_0$, $\mu_i^0 > 1 - \epsilon$ for $\gamma_0 = 1$ and small enough γ_j $(j = 1, 2, \cdots, m)$. Furthermore, $\hat{\lambda^0} > 1 - \epsilon$.*

Proof: From the dual problem of DFS, we have $\sum_{j=0}^{m} \gamma_j \mu_i^j = 1$. Set $\gamma_0 = 1$, then $\mu_i^0 = 1 - \sum_{j=1}^{m} \gamma_j \mu_i^j$. Since $x_i \to \hat{x}$, then $\mu_i^j \to \hat{\lambda^j}$ for $j = 1, 2, \cdots, m$ holds from Theorem 2.1 of Robinson (1974). So, $\mu_i^0 \to \hat{\lambda^0}$ since $\sum_{j=0}^{m} \gamma_j \hat{\lambda^j} = 1$.

For $j = 1, 2, \cdots, m$, since $\mu_i^j \to \hat{\lambda^j}$, i.e., $\forall\, \epsilon > 0$, there exists i_0 such that for all $i \geq i_0$, $|\mu_i^j - \hat{\lambda^j}| < \epsilon$, then, $\mu_i^j < \hat{\lambda^j} + \epsilon$. By selecting $\gamma_j = \frac{\epsilon}{(\hat{\lambda^j} + \epsilon) * m}$, we have $\sum_{j=1}^{m} \gamma_j \mu_i^j < \epsilon$. Thus, $\mu_i^0 > 1 - \epsilon$ for all $i \geq i_0$. So, $\hat{\lambda^0} > 1 - \epsilon$.

Theorem 3.3 *(Main Theorem). Under the assumptions (i) - (iv), for a fixed $\gamma_0 = 1$ and small enough γ_j $(j = 1, 2, \cdots, m)$, either Algorithm A stops after finite number of steps at the solution $\hat{x}$ or the infinite sequence $\{f^0(x_i)\}$ converges at least linearly to $f^0(\hat{x})$ with convergence ratio which is no more than a bound $\theta^A \in [0, 1)$, where:*

$$\theta^A = 1 - \frac{\bar{m}}{\bar{M}} \tag{3.33}$$

Proof: From Lemma 3.10, we have $\hat{\lambda^0} > 1 - \epsilon$. By Theorem 3.2, we have $\theta^A \leq 1 - \frac{\bar{m}}{\bar{M}}(1-\epsilon)$. Let $\epsilon \to 0$, then $\theta^A = 1 - \frac{\bar{m}}{\bar{M}}$.

Remark: During the discussion in this section, the matrix H in Algorithm A has been fixed. When H_i is used as each iteration i, the process of analyzing the convergence rate is similar if H_i is in a compact set of positive-definite matrices. Proof of the process is omitted.

4 COMPARISON OF CONVERGENCE RATES

Since Algorithm A includes Pironneau - Polak's MFD and Cawood - Kostreva's MFD, the convergence rate of Pironneau - Polak's MFD (Pironneau et al (1972) and Pironneau et al (1973)) and Chaney (Chaney (1976a) and Chaney (1976b)) can be considered as a special case that $H = I$ and $\gamma_j = 1$ $(j = 0, 1, 2, \cdots, m)$ and Korycki- Kostreva (Korycki et al (1996a)) as the case that $\gamma_j = 1$ $(j = 0, 1, 2, \cdots, m)$. In fact, these bounds are as follows:

- Pironneau - Polak's bound:

$$\theta^P = 1 - \frac{\min\{1, m^0(\hat{x})\}}{M^P}(\hat{\lambda}^0)^2$$

- Chaney's bound:

$$\theta^C = 1 - \frac{\bar{m}}{M^P}\hat{\lambda}^0$$

- Korycki - Kostreva's bound:

$$\theta^K = 1 - \frac{\bar{m}}{\bar{M}}\hat{\lambda}^0$$

- Kostreva - Chen's bound:

$$\theta^A = 1 - \frac{\bar{m}}{\bar{M}}$$

where $M^P \stackrel{\text{def}}{=} \max\{1, \max_{j=0,1,\cdots,m} \|\nabla^2 f^j(\hat{x})\|\}$. Chaney obtained θ^C by using different assumptions than (i)-(iii). Besides the assumptions (i) - (iii), we added a stricter but reasonable assumption that $\hat{x}$ is a regular point and strict complementary slackness holds at $\hat{x}$ to get our bound θ^A.

Theorem 4.1 *Let θ_I^K denote the bound θ^K above, θ_I denote the bound θ in Theorem 3.2, and θ_I^A denote the bound θ^A in Theorem 3.3 when H is an identity matrix. Under assumptions (i) - (iv), the following is true:*

$$\theta^P \geq \theta^C \geq \theta_I^K = \theta_I \geq \theta_I^A \tag{4.1}$$

Proof: If $H = I$, then $M^j(\hat{x})$ is the largest eigenvalue of $\nabla^2 f^j(\hat{x})$ and thus obviously $\|\nabla^2 f^j(\hat{x})\| \geq M^j(\hat{x})$ for all $j = 0, 1, \cdots, m$. So, $M^P \geq \bar{M}$. Moreover, to determine $\bar{M}$, we consider only indices of active constraints at $\hat{x}$, whereas in M^P all constraints contribute to the maximum. Hence, $\theta^C \geq \theta_I^K$. Since $\min\{1, m^0(\hat{x})\}\hat{\lambda}^0 \leq \min\{1, m^0(\hat{x})\hat{\lambda}^0\}$ and

$$\min_{\lambda\in\Lambda}\sum_{j=0}^{m} m^j(\hat{x})\lambda^j \geq \min_{\lambda\in\Lambda} m^0(\hat{x})\lambda^0 = m^0(\hat{x})\hat{\lambda}^0$$

then $\bar{m} \geq \min\{1, m^0(\hat{x})\}\hat{\lambda}^0$. Hence, $\theta^P \geq \theta^C$. Since $\hat{\lambda}^0 \leq 1$, then $\theta_I^K \geq \theta_I^A$. Therefore, (4.1) holds.

In Chen et al (1999b), we compared the number of iterations of Pironneau - Polak's MFD, Cawood - Kostreva's MFD and Algorithm A by numerical examples and those examples illustrated the theory before it was proved.

Lemma 4.1 *If* $\hat{H} = \nabla^2 f^0(\hat{x})$, *then* $m(\nabla^2 f^0(\hat{x})) = 1$ *and* $M(\nabla^2 f^0(\hat{x})) = 1$.

Proof: If we set $H = \nabla^2 f^0(\hat{x})$ then by the definitions of $m(\cdot)$ and $M(\cdot)$, and Propositions 2.2 and 2.3, it is clear that $m(\nabla^2 f^0(\hat{x})) = \eta_{min}(I) = 1$ and $M(\nabla^2 f^0(\hat{x})) = \eta_{max}(I) = 1$.

Theorem 4.2 *Suppose that all constraints in NLP (1.1) are linear, then* $\hat{H} = \nabla^2 f^0(\hat{x})$. *If this matrix is used in the modified MFD and an infinite sequence* $\{x_i\}$ *is produced, then*

(i) Under the assumptions (i) - (iii), the bound θ *on the rate of convergence is equal to* $1 - (\hat{\lambda}^0)^2$.

(ii) Under the assumptions (i) - (iv) and small enough γ_j $(j = 1, 2, \cdots, m)$, *the convergence is superlinear.*

Proof: In this case, $\nabla^2 f^j(\hat{x}) = 0$ $(j = 1, 2, \cdots, m)$. So, $m^j(\hat{x}) = 0$ and $M^j(\hat{x}) = 0$ for all $j = 1, \cdots, m$. So, $\bar{m} = \min\{1, \hat{\lambda}^0 m^0(\hat{x})\}$ and $\bar{M} = \max\{1, M^0(\hat{x})\}$. Let $\rho = \frac{\bar{m}}{\bar{M}}$. To maximize ρ, we certainly want to maximize $m^0(\hat{x})$ and at the same time minimize $M^0(\hat{x})$. The best we can hope for is the case when $m^0(\hat{x}) = M^0(\hat{x}) = 1$ which is actually achieved if we set $\hat{H} = \nabla^2 f^0(\hat{x})$. So, $\rho = \hat{\lambda}^0$ since $\hat{\lambda}^0 \leq 1$. By substituting ρ into the formula (3.28), we obtain that $\theta = 1 - (\hat{\lambda}^0)^2$, i.e., (i) holds. By (i) and Lemma 3.10, (ii) holds.

5 CONCLUSIONS

Based on the modified MFD - Algorithm A, we have analyzed the asymptotic rates of convergence of SQP-type MFD and its bound. This paper showed that SQP-type MFD is at least linearly convergent and induced a better bound on convergence ratio than the previous analysis. This analysis is an extension and a generalization of Pironneau - Polak's analysis and Korycki - Kostreva's analysis. Also, our results are asymptotic, similar to those of Pironneau - Polak, Chaney and Korycki - Kostreva. In addition, for the case that all constraints are linear, we showed that the modified MFD is superlinearly convergent under the assumptions.

Acknowledgments

The authors thank the referee for his careful reading and helpful comments.

References

Belegundu, A. D., Berke, L. and Patnaik, S. N. (1995), An optimization algorithm based on the method of feasible directions, *Structural Optimization*, Vol. 9, pp. 83-88.

Cawood, M. E. and Kostreva, M. M. (1994), Norm-relaxed method of feasible directions for solving nonlinear programming problems, *Journal of Optimization Theory and Applications*, Vol. 83, pp. 311-320.

Chaney, R. W. (1976a), On the Pironneau-Polak method of centers, *Journal of Optimization Theory and Applications*, Vol. 20, pp. 269-295.

Chaney, R. W. (1976b), On the rate of convergence of some feasible direction algorithms, *Journal of Optimization Theory and Applications*, Vol. 20, pp. 297-313.

Chen, X. and Kostreva, M. M. (1999a), A generalization of the norm-relaxed method of feasible directions, *Applied Mathematics and Computation*, Vol. 102, pp. 257-272.

Chen, X. and Kostreva, M. M. (2000), Methods of feasible directions: a review, Progress In Optimization, X.Q. Yang et al. (eds.), 205-219, Kluwer Academic publishers, Netherlands.

Ciarlet, G. P. (1989), *Introduction to Numerical Linear Algebra and Optimization*, Cambridge University press, Cambridge.

Huard, P. (1967), The Method of Centers, *Nonlinear Programming*, Edited by J. Abadie, North Holland, Amsterdam, Holland.

Korycki, J. A. and Kostreva, M. M. (1996a), Convergence analysis of norm-relaxed method of feasible directions, *Journal of Optimization Theory and Applications*, Vol. 91, pp. 389-418.

Korycki, J. A. and Kostreva, M. M. (1996b), Norm-relaxed method of feasible directions: application in structural optimization, *Structural Optimization*, Vol. 11, pp. 187-194.

Panier, E. R. and Tits, A. L. (1993), On combining feasibility, descent and superlinear convergence in inequality constrained optimization, *Mathematical Programming*, Vol. 59, pp. 261-276.

Pironneau, O. and Polak, E. (1972), On the rate of convergence of certain methods of centers, *Mathematical Programming*, Vol. 2, pp. 230-257.

Pironneau, O. and Polak, E. (1973), Rate of convergence of a class of methods of feasible directions, *SIAM Journal on Numerical Analysis*, Vol. 10, pp. 161-173.

Polak, E. (1971), *Computational Methods in Optimization*, Academic Press, New York.

Polak, E., Trahan, R. and Mayne, D. Q. (1979), Combined phase I - phase II methods of feasible directions, *Mathematical Programming*, Vol. 17, pp. 32-61.

Robinson, S. M. (1974), Perturbed Kuhn - Tucker points and rates of convergence for a class of nonlinear programming algorithms, *Mathematical Programming* Vol. 7, pp. 1-16.

Topkis, D. M. and Veinott, A. F. (1967), On the convergence of some feasible direction algorithms for nonlinear programming, *SIAM Journal on Control*, Vol. 5, pp. 268-279.

Vanderplaats, G. N. (1984a), *Numerical Optimization Techniques for Engineering Design*, McGraw Hill, New York.

Vanderplaats, G. N. (1984b), Efficient feasible directions algorithm for design synthesis, *AIAA Journal*, Vol. 22, pp. 1633-1640.

Vanderplaats, G. N. (1993), *DOT/DOC Users Manual*, Vanderplaats, Miura and Associates.

Wiest, E. J., Polak, E. (1992), A generalized quadratic programming - based phase I - phase II method for inequality constrained optimization, *Journal of Applied Mathematics and Optimization*, Vol. 26, pp. 223-252.

Zoutendijk, G. (1960), *Methods of Feasible Directions*, Elsevier Publishing Company, Amsterdam, Netherlands.

14 NONLINEAR LAGRANGIAN METHODS IN CONSTRAINED NONLINEAR OPTIMIZATION

Duan Li

Department of Systems Engineering and Engineering Management
The Chinese University of Hong Kong
Shatin, N. T., Hong Kong, China

and Xiaolin Sun

Department of Mathematics
Shanghai University, Shanghai, China

Abstract: Recent research has presented an extension from the traditional linear Lagrangian theory to nonlinear Lagrangian theory in order to achieve a guarantee of the identification of an optimal solution of the primal problem via dual search. This paper summarizes recent progress in new dual formulations with clear motivation and full geometric interpretation in order to better our understanding of the fundamental properties in constrained optimization and in the newly developed nonlinear Lagrangian duality theory.

Key words: Nonlinear constrained optimization, global optimal solution, saddle point, dual search, p-th power formulation, nonlinear Lagrangian.

X. Yang et al. (eds.), Optimization Methods and Applications, 267-277.

1 INTRODUCTION

The Lagrangian methods(Bertsekas (1982), Lasdon (1970), Luenberger (1984)), the penalty function methods (Courant (1962), Fiacco and McCormick (1968), Zangwill (1967)), and the successive quadratic programming method (Biggs (1975), Gill *et al* (1981), Han (1977), Powell (1983)), have been the most efficient solution algorithms in solving constrained optimization problems. In the convex situation, the existence of a saddle point guarantees the success of the dual search via sequential minimization of the Lagrangian function. In a presence of nonconvexity, however, the conventional dual search methods often fail to locate the global optimal solution of the primal problem. Recent research results in Goh and Yang (1997), Li (1995), Li (1997), Li and Sun (1999), Sun and Li (1999), Xu (1997), Yang and Li (2000) represent an extension from the traditional linear Lagrangian theory to nonlinear Lagrangian theory to guarantee the identification of an optimal solution of the primal problem via dual search. This paper investigates some fundamental properties in constrained nonlinear programming problems and summarizes recent progress in nonlinear Lagrangian formulations. Prominent features in both the theoretical achievements and computational implementation of the new dual formulations will be addressed in this paper.

Consider the following constrained optimization problem (P),

$$\begin{aligned} \min\ & f(x) \\ \text{s. t. } g_i(x) &\le b_i, \quad i=1,2,\ldots,m, \\ x &\in X, \end{aligned} \tag{1.1}$$

where $x \in R^n$, $f \in C^2$, $g_i \in C^2$, $i = 1,2,\ldots,m$, and X is a closed and bounded subset of R^n.

The perturbation function associated with problem (1.1) is defined by

$$w(y) = \min\{f(x) \mid g_i(x) \le y_i,\ i=1,2,\ldots,m,\ x \in X\}.$$

The domain of $w(y)$ is

$$F = \{y \mid \text{there exists } x \in X \text{ such that } g_i(x) \le y_i,\ i=1,2,\ldots,m\}.$$

Define set S in R^{m+1} as

$$S = \{(y_0, y) \mid y \in F,\ y_0 \ge w(y)\}.$$

Function $w(y)$ is nonincreasing with respect to each y_i. Assuming differentiability of $w(\cdot)$, we have

$$\frac{\partial w(y)}{\partial y_i} \le 0, \quad i=1,2,\ldots,m.$$

The primal-dual method has been one of the most efficient solution algorithms in solving problem (P). The classical Lagrangian function of (P) is

$$L(x,\lambda) = f(x) + \sum_{i=1}^{m} \lambda_i [g_i(x) - b_i],$$

where $\lambda = [\lambda_1, \lambda_2, \ldots, \lambda_m]' \in R^m_+$ is the Lagrangian multiplier vector. The dual search is performed to maximize the dual function $H(\lambda)$,

$$H(\lambda) = \min_{x \in X} L(x, \lambda).$$

The success of the primal-dual method resorts to the convergence of the sequential minimization of the Lagrangian function and has been limited largely to situations where there exists a saddle point (x^*, λ^*) that satisfies

$$L(x^*, \lambda) \leq L(x^*, \lambda^*) \leq L(x, \lambda^*)$$

for all $x \in X$ and all $\lambda \in R^m_+$.

The existence of the saddle point is closely related to the existence of a supporting hyperplane of the perturbation function $w(y)$ at $y = b$. Set S is convex if and only if F is convex and $w(\cdot)$ is a convex function over F. If $w(\cdot)$ is a convex function, there exists a supporting hyperplane of S at $[w(b), b]$ and for the optimal solution of problem (P), x^*, there is a Lagrangian multiplier vector λ^* such that (x^*, λ^*) is a saddle point of $L(x, \lambda)$. However, many non-convex optimization problems do not possess a saddle point. See Fig. 1.1 for geometrical interpretation.

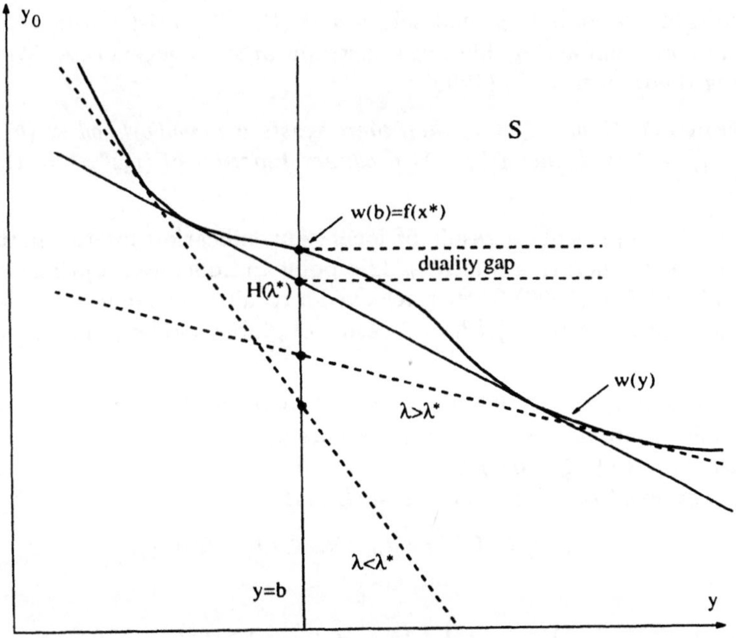

Figure 1.1 Nonconvex situation where there is no saddle point.

Failure of a linear Lagrangian in nonconvex situations is largely due to the nonexistence of a support to set S at $[w(b), b]$. Adopting some suitable nonlinear supports to set S, based on the property that perturbation function $w(y)$ is nonincreasing with respect to y, leads to the development of nonlinear Lagrangian theory. A major finding from recent study is that the existence of a saddle point is not an inherent property of (P). More specifically, a saddle point can be generated in a transformed new equivalent space to guarantee the success of a dual search.

2 P-TH POWER LAGRANGIAN METHOD

Without loss of generality, f, and g_i, $i = 1, 2, \ldots, m$, are assumed to be strictly positive for all $x \in X$. All constants b_i, $i = 1, 2, \ldots, m$, thus take strictly positive values. Define the mapping of S into the $\{(y_0)^p, y^p\}$ space as S_p,

$$S_p = \{[(y_0)^p, y^p] \mid y \in F,\ y_0 \geq w(y)\},$$

where notation y^p represents $[(y_1)^p, (y_2)^p, \ldots, (y_m)^p]$. For $y \in F$, the graph of $w(y)$ is converted to the following equivalent form in the $\{(y_0)^p, y^p\}$ space:

$$[w(y)]^p = \min\{[f(x)]^p \mid [g_i(x)]^p \leq (y_i)^p,\ i = 1, 2, \ldots, m, x \in X\}.$$

Define

$$\hat{w}(y^p) = [w(y)]^p.$$

Assume that the global optimal solution of (P), x^*, satisfies the second-order sufficient condition and no binding constraint at x^* is degenerate. We have the following theorem from Li (1995).

Theorem 2.1 *If $w(\cdot) \in C^2$, then there exists a neighborhood $N_\epsilon(b)$ of b and a finite $q_1 > 0$ such that $\hat{w}(y^p)$ is a convex function of $(y_i)^p$, $i = 1, 2, \ldots, m$, in $N_\epsilon(b)$ when $p \geq q_1$.*

Theorem 2.1 represents a result of local convexification for the perturbation function, thus generating a local saddle point in nonconvex optimization. Recently, Li and Sun (1999) further achieved a result of existence of a supporting hyperplane of S_p at $[\hat{w}(b^p), b^p]$ or a result of global saddle point in nonconvex optimization.

Theorem 2.2 *Let x^* be the unique global solution of (P) and $\lambda^* \in R^m_+$ be the Lagrangian multiplier vector at x^*. Assume that*

(i) $J(x^*) = \{j \mid \lambda^*_j > 0\} \neq \emptyset$;

(ii) *there exist positive ϵ_0 and ϵ_1 such that*

$$f(x) \geq f(x^*) + \epsilon_0, \quad \forall x \in Q_1 \cap Q_2(\epsilon_1),$$

where

$$Q_1 = \bigcup_{j \in J^-(x^*)} \{x \in X \mid g_j(x) > b_j\},$$

$$Q_2(\epsilon_1) = \bigcap_{j \in J(x^*)} \{x \in X \mid g_j(x) \leq b_j + \epsilon_1\},$$

$$J^-(x^*) = \{1, \ldots, m\} \setminus J(x^*).$$

Then there exists a $q_2 > 0$ and a multiplier vector λ_p^* satisfying

$$\hat{w}(y^p) \geq \hat{w}(b^p) - (\lambda_p^*)'(y^p - b^p)$$

for all $y \in F$ when $p \geq q_2$.

The implication of Theorems 2.1 and 2.2 is that instead of using the originally nonconvex optimization problem formulation in (P), the following equivalent p-th power formulation (Li (1995), Li (1997)) should be adopted to ensure the success of a dual search.

$$\begin{aligned} \min \quad & [f(x)]^p \\ \text{s. t.} \quad & [g_i(x)]^p \leq (b_i)^p, \quad i = 1, 2, \ldots, m, \\ & x \in X. \end{aligned} \tag{2.1}$$

The p-th power Lagrangian function is

$$L_p(x, \lambda) := [f(x)]^p + \sum_{i=1}^{m} \lambda_i \{[g_i(x)]^p - (b_i)^p\}.$$

Further, the dual function of (2.1) is defined as,

$$H_p(\lambda) = \min_{x \in X} L_p(x, \lambda).$$

It is highly possible that there exists no saddle point for the original nonconvex optimization formulation in (P). From Theorem 2.2, however, the existence of a saddle point is guaranteed for a general class of p-th power formulation in (2.1) with a sufficiently large value of p,

$$L_p(x^*, \lambda) \leq L_p(x^*, \lambda_p^*) \leq L_p(x, \lambda_p^*)$$

for all $x \in X$ and all $\lambda \in R_+^m$.

The perturbed dual function $H_p(\lambda, y) = \min[f(x)]^p + \lambda'([g(x)]^p - y^p)$ can be viewed as a nonlinear support of S. The dual problem is to maximize its intercept at the vertical axis $y = b$. The geometrical illustration of the p-th power formulation is showed in Fig. 2.1.

How large is large enough for p? The current result only offers theoretical lower bounds for p which are not implementable in the real calculation. Xu (1997) further extended the result in Li (1995). He showed that the same result can be obtained using p-th power transformation only on the constraints.

3 MINIMAX-TYPE LAGRANGIAN FUNCTION

Recognizing that every point on the perturbation function can be supported by a shifted cone obtained by shifting the $(m + 1)$ dimensional negative orthant in R^{m+1}, due to the nonincreasing property of the perturbation function, Goh and Yang (1997) introduces the following minimax-type Lagrangian function for (P):

$$L_{mm}(x, \theta) = \max\{f(x)/\theta, g_1(x)/b_1, \ldots, g_m(x)/b_m\}.$$

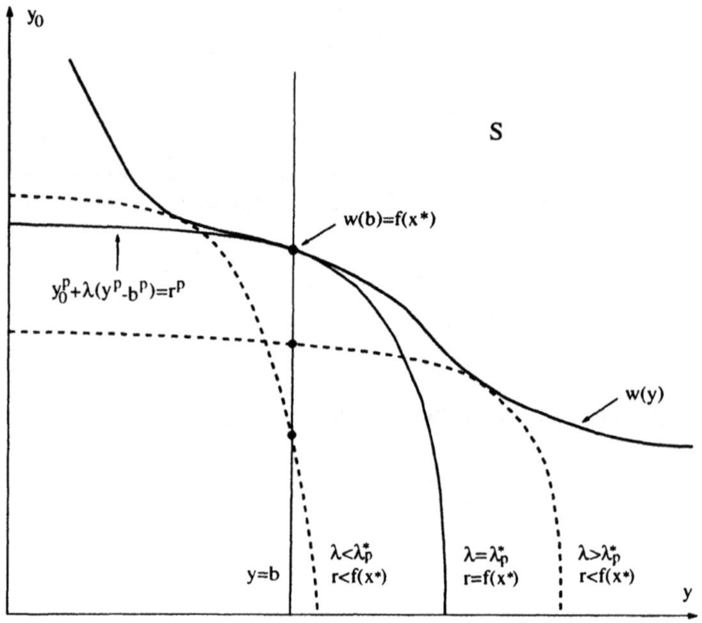

Figure 2.1 Geometrical interpretation of p-th power method.

The corresponding Lagrangian problem is:

$$\phi_{mm}(\theta) = \min_{x \in X} L_{mm}(x, \theta). \tag{3.1}$$

In the following sections, we assume that $\min_{x \in X} f(x) > 0$. Using function $\phi_{mm}(\theta)$, Goh and Yang (1997) derived a necessary and sufficient condition for optimality.

Theorem 3.1 *Let x^* be the global optimal solution of problem (P), and let $\theta^* = f(x^*)$. Then a solution x^0 solves (P) iff x^0 solves the following unconstrained problem*

$$\min_{x \in X} L_{mm}(x, \theta^*) = \min_{x \in X} \max\{f(x)/\theta^*, g_1(x)/b_1, \ldots, g_m(x)/b_m\}.$$

Nice properties of $\phi_{mm}(\theta)$ are found in Goh and Yang (1997).

Theorem 3.2 *Function $\phi_{mm}(\theta)$ has the following properties:*
(i) $\theta < f(x^*) \Rightarrow \phi_{mm}(\theta) > 1$;
(ii) $\theta > f(x^*) \Rightarrow \phi_{mm}(\theta) < 1$;
(iii) $\theta = f(x^*)$ *iff* $\phi_{mm}(\theta) = 1$;
(iv) $\phi_{mm}(\theta)$ *is a continuous and monotone nonincreasing function of* θ.

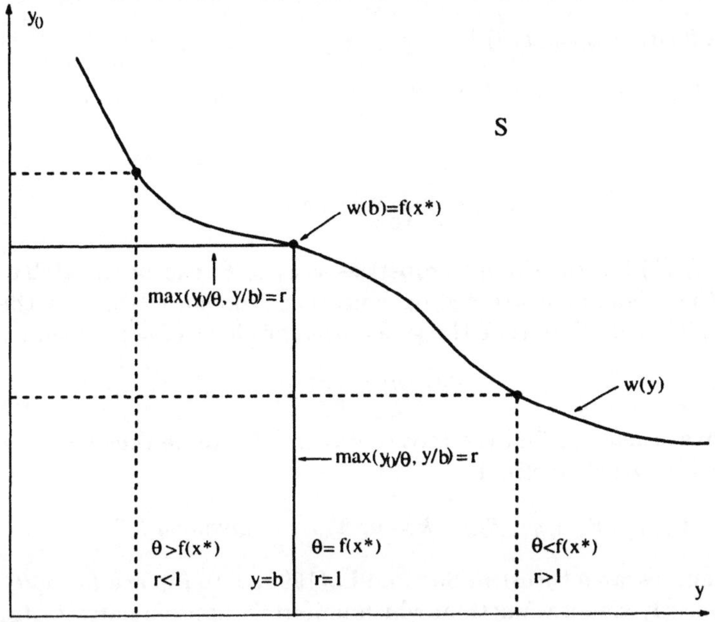

Figure 3.1 Properties of minimax-type supporting function $\max(y_0/\theta, y/b)$.

The geometrical interpretation of the above theorems for a single constraint case is illustrated in Fig. 3.1.

By theorem 3.2, the upper level problem can be reduced to a root-finding problem for $\phi(\theta) = 1$. Note that $\phi_{mm}(\theta)$ is a scalar monotone nonincreasing function of θ. When a root of $\phi(\theta) = 1$, θ^*, is found, the corresponding solution of $\min_{x\in X} L_{mm}(x, \theta^*)$ is the optimal solution of (P). Evaluating function $\phi_{mm}(\theta)$, however, involves solving the minimax-type Lagrangian problem in (3.1) which is of a nonsmooth structure. The lower-level problem in (3.1) is difficult to solve, even harder than (P) in most situations.

Yang and Li (2000) further extended the result in Goh and Yang (1997) by developing a general class of parametric monotone composition formulation that may have a smooth structure. The global optimal value of the original constrained optimization problem is shown to be the least root of the optimal value function of an auxiliary parametric optimization problem. The parametric optimization problem is formulated in such a way that it is a one-parameter optimization problem and its value function is a monotone composition function with respect to the original objective function and the constraints.

4 LOGARITHMIC-EXPONENTIAL LAGRANGIAN FUNCTION

A variant form of $L_{mm}(x,\theta)$ is

$$L_{le}(x,\theta) = \max\{f(x) - \theta, g_1(x) - b_1, \dots, g_m(x) - b_m\}. \tag{4.1}$$

Define

$$\phi_{le}(\theta) = \min_{x \in X} L_{le}(x,\theta).$$

Function $\phi_{le}(\theta)$ has the similar properties as $\phi_{mm}(\theta)$ except the global optimal value of the original constrained optimization problem (P) is now the unique root of $\phi_{le}(\theta) = 0$. Note that the generalized gradient of ϕ_{le} satisfies

$$\partial\phi_{le}(\theta) \subset [-1, 0].$$

An iterative scheme to find the root of $\phi_{le}(\theta) = 0$ can be thus developed using generalized Newton's method:

$$\theta_{k+1} = \theta_k + \phi_{le}(\theta_k), \quad k = 0,\ 1, \dots, \quad \text{given } \theta_0 \in R. \tag{4.2}$$

The main research thrust in Sun and Li (1999) is to replace $L_{le}(x,\theta)$ in (4.1) by a sufficiently smooth function, which uniformly approximates $L_{le}(x,\theta)$ over X. Corresponding parametric algorithms can then generate an approximate optimal global solution of (P), up to any predetermined degree of accuracy. Consider the following logarithmic-exponential function:

$$Q(x,\theta,p) = \frac{1}{p}\log[e^{(p(f(x)-\theta))} + \sum_{i=1}^{m} e^{(p(g_i(x)-b_i))}], \quad p > 0.$$

$Q(x,\theta,p)$ can be viewed as the p-th norm approximation of $L_{le}(x,\theta)$ when it is expressed to be the logarithm of the ∞-norm of $m+1$ dimensional vector $(e^{(f(x)-\theta)}, e^{(g_1(x)-b_1)}, \dots, e^{(g_m(x)-b_m)})$.

Now we consider the following unconstrained global optimization problem:

$$\psi(\theta,p) := \min_{x \in X} Q(x,\theta,p). \tag{4.3}$$

The following approximation properties hold when using the smooth auxiliary formulation to replace a nonsmooth one.

$$L_{le}(x,\theta) < Q(x,\theta,p) \le \frac{\log(m+1)}{p} + L_{le}(x,\theta),$$

$$\phi_{le}(\theta) < \psi(\theta,p) \le \frac{\log(m+1)}{p} + \phi_{le}(\theta).$$

Moreover, if f and g_is are twice continuously differentiable convex functions, then $Q(x,\theta,p)$ is also a convex function of x, for any $p > 0$.

The above properties of $Q(x, \theta, p)$ ensure that perturbation function $w(y)$ has a nonlinear smooth support at $y = b$ with the following form:

$$e^{p(y_0 - \theta)} + \sum_{i=1}^{m} e^{p(y_i - b_i)} = e^{pr}$$

for sufficiently large p, where θ can be viewed as an estimator of $f(x^*)$. See Fig. 4.1.

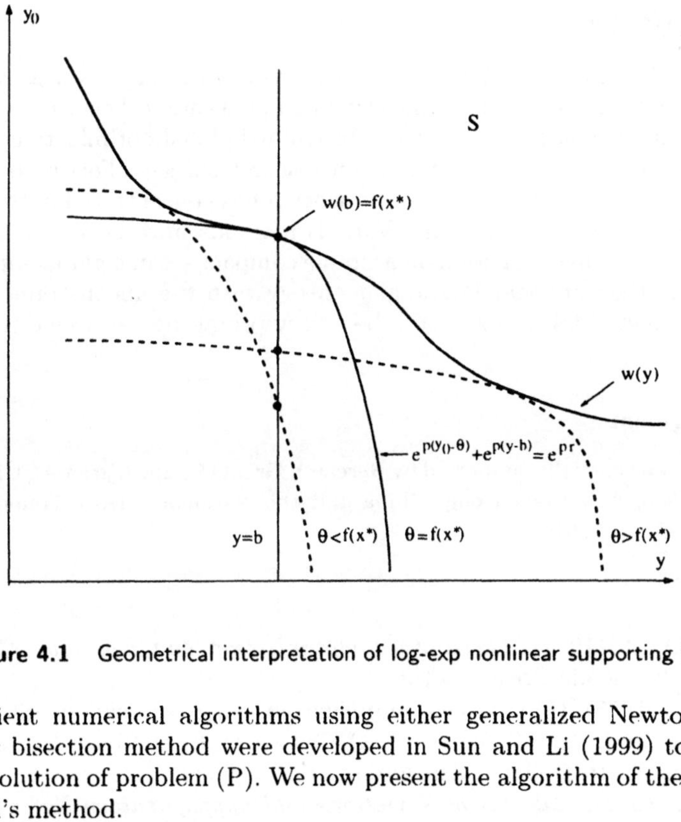

Figure 4.1 Geometrical interpretation of log-exp nonlinear supporting function.

Efficient numerical algorithms using either generalized Newton's method (4.2) or bisection method were developed in Sun and Li (1999) to search the global solution of problem (P). We now present the algorithm of the generalized Newton's method.

Algorithm (Generalized Newton's method)

Step 1. Choose a small positive $\epsilon > 0$ and a sequence $\{p_k\}$ with $p_k > 0$. Choose a $\theta_0 > f^*$. Set $k = 0$.

Step 2. Solve (4.3) with $\theta = \theta_k$ and $p = p_k$, return a solution x_k.

Step 3. If $-3\epsilon \le L_{le}(x_k, \theta_k) \le \epsilon$, set $x^* = x_k$, $\theta^* = \theta_k$, stop.

Step 4. Set

$$\theta_{k+1} = \theta_k + \left[1 + \frac{2\epsilon}{L_{le}(x_k, \theta_k)}\right] L_{le}(x_k, \theta_k),$$

$k := k + 1$, goto Step 2.

Theorem 4.1 *Assume that $p_k > \log(m+1)/\epsilon$ for each $k \geq 0$. The algorithm stops within finite number of iterations at an approximate global solution x^* of (P) satisfying x^* being feasible and $-4\epsilon < \phi_{le}(\theta^*) < 2\epsilon$.*

Compared with the traditional Lagrangian method where there are m multipliers, the logarithmic-exponential Lagrangian formulation only invokes a univariate auxiliary problem.

5 CONCLUSIONS

Nonlinear Lagrangian theory provides a new way of thinking to deal with constraints in the sense of global optimization and guarantees the success of a dual search. The implementation of the unconstrained global optimization problem at the lower level remains as an important research subject. Fortunately, much of the past research work in global optimization has concentrated in the area of unconstrained global optimization. Various methods have been developed for minimizing a nonconvex function on a simple compact set in a global sense. The integration of the nonlinear Lagrangian theory with the unconstrained global optimization methodologies seems to be a very promising research direction to pursue.

Acknowledgments

This research was partially supported by Research Grants Council, grants CUHK358/96P and CUHK4056/98E, Hong Kong, China and the National Science Foundation of China under Grant 79970107.

References

Bertsekas, D. P. (1982), *Constrained Optimization and Lagrange Multiplier Methods*, Academic Press, Boston.

Biggs, M. C. (1975), Constrained minimization using recursive quadratic programming, in: L. C. W. Dixon and G. P. Szergo, Eds., *Towards Global Optimization*, North-Holland, Amsterdam, pp. 341-349.

Courant, R. (1962), *Calculus of Variations and Supplementary Notes and Exercises*, Mimeographed notes, New York, New York University.

Fiacco, A. V. and McCormick, G. P. (1968), *Nonlinear Programming: Sequential Unconstrained Minimization Techniques*, John Wiley, New York.

Gill, P. E., Murray, W., and Wright, M. H. (1981), *Practical Optimization*, Academic Press, London.

Goh, C. J. and Yang, X. Q. (1997), A sufficient and necessary condition for nonconvex constrained optimization, *Applied Mathematical Letters*, Vol. 10, pp. 9-12.

Han, S. P. (1970), A globally convergent methods for nonlinear programming, *Journal of Optimization Theory and Applications*, Vol. 22, pp. 297-309.

Lasdon, L. S. (1970), *Optimization Theory for Large Systems*, MacMillan, London.

Li, D. (1995), Zero duality gap for a class of nonconvex optimization problems, *Journal of Optimization Theory and Applications*, Vol. 85, pp. 309-324.

Li, D. (1997), Saddle-point generation in nonlinear nonconvex optimization, *Nonlinear Analysis*, Vol. 30, pp. 4339-4344.

Li, D. and Sun, X. L. (1999), Existence of saddle point in nonconvex constrained optimization, submitted for publication.

Luenberger, D. G. (1984), *Linear and Nonlinear Programming*, Second Edition, Addison-Wesley, Reading, Massachusetts.

Powell, M. J. D. (1983), Variable metric methods for constrained optimization, in: A. Bachem, M. Grotschel and B. Korte, Eds., *Mathematical Programming: The State of the Art*, Springer, Berlin, pp. 288-311.

Sun, X. L. and Li, D. (1999), Value-estimation function method for constrained global optimization, *Journal of Optimization Theory and Applications*, Vol. 102, pp. 385-409.

Xu, Z. K. (1997), Local saddle points and convexification for nonconvex optimization problems, *Journal of Optimization Theory and Applications*, Vol. 94, pp. 739-746.

Yang, X. Q. and Li, D. (2000), Successive global optimization method via parametric monotone composition formulation, *Journal of Global Optimization*, Vol. 16, pp. 355-369.

Zangwill, W. I. (1967), Nonlinear programming via penalty functions, *Management Science*, Vol. 13, pp. 344-358.

15 PARALLEL ALGORITHMS FOR SOLVING LARGE-SCALE NONLINEAR OPTIMIZATION PROBLEMS

Paul Kang-Hoh Phua, Daohua Ming,
Weiguo Fan and Yan Zhang

School of Computing
National University of Singapore
Singapore 119260

Abstract: Multi-directional parallel algorithms for solving large-scale unconstrained optimization problems are proposed. Numerical results obtained from a broad class of test problems show that the average speedup factor achieved by our new algorithms is more than 300% (both in terms of total number of iterations and function/gradient evaluations) when they are compared with some well-known existing optimization codes. The maximum speedup factor obtained by our new algorithms can be as high as 25 times for some test problems. In terms of total CPU time required for solving all the problems, the average speedup factor obtained by our new algorithms is 27 over the well-known CONMIN package tested with a wide range of 56 test problems. In general, it is noticed that as the size and complexity of the problem increase, greater improvements and savings could be realized by our new algorithms. Our results also indicate that parallel algorithms are efficient and robust in solving large-scale nonlinear optimization problems, especially when they are applied in conjunction with vectorization techniques.

Key words: Nonlinear optimization, quasi-Newton methods, parallel algorithms.

X. Yang et al. (eds.), Optimization Methods and Applications, 279-293.

1 CONVENTIONAL METHODS FOR NONLINEAR OPTIMIZATION

1.1 Introduction to QN Methods

Many approaches for solving unconstrained optimization problems have been developed by researchers. Among these new techniques, the quasi-Newton (QN) methods are commonly used by practitioners. Assume that at the kth iteration, an approximation point x_k and an $n \times n$ matrix H_k are available, then the methods proceed by generating a sequence of approximation points via the equation

$$x_{k+1} = x_k + \alpha_k d_k, \tag{1.1}$$

where $\alpha_k > 0$ is the step-size which is calculated to satisfy certain line search conditions and d_k is an n-dimensional real vector representing the search direction. For QN methods, d_k is defined by:

$$d_k = -H_k g_k, \tag{1.2}$$

where $g_k = \nabla f(x_k)$ is the gradient vector of f(x) evaluated at point $x = x_k$.

One important feature of QN methods is the choice of the matrix H_k which is usually chosen to satisfy the QN equation

$$H_{k+1} y_k = \lambda_k \delta_k, \tag{1.3}$$

here, $\delta_k = x_{k+1} - x_k$, , and $\lambda_k > 0$ is a parameter.

Some commonly used QN methods are introduced as follows.

1.2 QN updates

One of the best known QN methods is the BFGS method proposed independently by Broyden (1970), Fletcher(1970), Goldfarb (1970) and Shanno (1970). The BFGS update is defined by the following equation

$$H_{k+1} = H_k - \frac{H_k y_k \delta_k^T + \delta_k y_k^T H_k}{\delta_k^T y_k} + \left(\lambda_k + \frac{y_k^T H_k y_k}{\delta_k^T y_k}\right) \frac{\delta_k \delta_k^T}{\delta_k^T y_k}, \tag{1.4}$$

where $\lambda_k \equiv 0$.

Based upon non-quadratic models, Biggs (1973) suggested a self-adjustable value for the parameter t_k and modified the BFGS update as below:

$$H_{k+1} = H_k - \frac{H_k y_k \delta_k^T + \delta_k y_k^T H_k}{\delta_k^T y_k} + \left(\lambda_k + \frac{y_k^T H_k y_k}{\delta_k^T y_k}\right) \frac{\delta_k \delta_k^T}{\delta_k^T y_k}, \tag{1.5}$$

where,

$$\lambda_k = \frac{1}{t_k}, \tag{1.6}$$

$$t_k = \frac{6}{\delta_k^T y_k}(f(x_k) - f(x_{k+1}) + \delta_k^T g_{k+1}) - 2, \tag{1.7}$$

The above update is referred as Biggs' BFGS update thereafter.

Davidon(1968) proposed a method that updates H_k by adding only a symmetric rank-one matrix. This symmetric rank-one (SR1) update is defined by

$$H_{k+1} = H_k + \frac{(\delta_k - H_k y_k)(\delta_k - H_k y_k)^T}{(\delta_k - H_k y_k)^T y_k}, \tag{1.8}$$

1.3 Optimization Codes

The BFGS methods were first implemented by Shanno and Phua (1976). This code, called MINIMI, is available in the ACM Mathematical Software Library. Incorporating the conjugate gradient algorithm into their earlier implementation of BFGS methods, Shanno and Phua (1980) developed the CONMIN optimization code which is currently available both in ACM Mathematical Software and IMSL libraries.

Based on the above two works, Buckley and Lenir (1985) developed the BBVSCG optimization code which offers users with the variable storage capability. Liu and Nocedal (1989) developed the LBFGS optimization code which modifies the BFGS methods with limited-memory capabilities. Another implementation of BFGS methods called E04DGE, is also available in the NAG library, NAG (1992).

2 PARALLEL QUASI-NEWTON ALGORITHMS

2.1 Introduction

We consider here the class of QN updates (see Shanno (1970)) given by

$$H^* = H + t\frac{\delta\delta^T}{\delta^T y} + \frac{[(1-t)\delta - Hy][(1-t)\delta - Hy]^T}{[(1-t)\delta - Hy]^T y}, \tag{2.1}$$

where t is a scalar parameter. In equation (2.1), the subscripts have been omitted by dropping (k) and replacing (k+1) with the subscript (*). Obviously, the BFGS and the SR1 updates defined by (1.4) and (1.8) correspond to $t = \infty$ and $t = 0$ respectively in (2.1). Consequently, we shall denote the SR1 and BFGS updates by $H^*(0)$ and $H^*(\infty)$ respectively. Analogously the Biggs' update given in (1.5) to (1.7) will be denoted by $H^*(t_k)$.

In practice, we have observed from our numerical experiments and results of many others that a particular QN algorithm may be 'effective' in solving certain types of minimization problems (Phua (1993)). For example, we have noticed that when the SR1 update without modifications is applied to solve a practical problem, it usually has higher efficiency than the other QN updates, such as BFGS (Phua and Chew (1992)), if it is able to solve a problem at all. On the other hand, it may encounter difficulties sometimes while other QN updates are able to solve the problem. Subsequent modifications made to this update may force it to solve the problem, but its efficiency will degenerate.

In order to incorporate the relative merits of different updates into the implementation, we propose parallel algorithms based on QN methods such that when these new algorithms are applied to solve practical problems, the parallel mechanisms will be able to explore different search directions generated by various QN updates during the minimization process. In addition, different search strategies will also be employed simultaneously in the process of locating the line minimum along each direction. These algorithms are described as follows.

2.2 PQN (Parallel quasi-Newton) algorithms

The proposed parallel quasi-Newton (PQN) algorithm consists of the following steps.

1. Initialization

   Let $k := 0$, and x_0 be the initial guess of the minimum and $H_0 = I$ be the identity matrix. Let $\epsilon > 0$ be the required accuracy.

2. Compute the function and gradient values at x_k

   Let $f_k := f(x_k)$ and $g_k := \nabla f(x_k)$

3. Compute the parallel search directions

   Let $m_1 > 0$ be the number of processors available for computing the search directions in parallel. Compute

$$d_k^{(j)} = -H_k(t_j)g_k, \quad j = 1, 2, \cdots, m_1 \tag{2.2}$$

   For example, t_j can be chosen as $0, 1, \infty$ in (2.1), or any other appropriate values.

4. Apply the parallel line search algorithm

   Call the line search routine PLS (see below) to perform the line search procedure in parallel along each search direction $d_k^{(j)}$, $j = 1, 2, \cdots, m_1$. Stop executing this routine once a line minimum α_k has been found satisfying the following Wolfe's conditions along any search direction $d_k^{(j)}$:

$$f(x_k + \alpha_k d_k^{(j)}) - f(x_k) \leq 0.0001 \times \alpha_k g_k^T d_k^{(j)} \tag{2.3}$$

   and

$$\nabla f(x_k + \alpha_k d_k^{(j)})^T d_k^{(j)} \geq 0.9 \times g_k^T d_k^{(j)} \tag{2.4}$$

   Let d_k^* be the search direction that α_k has been found successfully.

5. Compute the new point

   Let $x_{k+1} := x_k + \alpha_k d_k^*$ and $g_{k+1} := \nabla f(x_{k+1})$

6. Test for convergence

If $\|g_{k+1}\| \leq \epsilon \cdot \max\{1, \|x_{k+1}\|\}$, then stop; otherwise, proceed to Step 7.

7. Compute the new QN updates

Let $H_{k+1} = H_{k+1}(\infty)$. This is to update the approxiamte inverse Hessian matrix H_k by using the BFGS update defined in (1.5). Repeat the whole process from Step 2.

2.3 *Parallel Line Search Routine (PLS)*

The parallel line search algorithm works as follows. Let m_2 be the number of parallel processors available for locating the line minimum along a particular search direction $d_k^{(j)}$ $(j = 1, 2, \cdots, m_1)$. Let α_{max} be the maximum allowable step-size, and denote $\psi(\alpha) = f(x_k + \alpha d_k^{(j)})$. The main iteration of this routine consists of the following steps:

1. Choose the step sizes

Let

$$0 < \alpha^{(i)} < \alpha_{max} , \qquad i = 1, 2, \cdots, m_2$$

where $\alpha^{(1)} < \alpha^{(2)} < \cdots < \alpha^{(m_2)}$ are m_2 different approximately chosen step sizes. For instance, we may choose $\alpha^{(1)} = 0.5, \alpha^{(2)} = 1.0, \cdots, \alpha^{(m_2)} = \alpha_{max}$. Let Φ be the set of these step sizes.

2. Compute the function/gradient values concurrently

For $i = 1, 2, \cdots, m_2$, compute concurrently

$x_k^{(i)} = x_k + \alpha^{(i)} d_k^{(j)}, f_k^{(i)} = f(x_k^{(i)})$, and $g_k^{(i)} = \nabla f(x_k^{(i)})$.

3. Test for successful points

Let Φ^* be the set of step sizes $\alpha^{(i)}$ such that for each $\alpha^{(i)} \in \Phi^*$, $\alpha^{(i)}$ satisfies the Wolfe's conditions (2.3-(2.4). If $\Phi^* \neq \phi$ (empty set) and $\alpha_k^{(i)} \in \Phi^*$ is the stepsize which corresponds to the minimum functional value, that is,

$$\psi(\alpha_k^{(i)}) = \min_{\alpha^{(i)} \in \Phi^*} f(x_k + \alpha^{(i)} d_k^{(j)})$$

then set $\alpha_k := \alpha_k^{(i)}$ and return to the main PQN routine; Otherwise, proceed to Step 4.

4. Choose interpolation points

Let Φ^+ be the set of stepsizes such that for each $\alpha^{(i)} \in \Phi^+$, $\alpha^{(i)}$ satisfies

$${d_k^{(j)}}^T g_k^{(i)} > 0.$$

Let $\Phi^- = \Phi - \Phi^+$. Choose $\alpha_1 \in \Phi^-$ such that

$$\psi(\alpha_1) = \min_{\alpha^{(i)} \in \Phi^-} f(x_k + \alpha^{(i)} d_k^{(j)})$$

and choose $\alpha_2 \in \Phi^+$ such that

$$\psi(\alpha_2) = \min_{\alpha^{(i)} \in \Phi^+} f(x_k + \alpha^{(i)} d_k^{(j)})$$

If $\Phi^- = \phi$, then choose $\alpha_1 = 0$. If $\Phi^+ = \phi$, then choose $\alpha_2 = \alpha^{(m)}$, where $\alpha^{(m)} \in \Phi^-$ such that

$$\psi(\alpha^{(m)}) = \max_{\alpha^{(i)} \in \Phi^-} f(x_k + \alpha^{(i)} d_k^{(j)}).$$

5. Apply the cubic interpolation technique

Let $\varphi(\alpha)$ be the cubic polynomial passing the two points α_1 and α_2. Let α^* be the minimum of $\varphi(\alpha)$. Compute

$$\psi(\alpha^*) = f(x_k + \alpha^* d_k^{(j)})$$

and

$$h^* := \nabla\psi(\alpha^*)$$

If α^* satisfies the Wolfe's conditions (2.3-2.4), then set $\alpha_k = \alpha^*$ and return to the main PQN routine. Otherwise, if ${d_k^{(j)}}^T \cdot \nabla\psi(\alpha^*) > 0$, then replaces α_1 with α^*; if ${d_k^{(j)}}^T \cdot \nabla\psi(\alpha^*) \leq 0$, then replaces α_2 with α^*. Repeat Step 5.

3 VECTORIZATION AND FINE-TUNING TECHNIQUES

3.1 Introduction

To implement the proposed parallel QN methods, we have chosen three updates for computing the following QN directions in step 3 of the PQN routine:

$$d_k^{(1)} = -H_k(0) g_k, \tag{3.1}$$

$$d_k^{(2)} = -H_k(\infty) g_k, \tag{3.2}$$

$$d_k^{(3)} = -H_k(t_k) g_k, \tag{3.3}$$

They are the three search directions generate by SR1, BFGS and Biggs' updates as defined in (1.8), (1.4) and (1.5-1.7) respectively. Meanwhile, three line search steps are used for each of the three search directions ($m_2 = 3$) in the implementation of step 1 of the PLS (Parallel Line Search) routine. Here, $\alpha_1 = 0.5$, $\alpha_2 = 1$, and $\alpha_3 = 2$ are used. This particular implementation is referred to as PQN3_3 algorithm.

The algorithms were implemented on CRAY T94 whose two processors are insufficient for the parallel implementation. Thus the parallel version was converted into the sequential version called PQN33_S. In the sequential version, multi-updates and multi-line-search directions are still applied. However, they are executed on one processor in sequence rather than parallelly on a few processors.

3.2 Numerical Experiments on the FORTRAN90 Compilers

The Fortran90 installed on Cray T94 is equipped with four switches, namely *vector*, *scalar*, *inline* and *task*, for loop optimization, scalar optimization, subroutine inline, and task parallelization respectively. Each of the four options can be set as an integer from 0 to 3. Usually, value 0 means closing the switch, value 1 means low level usage of the switch, value 2 invokes moderate usage of the switch, and value 3 invokes aggressive usage of the switch. Different combinations of the four switches will be experimented.

To investigate the effects of different options of FORTRAN90 compiles on the computational efficiency of a program, a series of numerical experiments have been conducted with the PQN33_S on CRAY T94.

Table 3.1 shows the effects of different Fortran compiler switches in executing the program PQN33_S. Tests 1 to 4 are conducted by varying the *vector* switch from 0 to 3 while other switches are set at default. Test 1 with it vector0 failed to solve certain problems due to the fact that it exceeded the limit of function/gradient evaluations. Among tests 2 to 4, test 2 of *vector1* required the least CPU time.

Tests 5 to 8 are conducted by varying the *scalar* switch from 0 to 3. Also, test 5 of *scalar0* failed to solve certain problems for the same reason as test 1. Test 6 achieved the least CPU time compared with tests 7 and 8.

Tests 9 to 12 are conducted by varying the *inline* switch from 0 to 3. No significant differences are found in these experiments. This shows that the compile option of *inline* does not affect the executing of PQN33_S significantly. The *inline* option shows its effect in the group of tests 21 to 24 when *vector1* is used and *scalar* is set at default. Test 23 of (*vector1* + *inline2*) achieved the least CPU time, being 165 seconds faster than test 21 of (*vector1* + *inline0*).

In tests 13 to 16, we set the *vector* switch at value = 1, and vary the *scalar* switch from 0 to 3. Test 13 failed due to the fact that it exceeded the maximum allowable function/gradient evaluations. It is found that there are little differences among test 14 to 16. Analogously, in tests 17 to 20, we set the *vector* switch at value = 2, and vary the *scalar* switch from 0 to 3. All the four tests are completed. There are also little differences in this group. However, the average CPU time of tests 14 to 16 is 120 seconds less than the average CPU

Table 3.1 Numerical experiments on Fortran90 compile switches

Test No:	switches	CPU(sec.)	CPU-1000(sec.)
0	default	1109.9	109.9
1	*Vector0*	failed	NA
2	*Vector1*	980.7	-19.3
3	*Vector2*	1122.6	122.6
4	*Vector3*	1122.7	122.7
5	*Scalar0*	failed	NA
6	*Scalar1*	995.1	-4.9
7	*Scalar2*	1122.8	122.8
8	*Scalar3*	1108.5	108.5
9	*Inline0*	1122.7	122.7
10	*Inline1*	1123.6	123.6
11	*Inline2*	1122.6	122.6
12	*Inline3*	1124.5	124.5
13	*Vector1+Scalar0*	failed	NA
14	*Vector1+Scalar1*	1004.9	4.9
15	*Vector1+Scalar2*	983.4	-16.6
16	*Vector1+Scalar3*	1046.9	46.9
17	*Vector2+Scalar0*	1107.9	107.9
18	*Vector2+Scalar1*	1107.9	107.9
19	*Vector2+Scalar2*	1124.5	124.5
20	*Vector2+Scalar3*	1107.9	107.9
21	*Vector1+Inline0*	1048.8	48.8
22	*Vector1+Inline1*	981.8	-18.2
23	*Vector1+Inline2*	883.8	-116.2
24	*Vector1+Inline3*	894	-106
25	*Vector3+Inline0*	1123	123
26	*Vector3+Inline1*	1065	65
27	*Vector3+Inline2*	1065.2	65.2
28	*Vector3+Inline3*	1065.2	65.2

time of tests 17 to 20. This again shows that *vector1* is faster than *vector2* for the original PQN33_S code.

In tests 21 to 24, we set the *vector* switch at value = 1, and vary the *inline* switch from 0 to 3. Among these four tests, test 21 which requires 1048.8 seconds of CPU time is the slowest, and test 23 which requires 883.8 seconds of CPU time is the fastest.

In tests 25 to 28, we set the *vector* switch at value = 3, and vary the *inline* switch from 0 to 3. Test 25 which requires 1123 seconds of CPU time is the slowest, while tests 26 to 28 require almost the same CPU time of 1065 seconds.

Compared the group of tests 21 to 24 and the group of tests 25 to 28, it is once more shown that *vecotr1* is faster than *vector3* for the original PQN33_S code. The reasons behind these findings remain for further investigations.

In conclusion, various switches have different effects on the numerical performance of out program. For the original PQN33_S code, *vector1* is faster than both *vector2* and *vecotr3*. The compiler switch which achieved the least CPU time among our tests is the combination of *vecotr1* + *inline2* namely test 23.

3.3 Vectorization Techniques

To find out the utilization of CPU time by different routines in the PQN code, **FLOWVIEW** of the FORTRAN90 tool on the CRAY T94 is used to analyze the CPU time required by each routine. It is found that the subroutine **MVM** uses up to 93.44% of the CPU time in executing the program PQN33_S. The subroutine **MVM** is used to perform matrix operations for updating the search direction in the optimization process. The computation is performed using default switches in compiling the program in Fortran90. An example is shown as follows:

```
Subroutine mvm(n, h, x, y)
double precision h(n*n), x(n), y(n)
do i=1, n
  y(i) =0.0
  loc=i*(i-1)/2
  do j=1, i
     y(i)=y(i)+h(loc+j)*x(j)
     if (i.ne.j) y(j)=y(j)+h(loc+j)*x(i)
  end do
end do
end
```

Here, n is the dimension of the optimization problem, the $(n \times n)$ Hessian matrix H is stored with one dimensional array h(m), m = n(n+1)/2. As the Hessian matrix is symmetric, only the left-upper triangle of this matrix is stored for the calculation.

It is found that the subroutine can not be successfully vectorized as the inner loop contains a conditional **if** statement. To optimize the code, the routine is reconstructed as below:

```
Subroutine mvm(n, h, x, y)
double precision h(n*n), x(n), y(n)
do i=1,n
  loc=i*(i-1)/2
  y(i)=h(loc+i)*x(i)
  do j=1,i-1
     y(i)=y(i)+h(loc+j)*x(j)
  end do
  do j=1,i-1
     y(j)=y(j)+h(loc+j)*x(i)
  end do
end do
end
```

In addition to the above reconstruction of **MVM**, some other works have also been done to fine-tune the code. These include the replacing of nested structures of **if then else** statements with case statements. After fine-tuning, the routine **MVM** uses only 18.71% of the total CPU time.

After fine-tuning the code, we apply the same compiler option, it is found that the total CPU time for the execution of PQN33_S is reduced remarkably from some 1110 seconds to 115 seconds, which represents a speedup factor of 9.7 in average. The detailed results are shown in Table 3.2.

Table 3.2 Total CPU time required by PQN33_S before and after fine-tuning

CompileOption	*CPU(sec.) defore*	*CPU(Sec.) after*	*Speedupfactor*
default	1109.9	114.7	9.7
vector0	failed	failed	NA
vector1	980.7	112.5	8.7
vector2	1122.6	112.5	10
vector3	1122.7	113.2	9.9
scalar0	failed	3554.9	NA
scalar1	995.1	120	8.3
scalar2	1122.8	112.5	10
scalar3	1108.5	120	9.2
inline0	1122.7	113.7	9.9
inline1	1123.6	112.8	10
inline2	1122.6	112.8	10
inline3	1124.5	113.7	9.9

4 COMPUTATIONAL RESULTS

4.1 Test Problems and Convergence Criteria

To test the performance of the proposed parallel algorithms on CRAY T94, eleven standard functions are chosen. All the functions, except the Wood and Penalty II functions, are tested with variable dimensions of 20, 100, 200, 400, 800, and 1000 respectively. Wood 's function is tested for n = 4. Penalty Function II is tested for n = 20 and 50 respectively. Hence, a total of 57 problems are solved.

4.2 Analysis of Computational Results

Two serial optimization codes are selected for evaluation purposes. These are the CONMIN, developed by Shanno an Phua (1980), and the E04DGE of the NAG library, see NAG (1992). All codes are evaluated over the same set of 57 test problems. The CONMIN is executed on CRAY T94. As the NAG library is not available on CRAY T94, the E04DGE is executed on the CRAY J916 of the Computer Centre of NUS. In evaluating the performance of each optimization code, the criteria will be based on the total number of iterations, function/gradient evaluations, and the total CPU time required by the code in solving all the test problems. Computational results obtained by the above optimization codes, namely PQN33_S, CONMIN and E04DGE are summarized in Tables 4.1 - 4.2.

Table 4.1 Improvements and Speedups of PQN33_S over CONMIN and E04DGE

	In Terms of Iterations			In Terms of Function/ gradient evaluation		
Probms	*PQ/CN/E4*	*spu/C*	*spu/E*	*PQ/CN/E4*	*spu/C*	*spu/E*
Rosen -brock						
n=20	34/34/31	1.00	0.91	50/50/42	1.00	0.84
n=100	32/36/28	1.13	0.88	40/50/50	1.25	1.25
n=200	34/34/28	1.00	0.82	46/45/45	0.98	0.98
n=400	32/34/31	1.06	0.97	42/45/53	1.07	1.26
n=800	37/36/28	0.97	0.76	49/45/54	0.92	1.10
n=1000	34/35/26	1.03	0.76	46/46/48	1.00	1.04
Powell						
n=20	40/51/40	1.28	1.00	42/52/44	1.24	1.05
n=100	44/67/64	1.52	1.45	50/68/73	1.36	1.46
n=200	45/52/53	1.16	1.18	49/53/73	1.08	1.49
n=400	42/69/117	1.64	2.79	44/70/182	1.59	4.14
n=800	36/55/64	1.53	1.78	38/58/102	1.53	2.68
n=1000	38/49/222	1.29	5.84	41/53/364	1.29	8.88
Power						
n=20	41/280/100	6.83	2.44	42/281/105	6.69	2.50
n=100	73/867/201	11.88	2.75	74/868/225	11.73	3.04
n=200	90/1403/159	15.59	1.77	91/1404/186	15.43	2.04
n=400	112/2207/157	19.71	1.40	113/2208/191	19.54	1.69
n=800	141/3356/460	23.80	3.26	142/3357/473	23.64	3.33
n=1000	151/3857/183	25.54	1.21	152/3858/209	25.38	1.38
Watson						
n=20	86/115/709	1.34	8.24	92/119/715	1.29	7.77
n=100	154/186/1101	1.21	7.15	160/195/1138	1.22	7.11
n=200	147/228/1184	1.55	8.05	156/233/1257	1.49	8.06
n=400	221/252/1594	1.14	7.21	239/265/1749	1.11	7.32
n=800	265/298/2371	1.12	8.95	285/313/2656	1.10	9.32
n=1000	262/fail/2304	NA	8.79	278/fail/2678	NA	9.63
Broyden-Tridignl						
n=20	19/21/31	1.11	1.63	20/22/40	1.10	2.00
n=100	21/22/45	1.05	2.14	22/23/84	1.05	3.82
n=200	22/25/67	1.14	3.05	23/26/126	1.13	5.48
n=400	21/28/90	1.33	4.29	24/30/176	1.25	7.33
n=800	22/45/97	2.05	4.41	28/47/186	1.68	6.64
n=1000	23/60/98	2.61	4.26	31/65/187	2.10	6.03

(Remarks–PN/CN/E4=PQN3_S/CONMIN/E04DGE; spu/C=speedup of PQN33_S over CONMIN;spu/E=speedup of PQN33_S over E04DGE)

4.3 Comparisons, Discussions and Conclusions

Since CONMIN failed to solve the Watson function for n = 1000 due to the fact that the maximum allowable CPU time was exceeded, only the remaining 56 problems are considered for evaluating PQN33_S and CONMIN algorithms. To solve these 56 test problems, the total numbers of iterations performed by PQN33_S and CONMIN are 6211 and 20470 respectively, being a difference of 14259 iterations, which represents an overall speedup of 3.30. In terms of total

Table 4.1 *(continued)*

	In Terms of Iterations			In Terms of Function/ gradient evaluation		
Probms	*PQ/CN/E4*	*spu/C*	*spu/E*	*PQ/CN/E4*	*spu/C*	*spu/E*
Trigometry						
n=20	37/44/62	1.19	1.68	43/50/71	1.16	1.65
n=100	37/42/58	1.14	1.57	45/52/72	1.16	1.6
n=200	42/47/65	1.12	1.55	52/55/79	1.06	1.52
n=400	39/48/56	1.23	1.44	47/58/63	1.23	1.34
n=800	41/58/58	1.41	1.41	49/67/72	1.37	1.47
n=1000	42/56/60	1.33	1.43	50/63/65	1.26	1.3
Broyden -Toint						
n=20	28/36/61	1.29	2.18	29/37/72	1.28	2.48
n=100	41/51/50	1.24	1.22	42/52/96	1.24	2.29
n=200	38/47/58	1.24	1.53	39/48/117	1.23	3
n=400	44/922/85	20.95	1.93	45/926/174	20.58	3.87
n=800	1356/1693/1757	1.25	1.3	1368/1702/2109	1.24	1.54
n=1000	1714/2061/Fail	1.2	NA	1722/2071/Fail	1.2	NA
Wood						
n=4	15/23/37	1.53	2.47	24/29/44	1.21	1.83
Hilbert						
n=20	14/30/64	2.14	4.57	16/31/66	1.94	4.13
n=100	28/46/168	1.64	6	30/47/172	1.57	5.73
n=200	27/53/270	1.96	10	30/55/274	1.83	9.13
n=400	42/62/272	1.48	6.48	45/64/274	1.42	6.09
n=800	33/71/279	2.15	8.45	37/73/285	1.97	7.7
n=1000	39/70/344	1.79	8.82	44/72/350	1.64	7.95
Penalty Function I						
n=20	48/62/498	1.29	10.38	62/73/729	1.18	11.76
n=100	49/69/790	1.41	16.12	59/78/1230	1.32	20.85
n=200	52/59/140	1.13	2.69	60/72/147	1.2	2.45
n=400	48/69/113	1.44	2.35	58/80/115	1.38	1.98
n=800	60/58/220	0.97	3.67	71/71/222	1	3.13
n=1000	47/61/294	1.3	6.26	55/73/296	1.33	5.38
Penalty Function II						
n=20	93/307/181	3.3	1.95	117/359/254	3.07	2.17
n=50	100/523/113	5.23	1.13	113/1051/132	9.3	1.17
Grand Total:	6473/20470+F /17836+F	63.35	108.58	6861/21358+F /21091+F	65.37	113.51
Average		3.46	3.66		3.45	4.11

function/gradient evaluations, the savings gained by PQN33_S over CONMIN is 14775, which represents an overall speedup of 3.25.

Among these 56 problems, it is observed that different savings and speedup factors are obtained by PQN33_S over CONMIN for different test problems, as seen in Table 4.1. The maximum savings gained by PQN33_S happens for the Power function when n = 1000. In this case, a speedup factor of 25.3 is achieved

Table 4.2 CPU time and Speedup factors

Problem	*Dimension*	*PQN33_S*	*CONMIN*	*CPUSpeedup*
Rosenbrock	n=20	0.0055	0.0091	1.65
	n=100	0.0133	0.1105	8.31
	n=200	0.0292	0.3892	13.33
	n=400	0.0683	1.5094	22.1
	n=800	0.2358	6.3136	26.78
	n=1000	0.3172	9.5564	30.13
Powell	n=20	0.0058	0.0132	2.28
	n=100	0.0178	0.2081	11.69
	n=200	0.0387	0.6012	15.53
	n=400	0.0915	3.11	33.99
	n=800	0.2346	9.7404	41.52
	n=1000	0.3621	13.4911	37.26
Power	n=20	0.0057	0.0733	12.86
	n=100	0.0298	2.7414	91.99
	n=200	0.079	16.583	209.91
	n=400	0.2517	100.4295	399
	n=800	0.9481	605.3067	638.44
	n=1000	1.4871	1082.5679	728.0
Watson	n=20	0.0558	0.0468	0.84
	n=100	0.4099	0.7051	1.72
	n=200	0.8027	2.966	3.7
	n=400	2.7438	12.3283	4.49
	n=800	6.8846	49.2613	7.16
	n=1000	8.8798	78.4852	8.84
Bro-Tridiagonal	n=20	0.0026	0.0054	2.08
	n=100	0.0079	0.0664	8.41
	n=200	0.0178	0.2831	15.9
	n=400	0.0429	1.2353	28.79
	n=800	0.1352	7.9372	58.71
	n=1000	0.2069	16.5833	80.15
Trigometry	n=20	0.0173	0.016	0.92
	n=100	0.108	0.1669	1.55
	n=200	0.4224	0.6769	1.6
	n=400	1.3043	2.6732	2.05
	n=800	5.3572	12.6359	2.36
	n=1000	8.8715	18.951	2.14
Broyden-Toint	n=20	0.0129	0.0133	1.03
	n=100	0.0793	0.1837	2.32
	n=200	0.1511	0.5898	3.9
	n=400	0.365	43.8602	120.16
	n=800	25.6702	313.2293	12.2
	n=1000	42.7779	592.3554	13.85
Wood	n=4	0.0017	0.0024	1.41
Hilbert	n=20	0.003	0.0085	2.83
	n=100	0.0338	0.1534	4.54
	n=200	0.0849	0.6534	7.7
	n=400	0.3874	2.9269	7.56
	n=800	0.9415	13.1165	13.93
	n=1000	1.6714	20.1001	12.03
Penalty I	n=20	0.0081	0.0119	1.47
	n=100	0.0206	0.2209	10.72
	n=200	0.0468	0.6368	13.61
	n=400	0.1077	2.5616	23.78
	n=800	0.4039	13.5282	33.49
	n=1000	0.4577	16.8636	36.84
Penalty II	n=20	0.0175	0.0829	4.74
	n=50	0.0275	0.4873	17.72
	Total CPU:	114.7117	3079.3634	27.07

in terms of both total number iterations and total number of function/gradient evaluations.

E04DGE also failed to solve one problem, namely the Broyden-Toint function of n = 1000, due to internal errors occurred within the package. Therefore, only the remaining 56 problems are available for comparing the performance of PQN33_S and E04DGE. In solving this set of 56 test problems, the total number of iterations performed by PQN33_S and E04DGE are 4759 and 17836 respectively. Hence, a difference in performance of 13077 iterations, which represents an overall speedup factor of 3.75 is obtained by PQN33_S over E04DGE. In terms of total function/gradient evaluations, the savings gained by PQN33_S over E04DGE is 15952, which represents an overall speedup factor of 4.15. The highest speedup factor gained by PQN33_S over E04DGE are 16.12 (in terms of total number of iterations) and 20.85 (in terms of function/gradient evaluations) which happens for the case of Penalty I when n = 100.

Table 4.2 shows the amount of CPU time required by PQN33_S and CONMIN optimization codes in solving all the test problems. In terms of total CPU time required, the average speedup factors obtained by PQN33_S over CONMIN for the four problems is 183, and the maximum speedup factor is 728.

We have proposed here multi-directional parallel algorithms for solving large-scale unconstrained optimization problems based on quasi-Newton methods. Numerical results obtained from a broad class of test problems show that the average speedup factor achieved by our new algorithms is more than 300% (both in terms of total number of iterations and function/gradient evaluations) when they are compared with some well-known existing optimization codes. The maximum speedup factor obtained by our new algorithms can be as high as 25 times for some test problems.

In terms of total CPU time required for solving all the problems, the average speedup factor obtained by our new algorithms is 27 over the well-known CONMIN package tested with a wide range of 56 test problems. From our experiment, it is observed that the speedup factors gained from the parallelization of algorithms is about 3 times, since only three parallel search directions have been employed in our algorithms. On the other hand, the speedup factors from the parallelization and vectorization of the optimization codes is about 9 times. In some special cases, the maximum speedup factor gained by the new PQN33_S optimization codes is 728. These significant results are obtained due to fact that we have successfully parallelized and vectorized the optimization codes.

In general, it is noticed that as the size and complexity of the problem increase, greater improvements and savings could be realized by our new algorithms. Our results also indicate that parallel algorithms are efficient and robust in solving large-scale nonlinear optimization problems, especially when they are applied in conjunction with vectorization techniques.

References

Biggs M. C.(1973), A note on minimization algorithms which make use of non-quadratic properties of the objective function, *J. Inst. Maths Applics.*, pp. 337-338.

Broyden C. G.(1970), The convergence of a class of double-rank minimization algorithms", *J. Inst. Maths. Applics.*, pp. 76-90.

Byrd R. H., Liu D. and Nocedal J. (1992), On the behavior of Broyden's class of quasi-Newton methods, *SIAM J. Optimization*, Vol.2, pp. 533-557.

Davidon W.C.(1968), Variable metric method for minimization, *Computer Journal*, pp.406-410.

Fletcher R.(1970), A new approach to variable metric algorithm, *Computer Journal*, Vol.13, pp.392-399.

Goldfarb D.(1970), A family of variable-metric methods derived by variational methods, *Math. Comp.*, Vol. 24, pp. 23-26

Liu D. C. and Nocedal J.(1989) On the limited memory BFGS method ofr large scale optimization, *Mathematical Programming*, Vol.45, pp. 503-528

Nag (1992): *Fortran Library Reference manual (Mark14), Numerical Algorithms Group 3*, Nag Limited, Oxford OX2-8DR, United Kingdom.

Oren S.S. and Spedicato E.(1976), Optimal conditioning of self-scaling variable metric algorithms, *Math. Prog.*, pp. 70-90.

Phua K. H.(1993), Switching algorithms for quasi-Newton methods, *FSU-SCRI-93T-151*, Spuercomputer Computation Research Institute, Florida State University, 28 pp.

Phua K.H.(1996), Multi-directional parallel Algorithms for unconstrained optimization, *Optimization*, pp. 107-125.

Phua K.H. and Chew S.B.(1992), Symmetric rank-one update and quasi-Newton methods, in: K.H. Phua et al., eds., *Optimization Techniques and Applications, World Scientific, Singapore*, pp. 52-63.

Schnabel R.B.(1987), Concurrent function evaluations in local and global optimization, *Computer Methods in Applied Mechanics and Engineering*, pp. 537-552.

Shanno D.F.(1970), Conditioning of quasi-Newton methods for function minimization, *Mathematics of Computation*, pp.647-656.

Shanno D. F. and Phua K. H.(1976) Algorithm 500- minimization of unconstrained multivariate functions, *ACM Transactions on Mathematical Software*, Vol. 2(1)

Shanno D.F and Phua, K.H.(1980) Remark on algorithm 500, a variable method subroutine for unconstrained nonlinear optimization, *ACM Trans. Math. Software*, pp. 618-622.

van Laarhoven P.J.M.(1985), Parallel variable metric algorithms for unconstrained optimization, *Mathematical Programming*, pp. 68-81.

16 SECOND ORDER STRICT CONVERSE DUALITY IN NONLINEAR FRACTIONAL PROGRAMMING

Xin Min Yang,

Department of Mathematics
Chongqing Normal University
Chongqing, 400047, China

Kok Lay Teo and Xiaoqi Yang

Department of Applied Mathematics
The Hong Kong Polytechnic University
Kowloon, Hong Kong

Abstract: In this paper, two types of second order dual problems are introduced for a nonlinear fractional programming problem involving equality and inequality constraints. Four theorems on second order strict converse duality are proved under some generalized second order (F, ρ)-convexity assumptions.

Key words: Nonlinear fractional programming, second order duality, generalized second order convexity.

X. Yang et al. (eds.), Optimization Methods and Applications, 295-306.

1 INTRODUCTION AND PRELIMINARIES

Mangasaran (1975) formulated second order and higher order dual problems of a nonlinear program and established duality theorems under so-called inclusion conditions. Mond (1974) introduced another type of second order dual problems and obtained duality theorems using second order convexity conditions. Second order model and duality results in Mond and Weir (1981-1983) were extended to the invexity case in Bector et al (1991) and Hanson (1993). Mangasarian's second order dual model and results were generalized by Yang (1998) using second order comparative conditions.

Recently, in Zhang (1999) a second order generalized (F, ρ)-convexity was introduced and applied to obtain some second order weak and strong duality theorems for nonlinear optimization problems. However, to the best of our knowledge, second order duality results in fractional programming have not been investigated. In this paper, two types of second order dual problems are introduced for a nonlinear fractional programming problem involving equality and inequality constraints. Second order strict converse duality results are proved under some generalized (F, ρ)-convexity assumptions.

Consider the nonlinear fractional programming problem

$$\text{(FP)} \qquad \min_{x \in X} \quad \frac{f(x)}{g(x)}$$

$$\text{subject to} \quad h(x) \leq 0, \tag{1.1}$$

$$l(x) = 0, \tag{1.2}$$

where $X \subset \mathbb{R}^n$ is an open set and f, g, h and l are twice differentiable functions from X to $\mathbb{R}$, $\mathbb{R}$, $\mathbb{R}^m$ and $\mathbb{R}^k$ respectively. It is assumed that $g(x) > 0$, for any $x \in X$. The feasible set is denoted by

$$D = \{x | x \in X, h(x) \leq 0, l(x) = 0\}, \tag{1.3}$$

We present several definitions of second order convexity which will be used in the sequel.

Let $F : X \times X \times \mathbb{R}^n \longrightarrow \mathbb{R}$ be a sublinear functional in its third component and the function $\phi : X \longrightarrow \mathbb{R}$ be a real valued twice differentiable function, and let $\bar{x}$ be an element of X and $\nabla\phi(\bar{x})$ be the gradient vector of ϕ at $\bar{x}$, $\nabla^2\phi(\bar{x})$ the Hessian of ϕ at $\bar{x}$. Let $d(\cdot,\cdot) : \mathbb{R}^n \times \mathbb{R}^n \longrightarrow \mathbb{R}$, and $\rho \in \mathbb{R}$. Assume that $d(x, x) = 0$ for any x. The following definitions are due to Zhang (1999).

Definition 1.1 *Let $p \in \mathbb{R}^n$. The function ϕ is said to be second order (F, ϕ)-convex at $\bar{x}$ if, for any $x \in X$,*

$$\phi(x) \geq \phi(\bar{x}) + F(x, \bar{x}; \nabla\phi(\bar{x}) + \nabla^2\phi(\bar{x})p) - \frac{1}{2}p^T\nabla^2\phi(\bar{x})p + \rho d^2(\bar{x}, x).$$

Definition 1.2 *Let $p \in \mathbb{R}^n$. The function ϕ is said to be second order (F, ϕ)-quasiconvex at $\bar{x}$ if, for any $x \in X$ satisfying*

$$\phi(x) \geq \phi(\bar{x}) - \frac{1}{2}p^T \nabla^2 \phi(\bar{x})p,$$

we have

$$F(x, \bar{x}; \nabla\phi(\bar{x}) + \nabla^2\phi(\bar{x})p) \leq -\rho d^2(\bar{x}, x).$$

Definition 1.3 *Let $p \in \mathbb{R}^n$. The function ϕ is said to be second order weakly (F, ϕ)-quasiconvex at $\bar{x}$ if, for any $x \in X$ satisfying*

$$\phi(x) > \phi(\bar{x}) - \frac{1}{2}p^T \nabla^2 \phi(\bar{x})p,$$

we have

$$F(x, \bar{x}; \nabla\phi(\bar{x}) + \nabla^2\phi(\bar{x})p) \leq -\rho d^2(\bar{x}, x).$$

Definition 1.4 *Let $p \in \mathbb{R}^n$. The function ϕ is said to be second order (F, ϕ)-pseudoconvex at $\bar{x}$ if, for any $x \in X$ satisfying*

$$F(x, \bar{x}; \nabla\phi(\bar{x}) + \nabla^2\phi(\bar{x})p) \geq -\rho d^2(\bar{x}, x),$$

we have

$$\phi(x) \geq \phi(\bar{x}) - \frac{1}{2}p^T \nabla^2 \phi(\bar{x})p.$$

Definition 1.5 *Let $p \in \mathbb{R}^n$. The function ϕ is said to be second order strictly (F, ϕ)-pseudoconvex at $\bar{x}$ if, for any $x \in X$, $x \neq \bar{x}$ satisfying*

$$F(x, \bar{x}; \nabla\phi(\bar{x}) + \nabla^2\phi(\bar{x})p) \geq -\rho d^2(\bar{x}, x),$$

we have

$$\phi(x) > \phi(\bar{x}) - \frac{1}{2}p^T \nabla^2 \phi(\bar{x})p.$$

If $p = 0$, $F(x, \bar{x}; \nabla\phi(\bar{x}) + \nabla^2\phi(\bar{x})p) = \eta(x, \bar{x})^T \nabla\phi(\bar{x})$, for some vector function $\eta : \mathbb{R}^n \times \mathbb{R}^n \longrightarrow \mathbb{R}^n$, then above generalized convexity definitions reduce to generalized invexity properties in Kaul and Kaur (1985).

2 SECOND ORDER STRICT CONVERSE DUALITY I

Let I_s and J_s, $0 \leq s \leq r$, be partitions of the sets $\{1, 2, \cdots, m\}$ and $\{1, 2, \cdots, k\}$, respectively, and $r = \max\{r_1, r_2\}$, where r_1 and r_2 are the number of partitions of $\{1, 2, \cdots, m\}$ and $\{1, 2, \cdots, k\}$, respectively, with $I_s = \emptyset$ or $J_s = \emptyset$ when $s > \min\{r_1, r_2\}$. Here y_{I_s} is the vector consisting of the components y_i of vector y if $i \in I_s$. Similarly we define z_{I_s}, h_{I_s} and l_{I_s}.

Now by introducing an additional vector $p \in \mathbb{R}^n$, we can formulate the following second order dual problem (FD_1):

$$\max \quad \frac{f(u) + y_{I_0}^T h_{I_0}(u) + z_{J_0}^T l_{J_0}(u)}{g(u)} - \frac{1}{2} p^T \nabla^2 \frac{f(u) + y_{I_0}^T h_{I_0}(u) + z_{J_0}^T l_{J_0}(u)}{g(u)} p$$

subject to

$$\nabla\Big[\frac{f(u) + y_{I_0}^T h_{I_0}(u) + z_{J_0}^T l_{J_0}(u)}{g(u)} + y^T h(u) + z^T l(u) - y_{I_0}^T h_{I_0}(u) - z_{J_0}^T l_{J_0}(u)\Big] + \nabla^2\Big[\frac{f(u) + y_{I_0}^T h_{I_0}(u) + z_{J_0}^T l_{J_0}(u)}{g(u)} + y^T h(u) + z^T l(u) - y_{I_0}^T h_{I_0}(u) - z_{J_0}^T l_{J_0}(u)\Big]p = 0, \tag{2.1}$$

$$y_{I_s}^T h_{I_s}(u) + z_{J_s}^T l_{J_s}(u) - \frac{1}{2} p^T \nabla^2 [y_{I_s}^T h_{I_s}(u) + z_{J_s}^T l_{J_s}(u)] p \geq 0, \tag{2.2}$$

$$(1 \leq s \leq r), \tag{2.3}$$

$$y \geq 0, \tag{2.4}$$

where $u \in X$, $y \in \mathbb{R}^m$, $z \in \mathbb{R}^k$,

Remark 2.1 *If $p = 0$, $I_0 = \emptyset$, $J_0 = \emptyset$, $l(x) \equiv 0$, $I_1 = \{1, 2, \cdots, m\}$, $J_1 = \emptyset$, $I_i = J_i = \emptyset$, $(i = 2, 3, \cdots, r)$, then our dual model (FD_1) is reduced to a first order dual model in Craven and Mond (1999).*

In this section, we establish strict converse duality theorems between the dual fractional program (FD_1) and the primal fractional program (FP).

Theorem 2.1 *Let $(\bar{u}, \bar{y}, \bar{z})$ be a feasible solution of the dual fractional program (FD_1). Assume that the following conditions hold:*

(i) $\frac{f + \bar{y}_{I_0}^T h_{I_0} + \bar{z}_{J_0}^T l_{J_0}}{g}$ is second order (F, ρ)-pseudoconvex at $\bar{u}$ on the feasible region D;

(ii) $\bar{y}_{I_s}^T h_{I_s} + \bar{z}_{J_s}^T l_{J_s}$, $(1 \leq s \leq r)$ is second order (F, ρ_s)-quasiconvex at $\bar{x}$ on the feasible region D; and

(iii) $\rho + \sum_{s=1}^r \rho_s \geq 0$.

If the feasible solution $\bar{x}$ of the primal fractional program (FP) satisfies

$$\frac{f(\bar{x})}{g(\bar{x})} = \frac{f(\bar{u}) + \bar{y}_{I_0}^T h_{I_0}(\bar{u}) + \bar{z}_{J_0}^T l_{J_0}(\bar{u})}{g(\bar{u})} - \frac{1}{2} p^T \nabla^2 \Big[\frac{f(\bar{u}) + \bar{y}_{I_0}^T h_{I_0}(\bar{u}) + \bar{z}_{J_0}^T l_{J_0}(\bar{u})}{g(\bar{u})}\Big] p,$$

then $\bar{x}$ solves the primal fractional program (FP).

Proof. Assume that $\bar{x}$ does not solve (FP). Thus, there exists an x feasible to (FP) such that

$$\frac{f(x)}{g(x)} < \frac{f(\bar{x})}{g(\bar{x})}.$$

Since

$$\frac{f(\bar{x})}{g(\bar{x})} = \frac{f(\bar{u}) + \bar{y}_{I_0}^T h_{I_0}(\bar{u}) + \bar{z}_{J_0}^T l_{J_0}(\bar{u})}{g(\bar{u})} - \frac{1}{2}p^T \nabla^2 [\frac{f(\bar{u}) + \bar{y}_{I_0}^T h_{I_0}(\bar{u}) + \bar{z}_{J_0}^T l_{J_0}(\bar{u})}{g(\bar{u})}]p,$$

we have

$$\begin{aligned} \frac{f(x)}{g(x)} &< \frac{f(\bar{u}) + \bar{y}_{I_0}^T h_{I_0}(\bar{u}) + \bar{z}_{J_0}^T l_{J_0}(\bar{u})}{g(\bar{u})} \\ &- \frac{1}{2}p^T \nabla^2 [\frac{f(\bar{u}) + \bar{y}_{I_0}^T h_{I_0}(\bar{u}) + \bar{z}_{J_0}^T l_{J_0}(\bar{u})}{g(\bar{u})}]p. \end{aligned} \tag{2.5}$$

Because $\bar{y} \geq 0$, $h(x) \leq 0$ and $l(x) = 0$, we obtain

$$\bar{y}_{I_0}^T h_{I_0}(x) + \bar{z}_{J_0}^T l_{J_0}(x) \leq 0. \tag{2.6}$$

Because $g(x) > 0$, by (2.5) and (2.6), we have

$$\begin{aligned} \frac{f(x) + \bar{y}_{I_0}^T h_{I_0}(x) + \bar{z}_{J_0}^T l_{J_0}(x)}{g(x)} &< \frac{f(\bar{u}) + \bar{y}_{I_0}^T h_{I_0}(\bar{u}) + \bar{z}_{J_0}^T l_{J_0}(\bar{u})}{g(\bar{u})} \\ &- \frac{1}{2}p^T \nabla^2 [\frac{f(\bar{u}) + \bar{y}_{I_0}^T h_{I_0}(\bar{u}) + \bar{z}_{J_0}^T l_{J_0}(\bar{u})}{g(\bar{u})}]p. \end{aligned}$$

By the second order (F, ρ)-pseudoconvexity of $\frac{f + \bar{y}_{I_0}^T h_{I_0} + \bar{z}_{J_0}^T l_{J_0}}{g}$, we have

$$\begin{aligned} F(x, \bar{u}; \nabla [&\frac{f(\bar{u}) + \bar{y}_{I_0}^T h_{I_0}(\bar{u}) + \bar{z}_{J_0}^T l_{J_0}(\bar{u})}{g(\bar{u})}] \\ &- \nabla^2 [\frac{f(\bar{u}) + \bar{y}_{I_0}^T h_{I_0}(\bar{u}) + \bar{z}_{J_0}^T l_{J_0}(\bar{u})}{g(\bar{u})}]p) \leq -\rho d^2(x, \bar{u}). \end{aligned} \tag{2.7}$$

On the other hand, (1.1), (1.2), (2.3) and (2.4) yield

$$\begin{aligned} &\bar{y}_{I_s}^T h_{I_s}(x) + \bar{z}_{J_s}^T l_{J_s}(x) \leq 0 \\ &\leq \bar{y}_{I_s}^T h_{I_s}(\bar{u}) + \bar{z}_{J_s}^T l_{J_s}(\bar{u}) - \frac{1}{2}p^T \nabla^2 [\bar{y}_{I_s}^T h_{I_s}(\bar{u}) + \bar{z}_{J_s}^T l_{J_s}(\bar{u})]p, \quad (1 \leq s \leq r). \end{aligned}$$

Applying the second order (F, ρ_s)-quasiconvexity of $\bar{y}_{I_s}^T h_{I_s} + \bar{z}_{J_s}^T l_{J_s}$, we get

$$\begin{aligned} F(x, \bar{u}; \nabla [\bar{y}_{I_s}^T h_{I_s}(\bar{u}) + \bar{z}_{J_s}^T l_{J_s}(\bar{u})] + \nabla^2 [\bar{y}_{I_s}^T h_{I_s}(\bar{u}) + \bar{z}_{J_s}^T l_{J_s}(\bar{u})]p) \\ \leq -\rho_s d^2(x, \bar{u}), \qquad (1 \leq s \leq r). \end{aligned} \tag{2.8}$$

From (2.7), (2.8) and the sublinearity of F, we have

$$\begin{aligned} & F(x,\bar{u};\nabla[\frac{f(\bar{u})+\bar{y}_{I_0}^T h_{I_0}(\bar{u})+\bar{z}_{J_0}^T l_{J_0}(\bar{u})}{g(\bar{u})}+\sum_{s=0}^{r}\bar{y}_{I_s}^T h_{I_s}(\bar{u})+\sum_{s=0}^{r}\bar{z}_{J_s}^T l_{J_s}(\bar{u})] \\ + & \nabla^2[\frac{f(\bar{u})+\bar{y}_{I_0}^T h_{I_0}(\bar{u})+\bar{z}_{J_0}^T l_{J_0}(\bar{u})}{g(\bar{u})}+\sum_{s=0}^{r}\bar{y}_{I_s}^T h_{I_s}(\bar{u})+\sum_{s=0}^{r}\bar{z}_{J_s}^T l_{J_s}(\bar{u})]p) \\ \leq & -(\sum_{s=1}^{r}\rho_s+\rho)d^2(x,\bar{u}). \end{aligned}$$

By the condition (iii) and the above inequality, we obtain

$$\begin{aligned} & F(x,\bar{u};\nabla[\frac{f(\bar{u})+\bar{y}_{I_0}^T h_{I_0}(\bar{u})+\bar{z}_{J_0}^T l_{J_0}(\bar{u})}{g(\bar{u})}+\sum_{s=0}^{r}\bar{y}_{I_s}^T h_{I_s}(\bar{u})+\sum_{s=0}^{r}\bar{z}_{J_s}^T l_{J_s}(\bar{u})] \\ & +\nabla^2[\frac{f(\bar{u})+\bar{y}_{I_0}^T h_{I_0}(\bar{u})+\bar{z}_{J_0}^T l_{J_0}(\bar{u})}{g(\bar{u})}+\sum_{s=0}^{r}\bar{y}_{I_s}^T h_{I_s}(\bar{u})+\sum_{s=0}^{r}\bar{z}_{J_s}^T l_{J_s}(\bar{u})]p) < 0, \end{aligned}$$

which contradicts (2.1). The proof is complete.

Theorem 2.2 *Let $(\bar{u},\bar{y},\bar{z})$ be a feasible solution of the dual fractional program (FD_1). Assume that the following conditions hold:*

(i) $\frac{f+\bar{y}_{I_0}^T h_{I_0}+\bar{z}_{J_0}^T l_{J_0}}{g}$ is second order weakly (F,ρ)-quasiconvex at $\bar{u}$ on the feasible region D;

(ii) $\bar{y}_{I_{s0}}^T h_{I_{s0}}+\bar{z}_{J_{s0}}^T l_{J_{s0}}$ is second order strictly (F,ρ_s)-pseudoconvex at $\bar{u}$ with $(\bar{y}_{I_{s0}},\bar{z}_{I_{s0}})\neq 0$, $\bar{y}_{I_s}^T h_{I_s}+\bar{z}_{J_s}^T l_{J_s}$ $(1\leq s\leq r, s\neq s_0)$ is second order (F,ρ_s)-quasiconvex at $\bar{u}$; and

(iii) $\rho+\sum_{s=1}^{r}\rho_s\geq 0$.

If the feasible solution $\bar{x}$ of the primal fractional program (FP) satisfies

$$\frac{f(\bar{x})}{g(\bar{x})}=\frac{f(\bar{u})+\bar{y}_{I_0}^T h_{I_0}(\bar{u})+\bar{z}_{J_0}^T l_{J_0}(\bar{u})}{g(\bar{u})}-\frac{1}{2}p^T\nabla^2[\frac{f(\bar{u})+\bar{y}_{I_0}^T h_{I_0}(\bar{u})+\bar{z}_{J_0}^T l_{J_0}(\bar{u})}{g(\bar{u})}]p,$$

then $\bar{x}$ solves the primal fractional program (FP).

Proof. Assume that $\bar{x}$ does not solve (FP). Then, there exists a feasible solution x of (FP) such that $\frac{f(x)}{g(x)}<\frac{f(\bar{x})}{g(\bar{x})}$.

Since

$$\frac{f(\bar{x})}{g(\bar{x})}=\frac{f(\bar{u})+\bar{y}_{I_0}^T h_{I_0}(\bar{u})+\bar{z}_{J_0}^T l_{J_0}(\bar{u})}{g(\bar{u})}-\frac{1}{2}p^T\nabla^2[\frac{f(\bar{u})+\bar{y}_{I_0}^T h_{I_0}(\bar{u})+\bar{z}_{J_0}^T l_{J_0}(\bar{u})}{g(\bar{u})}]p,$$

the above inequality yields

$$\frac{f(x)}{g(x)}<\frac{f(\bar{u})+\bar{y}_{I_0}^T h_{I_0}(\bar{u})+\bar{z}_{J_0}^T l_{J_0}(\bar{u})}{g(\bar{u})}-\frac{1}{2}p^T\nabla^2[\frac{f(\bar{u})+\bar{y}_{I_0}^T h_{I_0}(\bar{u})+\bar{z}_{J_0}^T l_{J_0}(\bar{u})}{g(\bar{u})}]p.$$

By $g(x) > 0$, $\bar{y} \geq 0$, $h(x) \leq 0$ and $l(x) = 0$, we have

$$\frac{f(x) + \bar{y}_{I_0}^T h_{I_0}(x) + \bar{z}_{J_0}^T l_{J_0}(x)}{g(x)}$$
$$< \frac{f(\bar{u}) + \bar{y}_{I_0}^T h_{I_0}(\bar{u}) + \bar{z}_{J_0}^T l_{J_0}(\bar{u})}{g(\bar{u})} - \frac{1}{2} p^T \nabla^2 [\frac{f(\bar{u}) + \bar{y}_{I_0}^T h_{I_0}(\bar{u}) + \bar{z}_{J_0}^T l_{J_0}(\bar{u})}{g(\bar{u})}] p.$$

By the second order weak (F, ρ)-quasiconvexity of $\frac{f+\bar{y}_{I_0}^T h_{I_0} + \bar{z}_{J_0}^T l_{J_0}}{g}$, we get

$$F(x, \bar{u}; \nabla [\frac{f(\bar{u}) + \bar{y}_{I_0}^T h_{I_0}(\bar{u}) + \bar{z}_{J_0}^T l_{J_0}(\bar{u})}{g(\bar{u})}]$$
$$- \nabla^2 [\frac{f(\bar{u}) + \bar{y}_{I_0}^T h_{I_0}(\bar{u}) + \bar{z}_{J_0}^T l_{J_0}(\bar{u})}{g(\bar{u})}] p)$$
$$\leq -\rho d^2(x, \bar{u}). \tag{2.9}$$

From (1.1), (1.2), (2.3) and (2.4), we have

$$\bar{y}_{I_s}^T h_{I_s}(x) + \bar{z}_{J_s}^T l_{J_s}(x) \leq 0$$
$$\leq \bar{y}_{I_s}^T h_{I_s}(\bar{u}) + \bar{z}_{J_s}^T l_{J_s}(\bar{u}) - \frac{1}{2} p^T \nabla^2 [\bar{y}_{I_s}^T h_{I_s}(\bar{u}) + \bar{z}_{J_s}^T l_{J_s}(\bar{u})] p, \quad (1 \leq s \leq r).$$

By condition (ii), we have

$$F(x, \bar{u}; \nabla [\bar{y}_{I_{s_0}}^T h_{I_{s_0}}(\bar{u}) + \bar{z}_{J_{s_0}}^T l_{J_{s_0}}(\bar{u})] + \nabla^2 (\bar{y}_{I_{s_0}}^T h_{I_{s_0}}(\bar{u}) + \bar{z}_{J_{s_0}}^T l_{J_{s_0}}(\bar{u})) p])$$
$$< -\rho_{s_0} d^2(x, \bar{u}), \tag{2.10}$$

and

$$F(x, \bar{u}; \nabla [\bar{y}_{I_s}^T h_{I_s}(\bar{u}) + \bar{z}_{J_s}^T l_{J_s}(\bar{u})] + \nabla^2 [\bar{y}_{I_s}^T h_{I_s}(\bar{u}) + \bar{z}_{J_s}^T l_{J_s}(\bar{u})] p)$$
$$\leq -\rho_s d^2(x, \bar{u}), \quad 1 \leq s \leq r, s \neq s_0. \tag{2.11}$$

From (2.9), (2.10), (2.11) and the sublinearity of F, we give

$$F(x, \bar{u}; \nabla [\frac{f(\bar{u}) + \bar{y}_{I_0}^T h_{I_0}(\bar{u}) + \bar{z}_{J_0}^T l_{J_0}(\bar{u})}{g(\bar{u})} + \sum_{s=0}^{r} \bar{y}_{I_s}^T h_{I_s}(\bar{u}) + \sum_{s=0}^{r} \bar{z}_{J_s}^T l_{J_s}(\bar{u})]$$
$$+ \nabla^2 [\frac{f(\bar{u}) + \bar{y}_{I_0}^T h_{I_0}(\bar{u}) + \bar{z}_{J_0}^T l_{J_0}(\bar{u})}{g(\bar{u})} + \sum_{s=0}^{r} \bar{y}_{I_s}^T h_{I_s}(\bar{u}) + \sum_{s=0}^{r} \bar{z}_{J_s}^T l_{J_s}(\bar{u})] p)$$
$$< -(\sum_{s=1}^{r} \rho_s + \rho) d^2(x, \bar{u}). \tag{2.12}$$

By the condition (iii) and the above inequality, we obtain

$$F(x, \bar{u}; \nabla [\frac{f(\bar{u}) + \bar{y}_{I_0}^T h_{I_0}(\bar{u}) + \bar{z}_{J_0}^T l_{J_0}(\bar{u})}{g(\bar{u})} + \sum_{s=0}^{r} \bar{y}_{I_s}^T h_{I_s}(\bar{u}) + \sum_{s=0}^{r} \bar{z}_{J_s}^T l_{J_s}(\bar{u})]$$
$$+ \nabla^2 (\frac{f(\bar{u}) + \bar{y}_{I_0}^T h_{I_0}(\bar{u}) + \bar{z}_{J_0}^T l_{J_0}(\bar{u})}{g(\bar{u})} + \sum_{s=0}^{r} \bar{y}_{I_s}^T h_{I_s}(\bar{u}) + \sum_{s=0}^{r} \bar{z}_{J_s}^T l_{J_s}(\bar{u})) p]) < 0,$$

which contradicts (2.1). This shows that $\bar{x}$ solves (FP).

3 SECOND ORDER STRICT CONVERSE DUALITY II

In this section, we will introduce another second order dual model and establish strict converse duality results.

(FD_2):

$$\begin{aligned}
\max \quad & \lambda \\
\text{subject to} \quad & \nabla[f(u) + y^T h(u) + z^T l(u) - \lambda g(u)] \\
& \qquad + \nabla^2[f(u) + y^T h(u) + z^T l(u) - \lambda g(u)]p = 0, \quad (3.1) \\
& f(u) + y_{I_s}^T h_{I_s}(u) + z_{J_s}^T l_{J_s}(u) - \lambda g(u) \\
-\frac{1}{2}p^T \nabla^2[f(u) + & y_{I_s}^T h_{I_s}(u) + z_{J_s}^T l_{J_s}(u) - \lambda g(u)]p \geq 0, (1 \leq s \leq r), \quad (3.2) \\
& y_{I_s}^T h_{I_s}(u) + z_{J_s}^T l_{J_s}(u) \\
& \quad -\frac{1}{2}p^T \nabla^2[y_{I_s}^T h_{I_s}(u) + z_{J_s}^T l_{J_s}(u)]p \geq 0, (1 \leq s \leq r), \quad (3.3) \\
& y \geq 0, \quad (3.4) \\
& \lambda \geq 0, \quad (3.5)
\end{aligned}$$

where I_s, J_s, r, y_{I_s}, z_{J_s}, h_{I_s} and l_{J_s} are the same notations as defined in (FD_1).

Remark 3.1 *If $p = 0$, $I_0 = \emptyset$, $J_0 = \emptyset$, $l(x) \equiv 0$, $I_1 = \{1, 2, \cdots, m\}$, $J_1 = \emptyset$, $I_i = J_i = \emptyset$, $(i = 2, 3, \cdots, r)$, then our dual model (FD_2) is reduced to a first order dual model in Craven and Mond (1999).*

Now we prove results on strict converse duality theorems between the fractional program (FD_2) and the primal fractional program (FP).

Theorem 3.1 *Let $(\bar{u}, \bar{y}, \bar{z}, \bar{\lambda})$ be a feasible solution of the dual fractional program (FD_2). Assume that the following conditions hold:*

(i) $f + \bar{y}_{I_0}^T h_{I_0} + \bar{z}_{J_0}^T l_{J_0} - \bar{\lambda} g$ is second order (F, ρ)-pseudoconvex at $\bar{u}$ on the feasible region D;

(ii) $\bar{y}_{I_s}^T h_{I_s} + \bar{z}_{J_s}^T l_{J_s}$, $(1 \leq s \leq r)$ is second order (F, ρ_s)-quasiconvex at $\bar{u}$ on the feasible region D; and

(iii) $\rho + \sum_{s=1}^r \rho_s \geq 0$.

If the feasible solution $\bar{x}$ of the primal fractional program (FP) satisfies

$$\frac{f(\bar{x})}{g(\bar{x})} = \bar{\lambda},$$

then $\bar{x}$ solves the primal fractional program (FP).

Proof. Assume that $\bar{x}$ is not an optimal solution of (FP). Then, there would exist an $x \in D$ such that

$$\frac{f(x)}{g(x)} < \frac{f(\bar{x})}{g(\bar{x})}.$$

Based on above inequality and $\frac{f(\bar{x})}{g(\bar{x})} = \bar{\lambda}$, we have that

$$\frac{f(x)}{g(x)} < \bar{\lambda}.$$

Because $g(x) > 0$, the above inequality yields

$$f(x) - \bar{\lambda} g(x) < 0. \tag{3.6}$$

From (3.2), (1.1), (1.2) and (3.6), we have

$$f(x) - \bar{\lambda} g(x) + \bar{y}_{I_0}^T h_{I_0}(x) + \bar{z}_{J_0}^T l_{J_0}(x) < f(\bar{u}) - \bar{\lambda} g(\bar{u}) + \bar{y}_{I_0}^T h_{I_0}(\bar{u}) + \bar{z}_{J_0}^T l_{J_0}(\bar{u})$$
$$-\frac{1}{2} p^T \nabla^2 [f(\bar{u}) - \bar{\lambda} g(\bar{u}) + \bar{y}_{I_0}^T h_{I_0}(\bar{u}) + \bar{z}_{J_0}^T l_{J_0}(\bar{u})] p.$$

By the second order (F, ρ)-pseudoconvexity of $f - \bar{\lambda} g + \bar{y}_{I_0}^T h_{I_0} + \bar{z}_{J_0}^T l_{J_0}$, we have

$$F(x, \bar{u}; \nabla[f(\bar{u}) - \bar{\lambda} g(\bar{u}) + \bar{y}_{I_0}^T h_{I_0}(\bar{u}) + \bar{z}_{J_0}^T l_{J_0}(\bar{u})]$$
$$+\nabla^2 [f(\bar{u}) - \bar{\lambda} g(\bar{u}) + \bar{y}_{I_0}^T h_{I_0}(\bar{u}) + \bar{z}_{J_0}^T l_{J_0}(\bar{u})] p) < -\rho d^2(x, \bar{u}). \tag{3.7}$$

From $h(x) \leq 0, l(x) = 0, \bar{y} \geq 0$ and (3.3), we have

$$\bar{y}_{I_s}^T h_{I_s}(x) + \bar{z}_{J_s}^T l_{J_s}(x) \leq 0 \leq \bar{y}_{I_s}^T h_{I_s}(\bar{u}) + \bar{z}_{J_s}^T l_{J_s}(\bar{u})$$
$$-\frac{1}{2} p^T \nabla^2 [\bar{y}_{I_s}^T h_{I_s}(\bar{u}) + \bar{z}_{J_s}^T l_{J_s}(\bar{u})] p, \quad 1 \leq s \leq r. \tag{3.8}$$

Applying the second order (F, ρ_s)-quasiconvexity of $\bar{y}_{I_s}^T h_{I_s} + \bar{z}_{J_s}^T l_{J_s}$ $(1 \leq s \leq r)$, we get

$$F(x, \bar{u}; \nabla[\bar{y}_{I_s}^T h_{I_s}(\bar{u}) + \bar{z}_{J_s}^T l_{J_s}(\bar{u})] + \nabla^2 [\bar{y}_{I_s}^T h_{I_s}(\bar{u}) + \bar{z}_{J_s}^T l_{J_s}(\bar{u})] p)$$
$$\leq -\rho_s d^2(x, \bar{u}), \quad 1 \leq s \leq r. \tag{3.9}$$

By inequalities (3.7), (3.9) and the sublinearity of F, we have

$$F(x, \bar{u}; \nabla[f(\bar{u}) - \bar{\lambda} g(\bar{u}) + \bar{y}^T h(\bar{u}) + \bar{z}^T l(\bar{u})] + \nabla^2 [f(\bar{u}) - \bar{\lambda} g(\bar{u}) + \bar{y}^T h(\bar{u}) + \bar{z}^T l(\bar{u})] p)$$
$$< -(\rho + \sum_{s=1}^{r} \rho_s) d^2(x, \bar{u}).$$

By (iii) and the above inequality, we obtain

$$F(x, \bar{u}; \nabla[f(\bar{u}) - \bar{\lambda} g(\bar{u}) + \bar{y}^T h(\bar{u}) + \bar{z}^T l(\bar{u})] + \nabla^2 [f(\bar{u}) - \bar{\lambda} g(\bar{u}) + \bar{y}^T h(\bar{u}) + \bar{z}^T l(\bar{u})] p) < 0,$$

which contradicts (3.1). The proof is complete.

Theorem 3.2 *Let $(\bar{u}, \bar{y}, \bar{z}, \bar{\lambda})$ be a feasible solution of the dual fractional program (FD_2). Assume that the following conditions hold:*

(i) $f + \bar{y}_{I_0}^T h_{I_0} + \bar{z}_{J_0}^T l_{J_0} - \bar{\lambda} g$ is second order weakly (F, ρ)-quasiconvex at $\bar{u}$ on the feasible region D;

(ii) $\bar{y}_{I_{s_0}}^T h_{I_{s_0}} + \bar{z}_{J_{s_0}}^T l_{J_{s_0}}$ *is second order strictly* (F, ρ_s)*-pseudoconvex at* $\bar{u}$ *with* $(\bar{y}_{I_{s_0}}, \bar{z}_{I_{s_0}}) \neq 0$*;* $\bar{y}_{I_s}^T h_{I_s} + \bar{z}_{J_s}^T l_{J_s}$*,* $(1 \leq s \leq r)$ *is second order* (F, ρ_s)*-quasiconvex at* $\bar{x}$ *at* $\bar{u}$*,* $1 \leq s \leq r, s \neq s_0$*; and*

(iii) $\rho + \sum_{s=1}^{r} \rho_s \geq 0$.

If the feasible solution $\bar{x}$ *of the primal fractional program* (FP) *satisfies* $\frac{f(\bar{x})}{g(\bar{x})} = \bar{\lambda}$*, then* $\bar{x}$ *solves the primal fractional program* (FP).

Proof. For any $x \in D$, since $(\bar{u}, \bar{y}, \bar{z}, \bar{\lambda})$ is feasible to the dual fractional program (FD_2), it follows from (1.1), (1.2) and (3.3), that

$$\begin{aligned} &\bar{y}_{I_s}^T h_{I_s}(x) + \bar{z}_{J_s}^T l_{J_s}(x) \leq 0 \leq \bar{y}_{I_s}^T h_{I_s}(\bar{u}) + \bar{z}_{J_s}^T l_{J_s}(\bar{u}) \\ &\quad -\frac{1}{2} p^T \nabla^2 [\bar{y}_{I_s}^T h_{I_s}(\bar{u}) + \bar{z}_{J_s}^T l_{J_s}(\bar{u})] p, \quad 1 \leq s \leq r. \end{aligned}$$

From (ii), we have

$$\begin{aligned} &F(x, \bar{u}; \nabla[\bar{y}_{I_s}^T h_{I_s}(\bar{u}) + \bar{z}_{J_s}^T l_{J_s}(\bar{u})] + \nabla^2 [\bar{y}_{I_s}^T h_{I_s}(\bar{u}) + \bar{z}_{J_s}^T l_{J_s}(\bar{u})] p) \\ &\leq -\rho_s d^2(x, \bar{u}), \quad 1 \leq s \leq r, \end{aligned} \tag{3.10}$$

and

$$\begin{aligned} &F(x, \bar{u}; \nabla[\bar{y}_{I_{s_0}}^T h_{I_{s_0}}(\bar{u}) + \bar{z}_{J_{s_0}}^T l_{J_{s_0}}(\bar{u})] + \nabla^2 [\bar{y}_{I_{s_0}}^T h_{I_{s_0}}(\bar{u}) + \bar{z}_{J_{s_0}}^T l_{J_{s_0}}(\bar{u})] p) \\ &< -\rho_0 d^2(x, \bar{u}). \end{aligned} \tag{3.11}$$

From (3.10), (3.11) and the sublinearity of F, we have

$$\begin{aligned} &F(x, \bar{u}; \nabla[\sum_{s=1}^{r} \bar{y}_{I_s}^T h_{I_s}(\bar{u}) + \sum_{s=1}^{r} \bar{z}_{J_s}^T l_{J_s}(\bar{u})] \\ &+ \nabla^2 [\sum_{s=1}^{r} \bar{y}_{I_s}^T h_{I_s}(\bar{u}) + \sum_{s=1}^{r} \bar{z}_{J_s}^T l_{J_s}(\bar{u})] p) \leq -\sum_{s=1}^{r} \rho_s d^2(x, \bar{u}). \end{aligned} \tag{3.12}$$

From (3.12), (3.1), assumption condition (iii) and the sublinearity of F, we obtain

$$\begin{aligned} &F(x, \bar{u}; \nabla[\bar{y}_{I_0}^T h_{I_0}(\bar{u}) + \bar{z}_{J_0}^T l_{J_0}(\bar{u}) - \bar{\lambda} g(\bar{u})] \\ &+\nabla^2 [\bar{y}_{I_0}^T h_{I_0}(\bar{u}) + \bar{z}_{J_0}^T l_{J_0}(\bar{u}) - \bar{\lambda} g(\bar{u})] p) > -\rho d^2(x, \bar{u}). \end{aligned}$$

By the second order weak (F, ρ)-quasiconvexity of $f + \bar{y}_{I_0}^T h_{I_0} + \bar{z}_{J_0}^T l_{J_0} - \bar{\lambda} g$, we have

$$\begin{aligned} &f(x) + \bar{y}_{I_0}^T h_{I_0}(x) + \bar{z}_{J_0}^T l_{J_0}(x) - \bar{\lambda} g(x) \\ &\geq f(\bar{u}) + \bar{y}_{I_0}^T h_{I_0}(\bar{u}) + \bar{z}_{J_0}^T l_{J_0}(\bar{u}) - \bar{\lambda} g(\bar{u}) \\ &\quad -\frac{1}{2} p^T \nabla^2 [f(\bar{u}) + \bar{y}_{I_0}^T h_{I_0}(\bar{u}) + \bar{z}_{J_0}^T l_{J_0}(\bar{u}) - \bar{\lambda} g(\bar{u})] p. \end{aligned}$$

By (3.2) and the above inequality, we have

$$f(x) + \bar{y}_{I_0}^T h_{I_0}(x) + \bar{z}_{J_0}^T l_{J_0}(x) - \bar{\lambda} g(x) \geq 0, \forall x \in D.$$

It follows from $\bar{y}_{I_0}^T h_{I_0}(x) + \bar{z}_{J_0}^T l_{J_0}(x) \leq 0$ that

$$f(x) - \bar{\lambda} g(x) \geq 0, \forall x \in D.$$

Hence,

$$\frac{f(x)}{g(x)} \geq \bar{\lambda} = \frac{f(\bar{x})}{g(\bar{x})}, \forall x \in D.$$

This implies that $\bar{x}$ solves the primal fractional program (FP).

Acknowledgments

This research was supported in part by the National Natural Science Foundation of China and Competitive Research Grants of Hong Kong Polytechnic University.

References

Bector, C.R. (1973), Duality in nonlinear fractional programming, *Zeitschrift fur Operations Research*, Vol. 17, pp. 183-193.

Bector, C.R., Chandra, S. and Husain, I. (1991), Second order duality for a minimax programming problem, *Opsearch*, Vol. 28, pp. 249-263.

Craven, B. D. and Mond, B. (1999), Fractional programming with invexity, *Progress In Optimization: Contributions from Australasia*, Eberhard, A., Glover, B., Hill, R. and Ralph, D., Kluwer Academic Publishers, pp. 79-89.

Egudo, R. R. and Hanson, M. A. (1993), Second order duality in multiobjective programming, *Opsearch*, Vol. 30, pp. 223-230.

Hanson, M. A. (1993), Second order invexity and duality in mathematical programming, *Opsearch*, Vol. 30, pp. 313-320.

Jagannathan, R. (1973), Duality for nonlinear fractional programs, *Zeitschrift fur Operations Research*, Vol. 17, pp. 1-3.

Kaul, R. N. and Kaur, S. (1985), Optimality criteria in nonlinear programming involving nonlinear functions, *Journal of Mathematical Analysis and Applications*, Vol. 105, pp. 104-112.

Khan, Z. A. (1990), Converse duality in nonlinear fractional programming, *Asia-Pacific Journal of Operational Research*, Vol. 7, pp. 9-15.

Mahajan, D. G. and Vartak, M. N. (1977), Generalization of some duality theorems in nonlinear programming, *Mathematical Programming*, Vol. 12, pp. 293-317.

Mangasaran, O. L. (1975), Second and higher-order duality theorems in nonlinear programming, *Journal of Mathematical Analysis and Applications*, Vol. 51, pp. 607-620.

Mond, B. (1974), Second order duality for nonlinear programs, *Opsearch*, Vol. 11, pp. 90-98.

Mond, B. and Weir, T. (1981-1983), Generalized convexity and higher order duality, *Journal of Mathematical Sciences*, Vol. 16-18, pp. 74-94.

Schaible, S. (1976), Duality in fractional programming: A unified approach, *Operations Research*, Vol. 24, pp. 452-461.

Yang, X. Q. (1998), Second order global optimality conditions for convex composite optimization, *Mathematical Programming*, Vol. 81, pp. 327-347.

Zhang, J. (1999), Higher order convexity and duality in multiobjective programming problems, *Progress In Optimization: Contributions from Australasia*, Eberhard, A., Glover, B., Hill, R. and Ralph, D., Kluwer Academic Publishers, pp. 101-117.

III OPTIMIZATION APPLICATIONS

17 CHEBYSHEV OPTIMIZATION OF CIRCULAR ARRAYS

Mattias Dahl, Ingvar Claesson, Sven Nordebo and Sven Nordholm

Department of Telecommunications and Signal Processing
University of Karlskrona/Ronneby
Sweden

Abstract: The lobe design of a narrow–band antenna array is formulated as a complex Chebyshev optimization problem. Previously, the solution of similar problems have been approximated using various techniques including conventional finite–dimensional linear programming. In this paper, recently developed semi–infinite linear programming techniques employing simplex extension algorithms are used to design the required antenna array.

The proposed design technique is advantageous due to its numerical efficiency and its flexibility with respect to the design specification.

The design approach is applied to a numerical examples which deal with the design of a narrow–band circular and hexagonal antenna arrays for the far–field. The optimization technique is employed as a tool to investigate the effect of distributing the antenna elements around a planar shield.

Key words: Chebyshev optimization, semi-infinite linear programming, simplex extension algorithms, planar circular antenna array.

X. Yang et al. (eds.), Optimization Methods and Applications, 309-319.

1 INTRODUCTION

For antenna arrays as for digital FIR filters, the design problem is often cast as a finite–dimensional complex approximation problem as in Streit and Nuttall (1982); Chen and Parks (1987). Classical least squares approximation methods can in many cases be used to obtain a desired solution Veen and Buckley (1988). However, when the design specification is given as a bound on the complex design error, the problem is naturally converted to a uniform (Chebyshev) complex approximation problem.

It was shown in Streit and Nuttall (1982); Streit and Nuttall (1983) that the (non–linear) uniform complex approximation problem can be reformulated as an equivalent real semi–infinite linear program. However, in Streit and Nuttall (1982), the complex error was approximated by a finitization in order to solve the problem numerically.

The uniform complex approximation problem for the design of FIR filters has been intensively studied over the last few years, see e.g. Chen and Parks (1987); Alkhairy et al (1993); Burnside and Parks (1995). Earlier approaches such as in Chen and Parks (1987) approximates the optimum solution by finitization and employs finite–dimensional linear programming. However, it was shown in Anderson and Nash (1987) that the semi–infinite linear program corresponding to the uniform (real) approximation problem can be solved by numerically efficient simplex extension algorithms. These results were later exploited for the design of FIR filters and digital Laguerre filters networks with complex Chebyshev error criteria, see e.g. Burnside and Parks (1995); Nordebo and Zang (1999).

Finitization can in principle give an arbitrarily accurate approximation of the complex Chebyshev solution but becomes exceedingly memory intensive as the grid spacing decreases. The semi–infinite simplex extension is much more computational efficient since the constraint set can be represented in functional form rather than stored in memory as numerical values. Furthermore, the semi–infinite formulation deals directly with the true complex error and not an approximation thereof.

As a numerical example we consider the planar circular array defined by N sensor elements evenly distributed on a circle, and with the incident wave front propagating in the same plane as the array, see Fig. 1.1.

We consider the far–field and narrow–band case where the phase of the wave front is given by $e^{j(2\pi f_0 t - \mathbf{k}^T\mathbf{r})}$ where f_0 is the frequency, t is the time, $\mathbf{k} = -\frac{2\pi f_0}{c}(cos\varphi, \sin\varphi)$ is the wave vector, c is the speed of wave propagation, φ is the angle of incidence and $\mathbf{r}$ is the evaluated spatial point. The array elements have the spatial positions $\mathbf{r}_n = r(\cos(n\frac{2\pi}{N}), \sin(n\frac{2\pi}{N}))$ where $n = 0, \ldots, N-1$ and r is the radius of the array–circle.

With the phase center in the center of the array–circle, the array response is given by

$$H(\varphi) = \sum_{n=0}^{N-1} w_n^* a_n(\varphi) e^{j2\pi f_0 \frac{r}{c}\cos(n\frac{2\pi}{N} - \varphi)} \tag{1.1}$$

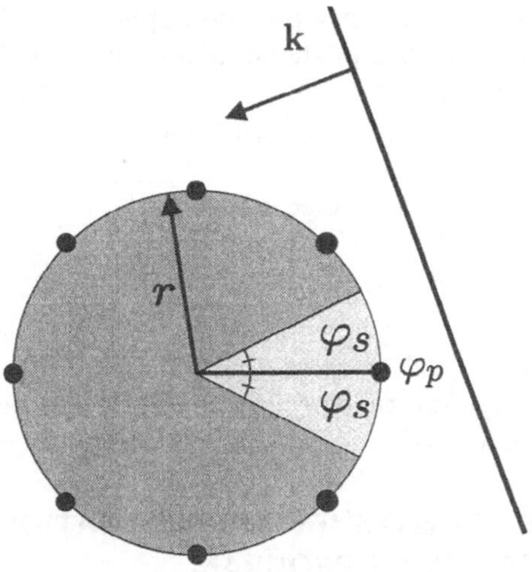

Figure 1.1 A planar circular array for the far–field where the • denotes the sensor elements. The passband is defined by the angle φ_p and the stopband (dark gray shaded) by the interval $[\varphi_p + \varphi_s, \varphi_p - \varphi_s + 2\pi]$, respectively.

where w_n are the complex array weights and $a_n(\varphi)$ are the responses of the individual sensors.

2 PROBLEM FORMULATION

The complex array response is given by

$$H(\varphi) = \sum_{n=0}^{N-1} w_n^* d_n(\varphi) = \mathbf{w}^H \mathbf{d}(\varphi) \qquad \varphi \in [-\pi, \pi] \tag{2.1}$$

where $\mathbf{w}$ is an $N \times 1$ array weight vector of complex coefficients w_n and $\mathbf{d}(\varphi)$ the corresponding array response vector of complex transfer functions $d_n(\varphi) = a_n(\varphi)e^{-j\mathbf{k}^T\mathbf{r}_n}$, $n = 0, \ldots, N-1$ (plane wave). It is assumed that the functions $d_n(\varphi)$ are continuous and linearly independent.

The narrow–band case often requires complex weights whereas the linear programming approach requires real variables. With a complex array weight vector $\mathbf{w} = \Re\{\mathbf{w}\} + j\Im\{\mathbf{w}\}$, the response in (2.1) can be redefined using real weights as

$$H(\varphi) = \mathbf{w}^H \mathbf{d}(\varphi) = \tilde{\mathbf{w}}^T \tilde{\mathbf{d}}(\varphi) \tag{2.2}$$

where $\tilde{\mathbf{w}}^T = [\Re\{\mathbf{w}^T\} \ \Im\{\mathbf{w}^T\}]$ and $\tilde{\mathbf{d}}^T(\varphi) = [\mathbf{d}^T(\varphi) \ -j\mathbf{d}^T(\varphi)]$. In this case, the number of real coefficients $\tilde{w}_n$ is $2N$.

The complex specification may be given in the form

$$\begin{cases} |H(\varphi)| \leq \varepsilon(\varphi) & \varphi \in \Phi_s = [\varphi_p + \varphi_s, \varphi_p - \varphi_s + 2\pi] \\ H(\varphi_p) = 1 \end{cases} \tag{2.3}$$

where $\varepsilon(\varphi)$ is a prescribed (strictly positive) upper bound. The first constraint in (2.3) specifies the sidelobes in the stopband Φ_s and the second defines the mainlobe in direction φ_p.

We now pose the following design criterion

$$\begin{cases} \min_{\tilde{\mathbf{w}}} \max_{\varphi \in \Phi_s} v(\varphi)|H(\varphi)| \\ \\ H(\varphi_p) = 1 \end{cases} \tag{2.4}$$

where $v(\varphi)$ is a strictly positive weighting function. Note that with the choice of weighting $v(\varphi) = 1/\varepsilon(\varphi)$, the specification (2.3) will be satisfied if and only if the optimum objective value in (2.4) is less than or equal to one.

3 THE COMPLEX CHEBYSHEV APPROXIMATION PROBLEM AS A SEMI–INFINITE LINEAR PROGRAM

According to the *real rotation theorem* Parks and Burrus (1987), a magnitude inequality in the complex plane can be expressed in the following equivalent form.

$$|z| \leq \delta \quad \Leftrightarrow \quad \Re\left\{ze^{j\theta}\right\} \leq \delta \qquad \forall \theta \in [0, 2\pi] \tag{3.1}$$

where z is a complex number and δ a real and positive number, see Fig. 3.1.

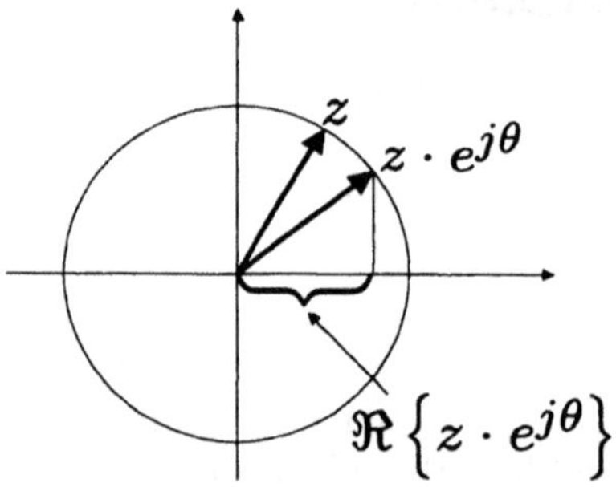

Figure 3.1 The *real rotation theorem* interpreted as a magnitude inequality in the complex plane.

Without loss of generality we assume from now on that the weighting function $v(\varphi) = 1$. By making use of (3.1), the nonlinear approximation problem (2.4) can be reformulated as the following continuous semi–infinite linear program

$$\begin{cases} \min \delta \\ |H(\varphi)| \leq \delta \\ H(\varphi_p) = 1 \end{cases} \Leftrightarrow \begin{cases} \min \delta \\ \Re\left\{H(\varphi) \cdot e^{j\theta}\right\} \leq \delta \\ H(\varphi_p) = 1 \end{cases} \tag{3.2}$$

where δ is an additional real variable, $\varphi \in \Phi_s$, $\theta \in [0, 2\pi]$. The linear program (3.2) is called (continuous) semi–infinite since the constraint set is infinite (uncountable) and the number of variables finite.

The problem (3.2) can now be expressed as

$$\left\{ \begin{array}{l} \min \mathbf{b}^T \mathbf{y} \\ \left[\begin{array}{cc} \Re\left\{\tilde{\mathbf{d}}^T(\varphi) \cdot e^{j\theta}\right\} & 1 \\ \Re\left\{\tilde{\mathbf{d}}^T(\varphi_p)\right\} & 0 \\ -\Re\left\{\tilde{\mathbf{d}}^T(\varphi_p)\right\} & 0 \\ \Im\left\{\tilde{\mathbf{d}}^T(\varphi_p)\right\} & 0 \\ -\Im\left\{\tilde{\mathbf{d}}^T(\varphi_p)\right\} & 0 \end{array} \right] \mathbf{y} \geq \left[\begin{array}{c} \mathbf{0} \\ 1 \\ -1 \\ 0 \\ 0 \end{array} \right] \end{array} \right. \Leftrightarrow \left\{ \begin{array}{l} \min \mathbf{b}^T \mathbf{y} \\ \mathbf{A}^T(\varphi, \theta)\mathbf{y} \geq c(\varphi, \theta) \end{array} \right. \tag{3.3}$$

where

$$\mathbf{y} = \left[\begin{array}{c} \tilde{\mathbf{w}} \\ \delta \end{array} \right], \quad \mathbf{b} = \left[\begin{array}{c} \mathbf{0} \\ 1 \end{array} \right] \tag{3.4}$$

In (3.3), $\mathbf{A}^T(\varphi, \theta)$ and $c(\varphi, \theta)$ are the rows of the left hand side constraint matrix and the elements of the right hand side constraint vector, respectively, defined for $(\varphi, \theta) \in \Phi_s \times [0, 2\pi] \cup \{\varphi_p\} \times \{1, 2, 3, 4\}$. The bold face $\mathbf{0}$ in the first entry of the right hand side vector in (3.3) denotes a zero function of φ and θ. Finally, we rewrite (3.3) as

$$\text{(D)} \left\{ \begin{array}{l} \min \mathbf{b}^T \mathbf{y} \\ \mathbf{A}_\alpha^T \mathbf{y} \geq c_\alpha \end{array} \right. \tag{3.5}$$

where α is an index belonging to a closed index set $\mathcal{A} \subset \mathcal{R}^k$ which has a one-to-one correspondence with the constraint rows of (3.3). The formulation in (3.5) is amenable of a solution due to the Caratheodory's theorem and the finiteness of related Lagrange multipliers Anderson and Nash (1987). Hence, (3.5) can be solved by using an extended simplex procedure as in Burnside and Parks (1995); Nordebo and Zang (1999). We refer to (3.5) as the dual formulation since it corresponds to the dual of a linear program in standard form Anderson and Nash (1987); Luenberger (1984).

4 SEMI–INFINITE LINEAR PROGRAMMING

The application of simplex extension algorithms for the solution of (3.2) is based on two important theorems given in Anderson and Nash (1987) concerning strong duality and finite dimensionality. These theorems will be quoted and explained below.

4.1 Duality and Dimensionality

The primal linear programming formulation corresponding to (3.5) is given by

$$(\mathrm{P}) \begin{cases} \max \int c_\alpha dx_\alpha \\ \int \mathbf{A}_\alpha dx_\alpha = \mathbf{b} \\ x_\alpha \geq 0 \end{cases} \tag{4.1}$$

where the maximization is with respect to the space of regular Borel measures x_α on $\mathcal{A}$ (dual of the space $\mathcal{C}$ of continuous functions c_α on $\mathcal{A}$), cf. Anderson and Nash (1987). The formulation (4.1) is said to be in standard form.

Finite dimensional linear programs satisfy a condition known as *strong duality*: If either of the primal and dual formulations has a finite optimum solution, so does the other, and the corresponding optimum values are equal Luenberger (1984). Strong duality does not necessarily hold for semi–infinite linear programs. Examples can easily be constructed for which there exists a *duality gap* between the optimal values of the dual programs in (3.5) and (4.1), cf. Anderson and Nash (1987).

Fortunately, the following result applies to the filter design problem (2.4) under consideration.

Theorem 4.1 *(Strong Duality) The semi–infinite linear program corresponding to the* uniform approximation problem *(3.2) satisfies strong duality.*

Theorem 4.1 is readily proven by applying theorem 4.4c in Anderson and Nash (1987).

The success of the semi-infinite linear programming approach relies on the property of strong duality, but also on the equally important *Dimensionality Theorem*:

Theorem 4.2 *(Finite Dimensionality) If there is a feasible solution to (P) in (4.1) with value z_0, then there is also a feasible solution of finite support (atomic measure) consisting of at most $2N+1$ points which achieves the same value.*

The dimensionality theorem (theorem 4.8 in Anderson and Nash (1987)) is based on an application of Caratheodory's theorem, a fundamental dimensionality result in convexity theory Rockafellar (1969). The implication of theorem 4.2 is extremely useful since it allows the simplex algorithm to work with finite basic feasible solutions in very much the same way as in the case of finite–dimensional linear programming. The primal formulation (4.1) can thus be rewritten as

$$(P) \begin{cases} \max \sum_{i=1}^{2N+1} c_{\alpha_i} x_{\alpha_i} \\ \sum_{i=1}^{2N+1} \mathbf{A}_{\alpha_i} x_{\alpha_i} = \mathbf{b} \\ x_{\alpha_i} \geq \mathbf{0} \\ \alpha_i \in \mathcal{A} \end{cases} \tag{4.2}$$

where the maximization is with respect to the points $\alpha_1, \ldots, \alpha_{2N+1}$ in $\mathcal{A}$.

The duality theorem enables us to solve the primal (P) in (4.2) rather than the dual (D) in (3.5). When the optimum solution to (P) is obtained we will also be in possession of the optimum solution to (D). The advantage of this procedure is that the primal (P) is in standard form for solution by the simplex algorithm employing a finite basis. The problem (P) is thus much easier to solve than (D).

5 BRIEF OUTLINE OF LINEAR PROGRAMMING

Given a *basis* $\mathbf{B} = [\mathbf{A}_{\alpha_1}, \ldots, \mathbf{A}_{\alpha_m}]$ where $\alpha_i \in \mathcal{A}$ for $i = 1, \ldots, m = 2N+1$ and a basic feasible solution $\mathbf{x}_B \geq \mathbf{0}$ satisfying $\mathbf{B}\mathbf{x}_B = \mathbf{b}$. The dimension of $\mathbf{B}$ and $\mathbf{x}_B$ are $m \times m$ and $m \times 1$, respectively.

Let the matrix $\mathbf{A}$ be partitioned as $\mathbf{A} = [\mathbf{B}\ \mathbf{N}]$ where $\mathbf{N} = (\mathbf{A}_\alpha)$ consists of those columns of $\mathbf{A}$ which are not included in $\mathbf{B}$. Define the corresponding partition of the vectors $\mathbf{c}^T = [\mathbf{c}_B^T\ \mathbf{c}_N^T]$ and $\mathbf{x}^T = [\mathbf{x}_B^T\ \mathbf{x}_N^T]$ (*basic* and *non–basic* variables). The primal cost is $\mathbf{c}^T\mathbf{x} = \mathbf{c}_B^T\mathbf{x}_B = \mathbf{c}_B^T\mathbf{B}^{-1}\mathbf{b}$.

The revised simplex algorithm proceeds with four basic steps as described below, cf. Luenberger (1984).

1. Calculate the dual variables by solving $\mathbf{B}^T\mathbf{y} = \mathbf{c}_B$, that is $\mathbf{y}^T = \mathbf{c}_B^T\mathbf{B}^{-1}$. The dual cost is $\mathbf{y}^T\mathbf{b} = \mathbf{c}_B^T\mathbf{B}^{-1}\mathbf{b} = \mathbf{c}^T\mathbf{x}$. If $\mathbf{y}$ is feasible for (D) then the current solution $\mathbf{x}_B$ is optimal for (P) according to the duality theorem. Stop.

2. Determine a column $\mathbf{A}_{\alpha'}$ to enter the basis by choosing an index α' so that $\mathbf{y}^T\mathbf{A}_{\alpha'} < c_{\alpha'}$ (constraint violation for (D) in (3.5)). The standard pivoting rule is to choose

$$\alpha' = \arg\max_{\alpha} \{c_\alpha - \mathbf{y}^T\mathbf{A}_\alpha\} \tag{5.1}$$

3. Determine a column $\mathbf{A}_{\alpha_j}$ to leave the basis by the ratio test

$$j = \arg\min_{i:d_i>0} \left\{ \frac{x_{Bi}}{d_i} \right\} \tag{5.2}$$

   where the x_{Bi}s are the elements of the basic feasible solution $\mathbf{x}_B$ and the d_is are the coordinates of the entering column in terms of the old basis. Thus, $\mathbf{B}\mathbf{d} = \mathbf{A}_{\alpha'}$ where $\mathbf{d} = (d_i)$.

4. Update the basis matrix $\mathbf{B}$ by replacing the column $\mathbf{A}_{\alpha_j}$ for $\mathbf{A}_{\alpha'}$.

The revised simplex algorithm as outlined above assumes that an initial basic feasible solution to (P) is available. This is phase two in the two–phase simplex algorithm Luenberger (1984). In phase one, a basic feasible solution to (P) is found by solving the so called artificial minimization problem

$$\begin{cases} \min \mathbf{1}^T\mathbf{z} \\ \mathbf{A}\mathbf{x} + \mathbf{z} = \mathbf{b} \\ \mathbf{x} \geq 0 \\ \mathbf{z} \geq 0 \end{cases} \tag{5.3}$$

where $\mathbf{z}$ is an $m \times 1$ vector of *artificial* variables and $\mathbf{1}$ is a vector of ones.

The formulation (5.3) is in standard form and can be solved by the simplex algorithm. The initial basis is the identity matrix $\mathbf{B} = \mathbf{I}$, and the initial basic feasible solution is $\mathbf{z} = \mathbf{b} \geq 0$.

If there is a basic feasible solution to (P), then the optimal solution to (5.3) has the objective value zero. Phase one is terminated when all artificial variables have left the basis.

Since the initial basic feasible solution $\mathbf{b}$ is highly degenerate, cycling may occur for this case. It is therefore advantageous to apply the following simple anticycling rule: once an artificial variable has left the basis, it cannot enter again.

5.1 The Extended Simplex Algorithm

Considering (5.1) and the definitions for $c(\varphi, \theta)$, $\mathbf{A}(\varphi, \theta)$ given in (3.3) we get the explicit relations for the variables (φ', θ') corresponding to the column entering the basis

$$(\varphi', \theta') = \arg\max_{\varphi,\theta} \left\{ c(\varphi, \theta) - \mathbf{y}^T \mathbf{A}(\varphi, \theta) \right\} \tag{5.4}$$

where

$$\varphi' = \arg\max_{\varphi} \left\{ |H(\varphi)| \right\} \tag{5.5}$$

$$\theta' = -\arg\left\{ H(\varphi') \right\} \tag{5.6}$$

The relations (5.5) and (5.6) display the infinite-dimensional nature of the simplex extension algorithm.

6 NUMERICAL EXAMPLE

As an application example we consider the planar circular array as defined in section 1 with the array response given by (1.1). The design is obtained by using the semi-infinite simplex algorithm as described in Nordebo and Zang (1999). The effect of using a circular shield (or 'ground-plane') as indicated in Fig. 6.1a is investigated. The sensor responses $a_n(\varphi)$ correspond either to isotropic elements or dipole elements due to the shielding. Fig. 6.1a shows a typical equiripple solution (with shield) using 25 active out of 48 elements. The sidelobe levels are superior in comparison with the unshielded array Fig. 6.1b (-30 dB vs. -13 dB).

The Semi-infinite design method described is very flexible and it is capable of design in a more complex antenna array situation. An example of this is the linear-hexagonal antenna in Figs. 6.1c–6.1d. The figures shows the corresponding array pattern using 3 front sections of 6 in the hexagonal. Each linear section consists of 8 isotropic dipole elements. Fig. 6.1c illustrates a typical equiripple solution for a single mainlobe in the 0° direction. Fig. 6.1d illustrates the performance for the 3(6)-linear–hexagonal antenna for the directions $\{0°, 5°, 10°, 15°, 20°, 25°, 30°, 35°\}$. The loss in stopband attenuation performance between the 0° and 30° design is about 3 dB. All antenna arrays are designed for the frequency $f_0 = 450$ MHz and the element spacing is $d = \lambda/2 \approx 0.3$ meter.

7 SUMMARY

The lobe design of a narrow-band antenna array is formulated as a complex Chebyshev optimization problem and solved by a semi-infinite linear programming technique. The advantages of semi-infinite simplex extension algorithms are: computational efficiency (the constraint set is represented in functional form), accuracy (phase- discretization is not necessary) and flexibility with respect to the design specification (auxiliary linear constraints are easily included).

Numerical examples are included to demonstrate the proposed design technique. The effect of employing a shield, or "ground–plane" in conjunction with a planar circular (or hexagonal) array is investigated. The following is concluded about the stopband attenuation: shielding increases the performance dramatically, the choice of circular or hexagonal geometry is indifferent, and the steering angle can be changed with insignificant loss in performance

References

Alkhairy, A. S. and Christian, K. G. and Lim, J. S. (1993), Design and characterization of optimal FIR filters with arbitrary phase, *IEEE Transactions on Signal Processing*, Vol. 41, no. 2.

Anderson, E. J. and Nash, P. (1987), *Linear Programming in Infinite-Dimensional Spaces*, New York: Wiley.

Burnside, D. and Parks, T. W. (1995), Optimal design of FIR filters with the complex chebyshev error criteria, *IEEE Transactions on Signal Processing*, Vol. 43, no. 3, pp. 605-616.

Chen, X. and Parks, T. W. (1987), Design of FIR filters in the complex domain, *IEEE Transactions on Acoustics, Speech, and Signal Processing*, Vol. ASSP-35, no. 2, pp. 144-153.

Luenberger, D. G. (1984), *Linear and Nonlinear Programming*, Addison-Wesley.

Nordebo, S. and Zang, Z. (1999), Semi-infinite linear programming: a unified approach to digital filter design with time and frequency domain specifications, to appear in *IEEE Transactions on CAS-II: Analog and Digital Signal Processing*.

Parks, T. W. and Burrus, C. S. (1987), *Digital Filter Design*, John Wiley & Sons, Inc.

Rockafellar, R. T. (1969), *Convex Analysis*, Princeton University Press, Princeton, N. J.

Streit, R. L. and Nuttall, A. H. (1983), A note on the semi-infinite programming approach to complex approximation, *Mathematics of Computation*, Vol. 40, no. 162, pp. 599-605.

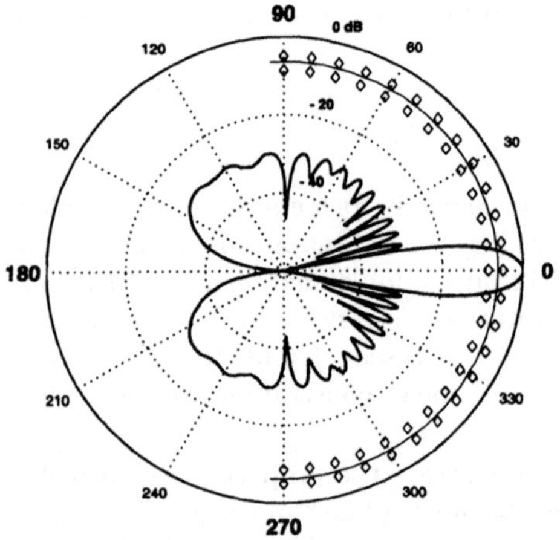

Figure 6.1a Circular shield and corresponding array pattern, using 25 of 48 elements, $f_0 = 450$ MHz, element spacing $d = \lambda/2$, $\varphi_s = 10°$, option: ground–plane, Look direction: $\varphi_p = 0°$.

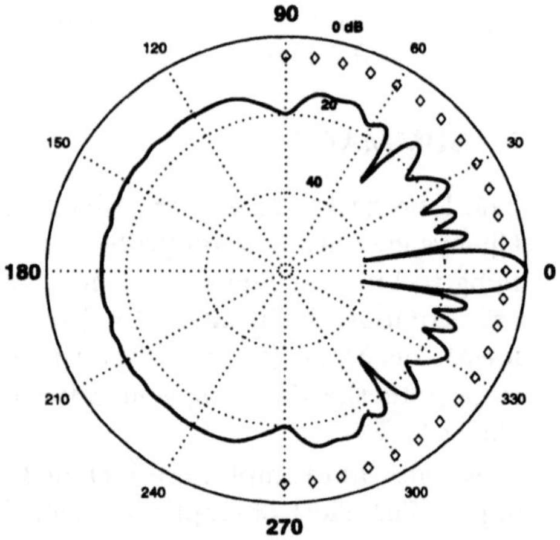

Figure 6.1b Same configuration as in (a) but <u>without</u> ground–plane.

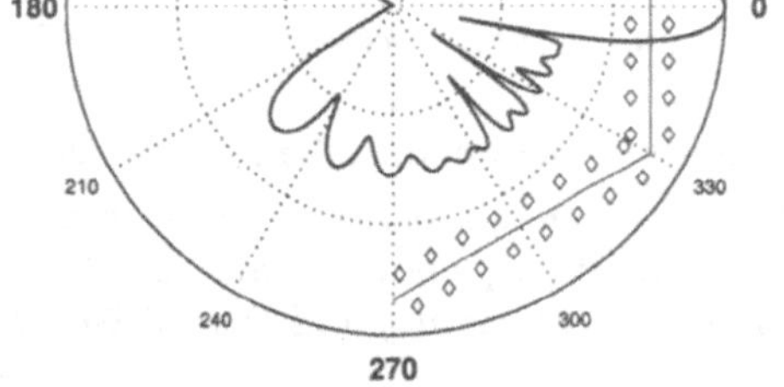

Figure 6.1c Linear-Hexagonal Antenna and the corresponding array pattern, using 3 sections of 6, $f_0 = 450$ MHz, element spacing $d = \lambda/2$, $\varphi_s = 10°$, option: ground–plane, Look direction: $\varphi_p = 0°$.

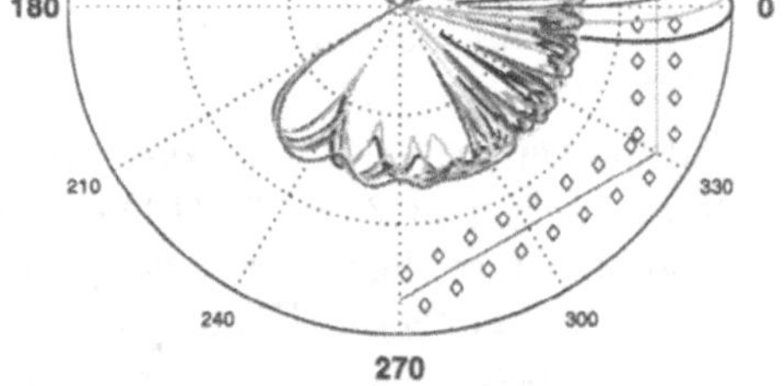

Figure 6.1d Same configuration as in (c) for look directions $\varphi_p = 0° \rightarrow 35°$.

Streit, R. L. and Nuttall, A. H. (1982), A general Chebyshev complex function approximation procedure and an application to beamforming, *Journal of the Acoustical Society of America*, Vol. 72, no. 1, pp. 181-190.

Van Veen, B. D. and Buckley, K. M. (1988), Beamforming: a versatile approach to spatial filtering, *IEEE ASSP Magazine*, pp. 4-24.

18 OPTIMUM POLE POSITION FOR DIGITAL LAGUERRE NETWORK WITH LEAST SQUARE ERROR CRITERION

H. H. Dam,

Australian Telecommunications Research Institute
Curtin University of Technology
Kent St., Bentley, WA 6102, Australia

Kok Lay Teo, Yanqun Liu

Department of Applied Mathematics
The Hong Kong Polytechnic University
Hong Kong

and S. Nordebo

Department of Signal Processing
University of Karlskrona/Ronneby
372 25, Ronneby, Sweden

Abstract: The digital Laguerre network has been studied as a means of implementing filters with a much smaller number of coefficients compared to Finite Impulse Response (FIR) filters. This paper considers the selection of the optimum pole for the Laguerre network. We propose a new effective algorithm for solving the optimum pole selection problem. Simulation results are presented, and comparison between the Laguerre structure and FIR filter structure are given.

Key words: Optimum pole, Laguerre structure, finite impulse response (FIR) filter, least square error, global optimization, frequency domain.

X. Yang et al. (eds.), Optimization Methods and Applications, 321-330.

1 INTRODUCTION

FIR filter structures are widely used in telecommunication systems due to their simplicity in construction, stability and cost–effectiveness. However, they usually require a large number of taps. On the other hand, it is known that Laguerre networks require fewer coefficients when compared with FIR filters, especially for specifications involving some over–sampling, see e.g Wahlberg (1991).

An important step in the application of the Laguerre network to filter design is to efficiently determine the best pole position for the Laguerre network. Recent approaches for solving this problem include finding all the stationary points for the Least Square Error (LSE) (Masnadi et al (1991), Masnadi-Shirazi et al (1995), Oliveira e Silva (1995)). The main difficulty is the possible existence of a large number of stationary points, and hence it can be very time consuming to obtain the global optimum solution.

In this paper, we consider the optimum pole selection problem for a digital Laguerre network structure with a frequency domain Least Square Error (LSE) criterion. The bridging method for global optimization problems developed in Liu (1998) is modified by incorporating a local search method such as the Golden Section method (Minoux (1986)). This yields an effective algorithm for locating the optimum pole with high accuracy. For illustration, the algorithm is applied to solve practical numerical examples.

2 PROBLEM FORMULATION

Let $\mathcal{H}_2$ denote the Hilbert space of all the complex–valued functions $F(z)$ which are analytic and square integrable outside the unit disc and equipped with the norm $\left(\frac{1}{2\pi}\oint |F(z)|^2 dz\right)^{1/2}$. Discrete time Laguerre functions form a complete orthogonal set with the following basis:

$$L_k(z,b) = \sqrt{1-b^2}\frac{(z^{-1}-b)^k}{(1-bz^{-1})^{k+1}} \tag{2.1}$$

where b is an unknown real pole value, $|b| \le 1$. Denote $z = e^{j\omega}$ on the unit circle. It is known that the sequence $\left\{L_k(e^{j\omega},b)\right\}_{k=0}^{\infty}$ forms an orthogonal basis in the Hardy space $\mathcal{H}_2$ (see McAulay et al (1971)), i.e.

$$\frac{1}{2\pi}\int_{-\pi}^{\pi} L_i(e^{j\omega},b)L_k^*(e^{j\omega},b)d\omega = \begin{cases} 1, & \text{if } i=k \\ 0, & \text{otherwise} \end{cases} \tag{2.2}$$

where $L_k^*(z,b)$ denotes the complex conjugate of $L_k(z,b)$. Thus, any desired frequency response $H_d(\omega)$ can be written as a linear combination of these basis functions

$$H_d(\omega) = \sum_{k=0}^{\infty} c_k(b)L_k(e^{j\omega},b), \tag{2.3}$$

where for each $k = 0, 1, 2, \ldots$

$$c_k(b) = \frac{1}{2\pi} \int_{-\pi}^{\pi} H_d(w) L_k^*(e^{j\omega}, b) dw.$$

Clearly, $c_k(b)$ can be written as:

$$c_k(b) = \frac{\sqrt{1-b^2}}{2\pi} \int_{-\pi}^{\pi} H_d(\omega) \frac{(e^{j\omega} - b)^k}{(1 - be^{j\omega})^{k+1}} d\omega. \tag{2.4}$$

These $c_k(b)$ are called the Laguerre's coefficients.

A finite Laguerre network with $L+1$ stages is realizable as shown in Figure 2.1. Its frequency response is obtained by truncating the infinite sequence in (2.3):

$$H(w) = \sum_{k=0}^{L} c_k(b) L_k(e^{j\omega}, b). \tag{2.5}$$

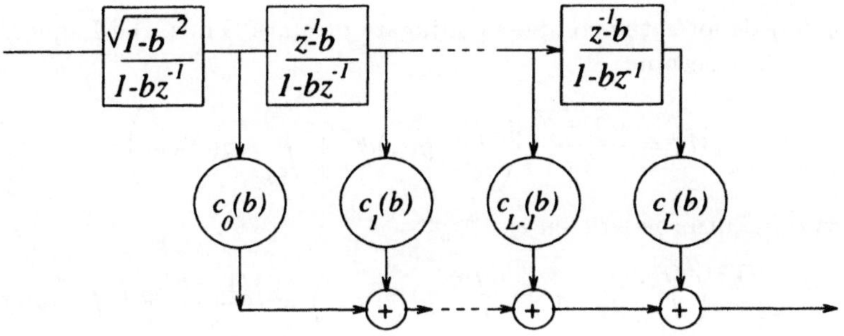

Figure 2.1 Laguerre filter network.

The cost function considered in this paper for determining the optimal Laguerre filter is the integral of the squared error between the frequency response $H(w)$ and a specified desired response $H_d(w)$. The cost function to be minimized is given by

$$\epsilon(b) = \frac{1}{2\pi} \int_{-\pi}^{\pi} |H_d(\omega) - H(\omega)|^2 d\omega,$$

and from (2.2)–(2.5), it can be written in a simpler form

$$\epsilon(b) = \frac{1}{2\pi} \int_{-\pi}^{\pi} |H_d(\omega)|^2 d\omega - \sum_{k=0}^{L} |c_k(b)|^2. \tag{2.6}$$

Here we considered only the case where the desired frequency response is a low–pass filter. Other cases such as high–pass and band–pass filters can be handled similarly.

The desired frequency response for a linear phase low–pass filter with constant *group delay* τ is:

$$H_d(\omega) = \begin{cases} e^{-j\omega\tau}, & \omega \in \mathcal{P} \\ 0, & \omega \in \mathcal{S} \end{cases}$$

where $\mathcal{P} = [-\omega_p, \omega_p]$ and $\mathcal{S} = \{[-\pi, -\omega_p], [\omega_p, \pi]\}$ are pass and stop bands, respectively. Thus, each coefficient $c_k(b)$ in (2.4) becomes

$$c_k(b) = \frac{\sqrt{1-b^2}}{2\pi} \int_{-\omega_p}^{\omega_p} e^{-j\omega\tau} \frac{(e^{j\omega} - b)^k}{(1 - be^{j\omega})^{k+1}} d\omega. \tag{2.7}$$

Denote

$$g(\omega) = e^{-j\omega\tau} \frac{(e^{j\omega} - b)^k}{(1 - be^{j\omega})^{k+1}}.$$

Obviously,

$$g(-\omega) = g^*(\omega),$$

where $g^*(\omega)$ denotes the complex conjugate of $g(\omega)$. Thus, the Laguerre coefficients in (2.7) become

$$c_k(b) = \frac{\sqrt{1-b^2}}{2\pi} \left(\int_0^{\omega_p} g(\omega) d\omega + \int_{-\omega_p}^{0} g(\omega) d\omega \right),$$

which can, in turn, be written as:

$$c_k(b) = \frac{\sqrt{1-b^2}}{2\pi} \left(\int_0^{\omega_p} g(\omega) d\omega + \int_0^{\omega_p} g^*(\omega) d\omega \right) = \frac{\sqrt{1-b^2}}{\pi} \Re \left(\int_0^{\omega_p} g(\omega) d\omega \right), \tag{2.8}$$

where $\Re\{.\}$ denotes the real value of the function inside. Therefore, the Laguerre coefficients obtained from (2.8) are real. Note that the integral involved in (2.8) is very complicated and it is very difficult to obtain the analytical solution. Thus, it is necessary to use numerical methods to obtain the value of the integration for each value of b. Details will be given in Section 4.

3 OPTIMIZATION METHOD

The problem of finding the optimum pole position for the error function (2.6), that is: $\min_{-1 \le b \le 1} \epsilon(b)$, has been a problem for many researchers. In Masnadi-Shirazi et al (1995) and Oliveira e Silva (1995), the global solution is obtained by finding all the stationary points of the error function $\epsilon(b)$ over $[-1, 1]$. More specifically, the first order derivative of LSE is taken. By using induction, it can be shown that this derivative depends only on $c_L(b)$ and $c_{L+1}(b)$, where $c_{L+1}(b)$ can be obtained from (2.4) for $k = L + 1$. More specifically,

$$\frac{d\epsilon(b)}{db} = -\frac{2(L+1)}{1-b^2} c_L(b) c_{L+1}(b).$$

For more information, see Masnadi-Shirazi et al (1995). The search for all the stationary points of LSE is thus equivalent to finding all the roots of the functions $c_L(b)$ and $c_{L+1}(b)$. Unfortunately, these functions are highly non-linear and analytical roots are very difficult to obtain. Thus, these functions are approximated in Masnadi-Shirazi et al (1995) and Oliveira e Silva (1995) by using the Taylor series expansion.

There are a number of limitations for finding the roots of a non-linear function. If the non-linear function is approximated by a polynomial, it must be a polynomial of high enough degree so as to achieve an acceptable accuracy. Thus, the calculation of all zeros of the approximating high order polynomial is required. This can be a very time consuming task and may yield inaccurate estimates.

In this paper, we propose a global optimization method for solving the non-linear optimization problem: $\min_{b\in[-1,1]} \epsilon(b)$. The method is based on a modification of the bridging method developed in Liu (1998) by incorporating a local search method such as the Golden Section method (Minoux (1986)). The main idea of the *bridging method* is to use the bridged functions in the search for the global minimum so that the method does not stuck at a local minimum point. The bridging method finds a local minimum point of the bridged function, starting from one end of the feasible region, and updates the bridged function every time a local solution is obtained until the global minimum of the original cost function is found. Since the method requires the continuous differentiability properties of the *bridged function*, a *smoothing procedure* is introduced in every iteration.

By incorporating a local search method (for example Golden Section Method), the bridging method can be improved so that no smoothing procedure is needed. Thus, it is applicable to cost functions which are only continuous instead of continuously differentiable. This method also works more efficiently, in that computing effort is substantially reduced. The algorithm is developed for a general one dimensional problems of the form

$$\textbf{Problem P:} \quad \begin{array}{l} \min f(x) \\ \text{subject to } x \in [x_1, x_2]. \end{array} \tag{3.1}$$

where $f(x)$ is a continuous function on $[x_1, x_2]$.

For any $x_0 \in [x_1, x_2]$, define the following minimization problem:

$$\textbf{Problem P}_r(x_0): \quad \min_{x\in[x_0,x_2]} b_{x_0}(x)$$

where $b_{x_0}(x) = \min\{f(x), f(x_0)\}$ is called the *bridged* function. Now, we describe the modified bridging algorithm as follows.

- *Step 0: Specify the step size h.*
- *Step 1: Initialize $x_0 = x_1$.*
- *Step 2: Find the left most local optimum solution x^* of $b_{x_0}(x)$. This is done by searching for the smallest value of i such that:*

$$b_{x_0}(x_0 + ih) < b_{x_0}(x_0 + (i+1)h), \quad x_0 + (i+1)h \le x_2. \tag{3.2}$$

Clearly,

$$b_{x_0}(x_0 + ih) \leq b_{x_0}(x_0 + (i-1)h). \tag{3.3}$$

If i does not exist, x^* is the global solution and stop the procedure. Otherwise, since $b_{x_0}(x_0) = f(x_0) \geq b_{x_0}(x_0 + h)$, it is clear that $i > 0$. By (3.2)-(3.3), there exists a local minimum in the interval $[x_o+(i-1)h, x_o+(i+1)h]$. Local minimization method such as Golden Section method can be used to locate this local minimum. We call this minimum x^* and go to the next step.

- *Step 3: Set $x_0 = x^*$. If $|x_2 - x^*| > h$ then return to Step 2, otherwise x^* is the global minimum and stop the procedure.*

Note that for a small step size h, the solution obtained from Step 2 is the global minimum solution of the function $f(x)$ over the interval $[x_0, x^*]$. Obviously, by introducing the *bridged function* at x_0, we can eliminate local minima of the original cost function $f(x)$ with values larger than $f(x_0)$.

Golden Section Search method is used for locating the minimum point in Step 3. The maximum error for the pole position using this method is $(0.618)^n \times 2h$ (Minoux (1986)), where h is the step size and n is the number of iterations for locating the local minimum. For this problem, the step size $h = 0.01$ is used. Thus, for the accuracy of up to 10^{-4}, 7–8 iterations are required in each minimization procedure.

The main advantage of this algorithm is that it only find a small number of useful local minima. Once a global solution is obtained, the procedure moves quickly through the rest of the interval with the step size h. Clearly, h needs to be small enough so as to avoid missing local minima between two consecutive points.

4 SIMULATION STUDIES

The above algorithm can be applied to the problem of selecting the optimum pole for the Laguerre network, namely, the selection of the value of the pole that yields the minimum LSE. Obviously, $c_k(b)$ in (2.8) is continuous with respect to the pole b. Thus, $\epsilon(b)$ is a continuous function in the interval $[-1, 1]$.

Numerical integration is used to obtain the LSE for each pole. Methods such as Simpson rule or Trapezoidal rule or Romberg integration can be used for this purpose (for details, see Atkinson et al (1989)). In this paper, we used Simpson rule with a partition of 100 points equally distributed in the interval $[0, \omega_p]$.

Example 1: Consider a low pass filter with $\omega_p = 0.25\pi$, i.e. $f_p = 0.125$ and the group delay $\tau_d = 16$. The total number of stages for the Laguerre structure is 18, i.e. $L = 17$.

Figure 4.1 plots the cost function, $10\log_{10}(\epsilon(b))$ [dB], as a function of b in $[-1, 1]$. From the plot, 8 local minima are obtained and the last one is the global solution. Once a local minimum is obtained, the *bridged function* $b_{x_0}(x)$

is modified. The value of the *bridged function* is taken as the minimum of the dash straight line (start from the local minimum) and the objective function.

Figure 4.2 plots the approximating polynomials of degree 49 and 50 for $c_{17}(b)$ and $c_{18}(b)$, with the dash and the solid curves respectively. This polynomials are obtained by using the Taylor series expansion in Masnadi-Shirazi et al (1995). The plot shows that there are about 30 zeros in the interval $[-1,1]$ for both polynomials. These zeros are matched with the local minima, maxima and saddle points of the original plot as shown in Figure 4.1. Hence, finding the positions of all these zeros as required by the methods proposed in Masnadi-Shirazi et al (1995) are time consuming.

Figure 4.3 and 4.4 plot the group delays in the pass band and the frequency responses, respectively, for the Laguerre structure and the FIR filter. The Laguerre structure with the optimum pole $b = 0.4016$ and FIR filter with 26 coefficients or $L = 25$ are indicated by the solid and dash lines, respectively. The group delays for both filters fluctuate around the desired group delay $\tau_d = 16$ with approximately 10% error. From the plots, there is a trade off between the frequency responses and group delay errors for the Laguerre structure and the FIR filter.

Example 2: The Laguerre structure has clearly advantages over the FIR structure when f_p is small, i.e. the filter has narrow pass band. Consider a low pass filter with $w_p = 0.15\pi$, i.e. $f_p = 0.075$ with the design group delay $\tau_d = 18$. Figure 4.5 plots the cost function $10\log_{10}\epsilon(b)$ [dB] for the Laguerre network with 21 coefficients or $L = 20$ using the bridging method. The Laguerre network with the minimum least square error has the optimum pole $b = 0.5927$. The FIR structure requires 48 coefficients, i.e. $L = 47$ for the same least square error.

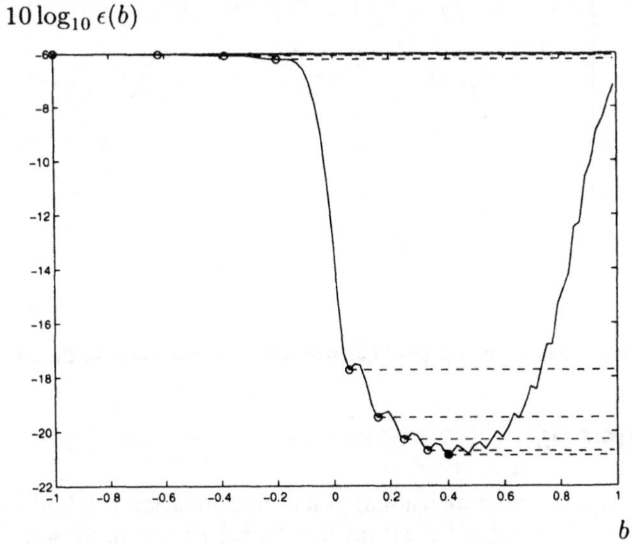

Figure 4.1 Plot of $10\log_{10}(\epsilon(b))$ for b in $[-1,1]$ for a low pass filter with $f_p = 0.125$.

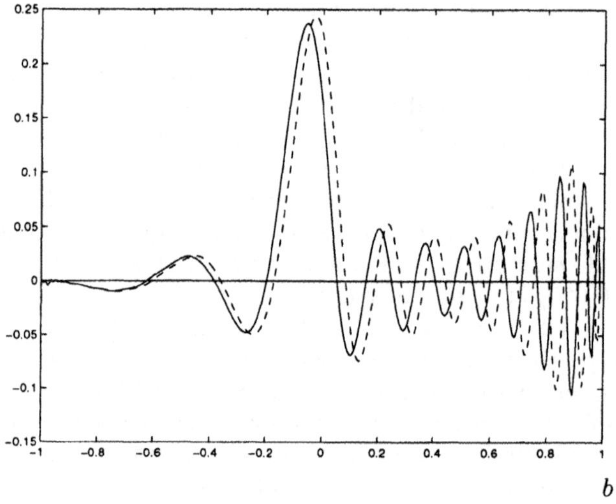

Figure 4.2 Plots of the two polynomials approximating $c_{17}(b)$ and $c_{18}(b)$

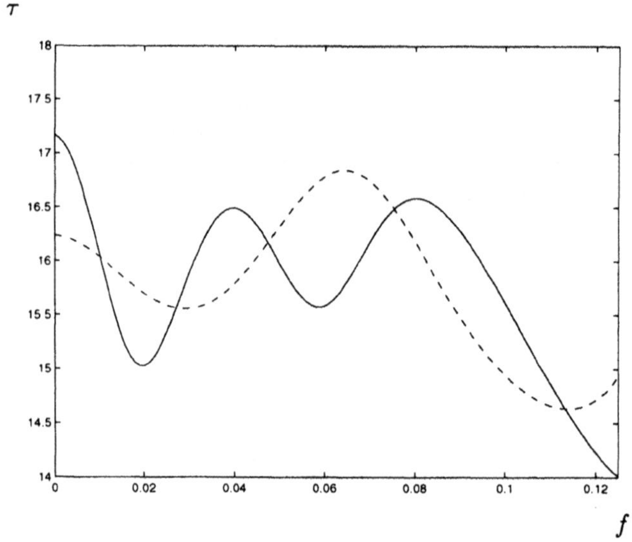

Figure 4.3 Group delay plot for the Laguerre filter and FIR filter in the pass band

5 CONCLUSIONS

In this paper, a new one dimensional global optimization method was proposed to obtain the optimum pole position for digital Laguerre network with Least Square error criterion in the frequency domain. This method does not require high order polynomial approximations and can locate the optimum pole position accurately and efficiently. Simulation results show that the Laguerre

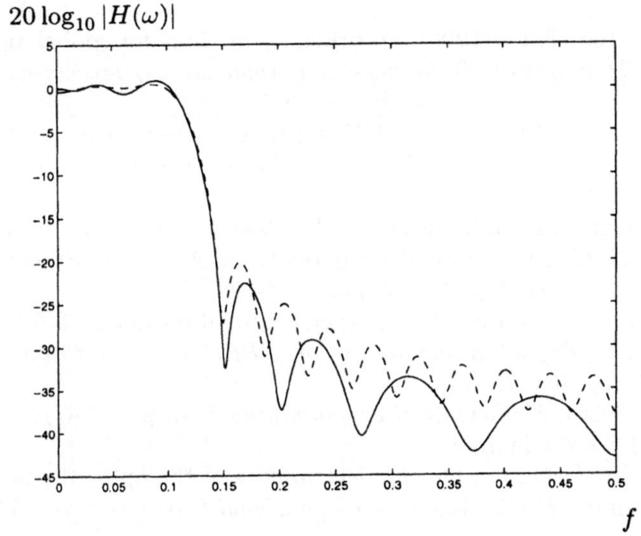

Figure 4.4 Frequency response plots for the Laguerre and FIR filters

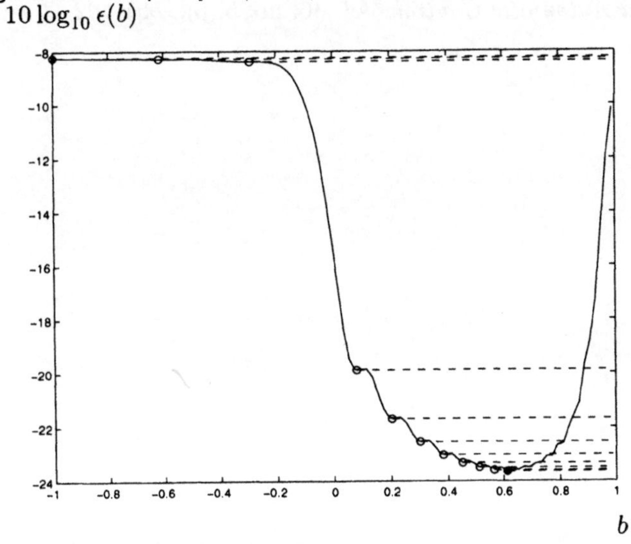

Figure 4.5 Plot of $10\log_{10}(\epsilon(b))$ for b in $[-1, 1]$ for a low pass filter with $f_p = 0.075$.

structure requires a smaller number of coefficients compared to the FIR filter structure, especially for specifications involving over sampling.

References

Atkinson, L.V., Harley, P.J. and Hudson, J.D. (1989), *Numerical Methods with Fortran 77, A Practical Introduction*; Wiley.

Liu, Y. and Teo, K.L. (1998), A bridging method for global optimization, *ICOTA (Optimization Techniques and Applications) proceeding*, Vol. 2, pp. 558-565.

Masnadi, M.A. and Ahmed, N. (1991), Optimum Laguerre networks for a class of discrete-time systems, *IEEE Transactions on Signal Processing*, Vol. 39, no. 9, pp. 2104-2108, Sep.

Masnadi-Shirazi, M.A. and Ghasemi, M. (1995), Laguerre digital filter design, *ICASSP, IEEE International Conference on Acoustics, Speech and Signal Processing - Proceedings*, Vol. 2. pp. 1284-1287, May.

McAulay, R.J. and Johnson, J.R. (1971), Optimal mismatched filter design for radar ranging detection and resolution, *IEEE Trans. Information Theory*, Vol. IT-17, pp. 696-701.

Minoux, M. (1986), *Mathematical Programming Theory and Applications*, John Wiley and Sons, Chixhester.

Oliveira e Silva, T. (1995), On the determining of the optimal pole position of Laguerre filters, *IEEE Transactions on Signal Processing*, Vol. 43, no. 9, pp. 2079-2087.

Wahlberg, B. (1991), System identification using Laguerre models, *IEEE Transactions on Automatic Control*, Vol. 36, no. 5, pp. 551-562.

19 A MARKOV MODEL FOR THE STOCHASTIC OPTIMAL CONTROL OF A SOLAR POWERED CAR

Phil Howlett

Centre for Industrial and Applicable Mathematics (CIAM)
University of South Australia
email p.howlett@unisa.edu.au

Abstract: The World Solar Challenge is a 3000km race for solar powered cars across the Australian continent from Darwin to Adelaide. Each car is powered by a panel of photovoltaic cells which convert solar irradiance to electrical power. The power can be used directly to drive the car or stored for later use. The solar radiation is modelled as a Markov process with a continuous state space. We use the Bellman Principle of dynamic programming to find an optimal control policy that maximises the expected value of the total distance to be travelled by the car on the remaining race days. We illustrate the general results with some basic examples.

Key words: Solar powered racing cars, stochastic optimal control, Markov processes, Bellman Principle.

X. Yang et al. (eds.), Optimization Methods and Applications, 331-341.

1 INTRODUCTION

It is said that a modern solar powered racing car can travel for over two kilometres at a speed of one hundred kilometres per hour on the amount of energy used by an electric toaster to toast one slice of bread. But this efficiency is borne of necessity. The cars must race using the energy they collect and so it is important to collect as much energy as possible. It is also important to use that energy efficiently. Studies by Gates and Westcott (1996), Howlett *et al* (1997) and Howlett and Pudney (1998) treated the solar car problem as a deterministic control problem where the solar radiation is known in advance. In each case the idea was to find driving strategies that maximised the daily distance travelled by the car. For both flat track and undulating track the optimal driving strategy was found to be a modified speedholding strategy.

There is no simple formula for the distance travelled in an optimal strategy but it is reasonable to assume that the distance travelled each day increases as the daily energy use increases. In fact we assume that the distance travelled is a known function of the energy used. Although the future solar radiation is not known we assume that solar radiation evolves in a probabilistic sense according to a known Markov transition probability. The state variables for this problem are the number of days remaining in the race, the amount of energy in the battery and the amount of solar energy collected on the previous day. The control variable is the amount of energy we decide to use on any given day. We formulate the following intuitive problem.

Problem 1.1 *For each allowable configuration of the state variables find a driving strategy that maximises the expected value of the total distance to be travelled by the solar car on the remaining race days.*

We use the Bellman Principle of dynamic programming to find a stochastic control policy that maximises the expected total distance travelled on the remaining race days. We will show, at least in some special cases, that a long term optimal strategy should use the same amount of energy each day. Since a long term strategy must be sustainable it is intuitively clear that the energy used each day must be equal to the average solar energy collected each day. The results in this paper extend the work in Boland *et al* 1999 to more general probability distributions.

2 FORMULATION

When energy r is supplied to the motor we assume that the distance travelled by the solar car is $x(r)$ where $x : [0,\infty) \mapsto [0,\infty)$ is strictly concave and increasing with $x(0) = 0$. We assume that $x' : (0,\infty) \mapsto (0,\infty)$ is continuous and decreasing with $x'(r) \downarrow 0$ as $r \uparrow \infty$ and that $x''(r) : (0,\infty) \mapsto (-\infty,0)$

is continuous. We assume that the solar energy collected each day is a random variable s which can take only non-negative values $s \in S = [0, M]$ with transition probabilities

$$P\{u, [0, v]\} = P\{s = u \to s \in [0, v]\}$$

for each $(u, v) \in S \times S$. Define $F_u : S \mapsto [0, 1]$ for each $u \in [0, M]$ by setting

$$F_u(v) = P\{u, [0, v]\}$$

for each $v \in S$. Note that F_u is monotone non-decreasing with bounded total variation and

$$\int_S dF_u(v) = 1$$

for each $u \in S$. We define value functions $W_0[u] : S \times [0, \infty) \to [0, \infty)$ for each $u \in S$ at the beginning of day 0, the last day of the race, as the expected distance to be travelled on day 0 given that the energy collected on day 1, the previous day, was u and that all energy will be used by the end of day 0. If b is the energy in the battery at the beginning of the last day then

$$W_0[u](b) = \int_S x(b + v_0) dF_u(v_0). \tag{2.1}$$

In general we write day n for the day that precedes day 0 by n and we wish to define value functions $W_n[u] : [0, \infty) \to [0, \infty)$ for each $u \in S$ given that the solar energy collected on day $n+1$ was u. We begin by defining auxilliary value functions $w_n[u] : [0, \infty) \times [0, \infty) \to [0, \infty)$ for each $u \in S$ at the beginning of day n given that the solar energy collected on day $n + 1$ was u by setting

$$w_n[u](b, r) = x(r) + \int_S W_{n-1}[v_n](b + v_n - r) dF_u(v_n). \tag{2.2}$$

The value $W_n[u](b)$ is obtained by maximising $w_n[u](b, r)$ over all possible values of r. We wish to find a control policy so that the value function is maximised. We will solve the following mathematical problem.

Problem 2.1 *For each state (n, u, b) find $r = \varphi_n[u](b)$ such that*

$$W_n[u](b) = \max_r w_n[u](b, r).$$

3 A RECURSIVE EQUATION FOR THE OPTIMAL CONTROLS

For each $u \in [0, M]$ we have

$$w_1[u](b, r) = x(r) + \int_S \left[\int_S x([b + v_1 - r] + v_0) dF_{v_1}(v_0) \right] dF_u(v_1).$$

We maximise $w_1[u](b,r)$ over all r by setting the partial derivative with respect to r equal to zero. If $r = \varphi_1[u](b)$ denotes the energy usage that gives the maximum of the auxilliary value function then we have the following result.

Lemma 3.1 *For each $u \in S$ the value $W_1[u](b)$ is given by the formula*

$$\begin{aligned} W_1[u](b) &= x(\varphi_1[u](b)) + \\ &\int_S \left[\int_S x([b + v_1 - \varphi_1[u](b)] + v_0) dF_{v_1}(v_0) \right] dF_u(v_1) \end{aligned}$$

where the optimal energy consumption $\varphi_1[u](b)$ satisfies the equation

$$x'(\varphi_1[u](b)) = \int_S \left[\int_S x'([b + v_1 - \varphi_1[u](b)] + v_0) dF_{v_1}(v_0) \right] dF_u(v_1).$$

We can also prove a general result.

Lemma 3.2 *For each integer $n > 1$ and each $u \in S$ the value $W_n[u](b)$ is given by the formula*

$$W_n[u](b) = x(\varphi_n[u](b)) + \int_S W_{n-1}[v_n](b + v_n - \varphi_n[u](b)) dF_u(v_n)$$

where the optimal energy consumption $\varphi_n[u](b)$ satisfies the recursive equation

$$x'(\varphi_n[u](b)) = \int_S x'(\varphi_{n-1}[v_n](b + v_n - \varphi_n[u](b))) dF_u(v_n).$$

Remark 3.1 *For convenience we will often write $\psi_n[u](b) = b - \varphi_n[u](b)$ for each natural number n and each $u \in [0, M]$.*

Proof of Lemma 3.2. By setting the partial derivative with respect to r of $w_2[u](b,r)$ equal to zero we obtain

$$x'(r) = \int_S W_1[v_2]'(b + v_2 - r) dF_u(v_2). \tag{3.1}$$

By differentiating the equation

$$\begin{aligned} W_1[v](b + v - r)) &= x(\varphi_1[v](b + v - r)) + \\ \int_S \Bigg[\int_S & x(\psi_1[v](b + v - r) + v_1 + v_0) dF_{v_1}(v_0) \Bigg] dF_v(v_1) \end{aligned}$$

with respect to r and by using Lemma 3.1 it follows that

$$W_1[v]'(b+v-r) = x'(\varphi_1[v](b+v-r)).$$

Substituting this expression into equation (3.1) and writing $r = \varphi_2[u](b)$ for the point where the maximum is attained gives

$$x'(\varphi_2[u](b)) = \int_S x'(\varphi_1[v_2](b+v_2-\varphi_2[u](b)))dF_u(v_2).$$

The general result can be proved in a similar manner by induction. We will suppose, for some integer $n > 1$, that

$$W_m[u](b) = x(\varphi_m[u](b)) + \int_S W_{m-1}[v_m](b+v_m-\varphi_m[u](b))dF_u(v_m) \quad (3.2)$$

and

$$x'(\varphi_m[u](b)) = \int_S x'(\varphi_{m-1}[v_m](b+v_m-\varphi_m[u](b)))dF_u(v_m) \quad (3.3)$$

for each integer $m < n$. We will also suppose that

$$W_m[v]'(b+v-r) = x'(\varphi_m[v](b+v-r)) \quad (3.4)$$

for all integers $m < n$. By setting the partial derivative with respect to r of $w_n[u](b,r)$ equal to zero we obtain

$$x'(r) = \int_S W_{n-1}[v_n]'(b+v_n-r)dF_u(v_n)$$

when $r = \varphi_n[u](b)$. Since the inductive hypothesis gives

$$W_{n-1}[v]'(b+v-r) = x'(\varphi_{n-1}[v](b+v-r))$$

we can now deduce that

$$x'(\varphi_n[u](b)) = \int_S x'(\varphi_{n-1}[v_n](b+v_n-\varphi_n[u](b)))dF_u(v_n). \quad (3.5)$$

We also know that $W_n[u](b) = w_n[u](b, \varphi_n[u](b))$ and hence

$$W_n[u](b) = x(\varphi_n[u](b)) + \int_S W_{n-1}[v_n](b+v_n-\varphi_n[u](b))dF_u(v_n). \quad (3.6)$$

By differentiating the equation

$$W_n[v](b+v-r) = x(\varphi_n[v](b+v-r)) + \int_S W_{n-1}[v_n](\psi_n[v](b+v-r)+v_n)dF_v(v_n)$$

with respect to r we get

$$W_n[v]'(b+v-r) = x'(\varphi_n[v](b+v-r))\varphi_n[v]'(b+v-r)+$$
$$\int_S W_{n-1}[v_n]'(\psi_n[v](b+v-r)+v_n)(1-\varphi_n[v]'(b+v-r))dF_v(v_n).$$

But we have established that

$$\begin{aligned} & x'(\varphi_n[v](b+v-r)) \\ &= \int_S x'(\varphi_{n-1}[v_n](\psi_n[v](b+v-r)+v_n)dF_v(v_n) \\ &= \int_S W_{n-1}[v_n]'(\psi_n[v](b+v-r)+v_n)dF_v(v_n) \end{aligned}$$

and hence we deduce that

$$W_n[v]'(b+v-r) = x'(\varphi_n[v](b+v-r)). \tag{3.7}$$

The equations (3.5), (3.6) and (3.7) show that the inductive hypothesis is also true for $m = n$. This completes the proof. □

4 THE PROPERTIES OF THE OPTIMAL CONTROLS

To determine the properties of the optimal controls we recall the necessary conditions given in Lemma 3.1. We can prove the following results.

Lemma 4.1 *For each $u \in S$ the value $\varphi_1[u](b)$ is uniquely defined with*

$$b/2 < \varphi_1[u](b) < b/2 + M$$

for all $b \in [0, \infty)$.

Proof of Lemma 4.1. Choose $b \in [0, \infty)$ and a fixed value of $u \in S$. Define a function $s_1 : [b/2, b/2+M] \mapsto (0, \infty)$ by the formula

$$s_1(r) = \int_S \left[\int_S x'((b+v_1-r)+v_0)dF_u(v_1) \right] dF_{v_1}(v_0)$$

for all $r \in [b/2, b/2+M]$. Clearly $s_1(b/2+M) < x'(b/2)$ and $s_1(b/2+M) > x'(b/2+M)$. Since $x'(r)$ is continuous there is at least one value $r_1 \in (b/2, b/2+M)$ such that $s_1(r_1) = x'(r_1)$. If $r > r_1$ then it is easy to show that $s_1(r) > s_1(r_1)$ and if $r < r_1$ then $s_1(r) < s_1(r_1)$. Therefore the value $r_1 = \varphi_1[u](b)$ is unique. Since

$$x'(\varphi_1[u](b)) = \int_S \left[\int_S x'(b+v_1-\varphi_1[u](b)+v_0)dF_u(v_1) \right] dF_{v_1}(v_0)$$

and since $x'(\cdot)$ is decreasing it follows that the argument on the left hand side is between the extreme values of the arguments on the right hand side. Thus

$$b - \varphi_1[u](b) < \varphi_1[u](b) < b + 2M - \varphi_1[u](b)$$

and the desired result follows from a simple rearrangement. □

Lemma 4.2 *For each $u \in [0, M]$ the function $\varphi_1[u] : [0, \infty) \mapsto (0, \infty)$ is differentiable with $0 < \varphi_1[u]'(b) < 1$ for all $b \in (0, \infty)$.*

Proof of Lemma 4.2. For $b < c$ we have

$$x'(\varphi_1[u](c)) - x'(\varphi_1[u](b)) =$$

$$\int_S \left[\int_S [x'(\psi_1[u](c) + v_1 + v_0) - x'(\psi_1[u](b) + v_1 + v_0)] dF_{v_1}(v_0) \right] dF_u(v_1).$$

¿From the mean value theorem we can find $\epsilon = \epsilon(b, c)$ between $\varphi_1[u](b)$ and $\varphi_1[u](c)$ and $\delta = \delta(b, c, v_1, v_0)$ between $\psi_1[u](b) + v_1 + v_0$ and $\psi_1[u](c) + v_1 + v_0$ so that the previous equation can be rearranged to give

$$\frac{\varphi_1[u](c) - \varphi_1[u](b)}{c - b} = \frac{\int_S \left[\int_S x''(\delta) dF_{v_1}(v_0) \right] dF_u(v_1)}{x''(\epsilon) + \int_S \left[\int_S x''(\delta) dF_{v_1}(v_0) \right] dF_u(v_1)}.$$

By taking the limit as $c \downarrow b$ we can see that

$$\varphi_1[u]'(b) = \frac{\int_S \left[\int_S x''(\psi_1[u](b) + v_1 + v_0) dF_{v_1}(v_0) \right] dF_u(v_1)}{x''(\varphi_1[u](b)) + \int_S \left[\int_S x''(\psi_1[u](b) + v_1 + v_0) dF_{v_1}(v_0) \right] dF_u(v_1)}$$

which implies $0 < \varphi_1[u]'(b) < 1$ for all $b \in [0, \infty)$. This completes the proof. □

We use induction on n and a similar argument to find corresponding properties for the functions $\varphi_n[u](\cdot)$ defined for each $n > 1$ and each $u \in [0, M]$ by the recursive equations of Lemma 3.2. We have the following two results.

Lemma 4.3 *For each $n > 1$ and each $u \in [0, M]$ the value $\varphi_n[u](b)$ is uniquely defined by the recursive equation of Lemma 3.2 with*

$$b/(n+1) < \varphi_n[u](b) < b/(n+1) + M$$

for all $b \in [0, \infty)$.

Lemma 4.4 *For each $n > 1$ and each $u \in [0, M]$ the function $\varphi_n[u] : [0, \infty) \mapsto (0, \infty)$ is differentiable with $0 < \varphi_n[u]'(b) < 1$ for all $b \in (0, \infty)$.*

5 SOME ELEMENTARY EXAMPLES

Example 5.1 *Consider the case where the function $F_u : [0, M] \mapsto [0, 1]$ is given by*

$$F_u(v) = \chi_{[u,M]}(v) = \begin{cases} 0 & \text{when } v < u \\ 1 & \text{when } v \geq u \end{cases}$$

This is actually a deterministic situation. The optimal strategy is given by the equations

$$\begin{aligned} x'(\varphi_1[u](b) &= \int_S \left[\int_S x'(b + v_1 - \varphi_1[u](b) + v_0) d\chi_{(v_1,M]}(v_0) \right] d\chi_{(u,M]}(v_1) \\ &= \int_S x'(b + 2v_1 - \varphi_1[u](b)) d\chi_{(u,M]}(v_1) \\ &= x'(b + 2u - \varphi_1[u](b)) \end{aligned}$$

and

$$\begin{aligned} x'(\varphi_n[u](b) &= \int_S x'(\varphi_{n-1}[v_n](b + v_n - \varphi_n[u](b))) d\chi_{(u,M]}(v_n) \\ &= x'(\varphi_{n-1}[u](b + u - \varphi_1[u](b))) \end{aligned}$$

for $n > 1$. Since x is strictly monotone increasing these equations are equivalent to the equations

$$\varphi_1[u](b) = b + 2u - \varphi_1[u](b)$$

and

$$\varphi_n[u](b) = \varphi_{n-1}[u](b + u - \varphi_n[u](b))$$

for $n > 1$. The solution to these equations is given by

$$\varphi_n[u](b) = b/(n+1) + u$$

for each $u \in [0, M]$ and each natural number n.

Example 5.2 *Consider the case where the function $F_u : [0, M] \mapsto [0, 1]$ is given by*

$$F_u(v) = \chi_{[M-u,M]}(v) = \begin{cases} 0 & \text{when } v < M - u \\ 1 & \text{when } v \geq M - u \end{cases}$$

Once again this is a deterministic situation. The optimal strategy is given by the equations

$$\begin{aligned} & x'(\varphi_1[u](b)) \\ &= \int_S \left[\int_S x'(\psi_1[u](b) + v_1 + v_0) d\chi_{(M-v_1,M]}(v_0) \right] d\chi_{(M-u,M]}(v_1) \\ &= \int_S x'(\psi_1[u](b) + M) d\chi_{(M-u,M]}(v_1) \\ &= x'(\psi_1[u](b) + M) \end{aligned}$$

and

$$\begin{aligned} x'(\varphi_n[u](b)) &= \int_S x'(\varphi_{n-1}[v_n](\psi_n[u](b) + v_n)) d\chi_{(M-u,M]}(v_n) \\ &= x'(\varphi_{n-1}[u](\psi_n[u](b) + M - u)) \end{aligned}$$

for $n > 1$. *Since* x *is strictly monotone increasing these equations are equivalent to the equations*

$$\varphi_1[u](b) = b + M - \varphi_1[u](b)$$

and

$$\varphi_n[u](b) = \varphi_{n-1}[u](b + M - u - \varphi_n[u](b))$$

for $n > 1$. *The solution to these equations is given by*

$$\varphi_{2p-1}[u](b) = b/(2p) + M/2$$

and

$$\varphi_{2p}[u](b) = b/(2p+1) + pM/(2p+1) - u/(2p+1)$$

for each $u \in [0, M]$ *and each natural number* p.

We need an elementary result in the next example.

Lemma 5.1 *Let* $\theta > 0$ *be a constant. The function* $f : (0, \infty) \mapsto (0, \infty)$ *defined by the formula*

$$f(r) = \left[\frac{\sinh r\theta}{r\theta}\right]^{1/r}$$

is strictly increasing.

Proof of Lemma 5.1. By differentiating we find that $f(r)$ is strictly increasing if and only if

$$r\theta \coth r\theta - \left[1 + \ln\left(\frac{\sinh r\theta}{r\theta}\right)\right] > 0.$$

If we define $g : [0, \infty) \mapsto (-\infty, \infty)$ by setting

$$g(s) = s \coth s - \left[1 + \ln\left(\frac{\sinh s}{s}\right)\right]$$

then it is clear that $g(0) = 0$ and that

$$g'(s) = \frac{1}{s} - \frac{s}{\sinh^2 s} > 0$$

for all $s > 0$. Now we must have $g(s) > 0$ for all $s > 0$ and hence $g(r\theta) > 0$ for all $r > 0$. This completes the proof. □

Example 5.3 *Consider the case where the distance function is*

$$x(r) = a[1 - e^{-kr}]$$

and the function $F_u : [0, M] \mapsto [0, 1]$ *is given by*

$$F_u(v) = (v/M).$$

This is a truly stochastic situation but is simplified by the uniformity of the probability distribution. At the first stage we have

$$x'(\varphi_1[u](b)) = \frac{1}{M^2} \int_S \left[\int_S x'(b + v_1 - \varphi_1[u](b) + v_0) dv_0 \right] dv_1$$

which can be rewritten explicitly in the form

$$ak \exp(-k\varphi_1[u](b)) = \frac{1}{M^2} ak \exp(-kb + k\varphi_1[u](b)) \frac{1}{k^2} [1 - \exp(-kM)]^2.$$

If we define $\theta = kM$ *and if* f *is the function defined in Lemma 5.1 then some elementary algebra can be used to show that*

$$\varphi_1[u](b) = b/2 + M/2 - c_1$$

where

$$c_1 = (1/(2k)) \ln f(1/2).$$

At stage $n > 1$ *the recursive equation is given by*

$$x'(\varphi_n[u](b)) = \frac{1}{M} \int_0^M x'(\psi_{n-1}[v_n](b) + v_n) dv_n$$

and if we make the inductive assumption that

$$\varphi_{n-1}[u](b) = b/n + M/2 - c_{n-1}$$

then it is easy to show that

$$\varphi_n[u](b) = b/(n+1) + M/2 - c_n$$

where

$$c_n = (n/(n+1)) \left[c_{n-1} + (1/(2nk)) \ln f(1/(2n)) \right].$$

It follows that $c_n < c_{n-1}$ *if and only if* $c_{n-1} > (1/(2k)) \ln f(1/2n)$. *Since*

$$c_1 = (1/(2k)) \ln f(1/2) > (1/(2k)) \ln f(1/4)$$

the inequality is true for $n = 2$. *If we suppose that* $c_{h-1} > (1/(2k)) \ln f(1/2h)$ *for each* $h < n$ *then*

$$\begin{aligned} c_{n-1} &= ((n-1)/n)\{c_{n-2} + (1/(2(n-1)k)) \ln f(1/2(n-1))\} \\ &> (1/(2k)) \ln f(1/2(n-1)) \\ &> (1/(2k)) \ln f(1/2n). \end{aligned}$$

Thus the inequality is also valid for $h = n$. *Since* $\{c_n\}$ *is positive and strictly decreasing* $c_n \downarrow c \geq 0$ *as* $n \uparrow \infty$. *By applying the recursive formula it is easy to show that*

$$c_n = (2/(n+1))c_1 + (1/2k(n+1))\left[\ln f(1/4) + \cdots + \ln f(1/2n)\right].$$

For each $\epsilon > 0$ *we can find* $N = N(\epsilon)$ *such that* $(1/2k)\ln f(1/2n) < \epsilon$ *for all* $n > N$ *and hence it follows that*

$$\begin{aligned} c_n &= (2/(n+1))c_1 + (1/2k(n+1))\left[\ln f(1/4) + \cdots + \ln f(1/2N)\right] \\ &\quad + (1/2k(n+1))\left[\ln f(1/2(N+1)) + \cdots + \ln f(1/2n)\right] \\ &\leq (2/(n+1))c_1 + (1/2k(n+1))\left[\ln f(1/4) + \cdots + \ln f(1/2N)\right] \\ &\quad + [n-(N+1)]\epsilon/(n+1). \end{aligned}$$

By taking the limit as $n \uparrow \infty$ *we see that* $c \leq \epsilon$. *Since* $\epsilon > 0$ *is arbitrary we conclude that* $c = 0$ *and that* $\varphi_n(b) \to M/2 = \bar{s}$ *as* $n \uparrow \infty$.

6 CONCLUSIONS

We have found a recursive equation to determine the optimal daily energy usage $\varphi_n[u](b)$ starting from any state (n, u, b). This recursive equation provides a much more effective calculation than the simple blind searches that are sometimes used to solve the Bellman equations. The elementary examples suggest that in many cases long term strategies do exist and that such strategies are independent of the present state. For an ergodic system we believe that the limit

$$\varphi[u](b) = \lim_{n \to \infty} \varphi_n[u](b)$$

exists and is well defined. Furthermore we conjecture that in such cases

$$\varphi[u](b) = \bar{s}$$

for all (u, b). This formula is certainly true for each of the elementary examples.

References

Boland, J., Gaitsgory, V., Howlett, P. and Pudney, P. J. (1999), Stochastic optimal control of a solar powered car, (to appear).

Gates, D.J. and Westcott, M.R. (1996), Solar cars and variational problems equivalent to shortest paths. *SIAM Journal on Control and Optimization*, Vol. 34, No. 2, pp. 428-436.

Howlett, P., Pudney, P., (1997), Tarnopolskaya, T. and Gates, D. (1997), Optimal driving strategy for a solar car on a level road. *IMA J. Maths. Applied in Business and Industry*, Vol. 8, pp. 59-81.

Howlett, P.G. and Pudney, P.J. (1998), An optimal driving strategy for a solar powered car on an undulating road. *Dynamics of Continuous, Discrete and Impulsive Systems*, Vol. 4, pp. 553-567.

20 THE PERSONNEL TASK SCHEDULING PROBLEM

Mohan Krishnamoorthy and Andreas T Ernst

CSIRO Mathematical & Information Sciences
Private Bag 10, Clayton South MDC VIC 3169, Australia
email: Mohan.Krishnamoorthy@cmis.csiro.au, Andreas.Ernst@cmis.csiro.au

Abstract: In this paper we introduce a group of problems called *Personnel Task Scheduling Problems* (PTSP). A PTSP is concerned with assigning a set of tasks with fixed start and end time to a heterogeneous workforce. Each person in the workforce works a shift with given start and end times as well as having skills that qualify the person to perform a subset of all the tasks. This type of problem arises in rostering applications. There are several variations depending on the objective and the exact nature of the constraints. The aim of this paper is to describe this class of problems together with some basic properties, to survey previous work in this area, and to introduce some data sets that may serve as a test bed for future developments in this area.

Key words: Personnel task scheduling, interval scheduling, fixed job scheduling, rostering.

X. Yang et al. (eds.), Optimization Methods and Applications, 343-368.

1 INTRODUCTION

In this paper, we introduce the personnel task scheduling problem (PTSP). The PTSP is a common problem that occurs in many practical instances. Production managers would face this problem on a routine basis while scheduling plant operations. In situations where specific tasks have already been assigned to machines over a specified planning horizon, decisions have to be made regarding the number of workers that are required and the machines to which each worker is assigned. This problem can be formulated as an instance of the PTSP. Typically however, this problem is faced by roster developers or shift supervisors in large service operations where the time at which a task has to be performed is determined by a fixed timetable or a set of appointments.

Consider a situation where a rosterer knows the start and end times of personnel rostered on a particular day. The task allocation problem is then one of optimally allocating a specified set of tasks to the available personnel. On a daily basis, the rosterer (or the shift supervisor) would need to allocate tasks, with specified start and end times, to available staff who have the requisite skills. Staff availability is indicated by personnel shifts (with specific start and end times, including overtime registrations).

We define the PTSP together with some of its properties in Section 2. Then in Section 3 a number of variants of the PTSP is given. We then review the literature in Section 4 and classify the work in terms of the PTSP variants. The need to study these types of problems is motivated through a number of different application areas that are described in Section 5. Finally we describe how a representative collection of test data sets can be generated in Section 6.

2 THE PTSP

In this section we give basic definitions and some properties of the PTSP. We also describe two special cases that can be solved easily. The PTSP can be described using the following definitions:

J The set of tasks that need to be assigned, $J = \{1, \ldots, n\}$.

W The set of workers that can perform tasks, $W = \{1, \ldots, m\}$.

s_j The start time of task $j \in J$.

f_j The finish time of task $j \in J$ $(f_j > s_j)$.

S_w The start time of shift $w \in W$.

F_w The finish time of shift $w \in W$ $(F_w > S_w)$.

P_j The set of personnel that can perform task $j \in J$ $(P_j \subseteq W)$.

T_w The set of tasks that worker $w \in W$ can perform $(T_w \subseteq J)$.

Workers in the above definitions, may also represent machines or processors. Note that the P_j's can be derived from the sets T_w or vice versa. In many cases these sets are obtained by considering a set of qualifications or skills Q. Each task requires a set of skills Q_j in order to perform that task, while each worker possesses a set of qualifications or skills Q_w. In order for a worker to perform a task (s)he needs to possess all of the necessary skills, hence $P_j = \{w \in W :$

$Q_w \supseteq Q_j\}$ and similarly $T_w = \{j \in J : Q_j \subseteq Q_w\}$. However for the remainder of this paper we shall generally ignore the set Q and only look at the P_j's and T_w's which could also have been arrived at in some other fashion.

The sets P_j and T_w can also include the shift time window constraints in addition to the skill considerations. For example $w \in P_j$ if $[s_j, f_j] \subseteq [S_w, F_w]$ and if worker w is qualified to perform task j. The sets T_w can be derived from the sets P_j through the relation $T_w = \{j \in J : w \in P_j\}$. Note that in the above description we do not take into account lunch breaks with flexible start and finish times however fixed lunch breaks can be included by removing conflicting jobs from T_w.

The *Personnel Task Scheduling Problem* (PTSP) is to find a *good* set of tasks for each worker to perform. The definition of what is a good (optimal) set of tasks depends on the objective function. Several different objective functions will be defined and discussed in Section 3. The main features that are present in all forms of the PTSP are that:

- Tasks must start at the given start times and hence have fixed finishing times.
- There is *no pre-emption*: once a task has been started by a worker, that person must complete the task.
- The assignment of tasks to workers must satisfy the skill requirements and respect the shift start and end times.

2.1 Conflict Graph

Define an interval graph, $G = (J, A)$, where J is the node set, one for each task in J and A is the arc set. Two nodes $j, k \in J$ are joined by an arc if the intervals $[s_j, f_j]$ and $[s_k, f_k]$ overlap. This graph G is called the *conflict graph* for the PTSP. The PTSP can now be described as a graph colouring problem. Each worker $w \in W$ corresponds to a colour and the aim is to assign a colour to each node of the graph such that: (a) no two adjacent nodes are coloured with the same colour and (b) the colour for each node j is chosen from P_j.

Let C be the set of *maximal cliques* in this interval graph. That is $C = \{K_1, \ldots, K_p\}$ consists of sets $K_t \subseteq J$ such that any two tasks in K_t overlap for some interval of time and K_t is maximal. It follows that no task in $J \setminus K_t$ will overlap with any one of the tasks in K_t. Obviously every node in a maximal clique must be assigned a different colour in any solution to the PTSP.

Unlike in general graphs, finding all maximal cliques in an interval graph is easy (ie can be performed in polynomial time). This involves sorting and a single pass over the tasks. For a detailed description of the algorithm see for example Gondran and Minoux (1984).

Usually we are not interested in the maximal cliques for the conflict graph G but in those of the subgraph G^w induced by considering only nodes in T_w for some $w \in W$. We denote by C^w the set of maximal cliques in G^w. Note that this cannot be arrived at by simply deleting elements from the maximal cliques in C. However each C^w can be calculated in $\mathcal{O}(|J|)$ for each worker w once the tasks have been sorted.

We illustrate these sets with a small example below. Assume a set of tasks with starting and ending times as shown in Figure 2.1. The resulting interval graph is as shown in Figure 2.2.

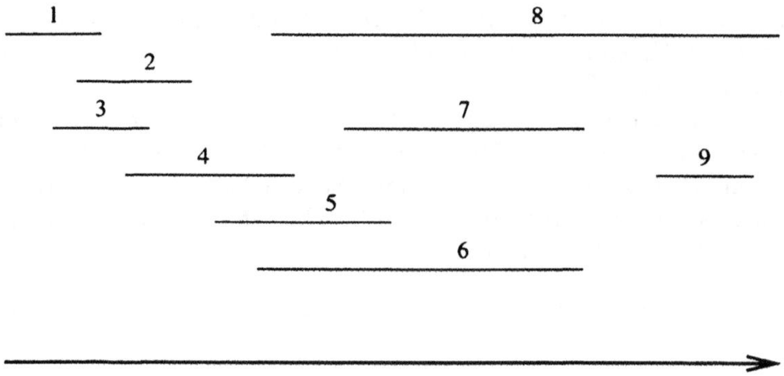

Figure 2.1 Set of tasks

Here the cliques are: $C = \{\{1,2,3\}\ \{2,3,4\}\ \{8,9\}\ \{4,5,6,8\}\ \{5,6,7,8\}\}$. In other words, it is not possible for a single worker to perform tasks 1, 2 and 3 simultaneously. Similarly, it is not possible for a single worker to perform tasks 5, 6, 7, and 8 simultaneously. Let us assume that, for a particular worker w, we have $T_w = \{2,4,6,8\}$. In other words, worker w in our example, can only perform tasks 2, 4, 6 and 8. Then, we can find $C^w = \{\{2,4\}\ \{4,6,8\}\}$.

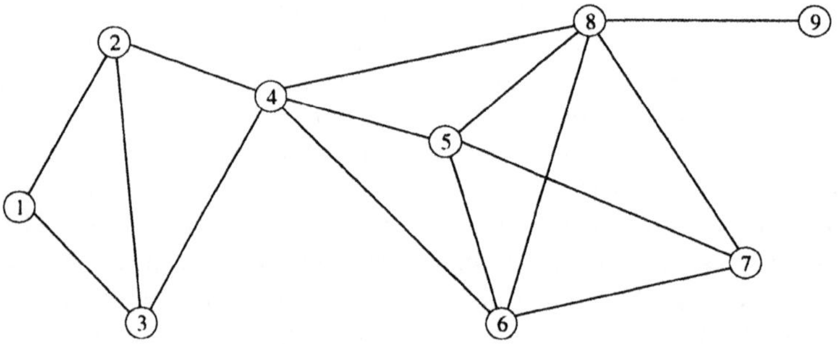

Figure 2.2 Interval graph

2.2 Identical machines

One very simple special case of the PTSP occurs where the aim is to find a feasible assignment of all jobs using the minimum number of people in a workforce in which all people have identical qualifications and shifts. In other words $T_w = J$ for all $w \in W$.

Obviously the maximum size of any clique in the conflict graph provides a lower bound on the number of people required. This lower bound is tight: it is possible to find a solution that requires no more people.

A polynomial time algorithm for finding this assignment works in a similar way to the algorithm for finding the set of maximal cliques in an interval graph and can be described briefly as:

- Sort all jobs in ascending order of start times;
- Assign each task $j \in J$ to the lowest indexed person $w \in W$ for which no conflict occurs with the set of tasks already assigned.

The complexity of this algorithm is $\mathcal{O}(|J|)$ plus sorting of the jobs. In fact a more general version of this problem can be solved in which there are setup (or turnaround) times between successive tasks. For more details see for example Ford and Fulkerson (1962).

2.3 The packing subproblem

Another simple version of the PTSP occurs when there is just a single worker or shift and not all jobs have to be performed. The aim in this simple PTSP variant is to maximise the value of tasks performed. This variant of the PTSP is akin to the knapsack problem in which a set of items need to be selected for packing in a knapsack in order to maximise total value subject to a simple capacity constraint. The difference in the PTSP packing problem is that instead of having a simple capacity constraint we require that the items packed do not overlap with each other (nor with the start or end of the shift).

Let v_j be the value of item $j \in J$. We will drop the w index as $|W| = 1$ in this section. The algorithm for finding a set of jobs $\bar{J} \subseteq J$ for which $\sum_{j \in \bar{J}} v_j$ is maximised involves the following steps:

1. Find the set of maximal cliques C in the conflict graph with $|C| = n$. For each clique $K_i \in C$, let t_i be a time when all of the tasks in K_i overlap. Also let $t_0 = -\infty$ and $t_{n+1} = \infty$.

2. Discretise the task start and finish time using the t_i: define $\bar{s}_j = i : t_i \leq s_j < t_{i+1}$ and similarly $\bar{f}_j = i : t_i \leq f_j < t_{i+1}$.

3. Solve the following dynamic programming recursion:

$$V(0) = 0 \qquad V(i) = \max_{j:\bar{f}_j = i} V(\bar{s}_j) + v_j$$

$$J(0) = 0 \qquad J(i) = \operatorname*{argmax}_{j:\bar{f}_j = i} V(\bar{s}_j) + v_j$$

The optimal solution to the PTSP packing problem has value $V(n)$ and uses jobs $J(n)$. Note that this algorithm could also be viewed as a shortest path problem in which each i represents a node and each task j an arc from $\bar{s}_j$ to $\bar{f}_j$.

3 VARIANTS OF THE PTSP

In order to define the basic PTSP and its variants we define the following decision variables. Let,

$$x_{jw} = \begin{cases} 1 & \text{if task } j \in J \text{ is assigned to worker } w \in W \\ 0 & \text{otherwise} \end{cases} \tag{3.1}$$

In this section, we use the above definitions, to define several variants of the PTSP and to present some Integer Programming (IP) formulations for these. The variants arise mainly out of different choices in objective function but also from possible complications in the constraints. In general each variant introduces a single feature which could occur together with other features.

In most of the PTSP problems the following family of constraints will be used to prevent overlapping jobs being assigned to the same person:

$$\sum_{j \in K} x_{jw} \leq 1 \qquad \forall w \in W,\ K \in C^w \tag{3.2}$$

This constraint simply states that in any maximal set of overlapping tasks K that could be performed by w, at most one task can be assigned to that person.

3.1 Nomenclature

In order to manage the large number of variants, we introduce a nomenclature to identify PTSP variants. This nomenclature takes the form `PTSP[T;S;Q;O]` where the slots T, S, Q and O refer to the *Task Type*, the *Shift Type*, the *Qualifications*, and the *Objective Function* respectively. Each slot can contain a list of appropriate features found in the PTSP variant. As we introduce variants in the remainder of this paper we will identify them in terms of this nomenclature. An overview of the possible range of values for each of the above attributes is given in the Table 3.1. Detailed explanations of each variant are given in the subsequent sections.

Table 3.1 allows the construction of a very large number of variants. It is not necessary that each attribute assumes only one value. For example we can envisage a complex variant of the PTSP in which there are multiple objectives. An example would be `PTSP[F;F;H;A-T-W]`. This problem has fixed contiguous tasks, fixed shift lengths, heterogeneous workforce and a linear combination of three different objectives: assignment cost, overtime cost and staff requirements. In general all variants of the PTSP will be identified by the quadruple of `[Tasks;Shifts;Qualifications;Objective]`, or `PTSP[T;S;Q;O]`. Obviously we cannot describe every possible combination of features. Hence we describe a simple basic version with the "default" values for the above four attributes.

Table 3.1 PTSP Nomenclature

Attribute	Feature	Description
T	F	Fixed contiguous tasks
	V	Variable task durations
	S	Split (non-contiguous) tasks
	C	Changeover times between consecutive tasks
S	F	Fixed, given shift lengths
	I	Identical shifts which are effectively of infinite duration
	D	Maximum duration without given start or end times
	U	Unlimited number of shifts of each type available
Q	I	Identical qualifications for all staff (homogeneous workforce)
	H	Heterogeneous workforce
O	F	No objective, just find a feasible schedule
	A	Minimise assignment cost
	T	Worktime costs including overtime
	W	Minimise number of workers
	U	Minimise unallocated tasks

In this paper we make `PTSP[F;F;H;F]` the default as described in Section 3.2. We then describe seven variants one for each of the significant features. Note that some of the above features have been included for completeness and to allow classification of PTSP variants that have been published previously in the literature. These "non-significant" features include S=I, S=D and Q=I.

3.2 *Feasible Assignment*

The simplest form of the PTSP is to just find a feasible assignment of tasks to people. The *Feasibility Problem* (FP) just requires that each task is allocated to a person that is suitably qualified with no person performing more than one task at a time. Furthermore tasks cannot be pre-empted, in other words once a task has been started by a person it must be completed by that person.

The `PTSP[F;F;H;F]` can be described mathematically as follows:

Problem PTSP[F;F;H;F]

$$\begin{array}{lrcll} & \multicolumn{4}{l}{\text{Find an assignment } x_{jw}} \\ \text{S.t.} & \sum_{w \in P_j} x_{jw} & = & 1 & \forall\, j \in J \\ \multicolumn{2}{l}{\text{overlapping constraint}} & & (3.2) & \\ & x_{jw} & \in & \{0,1\} & \forall\, j \in J,\ w \in P_j \end{array} \tag{3.3}$$

This formulation could be used in an IP solver to obtain feasible solutions to the problem, however the second set of constraints is not very tight. A better formulation, from the algorithmic point of view is described in Section 2. These constraints are sufficient though to exactly define the problem.

3.3 Minimum Assignment Cost

In problem FP it may be possible to find multiple feasible solutions. In this case it may be useful to be able to distinguish among these solutions by some criteria. The most obvious method for doing this is to attach a cost to each possible assignment. Let c_{jw} be the cost of assigning task j to person w. The problem to find a feasible assignment which minimises this cost is referred to as `PTSP[F;F;H;A]`.

Problem PTSP[F;F;H;A]

$$\begin{array}{lrcll} & \multicolumn{4}{l}{\min \sum_{j \in J} \sum_{w \in P_j} c_{jw} x_{jw}} \\ \text{S.t.} & \sum_{w \in P_j} x_{jw} & = & 1 & \forall\, j \in J \\ \multicolumn{2}{l}{\text{overlapping constraint}} & & (3.2) & \\ & x_{jw} & \in & \{0,1\} & \forall\, j \in J,\ w \in P_j \end{array} \tag{3.4}$$

This type of cost can be used to model a variety of factors that arise in different applications:

Preferences: Personnel might express preferences for performing particular tasks of types of tasks and also dislikes. This could be translated into a penalty for assigning a task to a person that is derived from that person's rating of the task. Hence the cost minimisation is actually a maximisation of staff "happiness".

Substitutions: The tasks may have a preference of person attached to them such that substitutions, while they may be possible, are not desirable. This is the case with the car rental application discussed in Section 5.2. Each booking is for a certain vehicle type so supplying a different vehicle type incurs a penalty that depends on the booking request (task) and the vehicle supplied (shift).

Overtime: Costs of this type can be used to provide a simple estimate of overtime costs. Each worker has a given start and end time spanning, say 12 hours. If the regular work time of this worker is only 8 hours, then any job assigned to the last 4 hours of the shift incurs a penalty consisting of the overtime rate times the duration of the job. Note that this is obviously a simplification of the real situation, a more accurate model is presented later.

Unassigned: In cases where not all tasks can be completed the objective could just be to minimise the number of tasks that are unassigned. This is achieved by setting $c_{jw} = -1$.

Task Priority: Each task may have an associated priority p_j depending on factors like the importance of the customer or the urgency of the job. Hence the objective may be to maximise the total priority of all the jobs completed (ie $c_{jw} = -p_j$). This type of objective occurs in the variant `PTSP[F;I;H;U-A]` studied by Kroon *et al.* (1995), who call it the *Operational Fixed Interval Scheduling Problem* (OFISP).

Idle time: If there are more jobs than can be completed, it may be desirable to keep the personnel (or machines) as busy as possible. This can be achieved by setting $c_{jw} = s_j - f_j$ to minimise idle time. Another way of looking at this cost function is that it maximises the total amount of work completed (measured by the duration of the jobs).

Contract Staff: Some service organisations deal with the fluctuations in load by using contract or casual staff. If this is the case then it is usually quite easy to find a solution in which all of the tasks are completed, however the aim is to minimise the cost of using these non-permanent staff. Again this type of objective can be modelled using a linear objective, for example by setting $c_{jw} = (M - r_w)(s_j - f_j)$. Here r_w is the pay rate which is small or zero for staff employed permanently and M is a suitably large constant to ensure that all of the c_{jw} are negative.

Naturally in many applications two or more of these factors should be considered. However it is not difficult to combine these different costs and still obtain a model that fits into the minimum assignment cost framework.

3.4 Working Time Minimisation

The previous discussions on costs showed that the linear cost PTSP, while already quite general is not always sufficient to model all of the objectives that may arise in real applications. The most important deficiency is that assignment objective does not include idle time explicitly. For example if a temporary staff member has to be called up to perform two tasks, they have to be paid not only for the duration of the two tasks, but also for all of the idle time between the two tasks. A similar problem occurs in the calculation of overtime. Hence the working time minimisation problem considers the case where the cost of a schedule includes a function of the working time. In many cases it is important to allow this function to be non-linear (though usually piecewise affine).

Let h_w be the number of hours worked by person $w \in W$, in other the time from the start of the first task allocated to w to the end of the last task. Furthermore let $C_w(h_w)$ be the cost for that person working the given hours. This type of model can represent a number of scenarios:

Casual Staff: If the workforce consists of a combination of permanent and casual staff then $C_w(h) = 0$ for the permanent staff (or some constant) and $C_w(h) = r_w h$ for casual staff where r_w is the rate of pay. The PTSP with worktime objective will then minimise the amount of money spent on hiring casual staff to cope with excess work.

Over Time: Where staff may work more than their regular hours but with additional hire rates of pay depending on the amount of overtime, the function C_w could be piecewise linear. For example:

$$C_w(h) = \begin{cases} 7r_w & \text{if } h \leq 7 \\ C_w(7) + 1.5r_w(h-7) & \text{if } 7 < h \leq 10 \\ C_w(10) + 2r_w(h-10) & \text{if } 10 < h \end{cases}$$

This allows for normal pay of work up to a complete shift, time-and-a-half and double-time, a common practice under Australian award rules.

Spread Time Constraints: It may simply be desirable to penalise excessive worktime. Hence C_w could just be a step function. This type of function captures the idea of fixed job scheduling with spread time constraints proposed in Fischetti *et al.* (1987), albeit as a soft constraint rather than a hard constraint.

Since the duration for which a person is working is part of the objective function, in this case it makes sense to have problems without prescribed shift start and end times, though this is not strictly necessary. Hence we obtain problem PTSP[F;I;H;T] which can be written mathematically as:

Problem PTSP[F;I;H;T]

$$\min \sum_{w \in W} C_w(h_w)$$

$$\text{S.t.} \quad \sum_{w \in P_j} x_{jw} = 1 \quad \forall\, j \in J$$

$$\text{overlapping constraint} \quad (3.2)$$

$$\max_{j \in T_w}\{f_j x_{jw}\} - \min_{j \in T_w}\{s_j x_{jw}\} = h_w \quad \forall\, w \in W \tag{3.5}$$

$$x_{jw} \in \{0,1\} \quad \forall\, j \in J,\ w \in P_j$$

$$h_w \geq 0 \quad \forall\, w \in W$$

Obviously the definition of h_w in the last constraint defaults to zero if $x_{jw} = 0$ for every $j \in T_w$. If an absolute limit on the number of working hours is required, beyond that required by any shift start and end times, then it be modelled easily by adding an upper bound to the h_w variables to give rise to problem PTSP[F;D;H;T].

Note that for all of the types of objective functions discussed above it is not difficult to transform the above non-linear programming formulation into equivalent MILP formulations. The crucial property required for this transformation is that the objective functions are convex and non-decreasing. Also if required it would be straight forward to combine this type of problem with an assignment cost to obtain problem PTSP[F;I;H;A-T].

3.5 *Workforce minimisation*

In this section we consider a planning problem, where the aim is to minimise the total number of staff required to perform all of the tasks on a typical day. We envisage the application of this model for a series of typical days, to derive a required size of the workforce. Alternatively we can use this model in situations where all of the work is performed by casual or contract staff. Typical instances of this application (in an Australian context) would be bank tellers and supermarket checkout operators.

Problem PTSP[F;F;H;W]

$$\min \sum_{w \in W} b_w \max_{j \in T_w} x_{jw}$$

$$\sum_{w \in P_j} x_{jw} = 1, \qquad \forall\, j \in J \tag{3.6}$$

$$\text{overlapping constraint} \qquad (3.2) \tag{3.7}$$

$$x_{jw} \in \{0,1\} \quad \forall\, j \in J,\ w \in P_j \tag{3.8}$$

Here b_w represents the cost of using worker w. In many applications $b_w = 1$ for all w so that the objective is just to minimise the total size of the workforce. Note that the above objective function is non-linear. However the problem can easily be rewritten as an integer linear program:

Problem PTSP[F;F;H;W]

$$\min \sum_{w \in W} b_w y_w$$

$$\sum_{w \in P_j} x_{jw} = 1, \qquad \forall\, j \in J$$

$$\sum_{j \in K} x_{jw} \leq y_w, \qquad \forall w \in W,\ K \in C^w$$

$$x_{jw} \in \{0,1\} \quad \forall\, j \in J,\ w \in P_j$$

$$y_w \geq 0 \qquad \forall\, w \in W$$

This type of problem includes the Tactical Fixed Interval Scheduling Problem (TFISP) introduced by Kroon *et al.* (1997). The TFISP is presented slightly differently in that there are no shift start and end times and that there is an "infinite" number of shifts available for each combination of qualifications. In other words the TFISP tries to find the mix of different type of personnel required to minimise the overall workforce. Hence the TFISP is called PTSP[F;U;H;W] in our nomenclature.

3.6 *Minimise Unallocated Tasks*

In some applications there may be more tasks than can be completed by the available workforce. If there is no possibility of increasing the workforce through the addition of temporary staff or contractors, then the problem becomes one of deciding on a set of tasks that can actually be performed with the available resources. Here we relax the constraint that every task must be assigned and replace it with the restriction that each task can be performed at most once.

Problem PTSP[F;F;H;U]

$$\begin{array}{lllll} & \min \quad \sum_{j \in J} \sum_{w \in P_j} 1 - x_{jw} & & & \\ \text{S.t.} & \sum_{w \in P_j} x_{jw} & \leq & 1 & \forall\, j \in J \\ & \text{overlapping constraint} & & (3.2) & \\ & x_{jw} & \in & \{0,1\} & \forall\, j \in J,\ w \in P_j \end{array} \tag{3.9}$$

The objective function could equivalently be written as one of maximising the number of tasks allocated. Another alternative is to associate a value with each task and try to maximise the value of the tasks performed. In other words we have a cost $c_{jw} = v_j < 0$ associated with each possible assignment and the objective function as for O=A. This problem is denoted by PTSP[F;F;H;U-A].

3.7 *Variable Task Lengths*

A quite different variation of the PTSP arises when the different possible workers that a task may be assigned to require different amounts of time to process the task. This scenario arises in particular when the work is done not by people but by machines, where newer or bigger machines may have significantly higher processing rates. It will still be assumed that the tasks have a fixed starting time. The task finishing times are denoted as f_j^w for each j and $w \in P_j$. Again many different objective functions are possible, including linear costs, work time minimisation, workforce minimisation or a combination of these. Here we just present a formulation for the basic problem of finding a feasible assignment:

Problem PTSP[V;F;H;F]

$$\begin{aligned} &\text{Find an assignment } x_{jw} \\ \text{S.t.} \quad &\sum_{w \in P_j} x_{jw} = 1 \qquad \forall\, j \in J \\ &\text{overlapping constraint} \qquad (3.2) \\ &x_{jw} \in \{0,1\} \quad \forall\, j \in J,\ w \in P_j \end{aligned} \tag{3.10}$$

Note that this formulation looks exactly like that for problem PTSP[F;F;H;F], however the differences arises in the overlapping of tasks which depends on the shift to which the tasks are to be assigned. In other words the set of cliques C^v and C^w may differ for two workers v and w even if $T_v = T_w$. Hence for integer programming approaches this variant may not be significantly different to the basic version of the PTSP, however it can make designing heuristic algorithms more complicated.

3.8 Non-Contiguous Tasks

Another variation which can significantly complicate the PTSP is if some tasks are non-contiguous. That is a task may involve an idle period during which the person assigned to this task may perform other work before returning to complete the original task. Equivalently this may be thought of as having some groups of tasks with the requirement that *all* of the tasks in the group have to be assigned to the same person. The formulation for a non-contiguous task problem in which just a feasible solution is required again looks exactly like that in Section 3.2 for the basic feasible assignment, however determining whether two tasks overlap in equation (3.2) can be more complicated. As for the variable task length variant, it is possible to consider a number of objective functions.

Applications of this type of problem occur in industries where an individual relationship must be maintained between a customer and their service provider. More concretely consider the health services provided by a hospital. Here the relationship between the doctor and the patient must be maintained over a series of tasks that could include a series of operations each of which is scheduled at fixed times in an operating theatre. Another application involving this type of tasks is described in Section 5.3.

3.9 Changeover Times

In all of the above PTSP variants it has been assumed that each new task can be scheduled independently of those that have occurred previously. However in some applications it may be necessary to include some changeover times t_{ij} for two successive tasks $i, j \in J$ assigned to the same person. This may be used to model setup times if the PTSP is applied to machines performing tasks. Another example occurs in the application to personnel scheduling if the tasks are to be performed at different locations and some allowance has to be made for the staff to travel between these in order to take up the next task.

It will be assumed that $t_{ik} \leq t_{ij} + t_{jk}$ for any $i, j, k \in J$. This simply eliminates a situation occurring that i and k can only be scheduled if j is also scheduled but not otherwise. Since this is not likely in any reasonable application of PTSP problems, it would seem unreasonable to require algorithms to cope with that case.

The formulation for the PTSP with changeover times looks like that for the feasible schedule problem however the conflict constraint is slightly more complicated:

Problem PTSP[C;F;H;F]

$$\text{Find an assignment } x_{jw}$$

$$\begin{array}{lrcll} \text{S.t.} & \sum_{w \in P_j} x_{jw} & = & 1 & \forall\, j \in J \\ & \sum_{j \in K} x_{jw} & \leq & 1 & \forall w \in W,\ K \in C^w \\ & x_{jw} & \in & \{0,1\} & \forall\, j \in J,\ w \in P_j \end{array} \tag{3.11}$$

In this formulation the sets $K \in C^w$ are harder to find since the conflict graph is no longer an interval graph. However it is not strictly necessary to find all maximal cliques. For example it is possible to define

$$K_j^w = \{i \in T_w : [s_j, f_j + t_{ji}] \cap [s_i, f_i + t_{ij}] \neq \emptyset\}.$$

By this definition the set of K_j^w for all $j \in T_w$ is certainly sufficient to include all clashes between jobs. However this definition may create some redundant overlapping constraints. Depending on the exact nature and structure of the changeover time data, it may be possible to find better cliques in a systematic manner. Alternatively a general heuristic could be employed on the clique graph.

The simplest case where all people are identical (ie `PTSP[C;I-U;I;F]`) is also known as Dilworth's problem. It is interesting to note that for Dilworth's problem a polynomial time algorithm exists. The algorithm can be thought of as a special case of an assignment problem (see Ford and Fulkerson (1962)).

In the definition of problem `PTSP[C;F;H;F]` we have not included any costs. In many applications it would be interesting to include a cost of performing a changeover. In particular we usually find that the cost of changeover depends on the time required. For example if the tasks are to be performed at geographically separated places, then the change over time and cost would depend on the time and cost required for staff to move from one place to another. Similarly if the shifts represent machines and the changeovers are required to reconfigure these for different types of tasks then the changeover cost would be derived from the labour cost of performing this change and hence again related to the duration. We have not included this as a separate cost function in the above list of features in order to limit the already very large set of variants of the PTSP. Furthermore adding this type of objective to the IP formulations presented above would make them significantly more complex and the LP relaxations in general quite weak.

4 LITERATURE REVIEW

In the literature, a similar problem to the PTSP is referred to as the fixed job schedule problem or a heterogeneous workforce scheduling problem, with constraints on working time. See, for example, Arkin and Silverberg (1987), Fischetti, Martello and Toth (1987, 1989, 1992) and Valls *et al.* (1996). Another similar problem has been introduced in Gondran and Minoux (1984) (page 476, Exercise 4). This involves identifying the minimum number of aircraft required to fly all the flights in a given planning horizon. We assume that none of the planes require any maintenance or repair and that the turnaround time between successive flights by the same plane is arbitrarily small. Each flight (task) has a start and end time and requires planes (staff) of a particular type (skill). Given that the planes (shifts) are available for specific durations, we are required to minimise the number of planes (staff). Thus, this problem is the `PTSP[F;F;H;W]` as defined above. A similar example has also been introduced in Gondran and Minoux (1984) (page 204–205, Example 23).

It has been shown by Arkin and Silverberg (1987) that the `PTSP[F;I;H;F]` is NP-complete (ie just finding a feasible solution with a heterogeneous workforce and fixed shifts) by reducing this variant to the 3-SAT problem. There are however algorithms that can solve special cases of this. Apart from the identical workers version or the single worker packing problem described in Section 2, there is also the case where there are just two types of qualification. Dondeti and Emmons (1992) present an algorithm based on solving a network flow problem for solving several variants of the PTSP provided that there are just two types of shifts with any number of jobs that can be done by either or both of the shift types. In other words the workforce is heterogeneous but only in a very limited way. The variants solved by Dondeti and Emmons (1992) include both the feasibility problem (O=F) and the workforce minimisation problem (O=W).

The variant `PTSP[F;I;H;W]` includes the model in Valls *et al.* (1996), where a heterogeneous workforce assignment problem is studied. In that paper the number of workers required to perform a machine load plan is minimised using a branch and bound algorithm. The authors use a colouring approach in which successively all possible ways of colouring all maximal independent node sets are explored together. This is combined with lower and upper bounding techniques to allow pruning of the search tree.

Different variants of PTSP with workforce minimisation objectives have been given in the literature. If the shifts are identical (S=I and Q=I) and if we were to minimise the number of machines that are used (O=W), the resulting `PTSP[F;I;I;W]` has been published under the name "*fixed job schedule problem*". This problem has been studied by Gertsbakh and Stern (1978) and by Fischetti *et al.* (1992). The fixed job schedule problem has been solved in $O(nlogn)$ time by Gupta *et al.* (1979) and also by Nakajima *et al.* (1982). Arkin and Silverberg (1987) provide an $O(n^2logn)$ algorithm for a problem similar to a variant of the fixed job schedule problem that would be called `PTSP[F;I;I;U-A]`. In other words they study a problem in which each job has a fixed start and end time and a value. They then minimise the value of the jobs left uncompleted on identical machines. Arkin and Silverberg (1987) show that the **PTSP[F;I;H;U-A]** is NP-complete, that is if each job has associated with it a subset of machines on which it can be processed. Note that a starting and finishing times can always be handled implicitly by restricting the allowable set of machines (shifts) for each task appropriately. Hence it appears that the crucial complication that makes the PTSP family of problems NP-complete is the qualifications which restrict the set of jobs that each machine can process. On the other hand for many problems with S=I and Q=I polynomial algorithms exist.

Fischetti *et al.* (1987) solve the fixed job schedule problem with spread time constraints. In this version of the problem we are given n tasks each with a start and end time (s_j, f_j). We are also given an unlimited number of identical processors ($S_w = -\infty$ and $F_w = \infty$). The objective is to minimise the number of machines used (O=W), subject to the additional constraint that the difference between the end of the last job and the start of the first job on every machine does not exceed the *maximum spread time.* This falls into the category **PTSP[F;D-U;H;W]**. Note that the most efficient method for formulating this is not to add variables h_w for the duration worked as in the objective $O = T$ but to change the definition of incompatibility between tasks so that two tasks j and k "overlap" (are incompatible) if the difference between the start time of the first and the end time of the second task is greater than the maximum spread time. Unfortunately this type of model would destroy the structure of the conflict graph described in Section 2. Fischetti *et al.* (1987) show that the fixed jobs scheduling problem with spread time constraints is NP-hard.

Fischetti *et al.* (1989) also introduce a variant of the fixed job schedule problem in which there are working time constraints. In this problem we are given n tasks with start and end times and an unlimited number of identical processors (S=I and Q=I). The objective is to perform all tasks with the minimum number of processors (O=W), such that no processor works, in total, for more than a given working time T. If we let $t_j = f_j - s_j$, then this problem can be formulated as **PTSP[F;I;I;W]** with the additional constraint

$$\sum_{j \in J} t_j x_{jw} \leq T \qquad \forall\, w \in W.$$

Hence this problem does not quite fit into the family of PTSP variants but is still closely related to them.

Kolen and Kroon study several variants of the PTSP under the name of class scheduling in a series of three papers (1991, 1992, 1993).The variants considered in these papers include:

- Feasible assignment with (a) qualifications but no shift start/end times (**PTSP[F;I;H;F]**)) and (b) fixed shift start and end times but no qualifications (**PTSP[F;F;I;F]**).
- Linear cost minimisation in which each job has a value v_j which is independent of the person performing the job. Again both shift start/end

times and qualifications are considered but not both for the same problem, ie variants PTSP[F;I;H;A], and PTSP[F;F;I;A]

- Workforce minimisation (Problem PTSP[F;I;H;W]) without shift start and end times,

It is interesting to note that while Arkin and Silverberg (1987) show that while problem PTSP[F;I;H;A-U] is NP-complete in general, they also present an dynamic programming algorithm that is polynomial in the number jobs for a fixed number of shifts provided the costs of performing a job is independent of the person it is assigned to. Kolen and Kroon (1993) study the analogous problem finding a feasible assignment for an arbitrary set of jobs to a given, fixed set of shifts, where there are shift start and end times but no qualification constraints (PTSP[F;F;I;F]). They show that this special case of the PTSP is already NP-complete (and the corresponding cost minimisation variant PTSP[F;F;I;U-A] NP-hard) by reduction to the Numerical three Dimensional Matching problem. Special cases of the PTSP without qualification constraints that are polynomialy solvable usually depend on the conflict graph of the *shifts* (rather than jobs) to have a very special structure. For example if none of the shifts overlap then the PTSP decomposes into a sequence of problems each with just one shift.

A quite different variant of task scheduling is introduced by Lipton and Tomkins (1994). They propose an on-line task scheduling problem in which the tasks arrive in chronological order of starting time and each have a value. The problem is to decide whether or not to schedule each of the tasks on a single machine such that the total value of all tasks completed is maximised. Algorithms for this problem and theoretical results on their expected performance are presented in Woeginger (1994) and Seiden (1998).

5 APPLICATIONS

5.1 *Rostering at an Airport*

Our motivation for carrying out research into this problem is derived from a consultancy that we undertook for a major international airline. In this consultancy project, we developed a rostering solution for the ground staff at a major international airport. See Dowling *et al.* (1997). In this study, we describe a human resource planning and scheduling system that assists in the rostering of approximately 500 staff (over a 35–day planning horizon) for the airport operations of a major international airline at one of the busiest international airports. Such rostering solutions consist, normally, of a two–phase approach. In the first phase, a master–roster or a planning–roster is created that satisfies all the labour constraints, work regulations and enterprise bargaining agreements. The second phase involves micro–planning on a day–to–day basis. Given the staff that are rostered on a particular day, the PTSP is to assign to each of them, specific tasks (each with a specific start and end time and each with specific skill requirements).

Thus, the PTSP is a day–to–day operational problem. We are given all the shifts that are operational on a particular day. Each operational shift has a start and end time. In addition, it is likely that some personnel have 'signed-up' (or, registered) for overtime. Such overtime registrations may be used if it is not possible to accommodate all the tasks within the normal shift durations. Each of the staff members have specific skills. We are also given the set of tasks that need to be allocated to personnel shifts. Tasks have specific start and end times and need to be performed by staff with specific attributes. The task times are derived from the airline timetables (arrivals and departures) and hence are known well in advance. A separate issue, which is ignored here, is disruption management on the day of operation when planes arrive or depart late or staff members are unexpectedly unavailable due to sickness. In the general case, we assume that it is not possible to preempt tasks once they are commenced. Then the PTSP is the problem of allocating *all* tasks to personnel shifts such that a person may only perform one task at a time.

A similar application, focusing on rostering airport maintenance staff is discussed by Dijkstra *et al.* (1991). This lead to the formulation of the OFISP (= `PTSP[F;I;H;U-A]`) described bu Kroon *et al.* (1995) and also the TFISP (Kroon *et al.* (1997), TFISP = `PTSP[F;I-U;H;W]`). Another scheduling application at airports that falls into the category of PTSP is the assignment of planes to gates (see Kroon *et al.* (1995) and references therein). Here the tasks are planes that need to be assigned to gates for the duration between arrival and departure. However not every gate is suitable for every type of plane (particularly at airports that handle a mixture of long-haul and short-haul, national and international flights). Furthermore if no gate is available it may be necessary to transfer passengers between the terminal and the plane by bus, which is obviously undesirable.

Finally it is worth noting that PTSP variants can also occur as subproblems when rostering staff that do not stay in one location, such as airline stewards or train drivers. In such rostering applications it is common to split the problem into two phases. The first phase roundtrips (also known as pairings or duties) are constructed which start and end at the same location, which is the home base for the personnel staffing this roundtrip. The first phase selects a minimum cost set of such pairings which covers each individual leg or trip exactly once. In the second phase a PTSP is solved to assign all of the roundtrips to the staff. Here the tasks are roundtrips and since each of these starts and ends at the same location there is no need to consider any geographic linking constraints.

5.2 Car rentals

A different application of this type of problem can be found in the vehicle rental business. Consider a car rental company that supplies vehicles to customers from a central depot. Here the 'personnel' are replaced by cars and the tasks are rental bookings. The company may have bookings for vehicles for several months in advance with each booking being for one or more days. Note that if each vehicle is only booked for a single day at a time then the vehicle schedule can be created by solving a separate assignment problem for each day. However this is not the case when bookings have longer durations.

Instead of skills required for each task there is the vehicle type promised to the customer. Note that is is usually possible to provide an 'upgrade' (ie a vehicle other than the one requested) to a customer although this is obviously not desirable. Furthermore there may not be a one to one correspondence between the bookings and vehicle types due to other factors. For example as part of a package deal the customer may be promised a "small car" with exactly specifying the brand. Another factor is that the rental company may make distinctions between vehicles that are invisible to the customer (eg based on the purchase date) that still impact on the scheduling (eg giving preference to using newer vehicles first).

This model can also take into account multiple depots in a city under the following assumptions on the operating policy of the rental company: Each vehicle has a 'home' depot and must start and end each booking at that location. Hence for bookings that start or end at a location other than the home location this involves a relocation at the start or end of the booking (and hence an additional cost) however this relocation would take 'no time'. Later in this paper we discuss a variant of the PTSP that relaxes this last assumption slightly.

A similar application is described briefly in Kroon *et al.* (1995) involving the assignment of holiday bungalows to vacationists. Again the tasks correspond to bookings and the costs are derived from the closeness of the match between the features requested in the booking and those provided by the bungalow actually assigned to it.

5.3 Classroom assignment

The last category of applications of the PTSP which has frequently been studied in its own right occurs in classroom assignment. Consider a school or university that has determined a timetable based on the courses selected by students, usually with a view to minimise conflicts among courses that require the same teacher or selected by the same students. The aim is now to assign the lessons to classrooms. There are obvious qualification constraints as many subjects require specialised equipment (eg laboratories, sports facilities) and depending on the number of students the room size has to be considered.

While many lessons are just for a single period, others may extend over two or more periods. In addition the problem may include constraints that require several classes for the same course to be assigned to the same room at different times of the week. This type of constraint leads to the non-contiguous task variant of the PTSP described in Section 3.8. An example of this type of classroom scheduling is described by Glassey and Mizrach (1986). This problem reduces to K-colourability of a graph and hence is NP-complete (see Carter and Tovey (1992)).

In this example there are obviously no start and end times for the "shifts" (rooms). The main aim would be to find a feasible assignment of the lessons to rooms. Hence the most obvious PTSP variant to apply here is `PTSP[S;I;H;F]`. However it may also be desirable to include some preference indication. For example while small English courses could be taught just about anywhere, the lecturers would obviously prefer them to be assigned to rooms in or near the English department.

The complication in this type of problem arises mainly due to the existence of lessons spanning two or more periods which need to be assigned a single room. If this feature is not present in the data set then the classroom assignment problem could be solved as series of independent single period assignment problems. A more comprehensive discussion of variants of the classroom assignment problem and which cases are polynomial solvable is given by Carter and Tovey (1992).

6 TEST DATA

The task of creating a representative collection of test problems for this class or problems is an almost impossible one. Nevertheless the authors believe that creating a reasonable library of test problems is a useful step towards the development and evaluation of algorithms for PTSP problems. Obviously, as with all NP hard problems, any heuristic or exact methods for solving PTSP problems must rely on exploiting the structure of the data in order to obtain good solutions in a reasonable amount of time. Hence it is meaningless to construct completely random test sets. On the other hand, given the wide range of possible application areas and the large number of variants it is not possible to provide sufficient data sets to reflect the typical structure of data for every instance of the PTSP that practitioners may come across. The choice of data sets presented in this section is heavily influenced by the authors experience and interest rostering for service industries.

Another choice that has to be made is the form in which data sets will be made available. The traditional method of providing text files containing data for individual problem instances on a ftp or web site is too limiting since it restricts the number and size of problem instances available. Hence in this paper we present algorithms for randomly generating problem instances with the type of structure that one might realistically expect to find in real world applications.

We present algorithms for generating each of property of a data set individually. This allows data sets for any variant of the PTSP to be generated by simply only including those features that are relevant. For each feature the parameters used to determine how data is generated are listed and then how these are used.

6.1 Shift times

Parameters: $|W|$ = number of shifts.

In many service industries, the demand for service (or the total amount of work to be done) follows a "double humped" pattern over the course of a day. Hence we will construct a set of shifts that will mirror this type of pattern. The following shifts are allowed:

	Name	Duration	Start time
E	Early Shift	7hrs	6am or 7am
M	Morning Shift	6hrs	8am or 9am
D	Day Shift	8hrs	10am or 11am
A	Afternoon S.	6hrs	2pm or 3pm
L	Late Shift	7hrs	3pm or 4pm

We will assume that all shifts are for a single day over the time period 6am to 8pm. All basic shifts will be eight hours long with start times at 6am, 9am or 12am. The reason for choosing this shift pattern is that it tracks reasonably well "double humped" demand pattern for service that is typical in many service industries. When generating $|W|$ shifts, each shift is assigned to one of the three types with equal probability.

Once a set of shifts is constructed this can be used to determine the supply of labour available. It will be assumed below that this supply correlates with the demand for labour to perform tasks. This is because in most applications, rosters are generated based on expected demand. For the test data sets though we will create this correlation in reverse by generating shifts first and then jobs.

6.2 Task times

Parameters: $|J|$ = number of jobs; T = tightness;

The aim in the method for generating tasks described here is that it produce a reasonable variety, however demand should roughly follow the same distribution as supply. Furthermore most, but not all, problems should be feasible. The number of jobs has been chosen as a parameter since it seems like a natural way to measure problem size while the tightness (density of packing) gives an idea of the difficulty. Variance of job length could be an additional parameter but has not been included for the sake of simplicity.

We use the following algorithm:

1. Assign each shift $\frac{|J|}{2|W|}$ of the shifts. Assign remaining jobs randomly to shifts (weighted by duration of shift so that longer shifts receive more jobs). This ensures a reasonably even distribution of jobs with some randomness and a minimum number of jobs per shift.

2. In each shift w split the duration $T * (F_w - S_w)$ among the jobs assigned to the shift using the algorithm: Each job is at least $T * (F_w - S_w)/10n$ long, where n is the number of jobs assigned to w. The remaining time $9T(F_w - S_w)/10$ is assigned to the jobs by generating a number uniformly in $[0, 1]$ for each job and scaling such that the sum of all job lengths is $T(F_w - S_w)$.

3. In each shift randomly distribute idle time $(1 - T) * (F_w - S_w)$ among the $n + 1$ spaces between the jobs and at the start and end of the shift.

Note that if there are no qualifications this would ensure that a feasible solution exists but given qualification constraints the problem may not be feasible.

6.3 Qualifications

Parameters: Q = number of skills;

The skills are divided into $Q/3$ categories with 3 levels for each skill category (say basic, intermediate and advanced). The point of these is that a person of the higher level always also possesses the lower skill category. This type of hierarchical structure not only reflects real life situations but can also make the PTSP slightly easier. Each person is randomly assigned skills in $\lceil\sqrt{Q/3}\rceil$ categories with probabilities of $\frac{1}{2}$, $\frac{1}{3}$ and $\frac{1}{6}$ of having the lowest to highest skill level in each category. Any person with a higher level of skill is automatically also assigned the lower skills in the same category.

Each job requires one, two or three different qualifications with equal probability. The actual skills are chosen by selecting first the category with uniform likelihood and then the level within the skill category with probability $\frac{1}{2}$, $\frac{1}{3}$ and $\frac{1}{6}$ as above. This ensures that no job requires at most one level of any skill category.

6.4 Assignment Costs

Parameters: C = cost type (1 or 2)

There is a plethora of different task-to-shift assignment costs that might arise in different applications. In order to keep the number of data sets manageable we will consider just two types of costs with significantly different properties:

1. Duration based: here we measure the cost of performing a job by the duration it requires as well as the cost of using a given person. Concretely we calculate the costs of performing a task j in shift w as

$$c_{jw} = (s_j - f_j)/s_w$$

   where s_w is the number of skills possessed by the person. The rational behind this costing is that longer jobs have a higher benefit (eg in a service industry earnings for completing a job would often be related to the length of the job). Hence $s_j - f_j$ gives a negative cost (or benefit). The quotient of s_w represents the fact that the more skills a person has the more expensive it is to employ that person. Hence the earnings are decreased. Note that this method of calculating the costs tends to make the problems easier by encouraging long jobs to be placed with inflexible (low-skilled) workers.

2. Priority based: in some cases we may simply want to prioritise tasks to be done. In this case the following cost could be used

$$c_{jw} = -(2^{p_j}),$$

where p_j is randomly selected in $0 \ldots 10$. The rational for this choice is that a task with priority $p+1$ is twice as important as one with priority p. Note that in this case the cost is not related to the task or shift characteristics. However the problem is made easier by having the assignment cost independent of the shift chosen. Hence one could view this as a two level problem of selecting a set of tasks to be completed such that the selected tasks can feasibly be performed by the available staff.

6.5 *Work Time Cost Function*

We simply provide a single function for the work time that captures most of the typical features that might be considered in this variant. Let s_w be the number of skills of w as defined above. We then define

$$C_w(h) = \begin{cases} s_w h & \text{if } h \leq 5 \\ C_w(5) + 1.5 s_w (h-6) & \text{if } 6 < h \leq 7 \\ C_w(6) + 2 s_w (h-7) & \text{if } 7 < h \leq 8 \\ C(8) + 10 s_w (h-8) & \text{otherwise} \end{cases}$$

This includes normal, time-and-a-half, and double-time rates of pay as well as imposing a spread time constraint, with a large penalty for exceeding 8 hours duration.

Optionally this variant could be solved without the shift start/end times defined in Section 6.1. In this case the PTSP actually determines optimal shifts for the staff in order to complete the given list of tasks with minimum wage costs.

6.6 *Workforce cost*

Parameters: B = cost type (1 or 2)
There are two different types of costs that can be used:

1. Unit cost: Simply minimise the total number of staff used ($b_w = 1$). This is the simplest cost of this type and has been considered in several papers in the literature.

2. Skill based cost: By using $b_w = s_w$, the number of skills of person w, we penalise the use of more expensive machines/staff. A cost of this type, where more powerful machines or highly skilled staff are more expensive, is more realistic in many applications. Note that this will make the problem harder since not using the more flexible staff makes it more difficult to find a feasible solution.

6.7 *Task Length Variation*

The task durations are varied such that more highly qualified people can perform tasks more rapidly. In particular we change the duration by ±20% based on the number of skills s_w of a worker with +20% for $s_w = 1$ and −20% if $s_w = Q$.

6.8 Combining Tasks

Parameters: X percentage of task segments that are part of non-contiguous tasks

The creation of data sets with non-contiguous tasks is performed by combining individual contiguous tasks defined previously. For these data sets we only allow tasks with a single break, although non-contiguous tasks with two or more breaks can easily be generated in a similar manner.

Randomly choose two contiguous tasks that have been "assigned" to the same shift in the generation of task times (Section 6.2) and that have at least one skill requirement in common. Combine these two tasks into a new non-contiguous task with two segments. The skill requirements of this task are set to the intersection of the skill requirements of the two individual tasks. Repeat this procedure until X percent of the segments are part of non-contiguous tasks.

Note that using this method it would generally not be possible to create a data set with $X = 100\%$ as many shifts would have an odd number of tasks assigned to them or they may have tasks with non-intersecting skill requirements. It is envisaged that in most applications these types of non-contiguous tasks are the exception rather than the rule.

6.9 Changeover Times

We generate changeover times for the test data set by assigning each task one of 10 locations (or task types) at random. The changeover time between two tasks of different locations (types) is given in the following table in minutes:

0									
11	0								
8	16	0							
11	6	13	0						
4	14	5	13	0					
9	2	15	7	13	0				
11	5	13	3	13	6	0			
5	10	12	11	8	9	12	0		
15	6	21	12	19	6	11	14	0	
7	9	11	8	9	8	10	4	14	0

Note that we only give the lower triangle as we assume that the changeover times are symmetric.

7 CONCLUSIONS

In this paper we presented several variants of the Personnel Task Scheduling Problem. We have shown that this class of models has many different applications and been studied under different names in very specific forms in the literature. Given the wide variety of applications the total amount of research in all aspects of this area is still very limited. This paper has set out to encourage further research of this class of problems in a more generic setting. To this end we have provided a comprehensive set of test problems which can easily be generated using simple rules. The test problems are designed to reflect real-world structure of typical PTSP problems while being simple enough to generate that they can easily be used as test problems by any researchers in this area.

Further research opportunities exist in using Constraint Logic Programming, Heuristic, and Integer Programming approaches for obtaining solutions to these problems. There is also a scope for a more complete classification of these problems into polynomialy solvable and NP-hard variants.

Acknowledgements

The authors have benefited in the writing of this paper from discussions about this problem with John Beasley, Stephen Bethune, Kim Marriott and Robert Storer.

References

Arkin, M. and Silverberg, B. (1987). Scheduling jobs with fixed start and end times. *Discrete Applied Mathematics*, Vol. 18, pp. 1–8.

Bazaara, M. S. and Goode, J. J. (1979). A survey of various tactics for generating lagrangian multipliers in the context of lagrangian duality. *European Journal of Operations Research*, Vol. 3, pp. 322–338.

Beasley, J. E. (1993). *Modern Heuristic Techniques for Combinatorial Problems, C.R.Reeves (Ed.).* Oxford: Blackwell Scientific Publications.

Carter, M. W. and Tovey, C. A. (1992). When is the classroom assignment problem hard? *Operations Research*, Vol. 40, pp. 28–39.

Dijkstra, M. C., Kroon, L. G., van Nunen, J. A. E. E. and Salomon, M. (1991). A DSS for capacity planning of aircraft maintenance personnel. *International Journal of Production Economics*, **23**, 69–78.

Dondeti, V. R. and Emmons, H. (1992). Fixed job scheduling with two types of processors. *Operations Research*, Vol. 40, pp. 76–85.

Dowling, D., Krishnamoorthy, M., Mackenzie, H. and Sier, D. (1997). Staff rostering at a large international airport. *The Annals of Operations Research*, Vol. 72, pp. 125–147.

Fischer, M. L. (1981). The Lagrangian relaxation method for solving integer programming problems. *Management Science*, Vol. 27, pp. 1–18.

Fischer, M. L. (1985). An applications oriented guide to lagrangian relaxation. *Interfaces*, Vol. 15, pp. 10–21.

Fischetti, M., Martello, S. and Toth, P. (1987). The fixed job schedule problem with spread-time constraints. *Operations Research*, Vol. 35, pp. 849–858.

Fischetti, M., Martello, S. and Toth, P. (1989). The fixed job schedule problem with working-time constraints. *Operations Research*, Vol. 37, pp. 395–403.

Fischetti, M., Martello, S. and Toth, P. (1992). Approximation algorithms for fixed job schedule problems. *Operations Research*, Vol. 40, pp. 96–108.

Ford, L. R. and Fulkerson, D. R. (1962). *Flows in Networks.* Princeton, New Jersey: Princeton University Press.

Geoffrion, A. M. (1974). Lagrangian relaxation for integer programming. *Mathematical Programming Studies*, Vol. 2, pp. 82–114.

Gertsbakh, I. and Stern, H. I. (1978). Minimal resources for fixed and variable job schedules. *Operations Research*, Vol. 26, pp. 68–85.

Glassey, C. R. and Mizrach, M. (1986). A decision support system for assigning classs to rooms. *Interfaces*, Vol. 16, pp. 92–100.

Gondran, M. and Minoux, M. (1984). *Graphs and Algorithms.* A Wiley–Interscience Publication. New York: John Wiley.

Gupta, U. I., Lee, D. T. and Leung, Y. T. (1979). An optimal solution for the channel assignment problem. *IEEE Transactions on Computing*, **C-28**, 807–810.

Held, M., Wolfe, P. and Crowder, H. P. (1974). Validation of subgradient optimisation. *Mathematical Programming*, Vol. 6, pp. 62–88.

Inc., C. O. (1994). *Using the CPLEX callable library (version 3.0).* CPLEX Optimization Inc., Suite 279, 930 Tahoe Blvd., Bldg 802, Incline Valley, NV 89451–9436,USA.

Kolen, A. W. J. and Kroon, L. G. (1991). On the computational complexity of (maximum) class scheduling. *European Journal of Operations Research*, Vol. 54, pp. 23–38.

Kolen, A. W. J. and Kroon, L. G. (1992). License class design: complexity and algorithms. *European Journal of Operations Research*, Vol. 63, pp. 432–444.

Kolen, A. W. J. and Kroon, L. G. (1993). On the computational complexity of (maximum) shift class scheduling. *European Journal of Operations Research*, Vol. 64, pp. 138–151.

Kroon, L. G., Salomon, M. and Wassenhove, L. N. V. (1995). Exact and approximation algorithms for the operations fixed interval scheduling problem. *European Journal of Operations Research*, Vol. 82, pp. 190–205.

Kroon, L. G., Salomon, M. and Wassenhove, L. N. V. (1997). Exact and approximation algorithms for the tactical fixed interval scheduling problem. *Operations Research*, Vol. 45, pp. 624–638.

Lipton, R. J. and Tomkins, A. (1994). Online interval scheduling. *Pages 302–311 of: Proc 5th Annu ACM SIAM Symp Discrete Algorithms.* ACM.

Nakajima, K., Hakimi, S. L. and Lenstra, J. K. (1982). Complexity results for scheduling tasks in fixed intervals on two types of machines. *SIAM Journal on Computing*, Vol. 11, pp. 512–520.

Seiden, S. S. (1998). Randomized online interval scheduling. *OR Letters*, Vol. 22, pp. 171–177.

Valls, V., Pérez, A. and Quintanilla, S. (1996). A graph colouring model for assigning a heterogenous workforce to a given schedule. *European Journal of Operations Research*, Vol. 90, pp. 285–302.

Woeginger, G. J. (1994). On-line scheduling of jobs with fixed start and end times. *Theoretical Computer Science*, Vol. 130, pp. 5–16.

21 ENVELOPE CONSTRAINED FILTER DESIGN: ROBUSTNESS ISSUES

B. Vo

Australian Telecommunications Research Institute
Curtin University of Technology
Australia

and A. Cantoni

Department of Electrical and Electronic Engineering
University of Western Australia
Australia

Abstract: In Envelope Constrained (EC) filter design, the effect of input noise is minimized subject to the constraint that the filter's response to a specified signal fits into a prescribed enevlope. It is invariably the case that the response of the optimal EC filter touches the output envelope at some instances. Consequntly, any disturbance in the prescribed input or implementation error could violate the output constraints. This paper presents a technique survey of progresses in EC filtering, including up-to-date results in EC filtering which is robust to input or implementation uncertainty.

Key words: Filter design, envelope constraints, semi-infinite programming, quadratic programming, robustness.

X. Yang et al. (eds.), Optimization Methods and Applications, 369-397.

1 INTRODUCTION

In signal processing the design of many filters can often be cast as a constrained optimization problem where the constraints are defined by the specifications of the filter. These specifications can arise either from practical considerations or from the standards set by certain regulatory bodies. For example in Telecommunication systems pulse shapes used in transmission systems are specified by recommendations issued by standard bodies (see e.g. Bell Communications (1988), CCITT (1984), Nobakt and Civanlar (1995), Lechleider (1990).

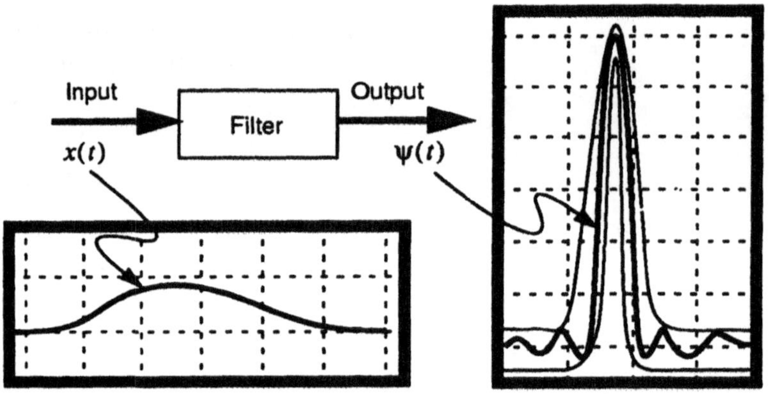

Figure 1. EC filtering

The (continuous-time) envelope-constrained filtering problem considers the design of a filter such that the noiseless response ψ to a specified excitation s fits into a prescribed envelope as shown in Figure 1. In addition various considerations lead us to seek those filters that possess the least output noise power.

McAulay and Johnson (1971) appears to be the first to consider the design of signals and filters subject to sidelobe constraints. The sidelobes are constrained approximately with inequalities at a finite number of points. Around the same time, Fortmann and Athans (1974) investigated similar signal and filter design problems with sidelobes constrained at an uncountable number of points. This approach was extended by Fortmann and Evans (1974), Evans (1975), to allow for time varying sidelobe constraints. This problem was tackled by the primal dual method, which involves transforming the constrained optimization problem into a non-smooth unconstrained dual problem (Fortmann and Evans (1974), Evans (1975)). The unconstrained problem is then discretized into a finite dimensional problem to be solved by a steepest ascent type algorithm (Fortmann and Evans (1974), Evans (1975)). The approximate solutions obtained by this technique are not guaranteed to satisfy the envelope constraints. Only in the limiting case i.e. with infinitesimal discretization that one can be certain the constraints are satisfied.

A discrete-time version of the EC filtering problem was formulated for FIR filters in Evans *et al* (1977). This problem can arise from discrete-time applications or as discrete-time approximation to the EC filtering problem. The primal dual algorithm was also used to solve this problem (Evans *et al* (1977)). One of the desirable feature of this algorithm is the simplicity of its on-line implementation. However, convergence can be quite slow. Recent approaches (Zheng *et al* (1995), Vo *et al* (1997) used penalty method to approximate this problem by an unconstrained problem. This is then solved by applying descent direction base methods such as SD and NR (Pshenichni and Danilin (1978), Dennis and Schnabel (1989), Dennis and Schnabel (1983)). These algorithms shows considerable improvements in the convergence speed. The discrete-time EC filtering problem was also generalized for IIR filters and these algorithms were extended to finite structured filters such as the discrete-time Laguerre network (Zang *et al* (1999)).

As for the continuous-time problem, not much progress had been reported for about two decades. Although Evans (1975) had suggested using finite filter structures to make the problem finite dimensional, the solution method still involves discretizing the dual problem and hence suffers from the associated drawback noted above. Advances in semi-infinite programming techniques have enabled the continuum of constraints in the EC filtering problem to be handled. In particular, the penalty approach (Vo *et al* (1995)) approximates the constrained problem by an unconstrained problem whose solution is guaranteed to satisfy the constraints. This technique was successfully applied to finite structured filter such as the analog Laguerre networks in Vo *et al* (1995). An H-infinity norm cost was consider for EC filters in Zang *et al* (1998), the rationale is to minimize the output noise for the worst case input noise. The most recent work used Caratheodory's dimensionality theorem to reduce the uncountable number constraints of the dual problem to a finite number (Tseng et al (1999)).

Apart from the progress in the optimization methods, the advances in the development of digital processors motivate the consideration of filter structures realized with digital components. While the discrete-time EC filter is well under control, its applications in a continuous time setting is not straight forward; after conversion to continuous time, the constraints are invariably violated. In Vo *et al* (1997) and (2000) the EC filtering problem was formulated for a hybrid filter composed of an A/D converter, a discrete-time filter, a D/A converter and a post-filter. It turns out that this problem assumes the same form as the analog EC filtering problem and hence the previously mentioned techniques can be applied to find the solution.

Assuming that the set of feasible filters does not contain the origin, i.e. no trivial solution, and since we seek feasible filters with smallest possible norm, it follows that the optimum filter always lies on the boundary of the feasible set. This means the response of the optimum filter to the prescribed input touches the output envelope at some points. Consequently, any disturbance in the prescribed input or implementation error could violate the output constraints.

The problem of designing filters which are robust to such disturbances is essential to the implementation of EC filters. Techniques for providing a guard band on the output mask has been addressed in Zang *et at* (1996). The EC with uncertain input (ECUI) filtering problem addresses the robustness to input disturbances by allowing for uncertainty in the input pulse. This is a generalized EC filtering problem where the input s is not specified exactly, but is known to lie within an input envelope. The filter is required to fit the response of all excitations within the input envelope into the output envelope. The ECUI problem was first addressed for FIR filters by Evans *et al* (1983). This problem was solved by combining the primal-dual algorithm with the Goldstein-Levitin-Poljak gradient projection method. In Teo *et al* (1994) it was shown that this problem is equivalent to a standard quadratic programming (QP) problem which can be solved efficiently by known techniques such as QP via active set strategy. However, there has been no study of this problem in a continuous-time setting until recently (Vo *et al* (2000)). For a technically oriented survey of the subject the reader is referred to Cantoni (1998).

In this paper, we present a unified version of the EC filtering problem and address the robustness issue for envelope constrained filters. The rest of the paper is divided into two sections. Section 2 presents the problem formulations for analog and hybrid filters and studies finite dimensional approximation to the problem while Section 3 addresses the robustness issue.

2 ENVELOPE CONSTRAINED FILTERING

Continuous-time filters can be realized by analogue or hybrid means. This section formulates the EC filtering problem for both realizations. For simplicity, we derive the cost function and the constraint set for analog filters first in Section 2.1. In Section 2.2 this process is repeated for hybrid filters. From an optimization view point both problems can be expressed in the same form and this is presented in Section 2.3. Inspecting the dual problem lead to interesting interpretation of the optimal filter configurations in terms of matched and whitening filters. In Section 2.4 we examine finite dimensional approximations and realizability by finite filter structures.

2.1 Analog filter

Let L_2 denote the Hilbert space of real Lebesque-square integrable functions on $\boldsymbol{R}$. For any interval $\Omega \subseteq \boldsymbol{R}$ let be $C(\Omega)$ the Banach space of bounded real continuous functions on Ω with norm $\|x\|_\infty = \sup_{t \in \Omega}|x(t)|$.

The notation Ξ_x is adopted to denote the mapping of u to the convolution product $x \otimes u$, i.e.

$$(\Xi_x u)(t) = (x \otimes u)(t) = \int_{\boldsymbol{R}} x(t-\lambda)u(\lambda)d\lambda. \tag{1}$$

The output constraints do not guarantee that the output is bounded everywhere. For stability, the output should be bounded at all times. The restriction that the filter impulse responses are Lebesque square integrable is not sufficient to guarantee bounded outputs. Boundedness of outputs also depends on the characteristics of the inputs. If the excitation x has finite energy, then for all filter impulse responses u in L_2, the response $x \otimes u$ is continuous and bounded at all times. This justifies the choice of the response space $C(\boldsymbol{R})$. The proof can be found in Lemma 1 (e) pp. 951 of Dunford and Schwartz (1963). While only a sufficient condition, it proves adequate for the problems of interest in this paper.

Output noise power

As discussed in previous sections, in many applications, it is desirable to minimize the noise enhancement due to the filter. Let ξ denote the noise component of the filter output, then $\xi = n \otimes u$. Assuming that the additive noise at the receiver input is zero-mean, stationary, Gaussian, with auto-correlation $R_{nn}(\tau)$, the output noise power is

$$f(u) = E[\xi^2(t)] = \langle u, Lu \rangle .$$

where $L: L_2 \to L_2$ is a non-negative self-adjoint linear operator defined by

$$(Lu)(t) = \int_{\boldsymbol{R}} R_{nn}(t-\lambda)u(\lambda)d\lambda$$

It can be shown that $\sqrt{f(u)}$ is a semi-norm or pseudo-norm on L_2. Consequently, the quadratic cost functional f is convex with respect to $u \in L_2$ although not necessarily strictly convex. Suppose that in addition to being additive, zero-mean, stationary and Gaussian, the input noise n is also *white*, i.e. the auto-correlation $R_{nn}(\tau) = N_0\delta(\tau)$, where $\delta(\tau)$ is a unit impulse. Then, it is easy to see that the output noise power $E[\xi^2(t)]$ is proportional to the square of the L_2-norm of the filter $E[\xi^2(t)] = N_0\|u\|_2^2$. The cost functional in this case is strictly convex.

2.2 Hybrid Filters

The EC filtering problem can be formulated for a hybrid filter composed of an A/D converter, a discrete-time filter, a D/A converter and a post-filter (Vo *et al* (2000)). The post-filter's function is to smooth the output of the D/A block. In practice, the post-filter is often implemented as a lowpass filter with cut-off frequency around half of the sampling frequency. Examples of these filters include Butterworth, Bessel, Chebyshev and elliptic filters. It is assumed that the incoming signal is sampled at or above the Nyquist rate. To simplify matters, quantization effects inherent in digital processes are neglected. This assumption results in the linearity of the system.

Suppose that the digital processor has a discrete-time impulse response $u \in l_2$, the space of square summable sequences. Then the output of the D/A converter when the system is excited by a continuous-time signal x can be modelled by

$$\sum_{i=-\infty}^{\infty} \sum_{j=0}^{\infty} x[(i-j)\tau]u(j)\Pi(t-i\tau) \tag{2}$$

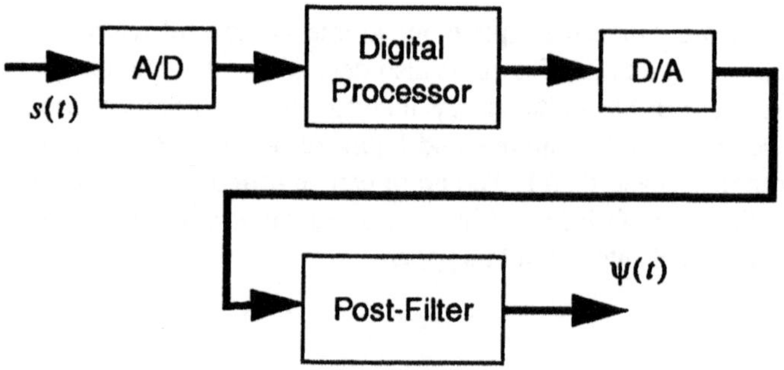

Figure 2. Configuration of system for digital processing of continuous-time signal

where τ is the sampling period of the A/D converter, Π is a rectangular pulse of length τ, i.e.

$$\Pi(t) = \begin{cases} 1, t \in [0, \tau] \\ 0, \text{otherwise} \end{cases}.$$

Let $\Lambda = \Pi \otimes h$ be the response of the post-filter h to the pulse Π, then the response of the hybrid filter to an input x is given by

$$\psi(u, t) = (\Xi_x u)(t) \equiv (\hat{\Xi}_x \Psi u)(t) \tag{3}$$

where $(\Psi u)(t) \equiv \sum_{j=0}^{\infty} \Lambda(t - j\tau)u(j)$ and $(\hat{\Xi}_x v)(t) \equiv \sum_{k=-\infty}^{\infty} v(t - k\tau)x(k\tau)$.

Output Noise Power

The output noise power of the system can be used as the cost functional for the filter optimization problem. Using (3), the output noise can be written as

$$\xi(t) = \sum_{i=-\infty}^{\infty} n(i\tau)(\Psi u)(t - i\tau)$$

Assuming the sequence $\{n(i\tau)\}_{i=-\infty}^{\infty}$ of noise samples is a stationary process with mean μ_n and auto-correlation $R_{nn}(k\tau)$, the mean $\mu_\xi(t)$ and auto-correlation $R_{\xi,\xi}(t + \lambda, t)$ of the output noise are given by

$$\mu_\xi(t) = \sum_{j=-\infty}^{\infty} \mu_n(\Psi u)(t - j\tau),$$

$$R_{\xi,\xi}(t + \lambda, t) = \sum_{j=-\infty}^{\infty} \sum_{k=-\infty}^{\infty} R_{nn}((j - k)\tau)(\Psi u)(t + \lambda - j\tau)(\Psi u)(t - k\tau).$$

It is clear from these expressions that the noise process ξ is not stationary. However, it can be easily verified that ξ is cyclo-stationary with period τ, i.e. $\mu_\xi(t)$ and

$R_{\xi,\xi}(t+\lambda,t)$ are periodic functions of t. Hence, the averaged output noise power $\bar{R}_{\xi,\xi}(0) \equiv (1/\tau)\int_0^\tau E[\xi^2(t)]dt$ is independent of time and is a well-defined cost

$$\begin{aligned} f(u) &= \frac{1}{\tau}\int_0^\tau \sum_{j=-\infty}^{\infty} \sum_{k=-\infty}^{\infty} R_{nn}[(j-k)\tau](\Psi u)(t-j\tau)(\Psi u)(t-k\tau)dt \\ &= \langle u, Lu \rangle \end{aligned} \tag{4}$$

where $L : l_2 \to l_2$ is a symmetric linear operator given by

$$(Lu)(l) = \sum_{m=0}^{\infty} u(m)\Gamma_{l-m}$$

in which

$$\Gamma_{l-m} = \frac{1}{\tau}\sum_{j=-\infty}^{\infty} \sum_{k=-\infty}^{\infty} R_{nn}[(j-k)\tau]\int_0^\tau \Lambda[t-(j+l-m)\tau]\Lambda(t-k\tau)dt .$$

When the noise is independent then, $R_{nn}(k\tau) = N_0\delta_k$ and by making the change of variable $\zeta = t - j\tau$, the cost functional can be rewritten as $(N_0/\tau)\int_{-\infty}^{\infty}[(\Psi u)(\zeta)]^2 d\zeta$. In this case, f is strictly convex, hence it defines a norm on l_2 and L is positive definite.

2.3 Problem statement

In both the analog and hybrid case, the ECUI filtering problem can be stated as an optimization problem in a Hilbert space H (H can be either L_2 or l_2)

$$\min f(u) = \langle u, Lu \rangle$$
$$\text{subject to } u \in H : |\Xi_s u - d| \le \varepsilon$$

where the partial ordering $\ge$ on $C(R)$, the space of continuous functions on R, is generated by the positive cone

$$P_C^+ = \{x \in C(R) : x(t) \ge 0, \forall t \in [0, T_c]\} \tag{5}$$

The interior of P_C^+ is $\{x \in C(R) : x(t) > 0, \forall t \in [0, T_c]\}$ and the boundary of P_C^+ consists of those $x \in P_C^+$ such that $x(t) = 0$ for some $t \in [0, T_c]$. It is easy to verify that P_C^+ is a closed convex cone with non-empty interior. Note that a non-empty interior is essential for the application of the Kuhn-Tucker theorem. However, positive cones corresponding to P_C^+ in $L_p(R)$ have no interior points (Luenberger (1969)).

Assumption: To eliminate output envelopes that permit the trivial solution $u = 0$, it is sufficient to assume the existence of an open subset I of $[0, T_c]$ such that $\varepsilon^+(t)\varepsilon^-(t) > 0, \forall t \in I$. This means that there is an open interval where both the upper and lower boundaries have the same sign. A necessary (but not sufficient) condition for the set $\mathcal{F}$ of feasible filters to have a non-empty interior is obviously $\varepsilon > 0$. This is assumed of the output mask. Because of the nature of the envelope constraints, it follows that if the tolerance ε is too small, there will be no feasible solutions at all. However, if there is a feasible solution, then an infinitesimal increase in ε allows the constraints to be satisfied with strict inequality.

Duality and optimal filter structures

The dual space $C^*(\boldsymbol{R})$ of $C(\boldsymbol{R})$ is well known to be represented by $M(\boldsymbol{R})$, the space of regular Borel measures on $\boldsymbol{R}$ where a linear functional $\phi \in C^*(\boldsymbol{R})$ corresponding to $m \in M(\boldsymbol{R})$ is given by $\phi(x) = m \bullet x = \int_{\boldsymbol{R}} x dm, x \in C(\boldsymbol{R})$. The ε-weighted norm of $m \in M(\boldsymbol{R})$ is given by the total variation $\|m\|_{\varepsilon} = (m_1 + m_2) \bullet \varepsilon$, where m_1 and m_2 are the positive and negative variations of m respectively (Fortmann and Athans (1974)).

The dual positive cone $P_M^{\oplus}$ of P_C^{+}, which generates the partial ordering on $M(\boldsymbol{R})$, consists of non-negative measures m on $\boldsymbol{R}$ which are zero outside $[0, T_c]$ i.e. $m \bullet x = \int_0^{T_c} x dm$ (see Fortmann and Athans (1974)). Consequently, we can regard $P_M^{\oplus}$ as the set of non-negative measures of $M[0, T_c]$. It is well known that $P_M^{\oplus}$ is closed (even if P_C^{+} is not).

It can be shown (Fortmann and Athans (1974), Vo and Cantoni (2000)) that the dual problem is the maximization of the dual cost functional η over $M[0, T_c]$, where

$$\eta(p) \equiv -\frac{1}{4}\langle L^{-1}\alpha(p), \alpha(p)\rangle - p \bullet d - \|p\|_{\varepsilon},$$

$$[\alpha(p)](\tau) = \int_0^{T_c} s(t-\tau)dp(t)$$

for the analog problem and

$$[\alpha(p)](j) = \int_0^{T_c} \sum_{i=0}^{\infty} s[(i-j)\tau]\Lambda(t-i\tau)dp(t)$$

for the hybrid case. Furthermore, the solution u^o to the EC filtering problem related to the dual solution p^o by

$$u^o = -\frac{1}{2}L^{-1}\alpha(p^o)$$

Examining the implications of L^{-1} and $\alpha(p^o)$ leads to the following interpretation of optimal EC filters; For the analog case, the EC filters consists of a whitening filter, a filter matched to the whitened input, followed by a pulse shaping filter determined by p^o as shown in Figure 3, (Fortmann and Athans (1974)). For the hybrid case, the optimal EC filter consists of a whitening filter, a filter matched to the whitened input, followed by a filter matched to the pulse response Λ, which is finally followed by a pulse shaping filter determined by p^o as shown in Figure 4 (Vo and Cantoni (2000)).

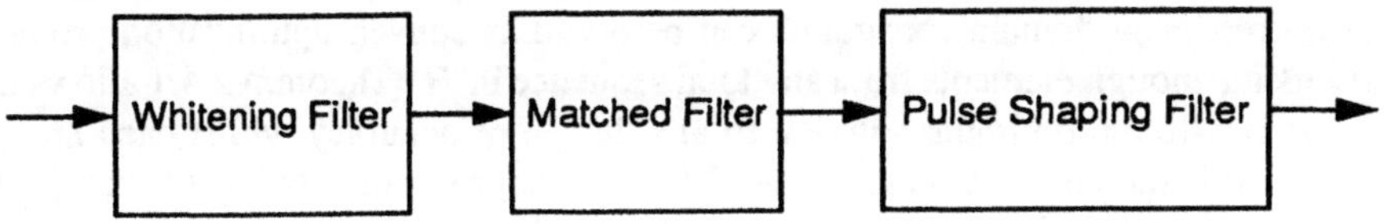

Figure 3. Configuration of optimal analog EC filter

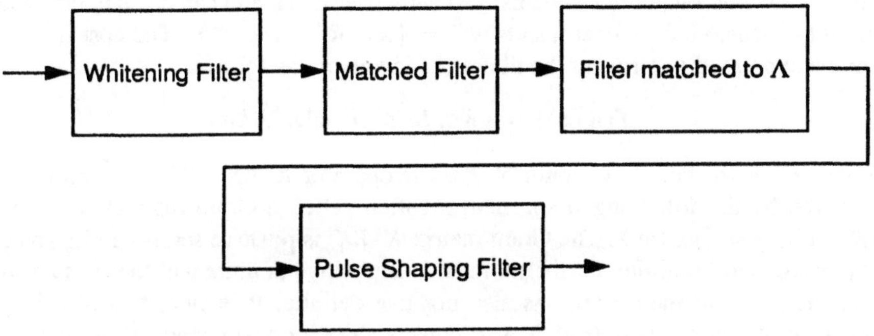

Figure 4. Configuration of optimal hybrid EC filter

2.4 Finite dimensional approximations

The optimum ECUI filter in L_2, for the analog problem or l_2 for the hybrid problem is only of theoretical interest since it is unlikely that these filters can be realized with finite circuitry without violating the constraints. Suppose that one seeks to realize such a filter by designing a network whose impulse response approximates that of the optimum filter in some sense, e.g. least squares, Pade. The important question is "would this approximation still satisfy the constraints?". Thus it would seem more appropriate to choose a particular finite-dimensional filter structure and then impose the constraints. This section addresses the issue of finite dimensional approximation. From a mathematical view point we have the following result, (Vo and Cantoni (2000)).

Theorem 2.4.1. *Let $\mathcal{F}$ be a closed and convex subset of a separable Hilbert space H with non-empty interior, and f a continuous and convex functional on H. Let u^o and u_n^o be minimizers of f on $\mathcal{F}$ and $\mathcal{F} \cap [\{v_i\}_{i=0}^{n-1}]$ respectively ($[\{v_i\}_{i=0}^{n-1}]$ denotes the span of $\{v_i\}_{i=0}^{n-1}$). If $\{v_i\}_{i=0}^{\infty}$ is total in H, then $f(u_n^o)$ monotonically decreases to $f(u^o)$ as n tends to infinity. Moreover, if f is an inner-product-induced norm, then* $\lim_{n \to \infty} \left\| u_n^o - u^o \right\| = 0$.

A large class of filtering problems including frequency domain constraints or time-frequency domain constraints can be posed as convex optimization problems. By using enough elements from any total sequence in H, Theorem 2.4.1 allows us to approximate the optimum solution to any degree of accuracy whilst each approximant still satisfies the constraints. With an appropriate choice of basis these sub-optimum solutions correspond to finite-structured filters that can be readily realized.

Let the finite-structured filters have impulse responses given by Ka, where $K : R^n \to H$ is a bounded linear operator defined by $Ka = v^T a$ with $\{v_i\}_{i=0}^{n-1}$ forming n elements of total sequence in H. The feasible region $\mathcal{F} \cap [\{v_i\}_{i=0}^{n-1}]$ is hence embedded in a finite-dimensional sub-space of H and can be characterized by the set of feasible filter coefficients $\mathcal{F}^n = \{a \in R^n: Ka \in \mathcal{F}\}$. The cost functional can be expressed in terms of the filter coefficients a, as follows

$$f(K(a)) = \langle Ka, LKa\rangle = a^{\dagger}K^{\dagger}LKa.$$

where $K^{\dagger}$ is the adjoint operator of K. The operator $K^{\dagger}LK : R^n \to R^n$ can be represented by the following $n \times n$ matrix (often called a Gram matrix) with entries $[K^{\dagger}LK]_{i,j} = \langle v_i, Lv_j\rangle$. The Gram matrix $K^{\dagger}LK$ is positive semi-definite since L is positive semi-definite. Furthermore if L is positive definite and the v_i's are linearly independent then $K^{\dagger}LK$ is also positive definite. It is clear that if f is convex/strictly convex then $f \circ K$ is also convex/strictly convex since K is linear. For analog filters

$$\langle v_j, Lv_k\rangle = \int_{\Omega_u}\int_{\Omega_u} R_{nn}(\kappa - \lambda)v_j(\lambda)v_k(\kappa)d\kappa d\lambda \tag{6}$$

and for hybrid filters,

$$\langle v_j, Lv_k\rangle = \sum_{l \in \hat{\Omega}_u}\sum_{m \in \hat{\Omega}_u} v_j(l)v_k(m)\Gamma_{l,m} \tag{7}$$

These finite-structured filters restrict the space of filter impulse responses to a finite dimensional subspace of H. That is, the optimization is carried out on this subspace rather than on the whole of H. The subspace spanned by $\{v_i\}_{i=0}^{n-1}$ is closed and convex, hence, the feasible region $\mathcal{F} \cap [\{v_i\}_{i=0}^{n-1}]$ is also closed and convex. Strict convexity of the cost functional ensures a unique optimum solution if an interior point exists. For a norm cost, it follows from Theorem 2.4.1 that the sub-optimum filters for the finite filter structures converge to the optimum filter as the filter structure grows.

Finite structure

In most practical analog realization K is specified by a finite set of orthonormal v_i's. The main reason for considering filter realization by orthonormal functions is that it is possible to generate and combine certain set of vectors using simple finite filter structures. The most notable examples are Laguerre and Legendre functions (Szego (1939), Makila (1990)).

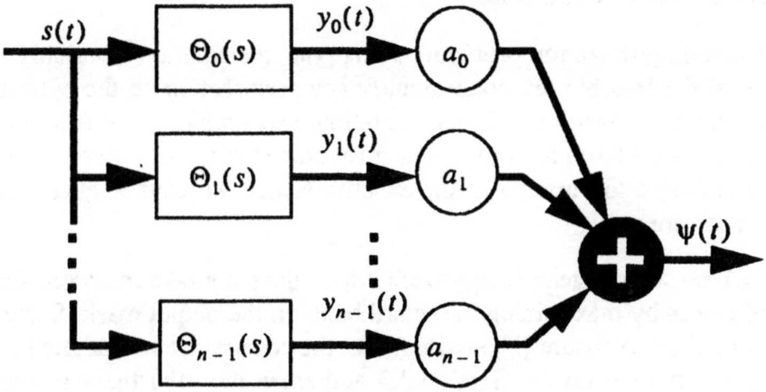

Figure 5. Parallel filter structure

There are two main techniques for building filters. The parallel structure is the simplest and most direct method (see Figure 5). With the transfer function for each block denoted by $\Theta_i(s)$, $i = 0, 1, \ldots, n-1$, the transfer function of the filter is $U(s) = \sum_{i=0}^{n-1} a_i \Theta_i(s)$. In this case, $v_i = \theta_i$, where θ_i is the inverse Laplace transform of Θ_i. The parallel structure is widely used for many filters employing orthonormal functions.

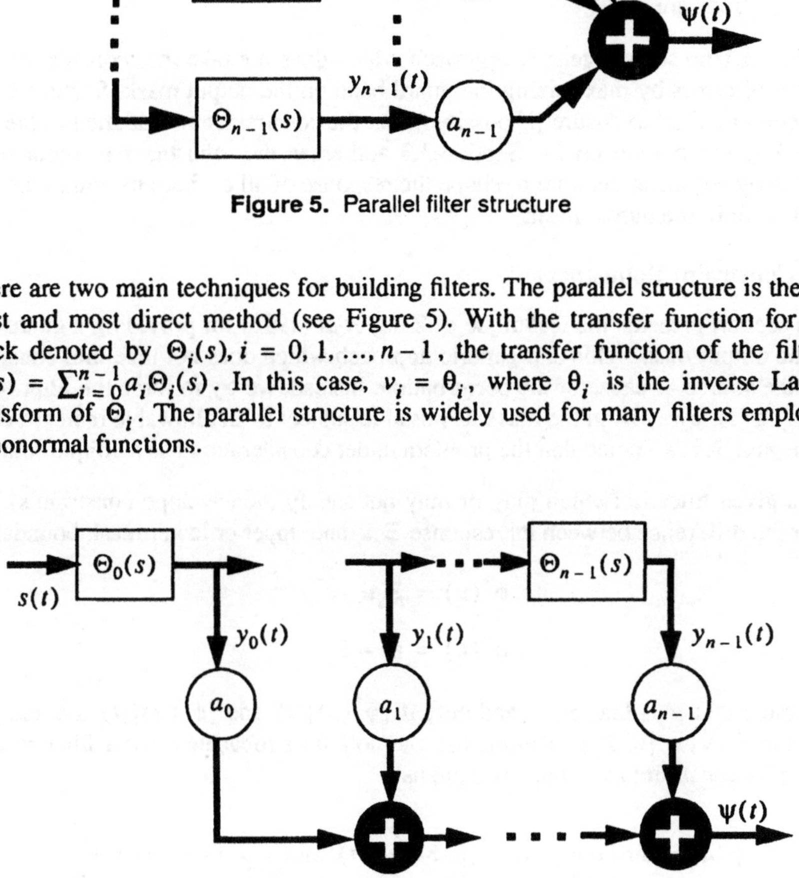

Figure 6. Transversal filter structure

The second method is the transversal structure, (see Figure 6). The best known examples of filters employing the transversal structure are Legendre or Laguerre filters or when the Θ_i's are pure delay elements. Let V_i denote the Laplace transform of v_i. Then the transfer function of the transversal filter is $U(s) = \sum_{i=0}^{n-1} a_i V_i(s)$ where $V_i(s) = \prod_{j=0}^{i} \Theta_j(s)$.

The digital component of the hybrid filters can also be realized in the same way.

3 ROBUSTNESS ISSUES

In constrained optimization problem of this type, the solution invariably lies on the boundary of the feasible set, consequently any perturbation on the optimal solution would violate the constraints. Implementation errors which arise from quantization effect (for digital filters) or component mismatch (for analog filters) means that we cannot implement the optimal filter exactly. Hence we cannot guarantee that the specifications are met.

Section 3.1 presents a generic approach which does not take into consideration the source of errors by maximizing the guard band on the output mask. Section 3.2, we present a method to ensure (if possible) that the constraints are satisfied in the presence implementation errors. Section 3.3 addresses the robustness to input disturbances by requiring the filter to shape the response of all excitations within the input mask to fit in the output mask.

3.1 Constraint Robustness

This section presents the technique of Zang *et al* (1996) for providing a guard band on the output mask. This is a generic approach which does not take into consideration the source or cause of the perturbation. Instead, we try to force the filter output to stay as far away from the mask as possible subject to an allowable increase on the noise gain. It is assumed that the problem under consideration has a unique solution.

For a given filter u (which may or may not satisfy the envelope constraints), consider the difference between its response $\Xi_s u$ and upper or lower mask boundaries.

$$\begin{aligned} \phi^+(u) &= \Xi_s u - \varepsilon^+ \\ \phi^-(u) &= \varepsilon^- - \Xi_s u \end{aligned} \tag{8}$$

It is clear that u is feasible if and only if $[\phi^+(u)](t)$ and $[\phi^-(u)](t)$ are non-positive for all $\forall t \in [0, T_c]$. To quantify the notion of robustness for a filter u let us define its constraint robustness margin as

$$\rho(u) \equiv \min\left\{\min_{t \in [0, T_c]}[-\phi^+(u)](t), \min_{[0, T_c]}[-\phi^-(u)](t)\right\} \tag{9}$$

$\rho(u)$ measures how far the filter output is from the envelope. Clearly u is a feasible point of the EC filtering problem if and only if $\rho(u) \geq 0$ (hence u is not feasible if and only if $\rho(u) < 0$). If $\rho(u) \geq 0$, then $\rho(u)$ is the smallest distance of the output $\Xi_s u$ from the boundary of the mask. Therefore we may say that the filter u is robust with constraint robustness margin $\rho(u)$. On the other hand if $\rho(u)$ is negative, then u is infeasible and $|\rho(u)|$ give the largest distance of the output $\Xi_s u$ from the boundary of the mask.

Let u^o denote the optimal solution of the EC filtering problem (note that $\rho(u^o) = 0$, since the optimal filter output touches the mask at some point). Hence, any filter with a positive robustness margin has a noise gain not less than $u^{\tilde{}}$ (otherwise would contradict the optimality of u^o). To achieve robustness, we must allow for the increase in noise gain and maximize the robustness margin as shown in the following constrained optimization problem

$$\max \rho(u)$$
$$\text{subject to } \|u\|_L^2 \leq (1+\delta)\|u^o\|_L^2 \qquad \text{(P.1)}$$

where $\delta > 0$ is a constant which specifies the allowable amount of increase of the output noise power. Let u^* denote an optimal solution of problem (P.1), then it is clear that $\rho(u^*) \geq 0$, since u^o satisfies the constraints of problem (P.1) and $\rho(u^o) = 0$.

Problem (P.1) can be converted to the following standard maximization problem

$$\max f(u, \sigma) = \sigma, (u, \sigma) \in L_2 \times \boldsymbol{R} \qquad \text{(P.2)}$$

subject to

$$g_1(u, \sigma) \equiv \sigma + \phi^+(u) \leq 0 \qquad (10)$$

$$g_2(u, \sigma) \equiv \sigma + \phi^-(u) \leq 0 \qquad (11)$$

$$g_3(u, \sigma) \equiv \|u\|_L^2 - (1+\delta)\|u^o\|_L^2 \leq 0 \qquad (12)$$

Intuitively, for positive σ the constraints (10) and (11) effectively shrinks the mask about the "desired" response d as shown in Figure 7. The larger σ is, the tighter the mask becomes. Conversely for negative σ, the mask is expanded about d. Thus, by maximizing σ, we force the filter output away from the boundary. Note that if (u^*, σ^*) is the solution of problem (P.2), then $\sigma^* \geq 0$ and the distance between the output wave form $[\Xi_s u^*](t)$ and the output mask is at least equal to $\rho(u^*)$. Problems (P.1) and (P.2) are equivalent in the sense that if (u^*, σ^*) is a solution to problem (P.2) then u^* is a solution to problem (P.1) and $\rho(u^*) = \sigma^*$. Conversely if u^* is a solution to problem (P.1) then $(u^*, \rho(u^*))$ is a solution of problem (P.2).

In practice, we may require a larger constraint robustness margin on certain intervals. In this case, a weighting function β can be used to specify the weightings in different time intervals. Define the weighted robustness constraints as

$$g_1(u, \sigma) \equiv \sigma\beta + \phi^+(u) \leq 0$$

$$g_2(u, \sigma) \equiv \sigma\beta + \phi^-(u) \leq 0$$

where β is a positive piecewise continuous weighting function. We normalize the weighting function so that $\max_{t \in [0, T_c]} \beta(t) = 1$. In this form, we can see that the envelope is also tightened, however, not by a constant amount but by a variable amount specified by the weighting function β. In other words β is used to specify

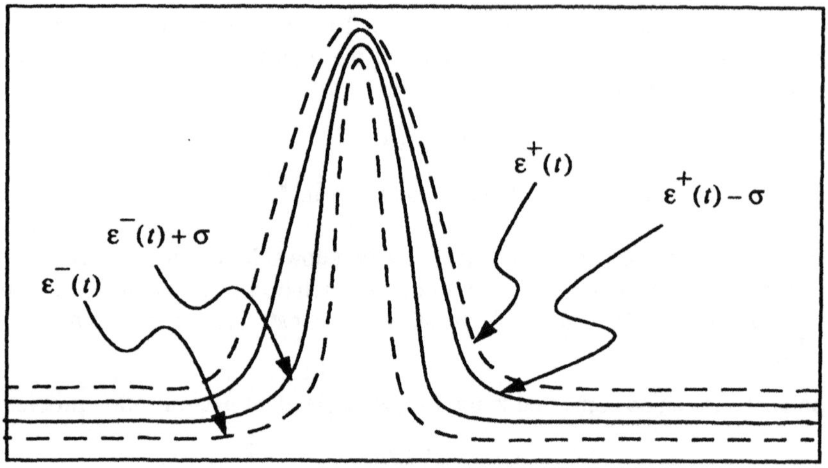

Figure 7. Weighted constraint robustness margin

the weightings in different time intervals. All previous results follows if we define the cost of problem (P.1) to be

$$\rho_\beta(u) \equiv \min\left\{ \min_{t \in [0, T_c]} \frac{[-\phi^+(u)](t)}{\beta(t)}, \min_{t \in [0, T_c]} \frac{[-\phi^-(u)](t)}{\beta(t)} \right\}$$

Numerical example with Laguerre filter

Consider the design of an equalization filter for a digital transmission channel consisting of a coaxial cable on which data is transmitted according to the DX3 standard (Bell Communications (1988), CCITT (1984)). The design objective is to find an equalizer which takes the impulse response of a coaxial with a loss of 30 dB at the Baud frequency and produces an output which lies within the DSX3 pulse template. For computational purpose, we consider the time domain constraints at discrete a finite number of points rather than the entire continuum.

In our numerical studies, we use 1024 points over an interval of $[0, 32\beta]$. Using a Laguerre filter with 14 coefficients and $p = 12$, we first solve the optimal EC filtering problem for which the optimal noise gain is $\|u^o\|^2 = \|a^o\|^2 = 54.2008$. The output mask, input signal and the output are shown in Figure 8. Figure 9 shows the frequency response of the unequalized cable and the equalized cable. To achieve maximum constraint robustness margin with $\delta = 1.5$, we solve the robust EC filtering problem for

$$\beta(t) = \begin{cases} 3\varepsilon(t), |\varepsilon(t)| \geq 0.05 \\ 0.035, |\varepsilon(t)| \leq 0.03 \\ \varepsilon(t), \text{elsewhere} \end{cases}.$$

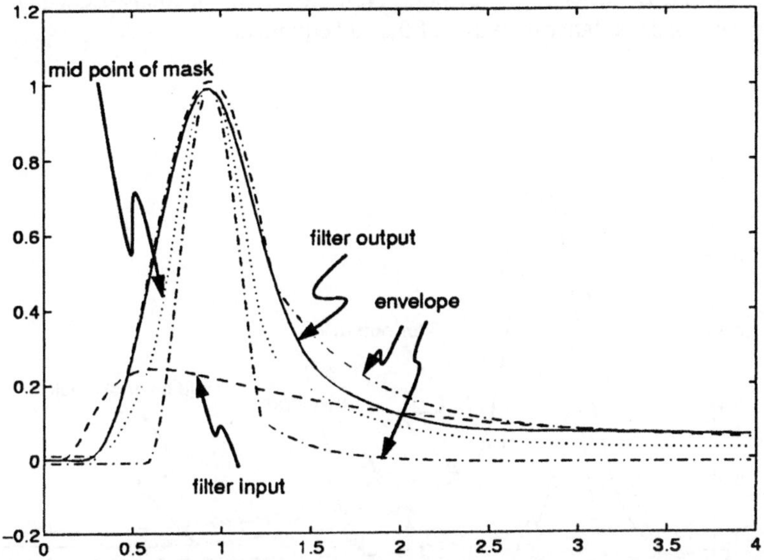

Figure 8. Laguerre filter - EC approach

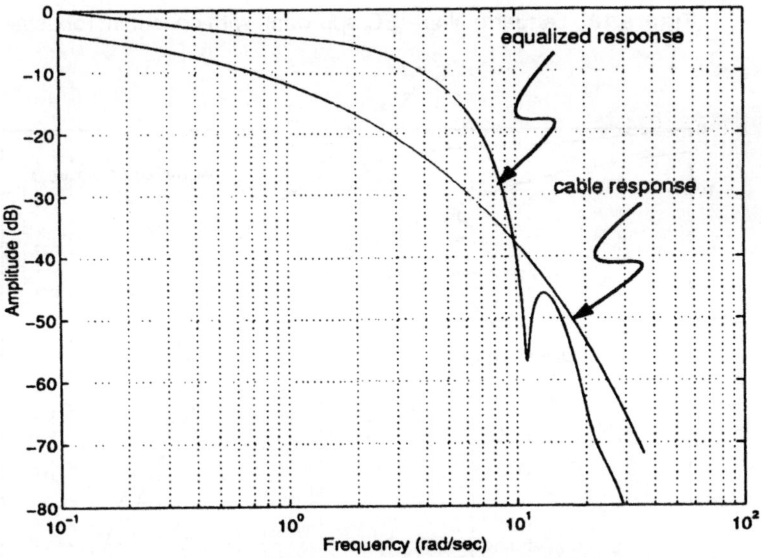

Figure 9. Magnitude response EC approach

The corresponding output is shown in Figure 10 and the frequency response is shown in Figure 11. From Figure 10, we see that the output is very close to the center of the output mask. The robustness to constraint violation of this filter in comparison to the optimal EC filter is self-evident. Note also that the robust-EC equalized frequency response is flatter than that of the EC equalized.

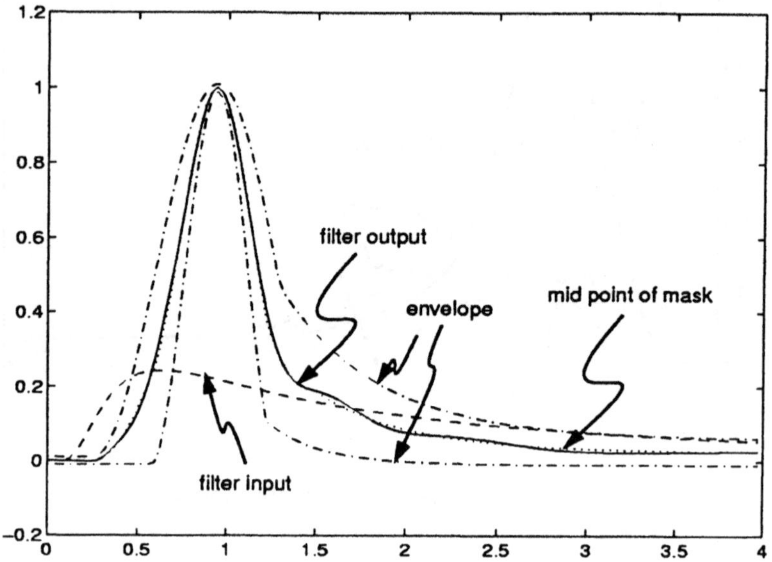

Figure 10. Laguerre filter - EC approach with constraint robustness

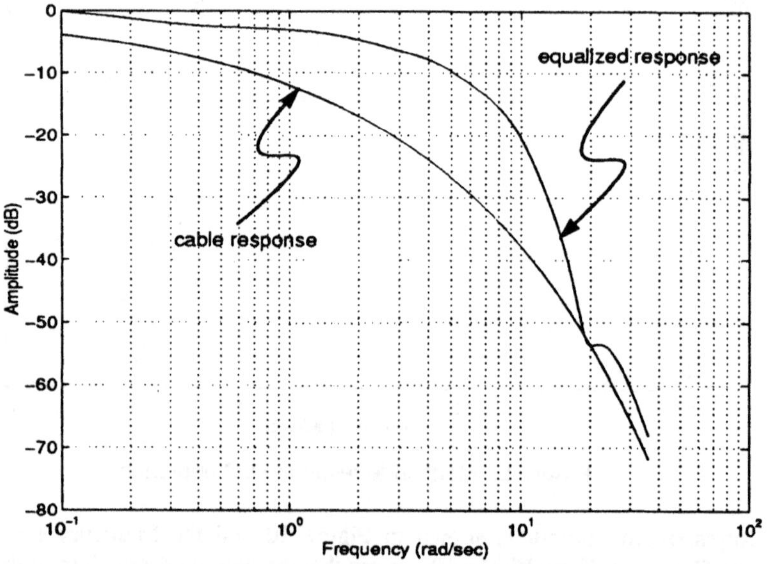

Figure 11. Magnitude response with constraint robustness

3.2 Envelope Constrained Filtering with Uncertain Implementation.

In practice it is often the case that a designed filter or system impulse response cannot be implemented exactly, there are always implementation errors. These errors can arise from component mismatch or quantization of filter coefficients. For example when digital techniques are used, the actual impulse response achieved differs from the desired value because of finite word lengths. As a result the constraints will invariably be violated. The robustness margin approach cannot guarantee feasibility in the presence of implementation errors. In this section, we present a method to determine filters (if they exists) which ensure that the constraints are satisfied in the presence of implementation errors.

It is assumed that the implementation errors are bounded in the sense that for a given filter u, we can achieve actual implementations that are known to be within δ of u, where δ is a bounded function. In other words, any implementation of u is known to belong to the set $\mathcal{V}(u, \delta)$ defined by

$$\mathcal{V}(u, \delta) \equiv \{v \in L_2 : |v - u| \leq \delta\}$$

where the partial ordering $\leq$ on L_2 is defined by $x, y \in L_2, x \leq y \Leftrightarrow \forall t \in R, x(t) \leq y(t)$. If δ is a constant, the region $\mathcal{V}(u, \delta)$ correspond to a ball centered at u with radius δ, if we use the sup-norm metric $\|v - u\| = \sup_{t \in R}|(v - u)(t)|$. Non constant δ reflects non-uniform weightings, regions with large δ mean large uncertainty and vice-versa.

In the EC filtering problem we seek filters u whose response to a prescribed signal s stays within some specified output envelope. With uncertainty in the implementations, we are interested in those u such that the response of any elements of $\mathcal{V}(u, \delta)$ to the signal s stay inside the output envelope. In both the analog and hybrid case, the set of feasible filters (in the filter space H which can be either L_2 or l_2) is given by

$$\mathcal{F} = \{u \in H : |\Xi_s v - d| \leq \varepsilon, \forall v \in H : |v - u| \leq \delta\}.$$

Although the same notation is used for both partial orderings, there should be no ambiguity if the signal space is taken into consideration. On L_2 the inequality is required to hold for the entire support, while on $C(R)$, the inequality only holds on $[0, T_c]$.

The u that solves this problem is such that any implementation v that is within δ of u will still force the filter output to be within the prescribed output mask. The description of the feasible region in terms of the set $\mathcal{V}(u, \delta)$ is adequate for characterizing the problem but it is not useful for computational purposes. To test the feasibility of a filter one would need to compute the response of every filter in $\mathcal{V}(u, \delta)$. There is no standard numerical techniques for handling problems with constraints of this form. The following result allows the constraint to be written in a more tractable form (Vo and Cantoni (2000)).

Lemma 3.2.1. *Suppose* X_1 *and* X_2 *are Banach spaces and each* X_j *is equipped with a partial ordering* $x \le y \Leftrightarrow x(t) \le y(t), \forall t \in \Omega_j$, *for all real-valued* x, y. *Let* $F : X_1 \to X_2$ *and* $|F| : X_1 \to X_2$ *be two linear operators defined by* $(Fx)(t) \equiv \int_{\Omega_1} x(\tau)f(t, \tau)d\tau$ *and* $(|F|x)(t) \equiv \int_{\Omega_1} x(\tau)|f(t, \tau)|d\tau$ *respectively. Then, for a given positive* $u \in X_1$, $|Fx - d| \le \varepsilon, \forall x: |x - s| \le u$ *is equivalent to* $|Fs - d| + |F|u \le \varepsilon$.

Identifying $F : H \to C(R)$ by $Fu = \Xi_s u$, and using Lemma 3.2.1, we have

$$|\Xi_s v - d| \le \varepsilon, \forall v: |v - u| \le \delta \text{ iff } |\Xi_s u - d| + |\Xi_s|\delta \le \varepsilon.$$

Thus to ensure robustness to implementation errors, we effectively tighten the output mask by reducing ε to $\varepsilon - |\Xi_s|\delta$. Of course if the uncertainty δ in the implementation is too large, the output mask is tighten beyond what would admit a feasible solution. We can determine from the implementation uncertainty δ whether the robustness can be achieved. For instance if there exists an interval where $(|\Xi_s|\delta)(t) \ge \varepsilon(t)$, then there is no feasible solution. Note that for the analog case the kernel of Ξ_s is $f(t, \tau) = s(t - \tau)$ while for the hybrid case the kernel is

$$f(t, l) = \sum_{k=0}^{\infty} \Lambda(t - kT_s - lT_s)s(kT_s).$$

Now we have an expression for the constraints which involves only d, ε, u, δ, and the input s. The EC filtering problem with uncertain implementation can now be written as an optimization problem in a Hilbert space H

$$\min f(u) = \langle u, Lu \rangle$$
$$\text{subject to } u \in H: |\Xi_s u - d| + |\Xi_s|\delta \le \varepsilon$$

This problem has the same form as the EC filtering problem. Here we are minimizing the cost (e.g. noise gain) of the theoretical filter u (i.e. without implementation error). The actual implementation $\hat{u}^o$ of the optimal theoretical filter u^o is not known exactly. The cost at $\hat{u}^o$ may be larger or smaller than that of u^o, but its response to the signal s still stays inside the output envelope.

Examples with Finite structured filters

For filters with finite structures $(F\boldsymbol{a})(t) \equiv \boldsymbol{y}^T(t)\boldsymbol{a}$, using Lemma 3.2.1, we have.

$$|\boldsymbol{y}^T(t)\hat{\boldsymbol{a}} - d| \le \varepsilon, \forall \hat{\boldsymbol{a}}: |\hat{\boldsymbol{a}} - \boldsymbol{a}| \le \delta \text{ iff } |\boldsymbol{y}^T(t)\boldsymbol{a} - d(t)| + |\boldsymbol{y}^T(t)|\delta \le \varepsilon.$$

As an example consider a pulse compression application using a hybrid configuration with a 27 tap FIR filter and a 6th order Bessel post-filter with cut-off frequency of $2\pi/\beta$ (β is the bit interval). The response of the system to a 13 bit Barker coded input is required to fit in the mask of Figure 12. The solution $\boldsymbol{a}^o$ has entries in the range $[-0.06, 0.06]$. For an 8-bit representation of $\boldsymbol{a}^o$ on an interval $[a, b]$, the bound ς on the quantization error is $(b - a)2^{-9}$. Using 8-bit representation of $\boldsymbol{a}^o$ on $[-0.065, 0.065]$, the bound on the quantization error is $\varsigma = 0.13 \times 2^{-9}$. Figure

12 shows the responses of the perturbations $\hat{a}$ within $\delta = \varsigma[1, \ldots, 1]^T$ of a^o in the sense that $|\hat{a} - a^o| \le \delta$.

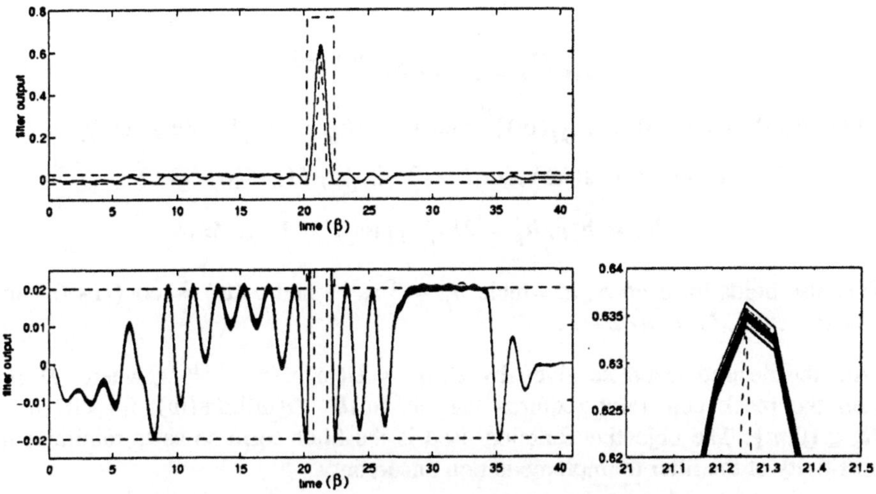

Figure 12. Response of perturbed filter

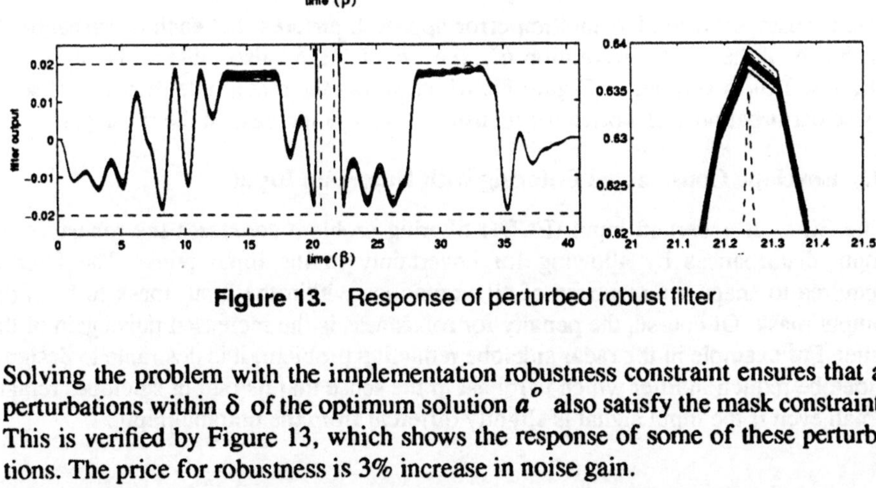

Figure 13. Response of perturbed robust filter

Solving the problem with the implementation robustness constraint ensures that all perturbations within δ of the optimum solution a^o also satisfy the mask constraints. This is verified by Figure 13, which shows the response of some of these perturbations. The price for robustness is 3% increase in noise gain.

To illustrate the application of Lemma 3.2.1 to problems other than EC filters, we consider the design of linear phase (FIR) filters with magnitude response specifications. For linear phase filters, the filter coefficients $\tilde{h}_i$, $i = 0, \ldots, n-1$ are symmetric about the middle coefficient, assuming n is odd. The frequency response is given by

$$H(e^{j\omega}) = c^T(\omega)he^{-j\omega(n-1)/2}$$

where $c(\omega) = [c_0(\omega), \ldots, c_M(\omega)]^T$ and $h = [h_0, \ldots, h_M]^T$ are given by

$$c_i(\omega) = \cos(i\omega), \text{for } i = 0, \ldots, M, M = (n-1)/2$$

$$h_0 = \tilde{h}_M, h_i = 2\tilde{h}_{M-i}, \text{for } i = 1, \ldots, M.$$

This also holds for even n, in which, $h_i = 2\tilde{h}_{M-i}$ and $c_i(\omega) = \cos((i+0.5)\omega)$, $i = 0, \ldots, M$, $M = n/2 - 1$.

With the desired response given by $d(\omega) = D(\omega)e^{-j\omega(n-1)/2}$, where $D \geq 0$, then the mask constraint requires that $|c^T(\omega)h - D(\omega)| \leq \varepsilon(\omega)$ for all ω on $\Omega_c \subseteq [0, \pi]$. The objective function used is the filter norm so as to minimize the sensitivity of the filter to implementation uncertainty.

For a 22-tap linear phase lowpass filter with

$$D(\omega) = \begin{cases} 1, \omega \in [0, 0.3\pi] \\ 0, \omega \in [0.45\pi, \pi] \end{cases}, \varepsilon(\omega) = \begin{cases} 0.050, \omega \in [0, 0.3\pi] \\ 0.025, \omega \in [0.45\pi, \pi] \end{cases},$$

the optimal solution h^o has entries take on value in the range $[-0.13, 0.67]$. For convenience we choose the interval $[-0.2, 0.8]$ and using 8-bit representation, hence $\varsigma = 2^{-9}$. Figure 14 shows the magnitude responses of perturbations $\hat{h}$ within $\delta = \varsigma[1, \ldots, 1]^T$ of h^o in the sense that $|\hat{h} - h^o| \leq \delta$. Note the constraint violation near 0, 0.15 and on the stopband.

The robustness to implementation error approach ensures that each perturbation $\hat{h}$ within δ of the optimum solution h^o, (i.e. $|\hat{h} - h^o| \leq \delta$) also satisfies the mask constraints. This is verified by Figure 15, which shows the magnitude response of some these perturbations. The price for robustness is a 4% increase in the noise gain.

3.3 Envelope Constrained Filtering with Uncertain Input

The EC with uncertain input (ECUI) filtering problem addresses the robustness to input disturbances by allowing for uncertainty in the input pulse. The filter is required to shape the response of all excitations within the input mask to fit in the output mask. Of course, the penalty for robustness is the increased noise gain of the filter. For example in the radar sidelobe reduction problem, it is desirable to design a sidelobe reduction filter which is robust in the sense that its output sidelobes remain small even if the input signal is slightly different from the nominal input.

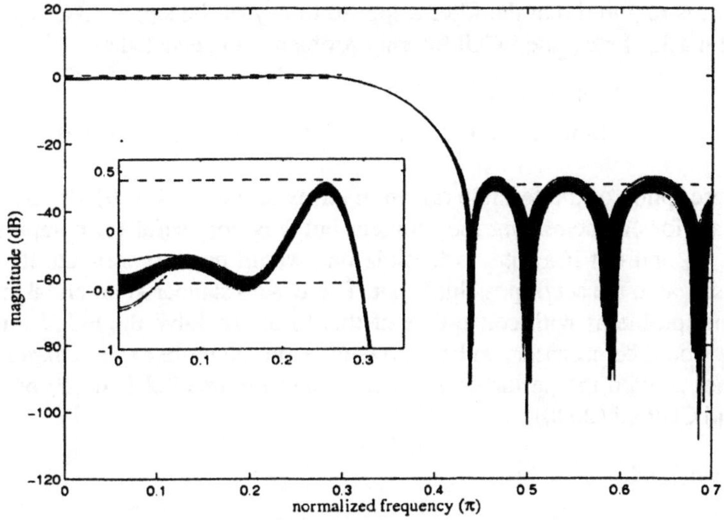

Figure 14. Magnitude response of perturbations

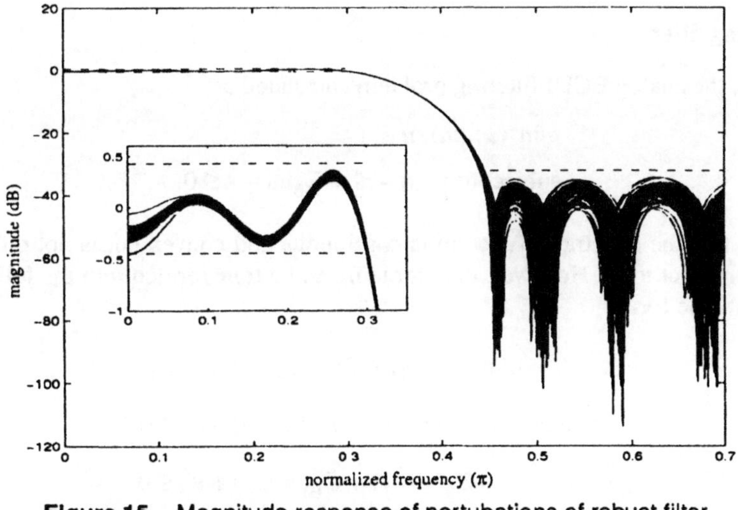

Figure 15. Magnitude response of pertubations of robust filter

Let the input envelope be described by upper and lower boundaries s^+, $s^- \in L_2$, then the set of allowable inputs is given by

$$\mathcal{S}(s, \theta) \equiv \{x \in L_2 : |x - s| \leq \theta\}$$

where $s \equiv 0.5(s^+ + s^-)$, $\theta \equiv 0.5(s^+ - s^-)$ and the partial ordering $\leq$ on L_2 is defined as before. It is assumed that $\theta \geq 0$ and s, θ are finite energy signals so that the set of allowable input is non-empty and bounded. In both the analog and hybrid cases, it is required that the filter response to any of the signals $\mathcal{S}(s, \theta)$ fits into the output mask. Hence, the ECUI filtering problem can be stated as

$$\begin{aligned} &\min f(u) = \langle u, Lu \rangle \\ &\text{subject to } u \in H : \left|\Xi_x u - d\right| \le \varepsilon,\ \forall x \in L_2 : |x - s| \le \theta \end{aligned}$$

The description of the feasible region in terms of the set $\mathcal{S}(s, \theta)$ of possible input is adequate for characterizing the problem but it is not useful for computational purposes. To find out if a filter is feasible one would need to compute its response to every signal in the set of possible input. There is no standard numerical technique for handling problems with constraints of this form. To solve the ECUI filtering problem, it would be necessary to have an equivalent but more explicit expression which does not contain the perturbed input x. Using Lemma 3.2.1, it can be shown that, (Vo and Cantoni (2000))

Theorem 3.3.1.

(i) $\left|\Xi_x u - d\right| \le \varepsilon,\ \forall x \in L_2 : |x - s| \le \theta$ *iff* $\left|\Xi_s u - d\right| + \Xi_\theta |u| \le \varepsilon.$

(ii) $\left|\hat{\Xi}_x \Psi u - d\right| \le \varepsilon,\ \forall x \in L_2 : |x - s| \le \theta$ *iff* $\left|\hat{\Xi}_s \Psi u - d\right| + \hat{\Xi}_\theta |\Psi u| \le \varepsilon.$

Analog filter

Thus, the analog ECUI filtering problem can stated as:

$$\begin{aligned} &\min \langle u, Lu \rangle,\ u \in L_2 \\ &\text{subject to } \left|\Xi_s u - d\right| + \Xi_\theta |u| - \varepsilon \le 0. \end{aligned} \qquad \text{(P.1)}$$

Note that the constraint function is continuous and convex but is not differentiable with respect to u. However, the problem can be transformed into the following differentiable form.

$$\begin{aligned} &\min \|x - y\|_L^2 + 2\rho\langle x, y \rangle,\ x, y \in L_2 \\ &\text{subject to } \Xi_s(x - y) + \Xi_\theta(x + y) - \varepsilon^+ \le 0 \\ &\qquad -\Xi_s(x - y) + \Xi_\theta(x + y) + \varepsilon^- \le 0 \\ &\qquad -x \le 0 \\ &\qquad -y \le 0 \end{aligned} \qquad \text{(P.2)}$$

where $\rho \in (0, 2\lambda_{min}(L))$.

The cost functional $f(x, y) \equiv \|x - y\|_L^2 + 2\rho\langle x, y \rangle$ is positive definite (Vo and CAntoni (2000)). Since the cost functional of (P.2) is strictly convex, the solution (x^o, y^o) is unique.

The ordered pair consisting of positive and negative parts of the solution $u^{\sim}$ to problem (P.1) is, in fact, a solution to problem (P.2). More importantly; if a solution (x^o, y^o) to problem (P.2) is found then we can obtain the solution u^o by the equation $u^o = x^o - y^o$. Note that x^o and y^o respectively may not necessarily be the positive and negative parts of u^o (Vo and Cantoni (2000)).

Hybrid filter

The hybrid ECUI filtering problem can be stated as:

$$\begin{aligned} &\min \langle u, Lu \rangle,\ u \in L_2 \\ &\text{subject to } |\hat{\Xi}_s \Psi u - d| + \hat{\Xi}_\theta |\Psi u| - \varepsilon \leq 0. \end{aligned} \tag{P.3}$$

The constraint function, similar to the analog case, is continuous, convex and non-differentiable. Again, this is generally a non-smooth optimization problem. However, this problem can not be transformed to a smooth form using the previous technique.

Finite dimensional analog filter

For finite filter structures of the form discussed in Section 2.4, we replace the filter impulse response u in problems (P.1) and (P.3) with Ka. The analog ECUI problem for finite filter structures also cannot be transformed using the previous technique. Supposing that the image $K(R^n)$ is closed under taking the operations of taking the positive and negative parts, then Theorem 3.3.1 still holds when we replace x and y in (P.1) and (P.2) by Kx and Ky respectively (Vo and Cantoni (2000)). In this case the problem can be transformed to the following finite-dimensional QP problem.

$$\begin{aligned} \min\ & z^\dagger H z,\ z \in R^{2n} \\ \text{subject to }\ & \Xi_s K[I_n, -I_n]z + \Xi_\theta K[I_n, I_n]z - \varepsilon^+ \leq 0 \\ & -\Xi_s K[I_n, -I_n]z + \Xi_\theta K[I_n, I_n]z + \varepsilon^- \leq 0 \\ & -K[I_n, 0_n]z \leq 0 \\ & -K[0_n, I_n]z \leq 0 \end{aligned}$$

where H is positive definite and is given by

$$H = \begin{bmatrix} K^\dagger LK & \rho K^\dagger K - K^\dagger LK \\ \rho K^\dagger K - K^\dagger LK & K^\dagger LK \end{bmatrix}. \tag{13}$$

If $z^* \in R^{2n}$ is the solution to the above problem then $a^* = [I_n, -I_n]z^*$ is the solution to the finite-dimensional analog ECUI filtering problem.

An example of a set of basis of functions which are closed under the operations of taking the positive and negative parts is the Walsh functions given by (Gabel and Roberts (1987))

$$v_0(t) = 1, t \in [0, T], \; v_1^0(t) = \begin{cases} 1, t \in [0, 0.5T) \\ -1, t \in (0.5T, T] \end{cases}$$

and for $k = 1, \ldots, 2^{m-1}$, $m = 1, 2, \ldots$

$$v_{m+1}^{(2k-1)}(t) = \begin{cases} v_m^{(k)}(t), t \in [0, 0.5T) \\ (-1)^{k+1} v_m^{(k)}(2t - T), t \in (0.5T, T] \end{cases}$$

$$v_{m+1}^{(2k)}(t) = \begin{cases} v_m^{(k)}(t), t \in [0, 0.5T) \\ (-1)^k v_m^{(k)}(2t - T), t \in (0.5T, T] \end{cases}$$

Consider the compression of a 13 bit barker coded signal with an allowable error of 3% as shown in Figure 16 (a unit on the time axis corresponds to one bit interval β .) We require all responses to signals within this input mask to fit in an output mask with allowable sidelobes of 0.03 and a mainlobe peak of 0.69 ± 0.075 as shown in Figure 17.

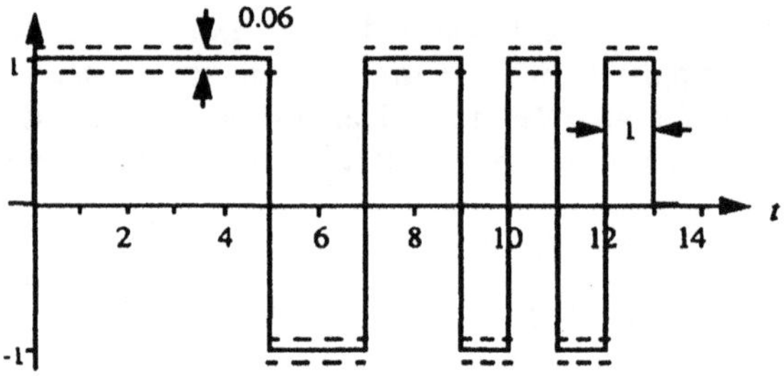

Figure 16. 13 -bit Barker-coded signal with input mask

Without loss of generality we consider the case where the input noise is white. For this example we use 32 Walsh functions, i.e. $m = 5$, $k = 16$. To match the bit interval we use $T = 32\beta$, where β is the bit interval. For computational purpose, the continuum of constraints is discretized at 64 constraints per bit interval.

This type of approximation results in a finite dimensional quadratic programming problem with a finite number of affine inequality constraints which can be easily dealt with by MATLAB routines. Figure 17 also shows the filter's responses to signals which were randomly perturbed about the nominal input but still fit inside the input mask. Note that these responses stay within the output envelope. Figure 18 plots the impulse response of the optimal ECUI which has a noise gain of $1.113 \times 10^{-3} T$, i.e. 0.0356.

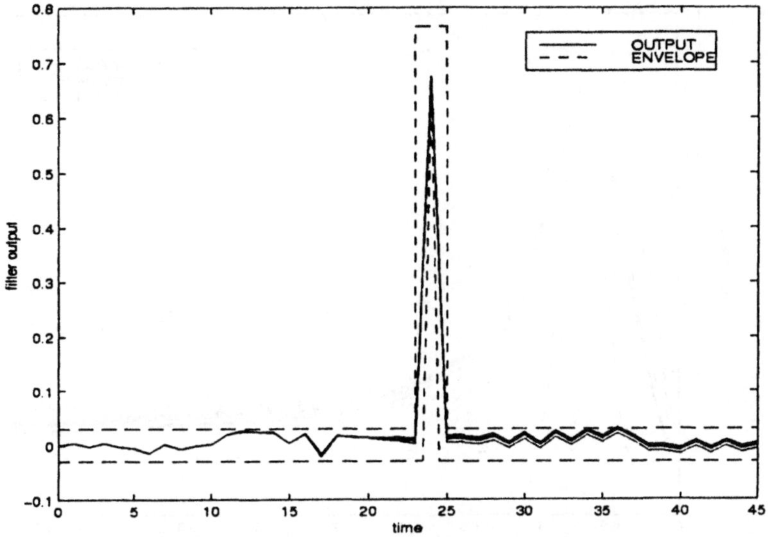

Figure 17. Pulse compression filter outputs

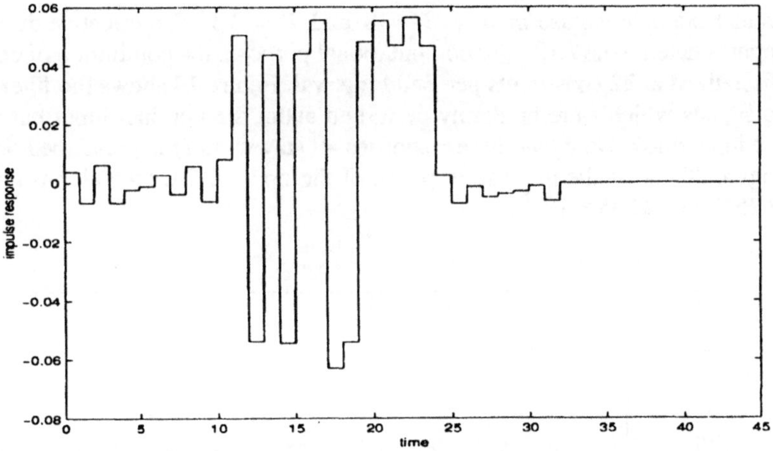

Figure 18. Impulse response of optimal ECUI pulse compression filter

The next example concerns the equalization of a digital transmission channel consisting of a coaxial cable operating at the DX3 rate (45Mb/s). The coaxial cable has a 30 dB attenuation at a normalized frequency of $2\pi/\beta$, where β is the baud interval (22.35×10^{-9} s). An equalizing filter is required to shape all signals within a 2% tolerance of the cable's impulse response so that it fits in the envelope given by the DSX3 pulse template (see Figure. 19). Note that the gap in the output mask is 0.1.

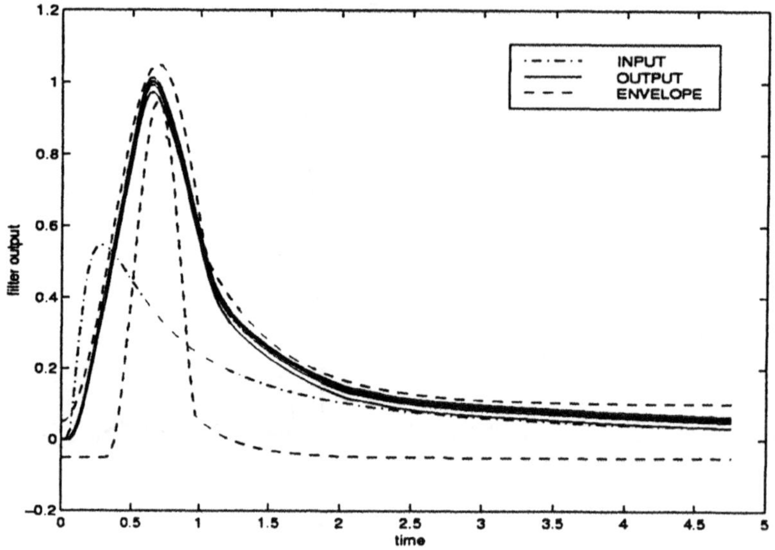

Figure 19. Signals for DSX3 example

In this example we use $m = 4$, $k = 8$ and $T = 4\beta$. The effective duration of the input is taken to be $20\,\beta$. For computational purpose, the continuum of constraints is discretized at 32 constraints per Baud interval. Figure 19 shows the filter's response to signals which were randomly perturbed about the nominal input but stay within the input mask. Note that these responses all stay within the prescribed output mask. Figure 20 shows the impulse response of the optimal filter which has noise gain of $2.25T$, i.e. 22.35×10^{-8}.

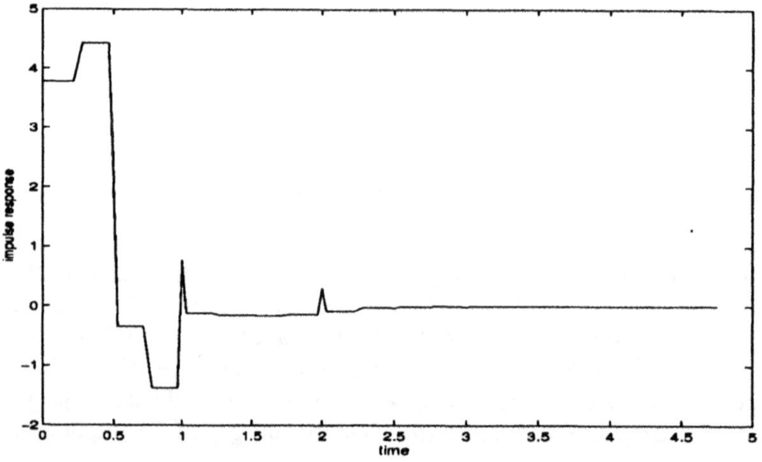

Figure 20. Impulse response of optimal ECUI equalization filter

4 CONCLUSIONS

We have seen how practical engineering requirements give rise to a raw engineering motivated optimization task. Considerations for the realizability of filter structures imposed extra constraints on the problem leading to finite dimensional problems. It has been shown that the finite dimensional solution converges to the optimum solution as the dimension increases. As with many engineering problems, the theoretical solutions are often not directly implementable, one has to move to the next stage to address the issue of robustness. A number of approaches for achieving robustness has been presented. The technique developed has been successfully applied to a range of signal processing problems namely equalizers for communications systems and pulse compression in sonar and radar ranging system.

References

Ahmed, K. M. and Evans, R. J. (1984), An adaptive array processor with robustness and broadband capabilities, *IEEE Trans. Antennas and Propagat.*, Vol. AP-32, pp. 944-950.

Bell Communications (1988), DSX-3 Isolated Pulse Template and Equations, *Technical Reference TR-TSY-000499*, Issue 2, pp. 9-17 .

Cantoni, A. (1988), Envelope-constrained filters - A practical application of optimization theory and algorithms. Key Note address, *ICOTA 98*, Perth, Australia, pp. 35-52.

CCITT (1984), Physical/Electrical characteristics of Hierarchical Digital Interfaces, *G.703*, Fascicle III.

Dennis, J. E. Jr. and Schnabel, R. B. (1989), A view of unconstrained optimization, *Handbooks in operations research and management science*, G. L. Nemhauser *et al.*, Eds., Elsevier Science Publishers B. V. (North-Holland) Vol. 1, pp. 1-72.

Dennis, J. E. Jr. and Schnabel, R. B. (1983), *Numerical methods for unconstrained optimization and nonlinear equations*, New Jersey: Prentice-Hall.

Dunford, N. and Schwartz, J. T. (1963), *Linear Operators*, Part II, Chapter XI, Interscience Publishers, Wiley.

Evans, R. J. (1975), *Optimal signal processing with constraints*, PhD thesis, Dept. of Electrical Engineering, University of Newcastle.

Evans, R. J., Cantoni, A., and Ahmed, K. M. (1983), Envelope-constrained filters with uncertain input, *Circuits System Signal processing*. Vol. 2, No. 2, pp. 131-154.

Evans, R. J., Cantoni, A. and Fortmann, T. E. (1977), Envelope-constrained filter, Part II, adaptive structures, *IEEE Trans. Information Theory*, Vol. IT-23, pp. 435-444.

Evans, R. J., Fortmann, T. E. and Cantoni A. (1977), Envelope-constrained filter, Part I, theory and applications, *IEEE Trans. Information Theory*, Vol. IT-23, pp. 421-434.

Fortmann, T. E. and Athans, M. (1974), Optimal filter design subject to output sidelobe constraints: Theoretical considerations, *Journal of Optimization Theory and Applications*, Vol. 14, No. 2, pp. 179-197.

Fortmann, T. E. and Evans R. J. (1974), Optimal filter design subject to output sidelobe constraints: Computational algorithm and numerical results, *Journal of Optimization Theory and Applications*, Vol. 14, No. 3, pp. 271-290.

Gabel, R. A.and Roberts, R. A. (1987), *Signal and Linear Systems*, 3rd Ed., John Wiley and Sons.

Lechleider, J. W. (1990), Pulse envelope for digital systems, *Proc. IEEE. Int. Conf. Communications*, Atlanta, GA, Vol.2, pp. 703-706.

Luenberger, D. G. (1969), *Optimization by vector space methods*, New York: Wiley.

Makila, P. M. (1990), Laguerre series approximation of infinite dimensional systems, *Automatica*, Vol. 26, No. 6, pp. 885-995.

McAulay, R. J. and Johnson, J. R. (1971), Optimal mismatched filter design for radar ranging detection and resolution, *IEEE Trans. Information Theory*, Vol. IT-17, pp. 696-701.

Nobakt, R. A. and Civanlar, M. R. (1995), Optimal pulse shape design for digital communication systems by projections onto convex sets, *IEEE Trans. Comm.*, Vol 43, pp. 2874-2877.

Pshenichni, B. N. and Danilin, Y. M. (1978), *Numerical methods in Extremal Problems*, Moscow: Mir Publishers.

Szego, G. (1939), *Orthogonal Polynomials*, American Mathematical Society Colloquium Publications Volume XXIII, American Mathematical Society Providence, Rhode Island.

Teo, K. L., Cantoni, A. and Lin, X. G. (1994), A new approach to the optimization of envelope-constrained filters with uncertain input, *IEEE Trans. Signal processing*, Vol. 2, No. 2, pp. 426-429.

Tsang, J., Zang, Z., Teo, K. L. and Cantoni, A. (1999), Dual approach to continuous-time Envelope Constrained Filter Design via Orthonormal Filters, submitted to *IEEE Trans. Signal processing*.

Vo, B. and Cantoni, A. (2000), Continuous-time Envelope Constrained filter Design via DSP approach, submitted to *IEE Proc. - Vision, Image and Signal Processing*.

Vo, B. and Cantoni, A. (2000), Continuous-time envelope constrained filter design with input uncertainty, (to be published in *IEEE Trans. Circuits & Systems I*),

Vo, B., Cantoni, A. and Teo, K. L. (1997), A penalty approach to iterative algorithms for envelope constrained filter design, *IEEE Trans. Signal Processing*, Vol. 45, No. 7, pp. 1869-1873.

Vo, B., Cantoni, A. and Teo, K. L. (1995), Computational methods for a class of functional inequality constrained optimization problem, *World Scientific Series in Applicable Analysis - Recent Trends in Optimization Theory and Applications*, R. P. Agarwal (Eds.), World Scientific Publishing (Singapore), Vol. 5, pp. 447-465.

Vo, B., Cantoni, A. and Teo, K. L. (1997), Envelope constrained filter with linear interpolator, *IEEE Trans. Signal Processing*. Vol. 45. No. 6, 1405-1415.

Vo, B., Zang, Z., Cantoni, A. and Teo, K. L. (1995), Continuous-time Envelope Constrained Filter Design via Orthonormal Filters, *IEE Proc. - Vision, Image and Signal Processing*, Vol. 142, No. 3, pp. 161-168.

Wood, L. C. and Treitel, S. (1975), Seismic signal processing, *Proc. IEEE* (Special issue on digital signal processing) Vol 63, pp. 649-661.

Zang, Z., Cantoni, A. and Teo, K. L. (1998), Continuous-Time Envelope-Constrained Filter Design via Laguerre Filters and H-infinity optimization methods, *IEEE Trans. Signal processing*, Vol. 46, No. 10, pp. 2601-2610.

Zang, Z., Cantoni, A., Vo, B. and Teo, K. L. (1996), Continuous-time envelope constrained filter design: Constraint robustness maximization, *4th IMA Conference on Mathematics in Signal Processing*, Warwick, UK, pp. 187-200..

Zang, Z., Vo, B., Cantoni, A. and Teo, K. L. (1999), Iterative algorithms for constrained recursive filter design via digital Laguerre networks, *IEEE Trans. Circuits & Systems, I*, Vol. 46, No. 11, pp. 1342-1348..

Zheng, W. X., Cantoni, A., Vo, B. and Teo, K. L. (1995), Recursive procedures for constrained optimization problems and its application in signal processing, *IEE Proc. - Vision, Image and Signal Processing*, Vol. 142, No. 3, pp. 161-168.

22 A ROBUST NUMERICAL ALGORITHM FOR THE OPTIMAL CONTROL OF HEAT TRANSFER IN THE CONTINUOUS CASTING OF STEEL

Y.H. Wu, M. Chuedoung and G. Zhang

School of Mathematics and Statistics
Curtin University of Technology
GPO Box U 1987, Perth 6845, Western Australia

Abstract: This paper concerns the optimal control of a two-phase heat transfer problem arising from the continuous steel casting process. The control action is performed via boundary conditions imposed on the computational region under consideration. Firstly, we introduce the continuous casting process and the equations governing the heat transfer and solidification of steel in the process. Then, an inverse problem associated with boundary controls is formulated mathematically and solved numerically. Finally, a numerical example is presented to demonstrate the application of the numerical algorithm for finding the optimal boundary cooling conditions for the casting.

Key words: Partial differential equations, heat transfer, optimal control, continuous casting.

X. Yang et al. (eds.), Optimization Methods and Applications, 399-412.

1 INTRODUCTION

Continuous casting is a very important and relatively modern heat extraction process for casting steel products such as blooms and billets. Figure 3.1 shows the essential features of the process. Molten steel is poured continuously into a water-cooled mould, where intense cooling causes a thin solidified steel shell to form around the edge of the steel. The solidified steel shell with a liquid pool in the center is then withdrawn from the bottom of the mould at a constant casting speed U. The product is supported by rollers after leaving the mould, cooled down by water sprays and then subsequently cooled by radiation only. This process is superior in many respects to the convention ingot casting and thus more and more percentage of steel in the world is produced by this process. In 1970, less than 5% of the crude steel produced in the world was continuously cast while this ratio has now exceeded 66% and is expected to increase continuously.

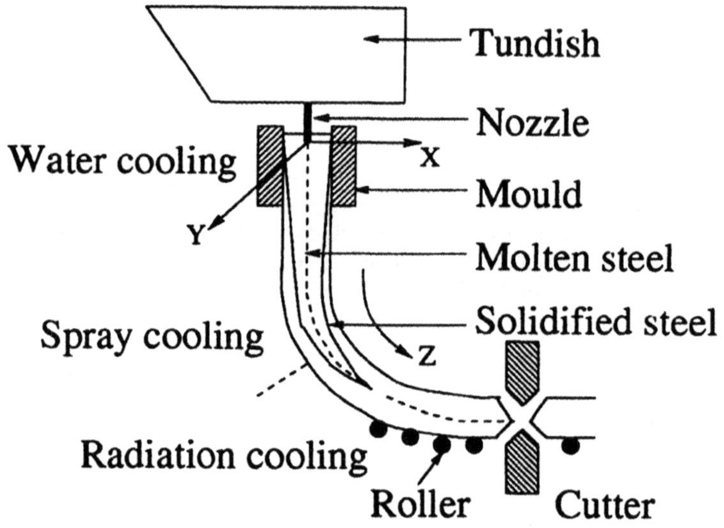

Figure 1.1 The continuous casting process and the coordinate system used.

In the casting process, sufficient heat must be removed from the steel to virtually complete solidification. Improper cooling strategy may change rapidly the surface temperature and consequently may cause different kind of surface and internal defects (Brimacombe *et al.* (1983), Schmidt & Fredriksson (1975) and Soejima *et al.* (1987)) and breakouts of molten steel from the bottom of the mould (Grill *et al.* (1976)). To avoid these problems, the strand has to be cooled down in such a way that the outcome is "optimal" in a certain sense. From metallurgical considerations, the engineer usually can provide information about the optimal behavior of the front of solidification. Thus, it is important that one is able to compute cooling rates that lead to a solidification front which is as close to the optimal one as possible (Upton *et al.* (1988), Engl and Langthaler (1978), Larrecq *et al.* (1983) and Neittaanmäki (1978)).

The problem of finding the optimal cooling strategy can be cast into an optimal control problem with constraints in state and control variables. The state variables include the temperature and the position of the interface between the molten steel and the solidified steel. The control variable is the boundary cooling rate which enters the mathematical model via the so-called average overall surface heat transfer coefficient between the casting surface and the environment. The state is governed by a heat transfer equation subject to a set of boundary conditions (Flint (1990), Hong *et al.* (1988), Laitinene & Neittaanmaki (1988) and Hill and Wu (1994)). The optimal control problem is : for a given optimal state (such as the optimal surface temperature or solidification profile), find the control such that the system state is as close to the optimal one as possible. This problem is also referred to as an inverse boundary value problem. While, the problem for finding the state corresponding to each specific control is referred to as a direct problem in this paper. In the following two sections, we present precisely the mathematical descriptions of the direct problem and the inverse problem respectively. In sections 4 - 6, a numerical algorithm for the solution of the inverse problem is presented and then, an example is given in section 7.

2 HEAT TRANSFER MODEL

The continuous casting process is a complex industrial process involving heat transfer with solidification and fluid flows. To formulate the problem mathematically, we choose a spatially fixed coordinate system O_{xyz} as shown in figure 3.1, with z representing the distance from the top surface. The equations governing the temperature and flow fields are the equations of momentum conservation, the continuity equation and the energy equation. Assume that the material is incompressible, then under the steady state condition, the governing equations are

$$\rho(\boldsymbol{u}.\nabla)\boldsymbol{u} = -\nabla P + \mu\nabla^2\boldsymbol{u}, \tag{2.1}$$

$$\nabla.\boldsymbol{u} = 0. \tag{2.2}$$

$$\rho c(T)(\boldsymbol{u}.\nabla)T = \nabla.(k(T)\nabla T) - \rho(\boldsymbol{u}.\nabla)H_L, \tag{2.3}$$

where P is the liquid pressure, ∇ denotes the gradient operator, T is the temperature, μ is the viscosity, $\boldsymbol{u} = (u, v, w)$ denotes the velocity vector in the Cartesian coordinate system $\boldsymbol{x} = (x, y, z)$ as shown in Figure 3.1, $\rho, c(T)$ and $k(T)$ denote density, specific heat and thermal conductivity respectively and $-\rho(\boldsymbol{u}.\nabla)H_L$ is a source term corresponding to latent heat being released due to solidification, and H_L represents the latent heat of the steel which in general can be expressed as a function of temperature, such as the linearly distributed function of temperature,

$$H_L = \begin{cases} L & if \quad T > T_L \\ L\dfrac{T - T_s}{T_L - T_s} & T_s \le T \le T_L \ , \\ 0 & T < T_s \end{cases} \tag{2.4}$$

where T_L, T_s and L are respectively the melting temperature, the solidification temperature and the latent heat of molten steel.

Equations (2.1)-(2.4) constitute a coupled system of five equations in term of five coordinate-dependent unknown functions(u, v, w, P and T) and in principle can be solved, but in practice, a great deal of computational effort is required. A rigorous study shows that the model can be simplified because

(a) relative to heat conduction in the x and y direction, conduction in the withdrawal (z) direction is small and can be neglected (Bland (1984));

(b) in both the solidified steel shell and the mould flux layer, the velocity components in the x and y directions are approximately equal to zero and hence the convective heat transfer in x and y directions for these regions can be neglected;

(c) the transfer of heat by convection in the liquid pool can be considered using an effective thermal conductivity (Lait *et al* (1988));

(d) the casting steel is withdrawn from the bottom of the mould continuously and hence the downward velocity component within the computational region is assumed to be equal to the constant casting speed U.

On using the above simplifications, the energy equation can be uncoupled from the system, thus

$$\rho U[c(T)\frac{\partial T}{\partial z} + \frac{\partial H_L}{\partial z}] = \nabla \cdot (k(T)\nabla T) \quad \text{in } \Omega \ , \tag{2.5}$$

where now ∇ denotes the two-dimensional (x, y) del-operator, and Ω is the horizontal cross-section of the casting steel.

To completely define the heat flow problem, boundary conditions must be specified. As heat flow from the strand surface to the environment takes place by convection and radiation, the boundary conditions on the standard surface are

$$-k\frac{\partial T}{\partial n} = \begin{cases} h(z)(T - T_w) & \text{in the mould region} \\ h(z)(T - T_w) + \sigma\varepsilon(T^4 - T_\omega^4) & \text{in the water-spray region} \ , \end{cases} \tag{2.6}$$

where $h(z)$, σ and ε are the surface heat transfer coefficient, the Stefan-Bolzmann constant and the emissivity of solid steel; T_w and T_∞ are respectively the water temperature and air temperature. The boundary condition on the top surface can be set up by assuming that the temperature distribution on the top surface $(z = 0)$ is uniform, namely

$$T(x, y, 0) = T_0 \quad \text{in } \Omega \tag{2.7}$$

where T_0 is assumed to be a constant. Further, since we are dealing with casting moulds which have a symmetric rectangular cross-section, the computational region can be limited to just one quadrant for three-dimensional analysis and thus an additional boundary condition is imposed on the planes of symmetry, namely

$$\frac{\partial T}{\partial n} = 0. \tag{2.8}$$

Equations (2.5)-(2.8) constitute a problem identical to the two-dimensional transient heat transfer problem with Neumann type boundary conditions and a discontinuous heat source H_L. The state variables to be determined are the temperature $T(x,y,z)$ and the location of a free boundary ($S(x,y,z) = 0$) - the interface between the molten steel and the solidified steel, or more specifically the surface where $T = T_s$. Therefore, our direct problem (state problem) associated with equations (2.5)-(2.8) can be stated as follows:

Given $h(z)$ and all other parameters, compute the temperature field $T(x,y,z)$ and consequently determine the free boundary $S(x,y,z)$ by solving equations (2.5)-(2.8).

3 STATEMENT OF THE OPTIMAL CONTROL PROBLEM

In the casting process, the control action is performed by means of data imposed on the fixed boundary $\partial\Omega \times [0, z_s]$ or part of the boundary. In this paper, we consider the control of the solidification process via the surface heat transfer coefficient $h(z)$ in boundary condition (2.6). More specifically, we want the solidification front $S(x,y,z)$ to be as close as possible to a given optimal solidification profile $\overline{s}(x,y,z)$.

Let U denote the set of admissible controls, namely

$$U = \{h(z) | h_{min} \leq h(z) \leq h_{max}\} , \tag{3.1}$$

where h_{min} and h_{max} are respectively the minimum and maximum possible heat transfer coefficients in the cooling zone. Let $\eta(s,z,h(z))$ and $\overline{\eta}(s,z)$ be respectively the actual and optimal thickness of the solidified steel shell corresponding to the boundary point(s,z) where s is the distance measured from a reference point A in a horizontal cross section as shown in Figure 3.1. Then, the optimal control problem is as follows:

$$\operatorname*{Inf}_{h(z) \in U} \left\{ J(h) = \int_0^{z_s} \int_{\partial\Omega} [\eta(s,z,h(z)) - \overline{\eta}(s,z)]^2 \, ds dz \right\} , \tag{3.2}$$

subject to the state problem consisting of equations (2.5)-(2.8).

This optimal control problem is also referred to as the inverse problem associated with equations (2.5) to (2.8).

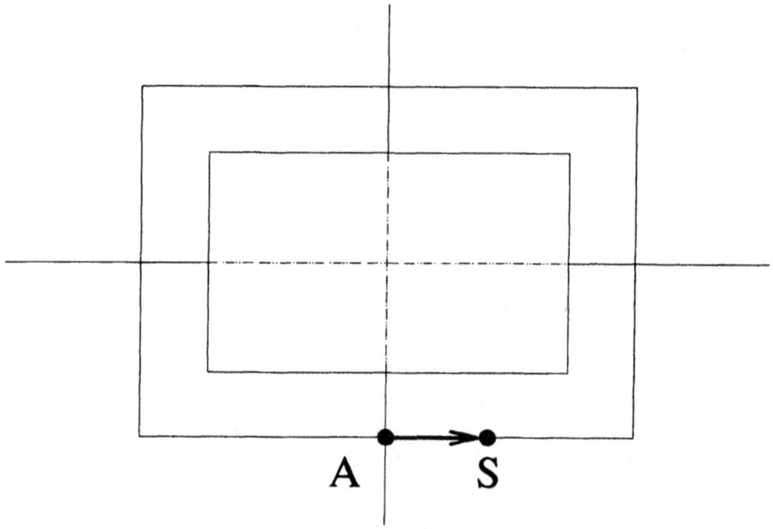

Figure 3.1 Sketch showing the definition of s.

In practice, boundary cooling conditions can be controlled usually only in such a way that the cooling region can be divided into a finite number (say P) of subregions and over each subregion $h(z)$ assumes a constant value. Thus the control variable $h(z)$ can be treated as a finite-dimensional variable and the cost function becomes:

$$J(\boldsymbol{h}) = \int_{0}^{z_s} \int_{\partial\Omega} [\eta(s, z, \boldsymbol{h}) - \overline{\eta}(s, z)]^2 \, dsdz\} \ , \tag{3.3}$$

where $\boldsymbol{h} = (h_1, h_2, \cdots, h_p)$ with h_j representing the value of $h(z_j)$. Hence our optimal control problem (inverse problem) can be stated as:

$$\min_{h_j \in U} \quad J(h_1, h_2, \cdots, h_p)$$

subject to the state problem consisting of equations (2.5) to (2.8).

4 NUMERICAL METHOD FOR THE OPTIMAL CONTROL PROBLEM

To solve the optimal control problem numerically, the integration region $\partial\Omega \times [0, z_s]$ is divided into N subregions and over each subregion, the integral is evaluated by the one-point quadrature rule. Thus, the cost function becomes:

$$J(\boldsymbol{h}) = \sum_{i=1}^{N} \triangle A_i [\eta_i(\boldsymbol{h}) - \overline{\eta_i}]^2 \ , \tag{4.1}$$

where η_i represents the value of η corresponding to the ith subregion, $\triangle A_i$ is the area of the i^{th}subregion. Let $\boldsymbol{\eta} = (\eta_1, \eta_2, \cdots, \eta_N)$ and $\overline{\boldsymbol{\eta}} = (\overline{\eta}_1, \overline{\eta}_2, \cdots, \overline{\eta}_N)$. Then, for $\triangle A_1 = \triangle A_2 = \cdots = \triangle A$, the problem is equivalent to the minimization of

$$J_1(\boldsymbol{h}) = (\boldsymbol{\eta} - \overline{\boldsymbol{\eta}})^T(\boldsymbol{\eta} - \overline{\boldsymbol{\eta}}). \tag{4.2}$$

Let h_j^0 be an initial estimate of the parameter $h_j, (j = 1, 2, 3, \cdots, P)$, and let $\boldsymbol{\eta}^0$ be the initial estimate of $\boldsymbol{\eta}$. Assume that for a small change of model parameters $\boldsymbol{\delta} = \boldsymbol{h} - \boldsymbol{h}^0$, the perturbation of the model response about $\boldsymbol{h}^0$ can be represented by the first order Taylor expansion

$$\boldsymbol{\eta} = \boldsymbol{\eta}^0 + \sum_{j=1}^{P} \frac{\partial \boldsymbol{\eta}}{\partial h_j} \bigg|_{\boldsymbol{h} = \boldsymbol{h}^0} (h_j - h_j^0) , \tag{4.3}$$

or in matrix notation

$$\boldsymbol{\eta} = \boldsymbol{\eta}^0 + A\boldsymbol{\delta} , \tag{4.4}$$

where A is the $N \times P$ Jacobian matrix of partial derivatives with elements

$$A_{ij} = \frac{\partial \eta_i}{\partial h_j}. \tag{4.5}$$

Then, the cost function $J_1(\boldsymbol{h})$ depends only on $\boldsymbol{\delta}$. Thus, our choice of perturbation in $\boldsymbol{h}$ will be made so as to minimize $J_1(\boldsymbol{\delta})$. Let $\boldsymbol{g} = \boldsymbol{\eta}^0 - \overline{\boldsymbol{\eta}}$. Then substituting (4.4) into (4.2) yields

$$J_1(\boldsymbol{\delta}) = [\boldsymbol{g} + A\boldsymbol{\delta}]^T[\boldsymbol{g} + A\boldsymbol{\delta}\ ]. \tag{4.6}$$

The minimization of $J_1(\boldsymbol{\delta})$ with respect to $\boldsymbol{\delta}$ requires that

$$\frac{\partial J_1(\boldsymbol{\delta})}{\partial \boldsymbol{\delta}} = 0. \tag{4.7}$$

Substituting (4.6) into (4.7) and carrying out the differentiation with respect to $\boldsymbol{\delta}$, we obtain the so-called "normal equation"

$$A^T A\boldsymbol{\delta} = -A^T\boldsymbol{g} , \tag{4.8}$$

which is known as the least squares solution or the Gauss-Newton solution. By repeatedly using (4.8), we can eventually obtain the optimal solution of $\boldsymbol{h}$ in the sense of minimizing the squares sum of errors.

A difficulty in using equation (4.8) is that when the matrix $A^T A$ becomes singular, the solution vector $\boldsymbol{\delta}$ tends to grow without bound. In order to reduce this difficulty, we use the method of damped least squares (Levenberg 1944 and Marquardt 1963) by imposing the constraining condition that the sum of the squares of the parameter change vector $\boldsymbol{\delta}$ is bounded by a finite quantity, say

δ_0^2. The effect of this constraint is to prevent unbounded oscillation in the solutions, i.e. to smooth the parameter change vector $\boldsymbol{\delta}$. Thus, we choose $\boldsymbol{\delta}$ to minimize a cost function $J_2(\boldsymbol{\delta}, \lambda)$ given by

$$J_2(\boldsymbol{\delta}, \lambda) = (\boldsymbol{g} + A\boldsymbol{\delta})^T(\boldsymbol{g} + A\boldsymbol{\delta}) + \lambda(\boldsymbol{\delta}^T\boldsymbol{\delta} - \delta_0^2) \ , \tag{4.9}$$

where λ is a Lagrange multiplier. Differentiation with respect to the vector $\boldsymbol{\delta}$ yields a modified form of the normal equation

$$(A^TA + \lambda I)\boldsymbol{\delta} = -A^T\boldsymbol{g} \ , \tag{4.10}$$

so that

$$\boldsymbol{\delta} = -(A^TA + \lambda I)^{-1}A^T\boldsymbol{g}. \tag{4.11}$$

For $\lambda = 0$, solution (4.11) reduces to the least squares solution. While for a very large value of λ, solution (4.11) becomes $\boldsymbol{\delta} = -\lambda^{-1}A^T\boldsymbol{g}$, which is in the steepest descent directions of $J_1(\boldsymbol{\delta})$. However, as λ becomes larger and larger, the elements of the parameter change vector $\boldsymbol{\delta}$ become smaller and smaller and a slow convergence rate can be expected. In the present work, λ is initially set to a large positive value so that the good initial convergence properties of the steepest descent method can come into play. Then, as this happens, λ is reduced by multiplying it by a constant factor less than one so that the least squares method may take over in the region closer to the solution. If divergence occurs during a given iteration, λ is divided once more by this factor until the error drops and convergence resumes.

5 NUMERICAL APPROXIMATION OF THE JACOBIAN MATRIX

One of the key steps for the numerical solution of the optimal control problem is the calculation of the Jacobian matrix A as defined in (4.5). To determine A, one needs to solve the direct boundary value problem defined by equations (2.5)-(2.8). Various numerical methods, such as the finite difference method (Pawlow *et al.* (1985), Voller and Prakash (1987)) and the finite element method can be used to solve the problem. In the present work, an efficient finite element method (Wu *et al.* (1994)) based on the enthalpy formulation is used for the solution of the direct boundary value problem.

Firstly, we multiply equation(2.5) in both sides by a weighting function ω and integrate over the relevant domain Ω to yield the following variational problem:

Find $T(x, y, z) \in H(\Omega) \times (0, Z)$ such that for all $\omega(x, y) \in H(\Omega)$

$$T|_{z=0} = T_0, \tag{5.1}$$

$$\left(\rho U\left(c\frac{\partial T}{\partial z}+L\frac{\partial H_L}{\partial z}\right),\omega\right)+A(T,\omega)=F(\omega), \tag{5.2}$$

where $H(\Omega)$ is the Sobolev space $W^{1,2}(\Omega)$ with norm $\|.\|_{1,2,\Omega}$, $(.)$ denotes the inner product on $L^2(\Omega), A(T,\omega)$ and $F(\omega)$ are defined respectively by

$$A(T,\omega)=\int_{\Omega} k\nabla T.\nabla\omega d\Omega+\int_{\partial\Omega_\omega} h_\infty T\omega ds, \qquad F(\omega)=\int_{\partial\Omega_\omega} h_\infty T_\infty \omega ds. \tag{5.3}$$

Then, we pose the variational problem in an N-dimensional subspace, namely approximating $T(x,y,z)$ and $\omega(x,y)$ by

$$T(x,y,z)=\sum_{j=1}^{N}\phi_j(x,y)T_j(z), \quad \omega(x,y)=\sum_{j=1}^{N}\phi_j(x,y)\omega_j, \tag{5.4}$$

where $\phi_1,\phi_2,\ldots,\phi_n$ are the basis function of the N-dimensional finite element subspace. Substituting (5.4) into equation (5.2) and noting that ω is arbitrary, we have for $i=1,2,\ldots,N$

$$\sum_{j=1}^{N}\left\{\rho U(c(T)\phi_j,\phi_i)\dot{T}_j+\rho UL(\phi_j,\phi_i)\dot{H}_j+A(\phi_j,\phi_i)T_j\right\}=F(\phi_i), \tag{5.5}$$

or in matrix form

$$C\frac{\partial \boldsymbol{T}}{\partial z}+D\frac{\partial \boldsymbol{H}_L}{\partial z}+K\boldsymbol{T}=\boldsymbol{F}. \tag{5.6}$$

where C and K are the assembled capacity matrix and conductivity matrix respectively, D is an assembled coefficient matrix of the latent heat term, $\boldsymbol{T}$ and $\boldsymbol{H}_L$ are the temperature and Latent heat vectors for all nodes. A relaxation iterative scheme (Wu *et al* (1994)) is then used to solve above problem for $\boldsymbol{T}$. Once $\boldsymbol{T}$ is determined, the solidification surface can then be determined as the surface where $T=T_s$. Hence, by solving the direct boundary value problem for different control variable h, we can determine the Jacobian matrix using equation (4.5).

6 NUMERICAL ALGORITHM

For any given $\overline{\eta}$ and tolerances ε_h and ε_η, the numerical algorithm used for solving the optimal control problem (inverse problem) is as follows.

step 0 Set $n = 0$, chose *scalar* < 1 and a positive value for λ and take an arbitrary starting point $\boldsymbol{h}^0$ with $h_j^0 \in U$.

step 1 Compute $\boldsymbol{\eta}(\boldsymbol{h}^0)$ by solving the direct problem and then calculate $\boldsymbol{g}^0$.

step 2 Calculate $\boldsymbol{\eta}(\boldsymbol{h}^n + I_j \Delta h_j)$ and then determine A^n by

$$A_{ij}^n = \frac{\partial \eta_i}{\partial h_j}\bigg|_{\boldsymbol{h} = \boldsymbol{h}^n} = \frac{\eta_i(\boldsymbol{h}^n + I_j \Delta h_j) - \eta_i(\boldsymbol{h}^n)}{\Delta h_j}$$

where I_j denotes the jth column of the $P \times P$ identity matrix I.

step 3 Solve the normal equation (4.10) for $\boldsymbol{\delta}^n$, namely

$$\boldsymbol{\delta}^n = -[(A^n)^T A^n + \lambda I]^{-1} (A^n)^T \boldsymbol{g}^n.$$

step 4 Set $\boldsymbol{h}^{n+1} = \boldsymbol{h}^n + \boldsymbol{\delta}^n$ and then solve the direct problem for $\boldsymbol{\eta}(\boldsymbol{h}^{n+1})$.

step 5 Check whether $\boldsymbol{h}^{n+1}$ satisfies the stop criterion for the optimal solution

$$(\|\boldsymbol{\delta}\| < \varepsilon_h \text{ and } \|\boldsymbol{\eta}(\boldsymbol{h}^{n+1}) - \overline{\boldsymbol{\eta}}\| < \varepsilon_\eta).$$

If yes then stop, otherwise proceed to step 6.

step 6 Check whether the squares sum of errors grows or not

$$(\| \boldsymbol{\eta}(\boldsymbol{h}^{n+1}) - \overline{\boldsymbol{\eta}}\| \geq \| \boldsymbol{\eta}(\boldsymbol{h}^n) - \overline{\boldsymbol{\eta}}\|).$$

If yes then let $\lambda := \dfrac{\lambda}{scale}$ and then return to step 3. Otherwise, let $n = n + 1$ and $\lambda = \lambda \times scale$ and then return to step 2.

7 AN ILLUSTRATIVE EXAMPLE

As an illustrative example, we consider a steel slab caster with depth $z_m = 0.8$ m, width $W = 1750$ mm and thickness 236 mm. The problem is considered as a plane problem and the calculation is limited to the vertical cross-section perpendicular to the broad face. The basic parameters are as follows: $z_s = 6.4$ m, $U = 1.3$ m/min, $\rho = 7800$ Kg/m^3, $C = 465 J/Kg$°C, $k = 35$ W/m°C, $L = 27200$ J/Kg, $T_s = T_L = 1495$°C, $T_w = T_\omega = 20$°C. The optimal control problem is : determine the surface heat transfer coefficient function $h(z)$ such that the solidification profile $\eta(z)$ of steel in the region $[0, z_s]$ is as close as

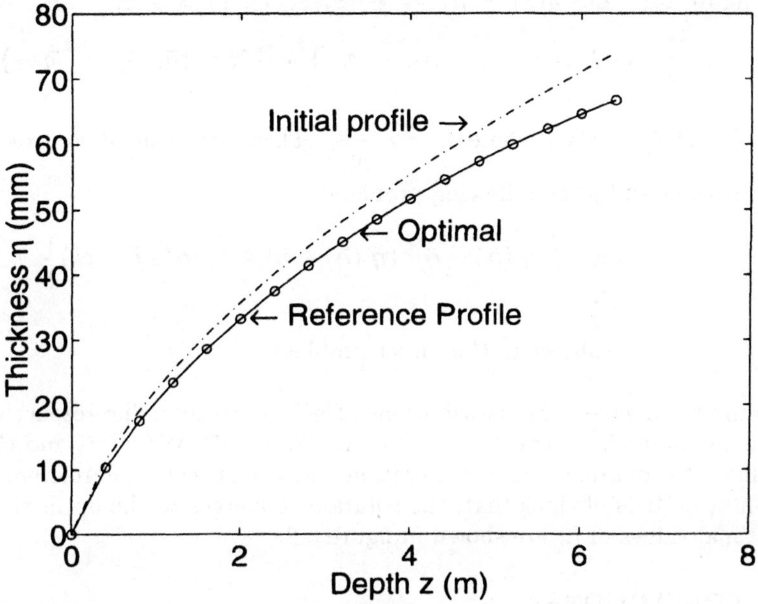

Figure 6.1 Thickness of solidified steel at different depth.

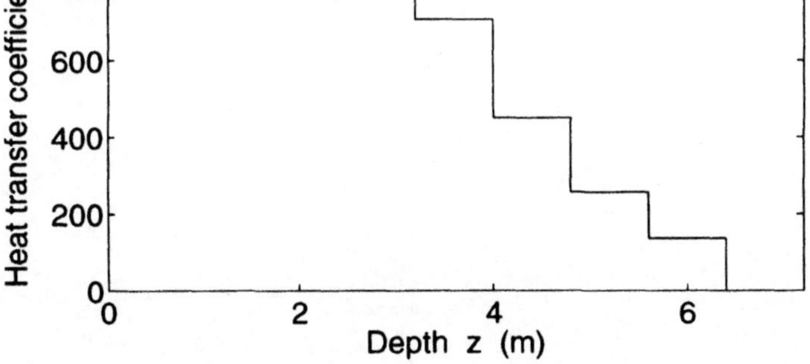

Figure 6.2 Distribution of optimal boundary heat transfer coefficient $h(z)$.

possible to the optimal profile given by the data set as shown by the points in figure 6.1. Let $h(z) = h_j \ \forall z \in (z_j, z_{j+1})$, $(j = 1, 2, \cdots, P)$, where $z_j = (j-1)\frac{z_s}{P}$, and let $\boldsymbol{\eta} = \{\eta_1, \eta_2, \cdots, \eta_N\}$ and $\overline{\boldsymbol{\eta}} = \{\overline{\eta}_1, \overline{\eta}_2, \cdots, \overline{\eta}_{N},\}$ with $\eta_i = \eta(z_i)$ and $\overline{\eta}_i = \overline{\eta}(z_i)$ where $z_i = i\frac{z_s}{N}$. Then the optimal control problem is approximated by the following problem:

$$\min_{\boldsymbol{h}} \left\{ (\boldsymbol{\eta}\ (\boldsymbol{h}) - \overline{\boldsymbol{\eta}})^T (\boldsymbol{\eta}\ (\boldsymbol{h}) - \overline{\boldsymbol{\eta}}) + \lambda(d\boldsymbol{h}^T d\boldsymbol{h} - \delta_0^2) \right\}$$

subject to the direct problem.

Figure 6.1 displays the distributions of $\eta(\boldsymbol{h})$ corresponding respectively to the starting point $\boldsymbol{h}^0 = (h_1, h_2, \cdots, h_8)$ with $h_j = 1300$ W/m^2 ^{o}C and the optimal values of $\boldsymbol{h}$ obtained after 25 iterations, with respect to the reference distribution $\overline{\eta}(z)$. It is obvious that, the solution converges to the optimal state. The optimal values of h_i are shown in figure 6.2.

8 CONCLUSIONS

The problem of finding the optimal boundary heat transfer condition for the continuous steel casting process was formulated in this paper as an inverse boundary value problem. A numerical algorithm has been developed to solve the problem and the numerical investigation shows that the numerical algorithm developed is efficient and robust. The algorithm can be used to determine the optimal boundary cooling conditions for different casting parameters.

References

Bland, D.R. (1984), Flux and the continuous casting of steel, *IMA J. of Appl. Math.*, Vol. 32, pp. 89-112.

Brimacombe, J.K., Samarasekera, I.V. and Laid, J.E. (eds.) (1983), *Continuous casting, Vol.*2nd*: Heat flow, solidification and crack formation*, Iron and Steel Society, AIME.

Engl, H.W., Langthaler, T. (1978), On an inverse problem for a nonlinear heat equation connected with continuous casting of steel, *Optimal control of differential equations* Birkhäuser Verlag, Boston.

Flint, P.J. (1990), A three dimensional finite difference model of heat transfer, fluid flow and solidification in continuous slab caster, *Proc* 73rd *ISS steelmakeing Conf.*, Detroit MI, March 25-28 .

Grill, A., Sorimachi, K. and Brimacombe, J.K. (1976), Heat flow, gap formation and break-outs in the continuous casting of steel slabs, *Met. Trans.* B, Vol. 7, pp. 177-189.

Hill, J.M. and Wu, Y.H. (1994), On a nonlinear Stefan problem arising in the continuous casting of steel, *Acta machanica*, Vol. 107, pp.183-198.

Hong, C.P., Umeda T. and Kimura, T. (1988), Numerical models for casting solidification: Part1, the coupling of the boundary element and finite difference methods for solidification problem; Part2, application of the boundary element method to solidification problems, *Metallurgical Transaction* B, Vol. 15, pp. 91-107.

Lait, J.E., Brimacombe, J.K. and Weinberg, F. (1988), Mathematical modelling of heat flow in the continuous casting of steel, *Ironmaking and Steelmaking*, Vol. 1, pp. 335-354.

Larrecq, M., Birat, J. P., Saquez, C. and Henry, J.(1983), Optimization of casting and cooling conditions on steel continuous caster; implementation of optimal strategies on slap and bloom caster, *IRSID*, ACI, Vol. 83, RE 1004.

Latinen, E., and Neittaanmaki, P. (1988), On numerical simulation of the continuous casting process, *J. Eng. Math.*, Vol. 22, pp. 335-354.

Levenberg, K.(1944), A method for the solution of certain nonlinear problems in least squares ,*Quarterly of Applied Mathematics*, Vol. 2, pp. 164-168.

Marquardt, D.W. (1963), An algorithm for least squares estimation of nonlinear parameters, *Journal of the Society of Industrial and Applied mathematics*, Vol.11, pp. 431-441.

Neittaanmäki, P. (1978), On the control of the secondary cooling in the continuous casting Process, *Optimal Control of Differential equations*, Birkhäuser Verlag, Boston.

Pawlow, I., Shindo Y. and Satawa, T.(1985), Numerical solution of multidimensional two - phase Stefan problem, *Numer. Funct. Anal. Optimiz.*, Vol. 8, pp. 55-84.

Schmidt, L, and Fredriksson, H. (1975), Formation of macro-segregation and centre-line cracks in continuous cast, *Ironmaking and Steelmaking*, vol. 2, pp. 61-67.

Soejima, T., Kitamura, M., Koyama, S. and Abu, J. (1987), *Surface quality of continuous cast slabs at Kobe Steel's Kakogawa Works*, Iron and Steel Society, AIME, USA, pp. 67-73.

Upton, E.A., Rao, T.R.S., Dauby, P.H. and Knechtges, R.C. (1988), Physical metallurgy and mathematical modelling as tools for continuous casting optimization at LTV steel, *Ironmaking and Steelmaking*, Vol. 15, pp. 51-57.

Voller, V.R. Prakash, C. (1987), A fixed grid numerical modelling methodology for convection-diffusion mushy region phase-change problems, *Int. J. Heat and Mass Transfer*, Vol. 30, pp. 1709-1719.

Wu, Y.H., Hill, J.M. and Flint, P. (1994), A novel finite element method for heat transfer in the continuous caster, *J. Austral. Math. Soc.* Ser. B, Vol. 35, pp. 263-288.

18. O. Maimon, E. Khmelnitsky and K. Kogan: *Optimal Flow Control in Manufacturing.* Production Planning and Scheduling. 1998 ISBN 0-7923-5106-1
19. C. Zopounidis and P.M. Pardalos (eds.): *Managing in Uncertainty: Theory and Practice.* 1998 ISBN 0-7923-5110-X
20. A.S. Belenky: *Operations Research in Transportation Systems:* Ideas and Schemes of Optimization Methods for Strategic Planning and Operations Management. 1998 ISBN 0-7923-5157-6
21. J. Gil-Aluja: *Investment in Uncertainty.* 1999 ISBN 0-7923-5296-3
22. M. Fukushima and L. Qi (eds.): *Reformulation: Nonsmooth, Piecewise Smooth, Semismooth and Smooting Methods.* 1999 ISBN 0-7923-5320-X
23. M. Patriksson: *Nonlinear Programming and Variational Inequality Problems.* A Unified Approach. 1999 ISBN 0-7923-5455-9
24. R. De Leone, A. Murli, P.M. Pardalos and G. Toraldo (eds.): *High Performance Algorithms and Software in Nonlinear Optimization.* 1999 ISBN 0-7923-5483-4
25. A. Schöbel: *Locating Lines and Hyperplanes.* Theory and Algorithms. 1999 ISBN 0-7923-5559-8
26. R.B. Statnikov: *Multicriteria Design.* Optimization and Identification. 1999 ISBN 0-7923-5560-1
27. V. Tsurkov and A. Mironov: *Minimax under Transportation Constrains.* 1999 ISBN 0-7923-5609-8
28. V.I. Ivanov: *Model Development and Optimization.* 1999 ISBN 0-7923-5610-1
29. F.A. Lootsma: *Multi-Criteria Decision Analysis via Ratio and Difference Judgement.* 1999 ISBN 0-7923-5669-1
30. A. Eberhard, R. Hill, D. Ralph and B.M. Glover (eds.): *Progress in Optimization.* Contributions from Australasia. 1999 ISBN 0-7923-5733-7
31. T. Hürlimann: *Mathematical Modeling and Optimization.* An Essay for the Design of Computer-Based Modeling Tools. 1999 ISBN 0-7923-5927-5
32. J. Gil-Aluja: *Elements for a Theory of Decision in Uncertainty.* 1999 ISBN 0-7923-5987-9
33. H. Frenk, K. Roos, T. Terlaky and S. Zhang (eds.): *High Performance Optimization.* 1999 ISBN 0-7923-6013-3
34. N. Hritonenko and Y. Yatsenko: *Mathematical Modeling in Economics, Ecology and the Environment.* 1999 ISBN 0-7923-6015-X
35. J. Virant: *Design Considerations of Time in Fuzzy Systems.* 2000 ISBN 0-7923-6100-8

Applied Optimization

36. G. Di Pillo and F. Giannessi (eds.): *Nonlinear Optimization and Related Topics*. 2000 ISBN 0-7923-6109-1
37. V. Tsurkov: *Hierarchical Optimization and Mathematical Physics*. 2000 ISBN 0-7923-6175-X
38. C. Zopounidis and M. Doumpos: *Intelligent Decision Aiding Systems Based on Multiple Criteria for Financial Engineering*. 2000 ISBN 0-7923-6273-X
39. X. Yang, A.I. Mees, M. Fisher and L.Jennings (eds.): *Progress in Optimization*. Contributions from Australasia. 2000 ISBN 0-7923-6286-1
40. D. Butnariu and A.N. Iusem: *Totally Convex Functions for Fixed Points Computation and Infinite Dimensional Optimization*. 2000 ISBN 0-7923-6287-X
41. J. Mockus: *A Set of Examples of Global and Discrete Optimization*. Applications of Bayesian Heuristic Approach. 2000 ISBN 0-7923-6359-0
42. H. Neunzert and A.H. Siddiqi: *Topics in Industrial Mathematics*. Case Studies and Related Mathematical Methods. 2000 ISBN 0-7923-6417-1
43. K. Kogan and E. Khmelnitsky: *Scheduling: Control-Based Theory and Polynomial-Time Algorithms*. 2000 ISBN 0-7923-6486-4
44. E. Triantaphyllou: *Multi-Criteria Decision Making Methods*. A Comparative Study. 2000 ISBN 0-7923-6607-7
45. S.H. Zanakis, G. Doukidis and C. Zopounidis (eds.): *Decision Making: Recent Developments and Worldwide Applications*. 2000 ISBN 0-7923-6621-2
46. G.E. Stavroulakis: *Inverse and Crack Identification Problems in Engineering Mechanics*. 2000 ISBN 0-7923-6690-5
47. A. Rubinov and B. Glover (eds.): *Optimization and Related Topics*. 2001 ISBN 0-7923-6732-4
48. M. Pursula and J. Niittymäki (eds.): *Mathematical Methods on Optimization in Transportation Systems*. 2000 ISBN 0-7923-6774-X
49. E. Cascetta: *Transportation Systems Engineering: Theory and Methods*. 2001 ISBN 0-7923-6792-8
50. M.C. Ferris, O.L. Mangasarian and J.-S. Pang (eds.): *Complementarity: Applications, Algorithms and Extensions*. 2001 ISBN 0-7923-6816-9
51. V. Tsurkov: *Large-scale Optimization – Problems and Methods*. 2001 ISBN 0-7923-6817-7

KLUWER ACADEMIC PUBLISHERS – DORDRECHT / BOSTON / LONDON

GPSR Compliance

The European Union's (EU) General Product Safety Regulation (GPSR) is a set of rules that requires consumer products to be safe and our obligations to ensure this.

If you have any concerns about our products, you can contact us on ProductSafety@springernature.com

In case Publisher is established outside the EU, the EU authorized representative is:

Springer Nature Customer Service Center GmbH
Europaplatz 3
69115 Heidelberg, Germany

Batch number: 10584518

Printed by Printforce, the Netherlands